Sensors, Focus on Tactile, Force and Stress Sensors

Edited by

Jose Gerardo Rocha

and

Senentxu Lanceros-Mendez

Sensors, Focus on Tactile, Force and Stress Sensors

Edited by Jose Gerardo Rocha and Senentxu Lanceros-Mendez

Republished by InTech

Janeza Trdine 9, 51000 Rijeka, Croatia

Technical Editor Teodora Smiljanic

Cover Designer Dino Smrekar

Additional hard copies can be obtained from orders@intechopen.com

Sensors, Focus on Tactile, Force and Stress Sensors,
Edited by Jose Gerardo Rocha and Senentxu Lanceros-Mendez
p. cm.
ISBN 978-953-7619-31-2

For EU product safety concerns:
IN TECH d.o.o., Prolaz Marije Krucifikse Kozulić 3, 51000 Rijeka, Croatia,
info@intechopen.com or visit our website at intechopen.com.

Preface

This decade has been called by many people as the decade of the sensors. With an enormous increase in the research and application of sensors in the last fifteen years, it can be considered that a revolution similar to the one of microcomputers in the decade of 1980 is in course. In the last times, we have witnessed enormous advances in sensor´s technology and more innovations are in the way. The sensitivity of the sensors is becoming higher, their dimensions lower, their selectivity better and their price lower. Some issues remain nevertheless constant: the basic principles used in the project of sensors and applications, once these principles are governed by the laws of the nature. However, through the times our appreciation, knowledge and mastering of these same laws has changed.

Among the existing sensors to measure the most diverse quantities, the tactile and force sensors are becoming more popular mainly, but not only, in the field of the robotic applications, where the machines are instructed to execute tasks more and more similar to the ones executed by human operators.

Tactile sensors are devices that measure the parameters related to the contact between the sensor itself and a certain object. This interaction is restricted to a well defined and usually small region. In contrast, the force and torque sensors normally measure the total forces and torques applied to an object.

Tactile sensors can be used to detect a wide range of stimulus: from the simple identification of a contact with a given object to a complete tactile image giving information on forces and shapes, for example. Usually, the active component of a tactile sensor is capable to feel and measure several properties, like contact forces, texture, impact, sliding and other contact conditions that can generate specific patterns of force and position. This information can be used to identify the state of the object handled by a manipulator, that is, its size, shape or if it is in the correct position, for example.

Once it does not exist a complete theory that describes the requirements of a robotic system in terms of tactile information, most of the knowledge in this area is produced from the study of the human tactile sensors and the way humans grasp and handle. From these studies, the investigators concluded that the function of grasping within the incorporation of tactile feelings requires several sensors, namely force sliding and even temperature knowledge. Moreover, the manipulator must have in its memory the right way to handle the object, that is, it must know a priori which are the sensations produced by the object, in order to handle it correctly.

This book describes some devices that are commonly identified as tactile or force sensors. It is achieved with different degrees of detail, in a unique and actual resource, the description of different approaches to this type of sensors. Understanding the design and the working principles of the sensors described here, requires a multidisciplinary background of electrical engineering, mechanical engineering, physics, biology, etc. It has been made an attempt to place side by side the most pertinent information in order to reach a more productive reading not only to professionals dedicated to the design of tactile sensors, but also all other sensor users, as for example, in the field of robotics. The latest technologies presented in this book, are more focused on information readout and processing: as new materials, micro and sub-micro sensors are available, wireless transmission and processing of the sensorial information, as well as some innovative methodologies for obtaining and interpreting tactile information are also strongly evolving.

This book is organized in twenty four chapters. In the first chapters, some considerations concerning tactile sensors and the way they must operate, as well as some examples of silicon sensors are presented. Then, tactile sensors of three and six axes are described. Some of them can measure, beyond the force, the slip. After that, several flexible sensors with anthropomorphous characteristics and with particularities resembling the human skin are reported. Finally, some methods of transmission and information processing, namely wireless and with more or less elaborated algorithms are described.

December, 2008

Editors

Jose Gerardo Rocha
and
Senentxu Lanceros-Mendez
University of Minho,
Portugal

Contents

1

How tactile sensors should be?

Satoshi Saga
Tohoku University
Japan

1. Introduction

Tactile sensation consists of sensory information at a contact status between human and the other environment. The contact status draws some physical phenomena. The tactile sensor has to record the sensory information, so the sensor should record these physical phenomena. The physical phenomena of the contact point are listed as follows; deformation, stress, temperature, and time variation of these information.
When human touch some environment the human finger will be deformed according to the pressed force and the reactive stress from the environment. The deformation and the stress are linked together and occur according to the Young's modulus and the Poisson's ratio of materials of the finger and the environment. If the materials can be assumed to be the total elastic body, the deformation and the stress can be linked by the linear elastic theory.
Because there exists no total elastic body, the link between the deformation and the stress is a little complex. The complexity is enhanced when the contact state is changed according to time. For example, the human moves his finger toward the environment or touch a vibratory environment, the environment may return the damper or mass property with the change of movement speed or acceleration. The most characteristic example is a dilatants phenomenon. A dilatants material is one in which viscosity increases with the rate of shear.
As a simple environment model, there exists such an impedance model;

$$F(x) = k(x - x_o) + d\frac{dx}{dt} + m\frac{d^2t}{dt^2} \tag{1}$$

By using this model the authors have proposed an environment recording system (Saga, et al. 2005). However the model is only for one point contact movement, so it cannot express the distribution of the deformation.
That is the reason why many sensors assume the materials as total elastic or rigid body and measure the deformation or stress by using some physical principles.
In the temperature domain, a governing physical equation is a diffusion equation. The key points of the thermal flow are the thermal difference between the finger and the environment, area distribution of contact surface, and thermal conductivities of both the finger and the environment. The existing thermal sensors are only measuring the current temperature. Neither contact area distribution nor thermal conductivities is measured. The lack of these information make the displaying of temperature difficult.

2. Tactile sensors in human

Human has some receptors beneath his/her skin. The known receptors are listed as follows; mechanoreceptors, nociceptors, thermal receptors, and muscle and skeletal mechanoreceptors. Each receptor has its own distribution and network; e.g. lateral inhibition. So the mapping and the network of the sensor is also important for tactile sensation.

2.1 Cutaneous receptors

First, there are some receptors in human skin (Kandel, et al. 2000) (Fig. 1). As mechanoreceptors there are Merkel cells, Meissner's corpscules, Pacinian corpuscles, and Ruffini endings. As nociceptors there are mechanical ones, thermal-mechanical ones, and polymodal ones. As thermal receptors there are cool receptors, warm receptors, heat nociceptors, and cold nociceptors. In addition, as muscle and skeletal mechanoreceptors, there are muscle spindle primary, secondary, Golgi tendon organs, joint capsule mechanoreceptors, stretch-sensitive free endings. By using these receptors human translate the physical phenomena to some electric signals.

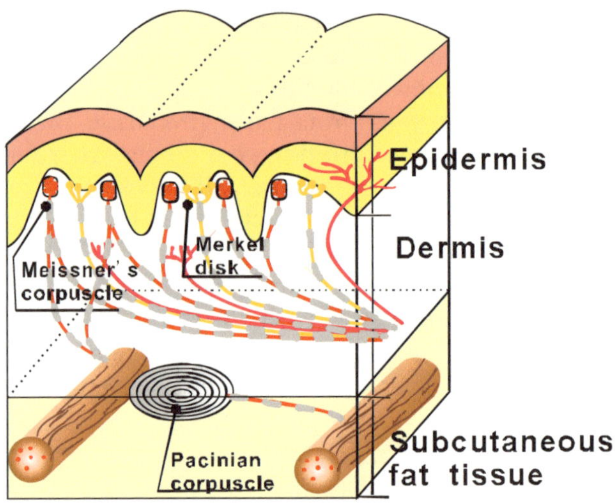

Fig. 1. Structure of skin (adapted from Kandel, et al. 2000)

Each receptor has its own unit density and responsibility. For example, the mechanoreceptors which measures mainly deformation and stress distributions have various densities and responsibilities. Merkel disks have its responsibility about 5 - 15Hz and has 70 units/cm square distribution, Meissner's corpuscles have its responsibility about 20 - 50Hz and has 140 units/cm square distribution, and Pacinian corpuscle have its responsibility about 60 - 400Hz and has 20 units/cm square distribution (Fig. 2).

These density and responsibility suggests that human processes the higher frequency signals with not so high density, but processes the lower frequency signals with high density.

2.2 Networks of receptors

In addition, each receptor has its own networks in the cortex, dorsal column nuclei, ventral posterior lateral nucleus of the thalamus, or cortex itself.

For example, there is convergent excitation, Surround inhibition, and lateral inhibition.

- Convergent excitation
- Surround inhibition
- Lateral inhibition

By using these networks parallel processing is exerted. Through the process the simple many signals became more extracted meaningful some signals.

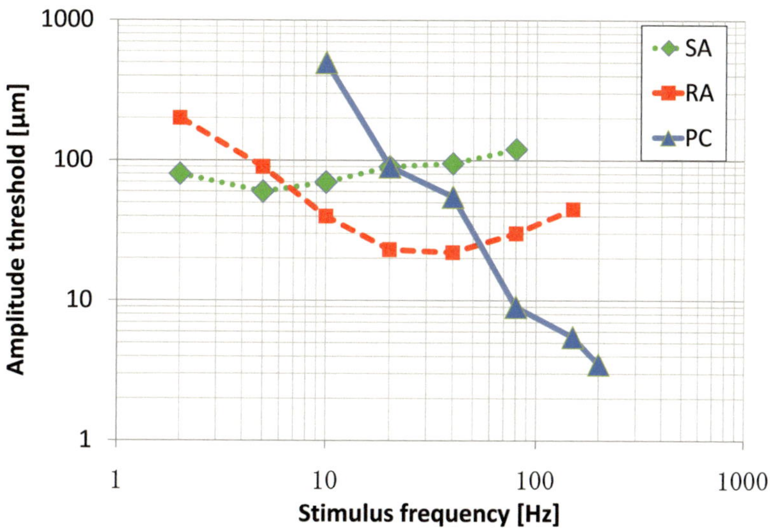

Fig. 2. Responsibility of each receptors (adapted from Freeman & Johnson, 1982)

2.3 Additional sensation

Furthermore, from clinical psychology's view, some sensations, such as pain, itchy, tickle, feel good, have their special dimension. Each of them is linked to one another, so the sensory information is more complex than what the conventional sensor can acquire. In order to detect and record and transmit tactile sensation of human, the tactile sensor should also have these complex sensitivities.

2.4 Feedbacks from cerebella

The complexities of these sensations are mainly caused by the cerebral feedbacks. These sensations are strongly affected by the emotion, knowledge or other information. These information also change the sensing ranges dynamically. In addition, as sensor hardware, the wirings of the sensors are also important for these sensations.

For example, the signals of pain sensation has time lag. These are the first pain and the second pain. The difference between the two is the transmitted path and the transmission speed. The first pain use A δ fiber which has myelin sheath, 13 - 22 µm gauge, and 70 - 120 m/s transmission speed, the other hand the second pain use C fiber which doesn't have myelin sheath, 0.2 - 1.0 µm gauge, and 0.2 - 2.0 m/s transmission speed.

By Melzack & Wall the gate control theory has been proposed according to these difference of transmission speed (Melzack & Wall, 1962). When the information is captured by the skin the signals are transmitted by between Aδ and C fiber and go into the spain. First, the signal going through Aδ fiber is transmitted toward the cerebellum. The arrival of the signal induces the search of memory. The processed information is transmitted to the T cell in the

spinal dorsal corn, and closes the gate of C fiber. Then the information of pain becomes difficult to be transmitted to the spine.
In tickle sensation, self tickling is not effective. This is because human uses his efferent copy in his tickle sensing. That is, the efferent copy is also a part of sensing information.

3. Conventional mechanical sensors using physical principles

3.1 Force sensors

Conventional tactile sensors have been created from some principles of physics. They record the phenomena of the contact status using some physical principles. In order to record the deformation or stress information, many tactile sensors have been developed. Strain gauge, piezoelectric effect, pressure sensitive rubber, diaphragm, photometric pressure gauge, and SAW force sensor, et al.

3.1.1 Strain gauge

A load cell usually uses a strain gauge. Through a mechanical arrangement, the force being sensed deforms a strain gauge. The strain gauge converts the deformation (strain) to electrical signals. The electrical signal output is typically in the order of a few millivolts and requires amplification by an instrumentation amplifier before it can be used.
A strain gauge takes advantage of the physical property of electrical conductance's dependency.

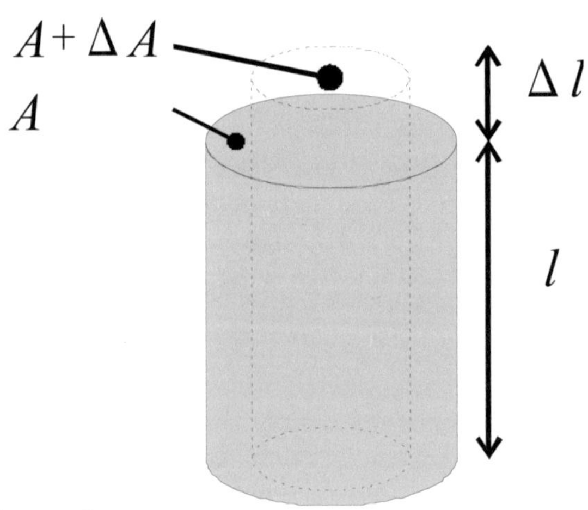

Fig. 3. Deformation of a conductor

$$R = \frac{\rho l}{A} \tag{2}$$

By differentiate the equation, we get

$$\frac{\Delta R}{R} = \frac{\Delta \rho}{\rho} + \frac{\Delta l}{l} - \frac{\Delta A}{A} \tag{3}$$

$$= (1 + 2\sigma)\frac{\Delta l}{l} \tag{4}$$

ρ: Resistance ratio,
R: Resistance value
l: Length
A: Cross section
σ: Poisson's ratio
This $(1+2\sigma)$ is called as gauge factor.

3.1.2 Piezoelectric device

A piezoelectric device uses a piezoelectricity effect. Piezoelectricity is the ability of some materials to generate an electric potential in response to applied mechanical stress. That is, this effect translates the strain information toward electric voltage. A PVDF also has a piezoelectricity. By using this characteristic some force sensors are created.

$$\frac{\Delta\rho}{\rho} = \pi E \frac{\Delta l}{l} \tag{4}$$

With the equation (2) and (4) we get;

$$\frac{\Delta R}{R} = (\pi + 1 + 2\sigma)\frac{\Delta l}{l} \tag{5}$$

π: Piezoresistance coefficient,
E: Young's modulus
This $(\pi+1+2\sigma)$ is called as gauge factor.

3.1.3 Pressure sensitive rubber

A pressure sensitive rubber has been developed for the sheet-switch of the electronic circuits, and has a unique property in that it conducts electric current only when compressed, and acts as an insulator when the pressure is released. This patented material is a composite of an elastomer and specially treated carbon particles, and is available in gray-black flexible sheet form, 0.5 mm in thickness.

3.1.4 Optical diaphragm

There is an interferometer sensor with optical diaphragm. Using the micro electro mechanical system technology the sensor has been made.

3.1.5 SAW force sensor

A SAW (Surface Acoustic Wave) force sensor measures the force in the frequency domain. If the force is applied to a SAW device, the phase shift occurs on the SAW signal. By recording the frequency shift the sensor can measure the force.

3.2 Thermometer

In general use, thermometer is not treated as tactile sensor. However temperature is also important information for tactile sensation. There are some contact type thermometers that are able to use as a tactile sensor; e.g. bi-metal, thermistor, thermocouple, thermal-diode, and optical fibers, et al.

3.2.1 Bi-metal

A bi-metal is a thermal dilation type sensor. This sensor is made of two kinds of metals which have different thermal dilation modulus. By the roll bonding of these metals this device can deform with thermal changes.

3.2.2 Thermistor

A thermistor is a type of resistor with resistance varying according to its temperature. With a first-order approximation, the relation depends on the equation;

$$\Delta R = k\Delta T \tag{6}$$

ΔR = change in resistance
ΔT = change in temperature
k = first-order temperature coefficient of resistance

3.2.3 Thermocouple

A thermocouple is a thermal electromotive force type sensor. When any conductor is subjected to a thermal gradient, it will generate a voltage. This phenomenon is known as Seebeck effect. Thermocouples measure the temperature difference between two points, not absolute temperature. In traditional applications, one of the junctions—the cold junction—was maintained at a reference temperature, while the other end was attached to a probe.

3.2.4 Thermal diode

A thermal diode is a semiconductor junction type device. The p-n junction has 1 - 2 mV/K voltage drop characteristic. By using this phenomenon the temperature is measured. By keeping the electric current to constant and by using the relation between the orthodromic voltage and current the sensor can measure the temperature.

3.2.5 Thermometer using optical fibers

An optical fiber can be used as a thermal sensor, too. There are two types of thermal sensor using an optical fiber. One is an interferometer type thermal sensor, and the other is a polarization type thermal sensor. An interferometer type sensor is using the phase shift against the thermal changes. A polarization type sensor is using double refraction characteristic and the refraction index is changed by the thermal changes. By monitoring the oblations the sensor can measure the thermal changes.

3.3 What is measured by conventional tactile sensors?

Conventional pressure /force/thermal sensors measure information by using some physical laws. These laws are mainly linked to electronic signals. Some of them use resistance shift and the others use electromotive force shift. This is because the signals are easily picked up by the electronic signals and integrated with other actuators. A few of them uses optical fibers for the safety of electric free system and for the accuracy of measuring. The sensors which use electronic signals has amplification problem in itself. This is because the acquired original electronic signals are often small and S/N ratio may be problem. The optical systems are free from these electronic amplification problems in itself.
Though the sensors are useful for the tactile sensor in part, the sensors are not designed for the tactile sensor. So there are some defects for tactile sensor. In the next session we talk about the defects of conventional sensors as tactile sensors.

4. Required ranges for tactile sensors

Previously discussed sensors have not enough ability for tactile sensor. This is partially because the range of the sensor is not enough. For the tactile sensing, we should not consider the sensor based on some physical principles but the sensor design based on the required functions. There are at least three lacked range, the lack of the spatial distribution, frequency distribution, and force distribution.

4.1 Spatial distribution

The spatial distribution means that the tactile sensation has two dimensional sensing distributions. The tactile itself is a boundary between human and the environment. Based on simple topology, the boundary of human whose body has three dimensions must be two dimensional distributions (Fig. 4). Many conventional sensors only measures force/temperature toward one point. This is because the sensor is not developed for the tactile sensor, and the applications often require only one point sensing information. However, in order to detect the changes of environment toward human, position information is also very much important. If there is some large pressure/thermal change information without position, human can detect the hazardous information but cannot understand which way he/she should escape (Fig. 5). In smaller range if there is rubbing movement of some object on the finger, human can detect some time varying information without position information (Fig. 6). Though he/she can detect the changing information, he/she cannot detect the direction of the movement of the object. Without the spatial information human cannot detect the changing direction of the signals.

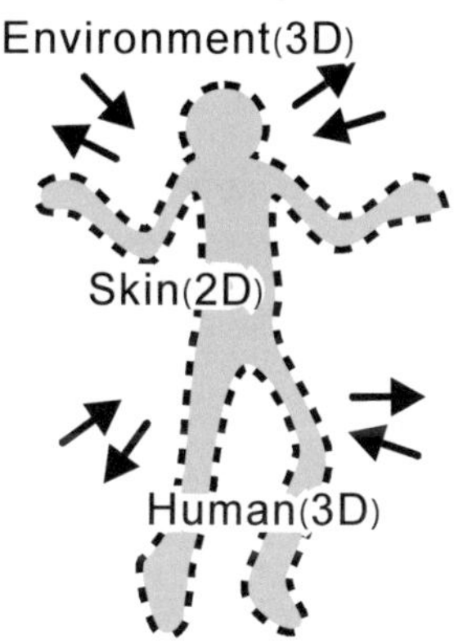

Fig. 4. Topology of human, environment and skin

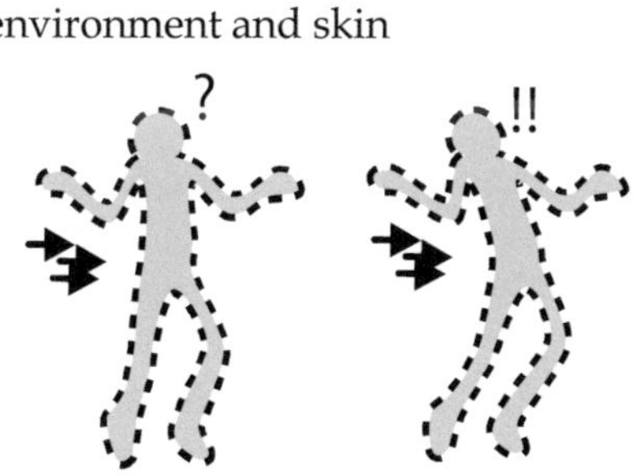

Fig. 5. With/without the distribution information (1): If he did not know the position of the hazardous information, he cannot understand which way to escape.

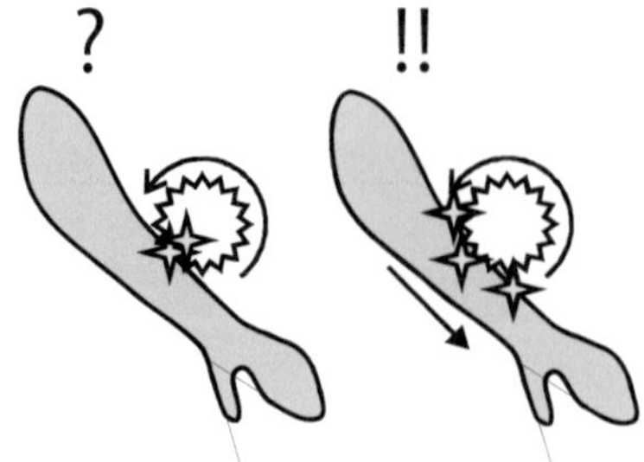

Fig. 6. With/without the distribution information (2): If he didn't acquire the position information, he cannot distinguish which movement occors.

Of course some distributed tactile sensor exists. For example, Pliance (Novel corp.), Tactilus seat type sensing (Sensor Products LLC.), and Flexi force (Nitta corp.). These are composed of small sensors unit and arranged in two dimensional arrays (Fig. 7). They can measure the distribution of added forces. Each of unit is independently connected. To analyse the contact state by many methods from the input data, the independency of the signal is useful. If we use these devise for tactile sensing application, the independency may cause some problems. In creating phase, the number of wirings may be a problem. In measuring phase, because of the independency there is no network for signal processing. In order to acquire the position or movement information, we have to integrate and analyse the information after acquiring the input.

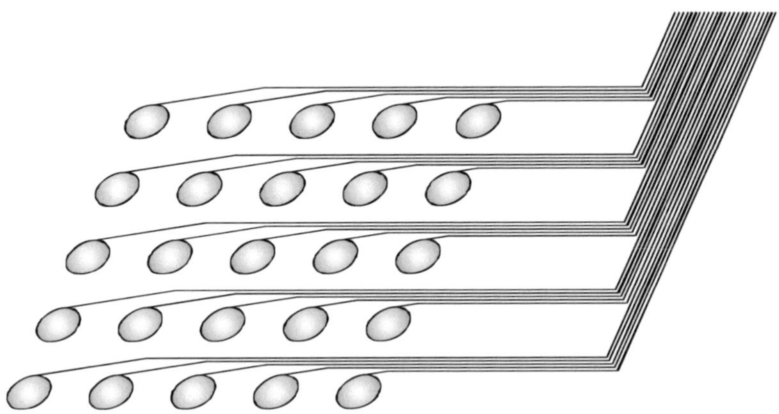

Fig. 7. Array type tactile sensor (ex. Pliance (Novel corp.), Tactilus (Sensor Procucts LLC.), Flexi Force (Nitta corp.))

4.2 Frequency distribution

The frequency distribution means that, from the figure 2, the receptors of human have their own responsibility toward frequency domain. For example, Merkel disks have their responsibility about 5 - 15 Hz, Meissner's corpuscles have about 20 - 50 Hz and Pacinian corpuscles have about 60 - 400 Hz. From this fact the tactile sensor should be able to measure at least the range of 0 - 400 Hz. Many kinds of conventional tactile sensors have such measurement range. So the frequency distribution seems sufficient for the tactile sensor. This is because the sensor often measures only one point. Sampling with few points will make the responsibility of the sensor faster.

4.2.1 Information transfer problem

If we treat the frequency distribution and the spatial distribution at the same time, the frequency distribution becomes difficult problem to solve. Though the ability of the sensors toward the frequency distribution is high, the multiple sensors arranged in two dimensional arrays requires some data collecting method, such as scanning or matrix switching.
Methods of acquiring two dimensional discrete data are critical for frequency distribution. Simple scanning method requires $n \times n$ ordered wirings and $n \times n$ switching device. Matrix switching method requires $2n$ ordered wirings and $n \times n$ switching device. Each method requires $n \times n$ ordered scanning speed. This is because each method aims to get all of the acquired information.

4.2.2 Imaging devices for information transfer

Here, the imaging device also has such switching technology, CMOS (Complementary Metal Oxide Semiconductor) imaging sensor and CCD (Charge coupled device) imaging sensor. CMOS imaging sensor is known as an active-pixel sensor (APS), also commonly written active pixel sensor. It is an image sensor consisting of an integrated circuit containing an array of pixel sensors, each pixel containing a photodetector and an active amplifier. There are many types of active pixel sensors including the CMOS imaging sensor used most commonly in web cameras. This imaging sensor is produced by a CMOS process, so it is also known as a CMOS imaging sensor. Because of its simplicity CMOS imaging device can realize block scanning. By separating the imaging area to some blocks, the device can scan each block simultaneously. In recent years Sony Inc. create fast scanning chip by specialized design (Barth, et al. 2007). CCD itself is an analog shift register, enabling electric charges to be transported through successive capacitors controlled by a clock signal. Charge coupled devices can be used as a form of memory or for delaying analog, sampled signals. By using this device CCD imaging sensor is created for serializing parallel analog signals.

4.2.3 Other novel devices for information transfer

Another communicating device, Two-Dimensional Signal Transmission (2DST), is developed by Shinoda, et al (Shinoda, et al. 2007). This device realizes the communication between each element without wiring them independently. The device is made from some layered conductive sheets and by using microwave confined around the surface it realize the low power and high security communication. This technology is designed for the use of tactile information, and now developing. By using the special transmitting protocol, this device may realize high speed transmission of information and compression technology.
Some methods are compressing sensing information without losing important aspects for tactile information. The soft tribo-sensor using PVDF Film is created by Jiang, et al (Jiang, et al. 1999). Though this sensor has only one dimensional measurement point, it can measure high frequency pressure change. By scanning the sensor itself on skin surfaces, it can measure the difference of them. Another thin and flexible tactile sensor is made of ordinal pressure-conductive rubber, though, the wirings of the sensor is very few (Shimojo & Ishikawa. 1990). Furthermore the sensor itself is flexible sheet. These sensors have been used as skins of some robots. The measurement information of the sensor is limited only the position of center of mass and the mass itself. By limiting the information the sensor require only four wirings for each area. For example, manipulation of some object with robot arm requires only this information. That is, the application decides the required information of the sensor, so the limitation of information matters little.

4.3 Force distribution

4.3.1 Range, dimension and material

The force distribution means that tactile sensor should measure the required force range and force vectors. Some sensors have their specific sensing ranges owing to their physical law.

If the physical law define the range which is different from humans' one, we have to prepare many types of sensing devices for the sufficient human sensing range. So the sensing method itself should not define the range. The ideal sensor should have enough range for human. However there is still no such sensor in the world. The second best sensor is a range changeable sensor by designing the sensing element without changing the sensing method.

The force vector information is also important. A force (F)/torque (T) toward one point has three dimensional components, F_x, F_y, F_z, T_x, T_y and T_z. Some conventional sensors can measure such information on only one point, but the spatial and force distribution occurs simultaneously. So the ideal sensor should measure the distribution of force vectors.

Additionally the sensor itself should have near or the same characteristics of material with the human. For example, Young's modulus, Poisson's ratio and friction coefficient are important aspects for the sensed information. This is because the stress and the strain are indivisible. So the sensing information between by strained sensing surface and by unstrained sensing surface is different. Furthermore, the friction coefficient also should be the same between the sensor and human. The most important thing in the tactile sensing is that the reproduction of the same contact state between the contact by the sensor and the contact by human skin.

4.3.2 Some devices for force distribution

Here Kamiyama, et al. proposed a tactile sensor called GelForce (Kamiyama, et al. 2005). The sensor is made of silicone rubber and imaging device. Inside the silicone rubber there is two layered marker patterns. By capturing the displacement of the markers with the imaging device, it can reconstruct the measured information. In the previous section the imaging device has well switching technology, so the use of imaging device is proper for the two dimensional sensing. The sensor can measure the force and spatial distribution simultaneously. It measures two dimensional distributions of three dimensional forces (x, y, z) and three dimensional torques (x, y, z). Furthermore the simplicity of the component of the sensor with silicone rubber, the responsibility can be changed easily. With this feature it realizes almost the same characteristics of material with human. The authors also research the similar sensor, named reflection-type tactile sensor, with the use of silicone rubber and imaging device (Saga, et al. 2007) (Fig. 8). The sensor can measure the displacement of the

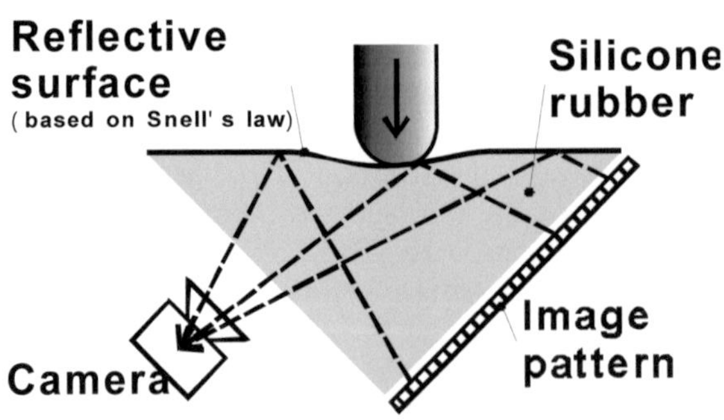

Fig. 8. Reflection type tactile sensor (Saga, et al. 2007)

sensor surface by using the total reflection of the contact surface. According to the use of the reflection image, the sensor realizes high resolution sensitivity and can detect 0.01 mm displacement of sensor surface. The same with the GelForce, the sensor can design the characteristics of the material.

4.4 Every distribution requires?

Though each of the distribution still cannot be combined now, the combination of these ranges will open the new sensing features (pain, itchy, tickle, and feel good) for the tactile sensor. Additionally, the important aspect of the tactile sensation is as follows. However we cannot compose perfect sensor with these distributions, we should well consider the application of the sensor and design it. Again, the application decides the required information of the sensor, so the limitation of information matters little.

5. Active touch for tactile sensors

The activeness of touch plays an important role (Gibson, 1962). This is a very much different thing from other sensation, such as vision and auditory. The tactile sensation uses not only the sensing information itself but also the efferent copy of arm/hand/finger movement.
In augmented reality researches, this sensing and efferent copy is more clearly examined. Nojima, et al. proposed the tool for augmented reality, SmartTool, by using a real time sensor and a haptic device (Nojima, et al. 2002). The sensor on the SmartTool measures the real environment, and the tool send the user the captured information through haptic sensation. The sensors are on the tool tip and it is the same point as the working point of the tool. Therefore, this device realizes "What the Sensors Detects is What the Tool Touches" (Fig 9, 10). So the sensing information and the efferent copy of user are integrated naturally. This discussion of active touch should be applied to tactile sensors.

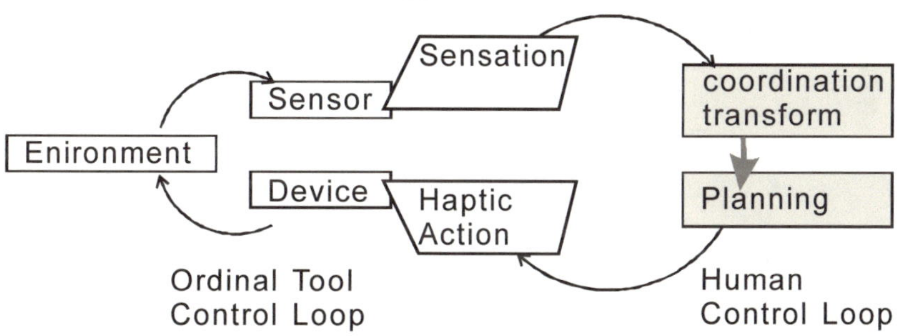

Fig. 9. Loop with ordinal tool: For human sensation and display is not the same point

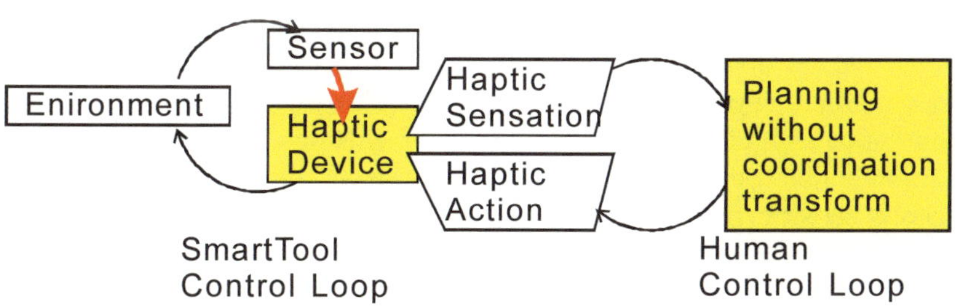

Fig. 10. Loop with SmartTool: For human sensation and display is the same point

5.1 Simultaneous sensing and display

In order to realize the active touch, sensing and display should be carried out simultaneously. (Here, the directions of sensing and display are to both environment and human in augmented reality. However to simplify the discussion we consider the direction only to human.) Furthermore, the most different thing of tactile sensation from other sensory information is their bilateral input/output. If someone touches something, the thing will always touch him/her. Touching and being touched occurs simultaneously. That is, the sensor itself has to be the display of tactile information simultaneously. In case of haptic device like SmartTools, the simultaneous sensing and display is realized by using rigid tool (Fig. 11). Because the rigid tool expands the force position to its body, the sensing point and displaying point can be separated.

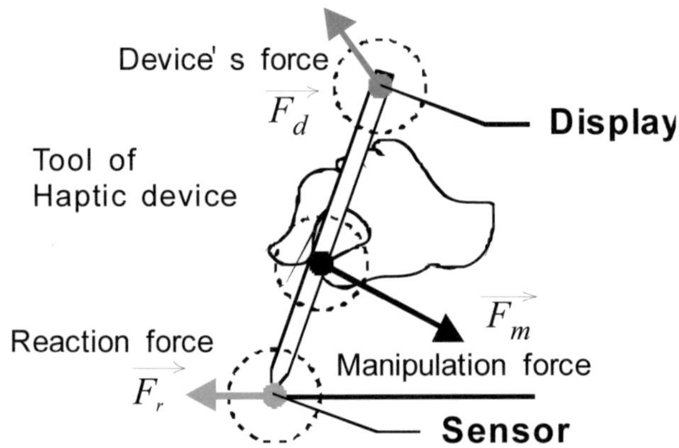

Fig. 11. Sensor & display of SmartTools

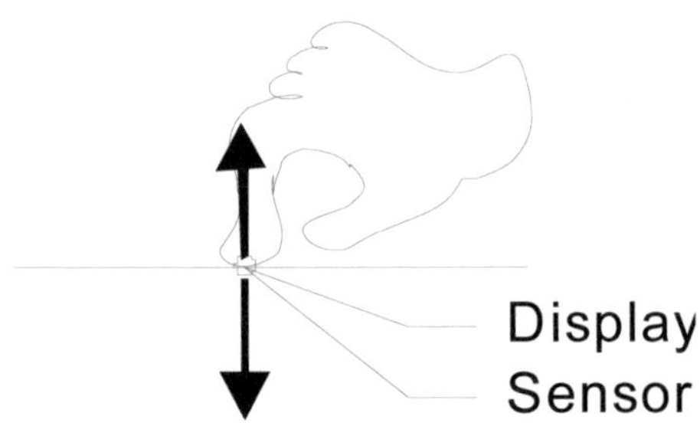

Fig. 12. Sensor & display of tactile device

However such devices are difficult to create for tactile device. This is because the contact phenomenon itself occurs in two dimensional surface. If sensor touched to some object from one side of the surface, there is no space left for the display (Fig. 12). The sensor side, there is a sensor arrays and the object side there is an object. If the sensor is created to be sparse and the displaying element is placed, the sensing and display can be realized at almost the same position simultaneously. However the resolution of sensing and displaying becomes sparse and the position is not "precisely" the same, but "almost" the same. The precise realization of simultaneous sensing and displaying is difficult in principle. So the different way of sensing technique should be realized.

In previous section, there is some new type of sensors using imaging device. These sensors are using the diffused light from the markers. So the sensing surface is free from some mechanical devices. Especially the reflection-type tactile sensor has only transparent silicone rubber at the sensor surface. The only important things are the transparency and the deformability. So the sensor can use the transparent functional fluid such as Magneto-Rheological fluid. It may realize the sensing and displaying simultaneously at precisely the same position.

5.2 Simultaneous sensing and display

The ideal tactile sensor may be like a mirror for contact object (Fig 13). The mirror can deform and change its hardness and contact with some object with ideal shape and softness. When the contact object changes its pressure, the mirror can change according to the change of pressure. In these days there is still no such device, but the development of such device will create the new world of communication.

Fig. 13. Ideal tactile sensor/display?

6. Acknowledgement

This work is partially supported by KAKENHI (19860012), Grant-in-Aid for Young Scientists (Start-up).

7. References

J. Barth, W. Reohr, P. Parries, G. Fredeman, J. Golz, S. Schuster, R. Matick, H. Hunter, C. Tanner, J. Harig, H. Kim, B. Khan, J. Griesemer, R. Havreluk, K. Yanagisawa, T. Kirihata & S. Iyer. (2007). A 500MHz Random Cycle 1.5ns-Latency, SOI Embedded DRAM Macro Featuring a 3T Micro Sense Amplifier. In *Digest of Technical Papers of*

IEEE International Solid-State Circuits Conference, IEEE, ISSN: 0193-6530, ISBN: 1-4244-0853-9

A. W. Freeman & K. O. Johnson. (1982). A model accounting for effects of vibratory amplitude on responses of cutaneous mechanoreceptors in macaque monkey. *Journal of Physiology*, Vol. 323, pp. 43-64, ISSN:0886-1714

J. J. Gibson. (1962). Observations on active touch. *Psychological Review*, Vol. 69, pp. 477-491, APA Journal, ISSN:0033-295X.

K. Kamiyama, K. Vlack, H. Kajimoto, N. Kawakami & S. Tachi. (2005). Vision-Based Sensor for Real-Time Measuring of Surface Traction Fields. *IEEE Computer Graphics & Applications Magazine*, Vol. 25, No. 1, pp. 68-75, IEEE, ISSN:0272-1716

Eric R. Kandel, James H. Schwartz & Thomas M. Jessell (ed.). (2000). *Principle of Neural Science*. Appleton & Lange, ISBN:978-0838577011, U.S.A.

R. Melzack & P. D. Wall. (1962). On the nature of cutaneous sensory mechanisms. *Brain*, Vol. 85, pp. 331-356, Oxford Journal, ISSN:0006-8950

T. Nojima, D. Sekiguchi, M. Inami & S. Tachi. (2002) The SmartTool: A system for augmented reality of haptics. In *Proceedings of IEEE Virtual Reality*, IEEE, ISBN: 978-0769514925, U.S.A.

S. Saga, K. Vlack, H. Kajimoto & S. Tachi. (2005). Haptic Video, *In Conference Abstracts and Applications of SIGGRAPH2005*, ACM SIGGRAPH, Los Angeles.

S. Saga, H. Kajimoto & S. Tachi. (2007). High-resolution Tactile Sensor using the Deformation of a Reflection Image. *Sensor Review*, Vol. 27, pp. 35-42, Emerald Group Publishing Limited, ISSN: 0260-2288.

H. Shinoda, Y. Makino, N. Yamahira & H. Itai. (2007). Surface Sensor Network Using Inductive Signal Transmission Layer. In *Proceedings of 4th International Conference on Networked Sensing Systems*, ACM SIGMOBILE, ISBN: 1-59593-343-3, U.S.A.

M. Shimojo & M. Ishikawa. (1990). Thin and Flexible Position Sensor. *Journal of Robotics and Mechatronics*, Vol. 2, No. 1, pp. 38-41, Fuji Technology Press, ISSN: 0915-3942.

2

Torque Sensors for Robot Joint Control

Dzmitry Tsetserukou and Susumu Tachi
University of Tokyo
Japan

1. Introduction

In the field of service robotics, there is a growing need for robots capable of physical interaction with humans to assist with daily life tasks. The desired coexistence of robotic systems and humans in the same physical domain (sharing the same workspace and actually cooperating in a physical manner) poses very fundamental problem of ensuring safety to the user and robot. Even without wrong programming, a robot, moving freely in a human environment, is potentially dangerous because of its large moving masses, powerful actuators, and unpredictably complex behavior. Design and programming of the robots exhibiting intrinsically safe behavior in a human domain are great challenges in robotics because such robots have to deal with unstructured time-varying environment. Several humanoid robots aimed at integration into people environment were developed (Sakagami et al., 2002), (Kaneko et al., 2004). However, despite the splendid means for sensing the environment (visual, audio, and haptics), the 6-axis force/torque sensors attached at the tip of the robot arm and a stereo vision system which is slow to track the changing environment in real-time, are only the abilities to anticipate and handle the collision. The rest parts of the robot body (forearm, elbow, upper arm, shoulder, and torso) are presenting the significant danger not only for human being, but also for the robot structure itself.

Effective methods on enhancement of contact detection ability of manipulator were reported. To avoid collisions in time-varying environment, Lumelsky & Cheung (2001) proposed to cover manipulator with a sensitive skin capable of detecting nearby objects. Mitsunaga et al. (2006) progressively improved the tactile ability of the robot through covering its entire body with piezoelectric-film-based tactile sensors. Since this device integrates a huge amount of small sensors incorporated into soft layer and requires the complicated wiring and signal processing hardware, it has high cost and reliability issues. The high-speed vision system attached to the robot arm aimed at real-time collision avoidance (Morikawa et al., 2007) presumes usage of expensive detectors, complex signal processing techniques, and issues of self-body extraction from the camera view area.

Is should be noted, that such tasks under human supervision as transporting the object, leading the robot tip via force-following, performing the assembling tasks, require the processing algorithm of contact state. Finding the technical solution for trade-off between performance and safety is the target of a new manipulation technology. To cope with this issue, an active compliance control implying fast joint torque controlling based on measuring the applied external torque in each joint was developed. The first embodiment of torque measurement is the integration of a torque sensor into each joint of the manipulator.

The impedance control generates compliant trajectory based on information on the measured external torque. Such approach has two main advantages: (1) sensor detects not only forces applied to the hand but also those exerted at other points on the manipulator, (2) and allows to increase the performance of fast movements by active vibration damping. Several attempts have been made by researchers to improve joint torque control. Wu & Paul (1980) proposed a simple, wide bandwidth, torque servo system using strain-gauge-based joint torque sensor. The torque-controlled lightweight robot arm with high load to weight ratio was developed by Hirzinger et al. (2001). Each joint of the arm is facilitated with strain-gauge-based torque sensor, position sensor and piezzo-brake. Sakaki & Iwakane (1992) proposed to use a compact motor drive with embedded magnetostrictive torque sensor for impedance control of the robot arm. The approach of torque measurement through the elasticity of the harmonic drive flexsplines allows keeping the same stiffness and mechanical structure of the robot (Hashimoto et al., 1993), (Golder et al., 2001). This method requires the strain gauges to be installed on the flexsplines. The crucial shortcomings of the torque measurement approaches mentioned above will be discussed in Section 2.

The alternative method for increasing the safety level of robot arms interacting with humans is intentionally introducing compliance at the mechanical design level (passive control). The main idea here is decoupling the rotor inertia from the link inertia through use of passive elasticity (Bicchi & Tonietti, 2004). However, the robot control is complicated by many unknown parameters (e.g. actuator stiffness and damping). Furthermore, compliant transmission negatively affects the performance in term of increased oscillations and settling time.

To realize the safe physical contact of entire robot arm structure with human and to guarantee the collision avoidance, our primary idea is concentrated on the design of a whole-sensitive robot arm (by using distributed torque sensors in each joint). When contact with environment occurs, manipulator automatically generates compliant motion according to the measured external torque and time derivative of torque.

The Chapter is devoted to the design of new optical torque sensors and robot joint control system for safe physical interaction with surroundings.

2. Background of torque measurement techniques

The torque transducer on the motor shaft includes a load cell structure supporting the load and deflecting a known amount in response to applied torque. Deformation of the elastic element is measured in different ways as follows:

1. *Electrically.*
2. *Based on electromagnetic phenomena.*
3. *Optically.*

These approaches include: 1) noncontact transducers measuring relative displacement by light radiation detection, Hall effect, Faraday's law, and 2) contacting transducers measuring strain by means of strain gauges, potentiometers, piezoelectric effect, optical fibre.

2.1 Electrical methods

Electrical measurement is essentially associated with strain gauges, capacitive, and piezoelectric sensing. Strain gauge operation is based on variations in electrical resistance with strain. When force is applied, strain changes the electrical resistance of gauges proportionally to the load. Silicon semiconductor strain gauges are often used due to high sensitivity. Strain in silicon causes its bulk resistance to change, producing a signal 75 times stronger than conventional foil gauges where resistance changes are only due to dimensional changes in the resistor. Despite such benefits of strain gauges as high linearity, about 0.03%-2.5% of full scale (FS), high resolution of 1-3 mV/V, their maximum allowable strain is close to their breaking point. To guarantee overload protection of transducers, mechanical stops limiting deflections of flexures are necessary. Very stiff sensors may only deflect a few ten-thousandths of a millimeter. Production of limit stops with such small clearances is very difficult. Strain-gauge-based torque/force sensors are greatly subjected to radial and other force components. Semiconductor and foil gauges require elaborate process for attachment by a specialist. Another shortcoming of these sensors is their high sensitivity to electrical noise and temperature (Westbrook & Turner, 1994).

Piezoelectric torque sensors are similar in operation to strain gauges and based on the phenomenon, in which a crystal becomes electrically charged under the action of mechanical stress. High stiffness and strength enable sensors to be directly inserted into the torsion member. An example of the piezoelectric effect is the invention (Kovacich et al., 2002), where authors exploited changes in the resonant frequency of the piezoelectric element as a measure of the strain, to which the torsion member is subjected. Extremely high accuracy (0.03% of FS) and high signal output are their main advantages. Drawbacks restricting their application are high cost and nonlinear output.

Many torque transducers are based on measuring the relative angle between the two ends of the torsion bar. This principle was realized in the differential capacitive sensor for measurement of the relative angle (Fulmek et al., 2002). The transducer is noncontact, robust, and compact. Two rotatable electrodes are placed between two sensor plates. The relative angle between the two rotors and the absolute position of the rotor blades are calculated from measurement of capacitive coupling between different transmitting stator segments and a single receiving electrode. Its drawback is high sensitivity to radial displacement and high cost. The relationship between the capacitor capacity and permittivity of the dielectric material between the capacitor plates also was used in patent (Madni et al., 2004). In this invention, the apertured metal cage shielding a dielectric rotor is placed between capacitor plate rings fixed on opposite sides of a torsion bar. The relative rotation of the apertured conductive plates and the dielectric rotor changes the overall differential capacitance of the system in proportion to torque.

2.2 Electromagnetic phenomena

Faraday's law (inductive sensors), magnetostriction, and magnetoelastic effect are used in electromagnetic sensors. Vischer & Khatib (1995) used linear variable differential transformers (LVDT) in the torque sensor. The hub-spoke structure of this torque sensor is given in Fig. 1 (reference numeral 1 denotes a flexible beam, 2 denotes a LVDT).

The main advantage of LVDT is their high degree of robustness, remarkable resolution of about 0.1μm, good accuracy (0.01-0.3%), and easy installation and calibration. High reliability is derived from their operation principle based on magnetic transfer eliminating physical contact across the sensing element. The strong relationship between core position and output voltage of secondary coils yields excellent resolution. Inductive sensors suffer from reduction in signal at very low frequencies, and they are affected by electromagnetic noise. The smallest LVDT made by the Lucas Schaevitz in XS-B series weighs only 4.36 g and has an outer diameter of 4.77 mm, but its length of 22.4 mm complicates compact overall dimensions of the torque transducer.

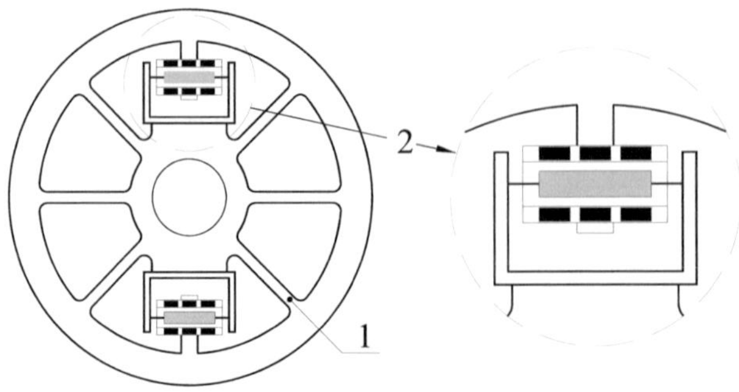

Fig. 1. Six-beam torque sensor with LVDT transducer (Vischer & Khatib, 1995)

The effect, in which stress applied to the material causes a change in its magnetization, is known as the Villari effect or magnetostriction phenomena. Magnetostrictive torque sensors consist of rotational shaft, having a magnetostrictive metal layer of a chevron configuration formed on the shaft, an exiting coil, and a pickup coil for detecting the magnetic property change of the magnetostrictive layer. When AC current is applied to the exciting core, the magnetostrictive layer is excited. Torque applied to the magnetostrictive element generates stress. Permeability is changed by the Villari effect, and inductive output is generated in the pickup coil with torsional load (Shinoura, 2003). Advantages of the sensor are nonphysical contact between the shaft and housing, and high torsional stiffness. Drawbacks are complicated manufacturing, bulky heavy structure, need for a robust magnetic shield, and insufficient performance (linearity of 3-5% FS, hysteresis of 2-3%, resolution of 10 mV/V) (Ishino et al., 1996).

2.3 Optical approaches

A light source, photosensor, and solid object modifying the amount of light incident on the optical detector are necessary to measure displacement between unmovable and flexible parts of the optical sensor. Photosensors have such drawbacks as nonlinearity and temperature sensitivity, but they are considerably more reliable, cheap, and simplified in design than other sensors. A displacement is detected by interrupting light between source and detector, changing the intensity of reflected light, or the relative movement of source and detector. Hirose & Yoneda (1990) have significantly contributed to research on the optical force/torque sensors. They proposed using a split photosensor to detect displacement of the light source (LED) caused by applied force in two directions. This sensor is shown in Fig. 2 (1 denotes a force loaded adapter plate, 2 denotes a load cell (elastic element), 3 denotes a photodetector, 4 denotes a LED, 5 denotes an installation adapter plate).

When force is applied to the plate 1, elastic solid body 2 deflects LED light incident on photosensor 3. Thus, magnitude of the photodetector output responds on exerting force/moment. In cooperation with the Minebea Co., OPFT series of 6-axis optical force/torque sensors were manufactured (Minebea). Compared to conventional strain-gauge-based transducers, they are more compact, lightweight, and cheap, but they have complicated calibration due to nonlinear output, require application of DSP for real-time computation of measured force, and have an average accuracy of 5% FS.

At Nara Institute of Science and Technology, a 6-axis optical force/torque sensor was developed for fMRI application (Takahashi et al., 2003). The sensor was made from acrylic resin to eliminate any metal sensor components that generate fMRI signal noise. The layout of the transducer is shown in Fig. 3. The elastic frame has a Y-topology with S-shaped beams to enable 6-DOF displacement of the sensing face. Force exerted by users deflects the elastic frame, altering the intensity of light falling on optical fiber. The sensor provides accuracy of 2.65% for measurement of moment M_Z. The transducer is complicated and intended only for narrow applications.

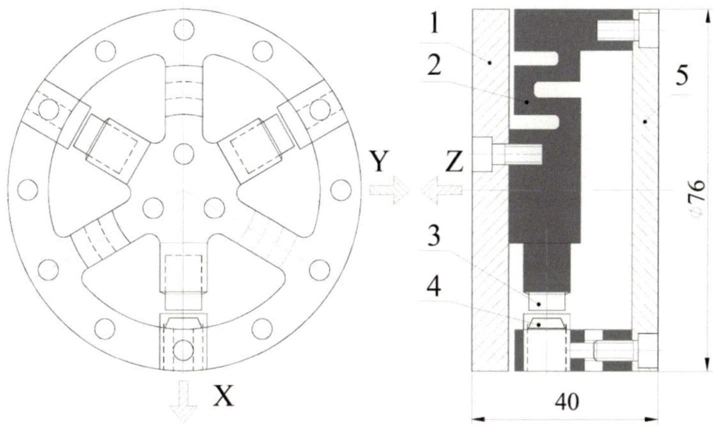

Fig. 2. Layout of the 6-axis optical force/torque sensor (Hirose & Yoneda, 1990)

The idea of the patent (Okutani & Nakazawa, 1993) is calculation of torque by measuring the angle of twist of the torsion shaft through detection of differences in the rotation position of disks at opposite sides of the torsion shaft using an optical encoder. The encoder-type torque sensor suffers from large overall dimensions and requires very precise relative installation of apertured disks. To overcome this, the author (Horton, 2004) invented a torque sensor including a source of optical radiation, a two-dimensional array of a radiation detector and two modulating apertured disks placed between the source and detector. When torque is applied to the shaft, the relative position between disks alters overlapping slots and hence the size of the apertures, which control the pattern of light incident upon the optical detector.

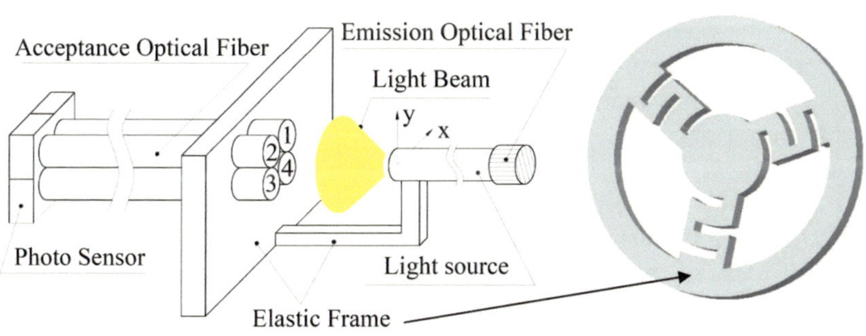

Fig. 3. Layout of the 6-axis optical force/torque sensor (Takahashi et al., 2003)

The 6-axis force/torque converter using LEDs and photodiodes is described in (Hilton, 1989).

3. Development of a new optical torque sensor

3.1 Senor specifications

The specific application (i.e. robot arm interacting with human beings in a safe manner) introduces special requirements to the design of torque sensor. New anthropomorphic robot arm iSoRA (intelligent **So**ft **R**obot **A**rm) enabling integration of torque sensors was developed. The 3D model of the robot arm, cross-section of the shoulder joint with built-in torque sensors, and frame assignments are shown in Fig. 4.

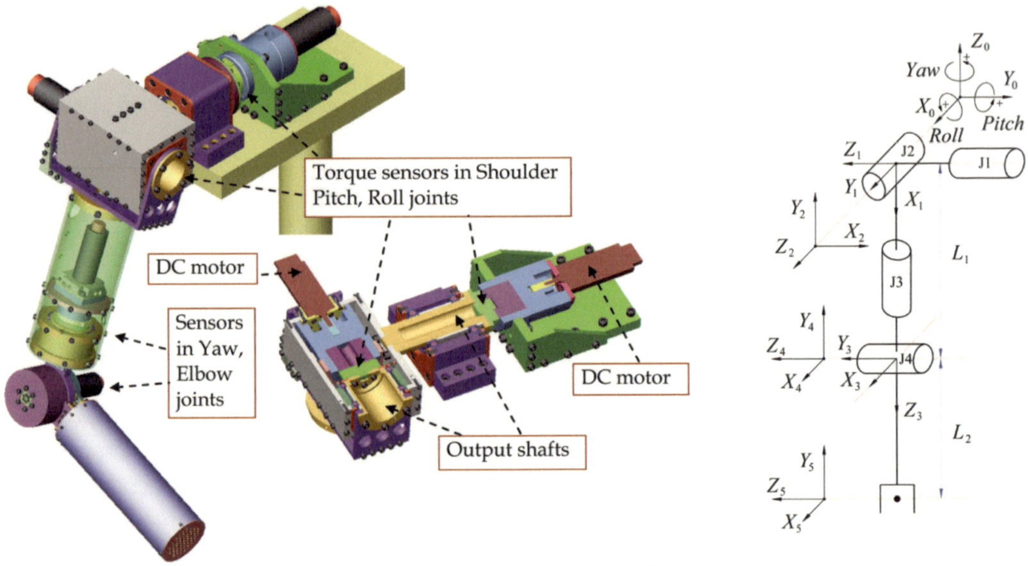

a) Robot arm b) Cross section of the robot shoulder c) Frame assignments

Fig. 4. 3D model of the new anthropomorphic robot arm

The list of desirable properties of torque sensor aimed at integration into anthropomorphic robot arm is given below.

1. Addition of the torque sensor to a robot joint should not require redesign of the joint and should cause minimal modification in kinematics and dynamics. Therefore, lightweight sensor with small width is preferable.
2. Noise created by current passing through the DC motor at each joint of the robot arm should not affect the sensing element, and usage of electromagnetic sensors should be avoided in this application.
3. The angle of twist of the movable part of the transducer at the maximum expected external torque should enable exploiting as wide range of detector sensitivity as possible to achieve high signal-to-noise ratio and sensor resolution.
4. Torsional stiffness of the sensor should not considerably reduce the natural frequency of the robot arm and deteriorate the position accuracy. This introduces the trade-off in stiffness maximization while maintaining high sensitivity.
5. Another difficulty in design of the torque sensor is hysteresis elimination. Most metals used as flexures have very little hysteresis. Bolted, press fit, and welded joints near the flexure introduce hysteresis. Hence, mechanical structure should be machined from a single piece of metal.

6. Influence from any of the nontorsional components of load should be canceled to guarantee precise measurement of torque T that is moment around Z-axis M_Z in 6-axis sensors.
7. Behavior of the sensing element output and mechanical structure should be as close to linear as possible.
8. Simple to manufacture, low-cost, and robust.

Optical approaches of torque measurement satisfy the demands of compact sizes, light in weight, and robustness. The small influence of electrical noise created by DC motors on output signal of optical sensors results in high signal-to-noise ratio and sensor resolution. Therefore, we decided to employ this technique to measure torque in robot joints.

3.2 Design of new optical torque sensors

The novelty of our method is application of the ultra-small size photointerrupter (PI) as sensitive element to measure relative motion of sensor components. The relationship between the output signal and position of the shield plate for RPI-121 (ROHM) is shown in Fig. 5. The linear section of the transferring characteristic corresponding approximately to 0.2 mm is used for detection of the relative displacement of the object. The dimensions of the photointerrupter (RPI-121: 3.6 × 2.6 × 3.3 mm) and weight of 0.05 g enable realization of compact sensor design.

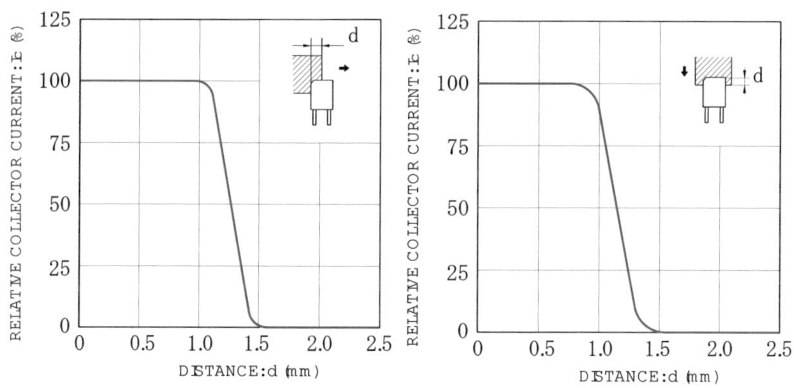

Fig. 5. Relative output vs. Distance (ROHM)

Two mechanical structures were realized to optimize the sensor design: the in-line structure, where detector input and output are displaced axially by the torsion component; and the "in plane" one, where sensor input and output are disposed in one plane and linked by bending radial flexures. The layout of the in-line structure based on a spring with a cross-shaped cross section is shown in Fig. 6. This spring enables large deflections without yielding. The detector consists of input part 1, output part 2, fixed PI 3, shield 4, and cross-shaped spring 5. The operating principle is as follows: when torque T is applied to the input shaft, the spring is deflected, rotating the shield 4. Shield displacement is detected by the degree of interruption of infrared light falling on the phototransistor. The magnitude of the PI output signal corresponds to the applied torque. The "in plane" arrangement of the load cell was designed to decrease the sensor thickness and, therefore, to minimize modification in dimensions and weight of robot joint.

The layout of the structure having hub and three spokes (Y-shaped structure) and 3D 3D assembly model are shown in Fig. 7. The detector consists of inner part 1 connected by flexure 3 with outer part 2, fixed PI 5, slider with shield plate 4, and screws 6. When torque is applied, radial flexures are bent. The shield is adjusted by rotating oppositely located screws 6. The pitch of screws enables smooth movement of the slider along with the shield plate.

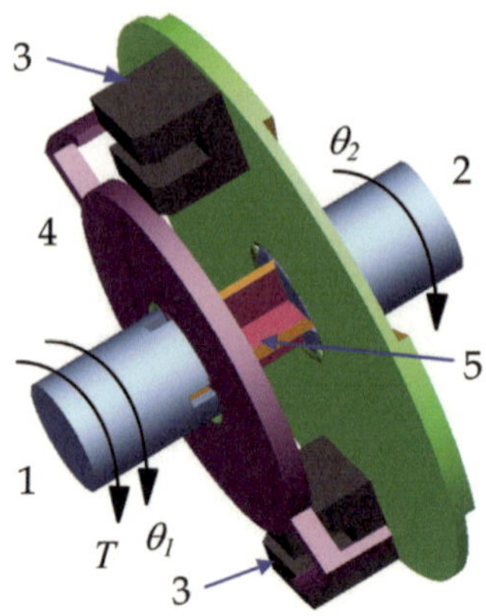

Fig. 6. Construction of the optical torque sensor

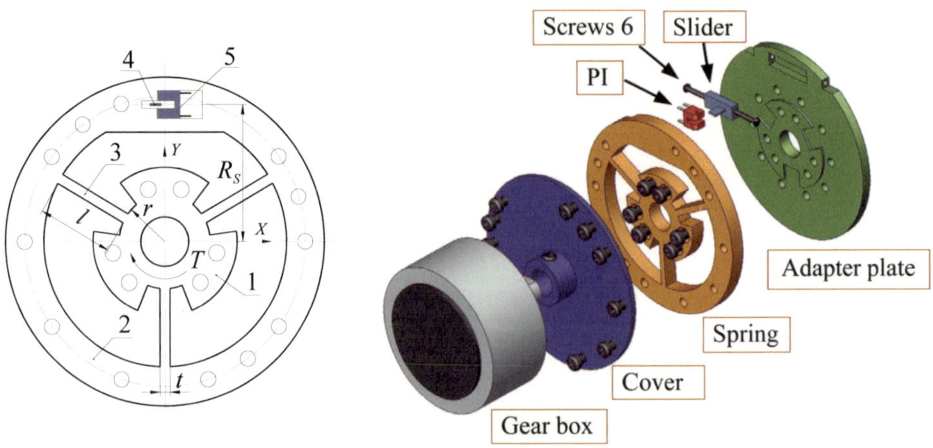

Fig. 7. Layout of hub-spoke spring and position regulator

The relationship between the applied torques to robot arm structure and the resultant angles of twist for the case of linear elastic material is as follows:

$$T = k\theta = k\left(\theta_{in} - \theta_{out}\right), \tag{1}$$

where $T = [\tau_1, \tau_2, ..., \tau_n]^T \in R^n$ is the vector of applied joint torques (Nm); $k = [k_1, k_2, ..., k_n]^T \in R^n$ is the vector of torsional stiffness of the flexures (Nm/rad); $\theta = [\theta_1, \theta_2, ..., \theta_n]^T \in R^n$ is the vector of angles of twist (rad), θ_{in} is the vector of angles of input shaft rotation; θ_{out} is the vector of angles of output shaft rotation.

Since the angle of twist is fairly small, it can be calculated from the displacement of the shield in tangential direction Δx, then Eq. (1) becomes:

$$T = k\Delta x / R_s\,, \tag{2}$$

where R_S is the vector of distances from the sensor axis to the middle of the shield plate in radial direction.

Sensor structure rigidity can be increased by introducing additional evenly distributed spokes (Nicot, 2004). The torsional stiffness of this sensor is derived from:

$$k = 4NEI\left(\frac{1}{l} + \frac{3r}{l^2} + \frac{3r^2}{l^3}\right), \tag{3}$$

where N is the number of spokes, l is the spoke length, E is the modulus of elasticity, r is the inner radius of the sensor (Vischer & Khatib, 1995).

The moment of inertia of spoke cross section I is calculated as:

$$I = \frac{bt^3}{12}, \tag{4}$$

where b is the beam width, t is the beam thickness.

The sensor was designed to withstand torque of 0.8 Nm. The results of analysis using FEM show von Mises stress in MPa under a torque T of 0.8 Nm (Fig. 8a), tangential displacement in mm (Fig. 8b), von Mises stress under a bending moment M_{YZ} of 0.8 Nm (Fig. 8c), and von Mises stress under an axial force F_Z of 10 N (Fig. 8d).

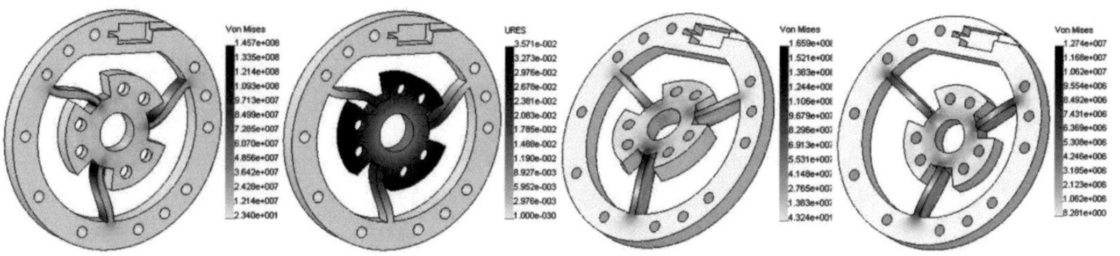

a) b) c) d)

Fig. 8. Results of analysis of hub-spoke spring using FEM

The maximum von Mises stress under torque T of 0.8 Nm equals $\sigma_{MaxVonMises} = 14.57 \cdot 10^7$ N/m^2 < $\sigma_{yield} = 15.0 \cdot 10^7$ N/m^2. The angle of twist of 0.209° is calculated from the tangential displacement. The ability to counteract bending moment is estimated by the coefficient:

$$K_{TM} = \frac{\sigma_{MaxVonMises(T)}}{\sigma_{MaxVonMises(M_{YZ})}}. \tag{5}$$

The hub-spoke spring coefficient K_{TM} equals 0.878. To estimate the ability to counteract axial force F_Z, the same approach is applied:

$$K_{TF} = \frac{\sigma_{MaxVonMises(T)}}{\sigma_{MaxVonMises(F_Z)}}. \tag{6}$$

After substitution of magnitudes, we calculate $K_{TF} = 11.44$. Our sensor was machined from one piece of brass using wire electrical discharge machining (EDM) cutting to eliminate hysteresis and guarantee high strength (Fig. 9). In this sensor, the ultra-small photointerrupter RPI-121 was used. We achieved as small thickness of the sensor as 6.5 mm.

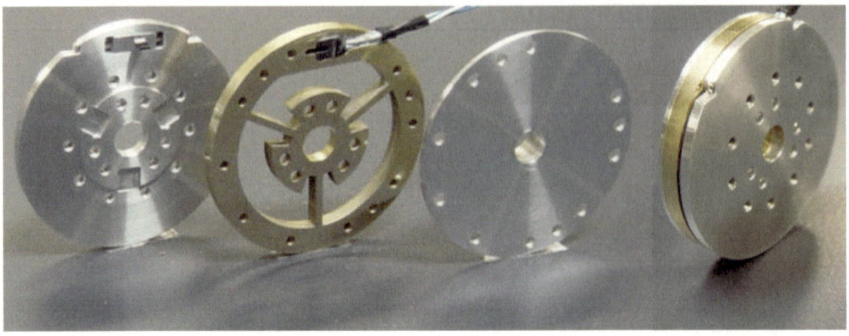

Fig. 9. Optical torque sensor with hub-spoke-shaped flexure

The ring-shaped spring was designed to extend the exploiting range of the PI sensitivity while keeping same strength and outer diameter. Layout and 3D assembly model of the developed optical torque sensor are shown in Fig. 10 (1 designates a shield plate, 2 designates a PI RPI 131, 3 designates a ring-shaped flexure). The flexible ring is connected to the inner and outer parts of the sensor through beams. Inner and outer beams are displaced with an angle of 90° that enables large compliance of the ring-shaped flexure.

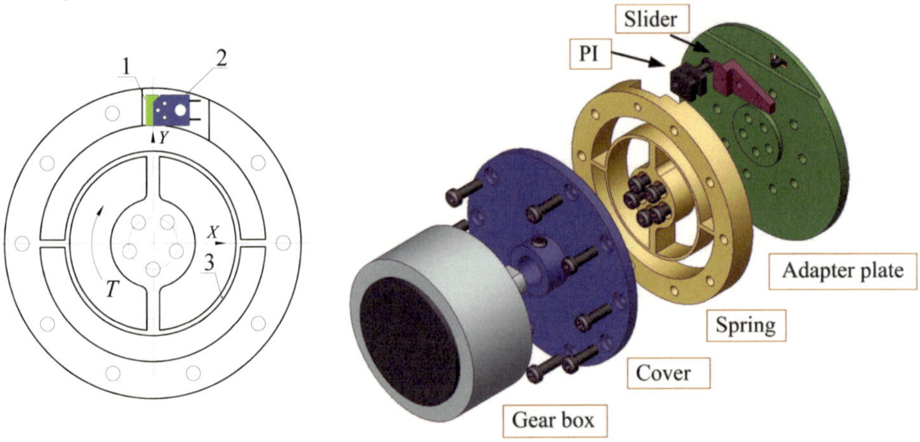

Fig. 10. Ring-shaped topology of the spring

The results of analysis using FEM show von Mises stress in MPa under torque T of 0.8 Nm (Fig. 11a), tangential displacement in mm (Fig. 11b), von Mises stress under bending moment M_{YZ} of 0.8 Nm (Fig. 11c), and von Mises stress under axial force F_Z of 10 N (Fig. 11d).
The maximum von Mises stress under torque T of 0.8 Nm equals $\sigma_{MaxVonMises} = 8.74 \cdot 10^7$ N/m^2 $< \sigma_{yield} = 8.96 \cdot 10^7$ N/m^2. Given structure provides the following coefficients: $K_{TM} = 0.217$, $K_{TF} = 3.56$, and angle of twist θ of 0.4°. Thus, the ring-shaped structure enables magnifying the angle of twist deteriorating the degree of insensitivity to bending torque and axial force. This structure was machined from one piece of aluminium A5052. The components and assembly of the optical torque sensor are shown in Fig. 12. The sensor thickness is 10 mm. The displacement of the shield is measured by photointerrupter RPI-131. The shortcomings of this design are complicated procedure of adjusting the position of the shield relatively photosensor and deficiency of the housing to prevent the optical transducer from damage.

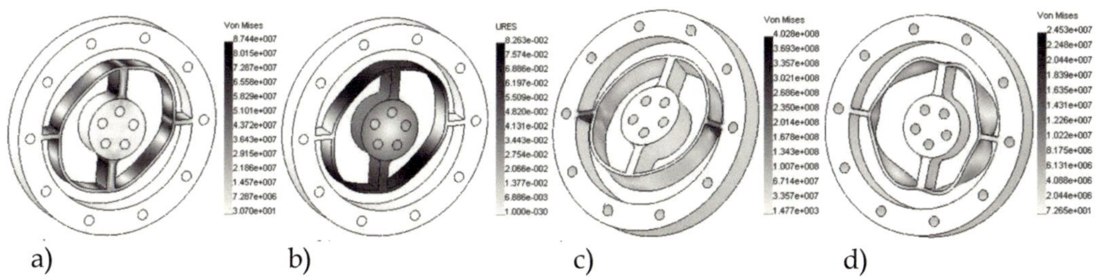

Fig. 11. Result of analysis of ring-shaped flexure using FEM

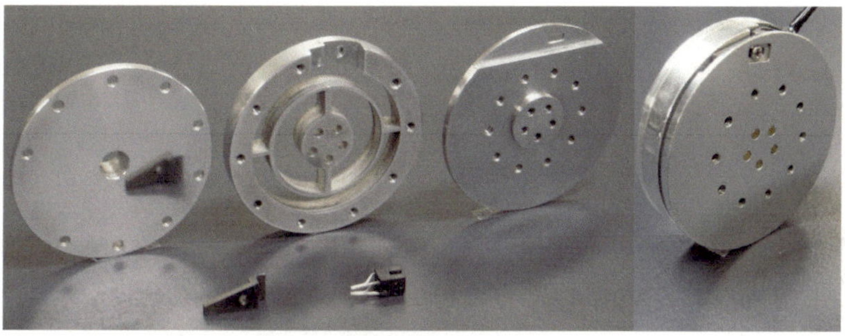

Fig. 12. Optical torque sensor with ring-shaped flexure

The sensor with a ring topology was modified. The layout of the detector with semicircular flexure and 3D model are given in Fig. 13 (1 designates a shield, 2 designates a PI RPI-121, 3 designates a semicircular flexure).

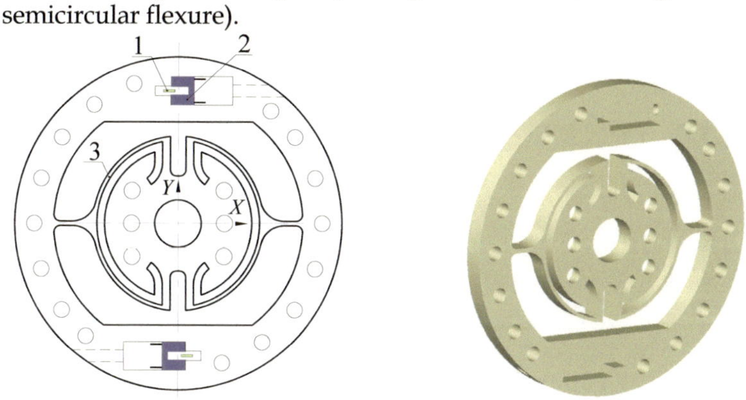

Fig. 13. Semi-ring-shaped spring

The results of analysis using FEM show von Mises stress in MPa under torque T of 0.8 Nm (Fig. 14a), tangential displacement in mm (Fig. 14b), von Mises stress under bending moment M_{YZ} of 0.8 Nm (Fig. 14c), and von Mises stress under axial force F_Z of 10 N (Fig. 14d).

The maximum von Mises stress under maximum loading is less then yield stress $\sigma_{Max\,VonMises}=14.94 \cdot 10^7$ N/m^2 < σ_{yield} = $15.0 \cdot 10^7$ N/m^2. The semicircular flexure provides the following coefficients: K_{TM} = 0.082, K_{TF} = 2.83, and angle of twist θ of 0.39°. This structure was machined from one piece of brass C2801. The sensor is 7.5 mm thick. Its drawback is high sensitivity to bending moment. Components and assembly of this optical torque sensor are shown in Fig. 15.

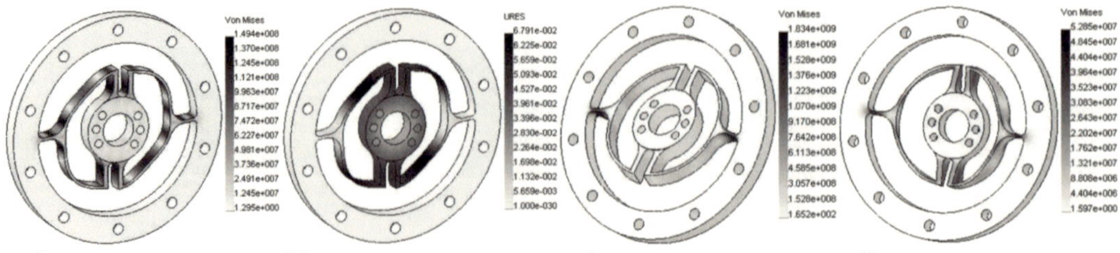

a) b) c) d)

Fig. 14. Results of analysis of semi-ring-shaped spring using FEM

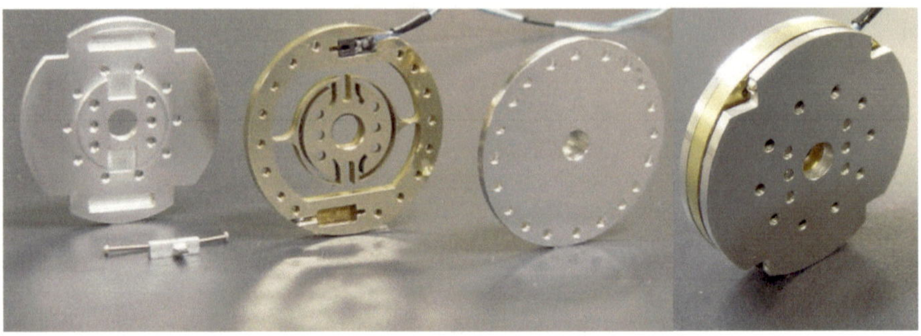

Fig. 15. Optical torque sensor with semi-ring-shaped flexure

In the test rig for calibrating the optical sensor (Fig. 16), force applied to the arm, secured by screws to the rotatable shaft, creates the loading torque. Calibration was realized by incrementing the loading weights and measuring the output signal from the PI. Calibration plots indicate high linearity of the sensors output signal.

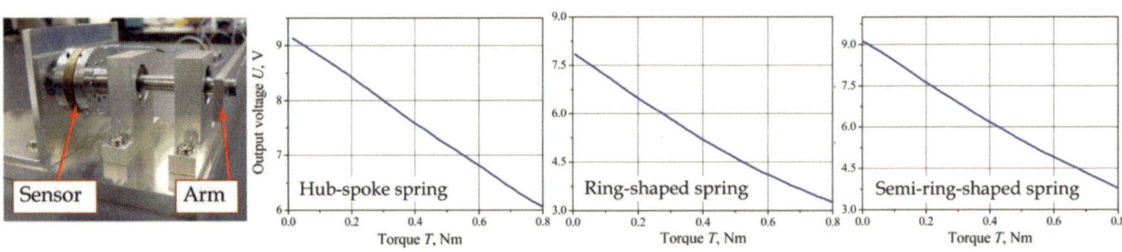

a) b) c) d)

Fig. 16. Test rig and calibration result

Technical specifications of optical torque sensors are listed in Table 1.

The technical specifications of 6-axis force/torque sensors with a similar sensing range of torque around Z-axis are listed in Table 2 (ROHM), (ATI), (BL AUTOTEC).

The spoke-hub topology enables a compact and lightweight sensor. The large torsional stiffness does not considerably deteriorate the dynamic behavior, but diminishes PI resolution. The semicircular spring has high sensitivity to bending moment and axial force and small natural frequency. As regards the ring-shaped flexure, it provides wide torsional stiffness with high mechanical strength. The main shortcoming of this topology is high sensitivity to bending moment. Nevertheless, this obstacle is overcome through realization of a simple supported loading shaft of the robot joint. In the most loaded joints, e.g.

shoulder, such material as hardened stainless steel can be used for elastic elements to keep sensor dimensions the same. Compared to strain-gauge-based sensor ATI Mini 40, our optical sensors have small torsional stiffness and low factor of safety. However, such advantages of designed sensors as low cost, easy manufacture, immunity to the electro-magnetic noise, and compactness make them preferable for torque measurement in robot arm joints. The linear transfer characteristic of the PI simplifies calibration of the sensor. Because of sufficient stiffness, high natural frequency, small influence of bending moment and axial force on the sensor accuracy, the hub-spoke spring as deflecting part of the optical torque sensor was chosen. Four torque sensors for integration into robot joints were manufactured and calibrated (Tsetserukou et al., 2007). The sensors were installed between the harmonic drives and driven shafts of the robot joints.

Sensor	Hub-spoke spring	Ring-shaped spring	Semicircular spring
Spring member material	Brass C2801	Aluminium A5052	Brass C2801
Photointerrupter type	RPI-121	RPI-131	RPI-121
Load capacity, Nm	0.8	0.8	0.8
Torsional stiffness, Nm/rad	219.8	115.86	116.99
Natural frequency, kHz	5.25	2.7	1.37
Factor of safety	1.0	1.0	1.0
Outer diameter, mm	42	42	42
Thickness, mm	6.5	10.0	7.5
Sensor mass, g	34.7	28.7	36.8

Table 1. Technical specifications

Sensor	ATI Mini 40 Hub-spoke spring	BL Autotec Mini 2/10 Hub-spoke spring	Minebea OPFT-50N Hub-spoke spring
Spring member material	Hardened stainless steel	Stainless steel	Aluminium
Sensing element	Silicon strain gauge	Strain gauge	LED-Photodetector
Sensing range MZ, Nm	1.0	1.0	2.5
Torsional stiffness Z-axis, Nm/rad	4300	-	-
Natural frequency, kHz	3.2	-	-
Accuracy, %	-	1.0	5.0
Factor of safety	5.0	5.0	1.5
Outer diameter, mm	40	40	50
Thickness, mm	12.25	20	31.5
Sensor mass, g	50	90	133

Table 2. Technical specifications of the 6-axis force sensors

4. Robot arm control

4.1 Joint impedance control

The dynamic equation of an n-DOF manipulator in joint space coordinates (during interaction with environment) is given by:

$$M(\theta)\ddot{\theta} + C(\theta,\dot{\theta})\dot{\theta} + \tau_f(\dot{\theta}) + G(\theta) = \tau + \tau_{EXT}, \tag{7}$$

where $\theta, \dot{\theta}, \ddot{\theta}$ are the joint angle, the joint angular velocity, and the joint angular acceleration, respectively; $M(\theta) \in R^{n\times n}$ is the symmetric positive definite inertia matrix; $C(\theta,\dot{\theta}) \in R^n$ is the vector of Coriolis and centrifugal terms; $\tau_f(\dot{\theta}) \in R^n$ is the vector of actuator joint friction torques; $G(\theta) \in R^n$ is the vector of gravitational torques; $\tau \in R^n$ is the vector of actuator joint torques; $\tau_{EXT} \in R^n$ is the vector of external disturbance joint torques.

People can perform dexterous contact tasks in daily activities, regulating own dynamics according to time-varying environment. To achieve skillful human-like behavior, the robot has to be able to change its dynamic characteristics depending on time-varying interaction forces. The most efficient method of controlling the interaction between a manipulator and an environment is impedance control (Hogan, 1985). This approach enables to regulate response properties of the robot to external forces through modifying the mechanical impedance parameters. The graphical representation of joint impedance control is given in Fig. 17.

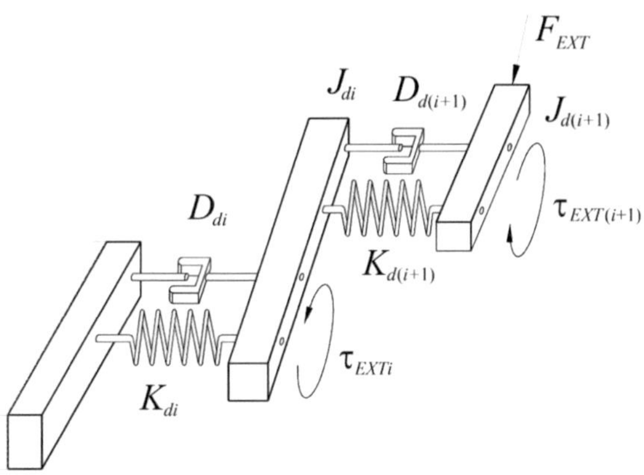

Fig. 17. Concept of the local impedance control

The desired impedance properties of *i-th* joint of manipulator can be expressed as:

$$J_{di}\Delta\ddot{\theta}_i + D_{di}\Delta\dot{\theta}_i + K_{di}\Delta\theta_i = \tau_{EXTi};\ \Delta\theta_i = \theta_{ci} - \theta_{di}, \tag{8}$$

where J_{di}, D_{di}, K_{di} are the desired inertia, damping, and stiffness of *i-th* joint, respectively; τ_{EXTi} is torque applied to *i-th* joint and caused by external forces, $\Delta\theta_i$ is the difference between the current position θ_{ci} and desired one θ_{di}. The state-space presentation of the equation of local impedance control is written as follows:

$$\begin{bmatrix}\Delta\dot{\theta}_i \\ \dot{v}_i\end{bmatrix} = \begin{bmatrix} 0 & 1 \\ -K_d/J_d & -D_d/J_d \end{bmatrix}\begin{bmatrix}\theta_i \\ v_i\end{bmatrix} + \begin{bmatrix} 0 \\ 1/J_d \end{bmatrix}\tau_{EXTi}(t), \tag{9}$$

or:

$$\begin{bmatrix} \Delta\dot{\theta}_i \\ \dot{v}_i \end{bmatrix} = A \begin{bmatrix} \theta_i \\ v_i \end{bmatrix} + B\tau_{EXTi}(t), \tag{10}$$

where the state variable is defined as $v_i = \Delta\dot{\theta}_i$; A, B are matrices. After integration of Eq. (10), the discrete time presentation of the impedance equation is expressed as:

$$\begin{bmatrix} \Delta\theta_{k+1} \\ \Delta\dot{\theta}_{k+1} \end{bmatrix} = A_d \begin{bmatrix} \Delta\theta_k \\ \Delta\dot{\theta}_k \end{bmatrix} + B_d T_{EXT(k)}. \tag{11}$$

To achieve the fast non-oscillatory response on the external force, we assigned the eigenvalues λ_1 and λ_2 of matrix A as real and unequal $\lambda_1 \neq \lambda_2$. By using Cayley-Hamilton method for matrix exponential determination, we have:

$$A_d = e^{AT} = \frac{1}{\lambda_1 - \lambda_2} \begin{bmatrix} e^{\lambda_2 T}\lambda_1 - e^{\lambda_1 T}\lambda_2 & e^{\lambda_1 T} - e^{\lambda_2 T} \\ -b\left(e^{\lambda_1 T} - e^{\lambda_2 T}\right) & e^{\lambda_2 T}\left(\lambda_1 + a\right) - e^{\lambda_1 T}\left(\lambda_2 + a\right) \end{bmatrix}, \tag{12}$$

$$B_d = \left(A_d - I\right)A^{-1}B = -\frac{c}{b\left(\lambda_1 - \lambda_2\right)} \begin{bmatrix} e^{\lambda_2 T}\lambda_1 - e^{\lambda_1 T}\lambda_2 - \left(\lambda_1 - \lambda_2\right) \\ -b\left(e^{\lambda_1 T} - e^{\lambda_2 T}\right) \end{bmatrix}, \tag{13}$$

where T is the sampling time; coefficients a, b, and c equal to D_d/M_d, K_d/M_d, and $1/M_d$, respectively; I is the identity matrix.
The eigenvalues λ_1 and λ_2 can be calculated from:

$$\lambda_1 = \frac{-a + \sqrt{a^2 - 4b}}{2}; \; \lambda_2 = \frac{-a - \sqrt{a^2 - 4b}}{2}. \tag{14}$$

The value of contact torque τ_{EXTi} defines the character of joint compliant trajectory $\Delta\theta_i, \Delta\dot{\theta}_i$. In addition to contact force, torque sensor continuously measures the gravitational, inertial, friction, Coriolis, and centrifugal torques (Eq. (7)). The plausible assumptions of small speed of joint rotation and neglible friction forces allow us to consider only gravitational torques. To extract the value of the contact force from sensor signal, we elaborated the gravity compensation algorithm.

4.2 Gravity compensation

In this subsection, we consider the problem of computing the joint torques corresponding to the gravity forces acting on links with knowledge of kinematics and mass distribution. It is assumed that due to small operation speed the angular accelerations equal zero. The Newton-Euler dynamics formulation was adopted. In order to simplify the calculation procedure, the effect of gravity loading is included by setting linear acceleration of reference frame ${}^{0}\vartheta_0 = G$, where G is the gravity vector. First, link linear accelerations ${}^{i+1}\dot{\vartheta}_{C_{i+1}}$ of the center of mass (COM) of each link are iteratively computed from Eq. (15). Then, gravitational forces ${}^{i+1}F_{i+1}$ acting at the COM of the first and second link are derived from Eq. (16):

$$ {}^{0}\dot{\vartheta}_{0} = g\hat{Z}_{0}; \quad {}^{i+1}\dot{\vartheta}_{C_{i+1}} = {}^{i+1}R_{i}\,{}^{i}\dot{\vartheta}_{i} \tag{15} $$

$$ {}^{i+1}F_{i+1} = m_{i+1}\,{}^{i+1}\dot{\vartheta}_{C_{i+1}}, \tag{16} $$

where m_{i+1} is mass of the link i+1, ${}^{i+1}R_i$ is matrix of rotation between successive links calculated using Denavit-Hartenberg notation.
While inward iterations, we calculate force ${}^{i}f_i$ (Eq. (17)) and moment ${}^{i}n_i$ (Eq. (18)) acting in the coordinate system of each joint. In the static case, the joint torques caused by gravity forces are derived by taking Z component of the torque applied to the link (Eq. (19)).

$$ {}^{i}f_{i} = {}_{i+1}^{\;\;i}R\,{}^{i+1}f_{i+1} + {}^{i}F_{i} \tag{17} $$

$$ {}^{i}n_{i} = {}_{i+1}^{\;\;i}R\,{}^{i+1}n_{i+1} + {}^{i}P_{C_i} \times {}^{i}F_{i} + {}^{i}P_{i+1} \times {}_{i+1}^{\;\;i}R\,{}^{i+1}f_{i+1} \tag{18} $$

$$ \tau_{gi} = {}^{i}n_{i}^{T}\,{}^{i}\hat{Z}_{i}, \tag{19} $$

where ${}^{i}P_{C_i}$ is vector locating the COM for the *i-th* link, ${}^{i}P_{i+1}$ is vector locating the origin of the coordinate system i+1 in the coordinate system i.
The application of the algorithm for robot arm iSoRA results in the equation of gravitational torque vector:

$$ G(\theta) = \begin{bmatrix} \tau_{g1} \\ \tau_{g2} \\ \tau_{g3} \\ \tau_{g4} \end{bmatrix} = \begin{bmatrix} L_{M2}m_2\left(s_1c_2c_4 + c_1c_3s_4 + s_1s_2s_3s_4\right) + \left(L_{M1}m_1 + L_1m_2\right)s_1c_2 \\ L_{M2}m_2\left(c_1s_2c_4 - c_1c_2s_3s_4\right) + \left(L_{M1}m_1 + L_1m_2\right)c_1s_2 \\ L_{M2}m_2\left(-c_1s_2c_3s_4 - s_1s_3s_4\right) \\ L_{M2}m_2\left(-c_1s_2s_3c_4 + s_1c_3c_4 + c_1c_2s_4\right) \end{bmatrix} \begin{bmatrix} g \\ g \\ g \\ g \end{bmatrix}, \tag{20} $$

where τ_{gi} is the gravitational torque in *i-th* joint; m_1 and m_2 are the point masses of the first and second link, respectively; L_{M1} and L_{M2} are the distances from the first and second link origins to the centers of mass, respectively; L_1 is the upper arm length; c_1, c_2, c_3, c_4, s_1, s_2, s_3, and s_4 are abbreviations for $\cos(\theta_1)$, $\cos(\theta_2)$, $\cos(\theta_3)$, $\cos(\theta_4)$, $\sin(\theta_1)$, $\sin(\theta_2)$, $\sin(\theta_3)$, and $\sin(\theta_4)$, respectively.
The experiment with the fourth joint of the robot arm was conducted in order to measure the gravity torque (Fig. 18a) and to estimate the error by comparison with reference model (Fig. 18b).
As can be seen from Fig. 18, the pick values of the gravity torque estimation error arise at the start and stop stages of the joint rotation. The reason of this is high inertial loading that provokes the vibrations during acceleration and deceleration transients. This disturbance can be evaluated by using accelerometers and excluded from further consideration. The applied torque while physical contacting with environment is derived by subtraction of gravity term $G(\theta)$ from the sensed signal value. Observing the measurement error plot (Fig. 18b), we can assign the relevant threshold of 0.02 Nm that triggers control of constraint motion.

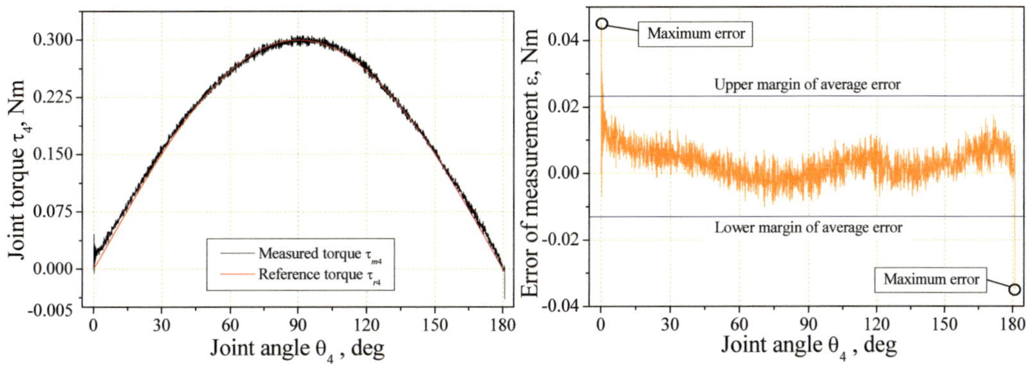

a) b)

Fig. 18. Experimental results of gravity torque measurement

4.3 Experimental results of joint admittance control

To improve the service task effectiveness, we decided to implement admittance control (Fig. 19). In this case, compliant trajectory generated by the impedance controller is traced by the PD control loop. Thus, inherent dynamics of the robot does not affect the performance of the target impedance model. We adopted K_d = 29 (Nm/rad), D_d = 6.9 (Nm·s/rad), J_d = 0.4 (kg·m²), to achieve closed to critical damped response and sufficient for safe interaction compliance.

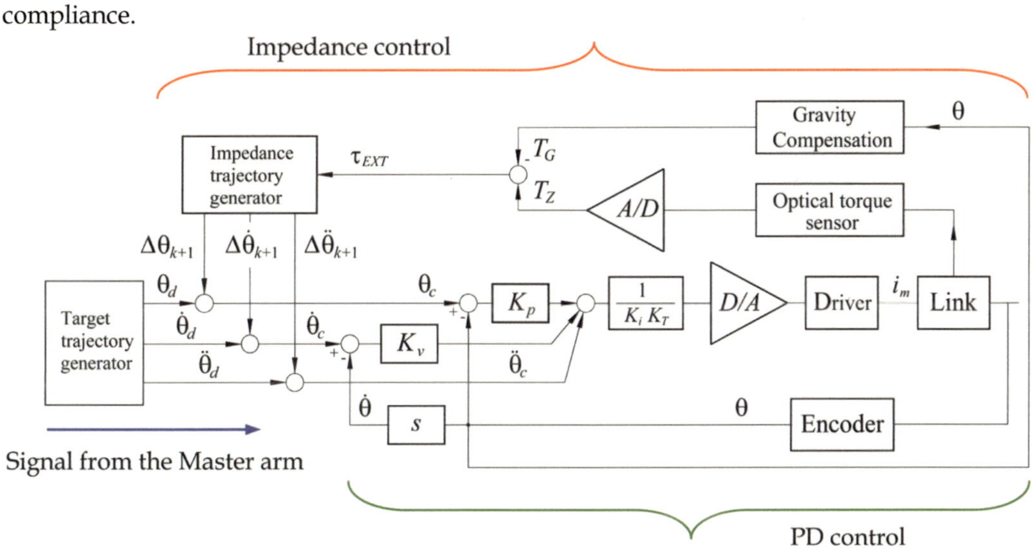

Fig. 19. Block diagram of joint admittance control

To verify the theory and to evaluate the feasibility and performance of the proposed controller, the experiments were conducted with developed robot arm. During the experiment, robot forearm was pushed several times by human in different directions with forces having different magnitude. The experimental results for the elbow joint - applied torque, angle generated by impedance controller, measured joint angle, and error of joint angle in the function of time - are presented in Fig. 20, Fig. 21, Fig. 22 and Fig. 23, respectively.

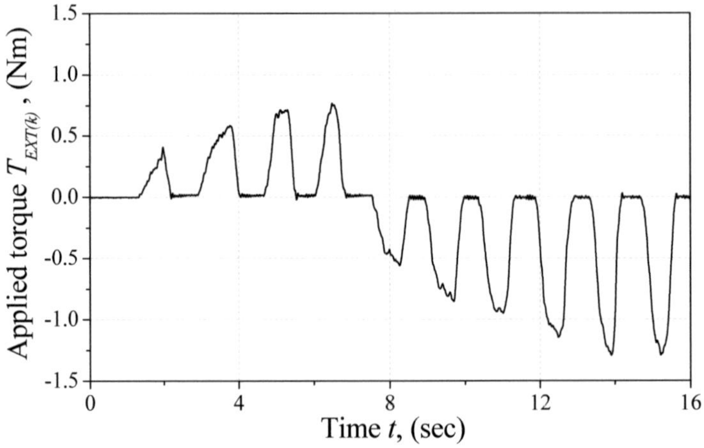

Fig. 20. External torque

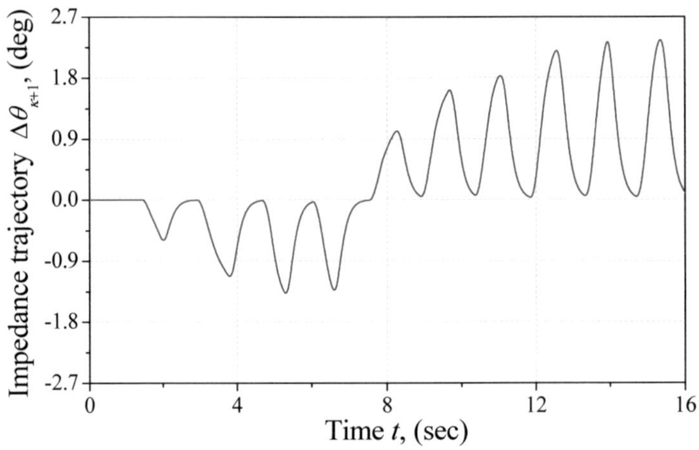

Fig. 21. Impedance trajectory

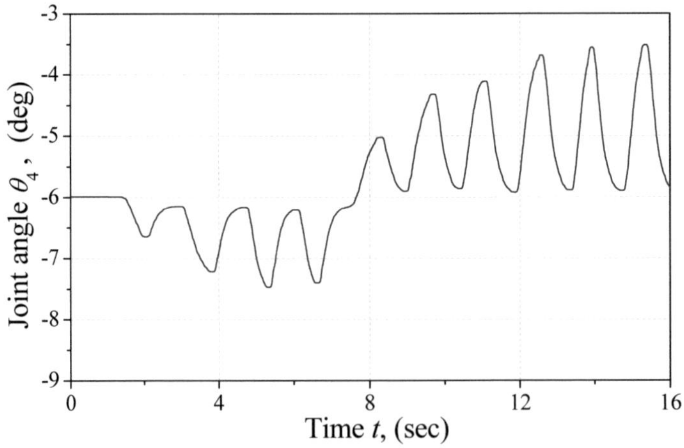

Fig. 22. Measured joint angle

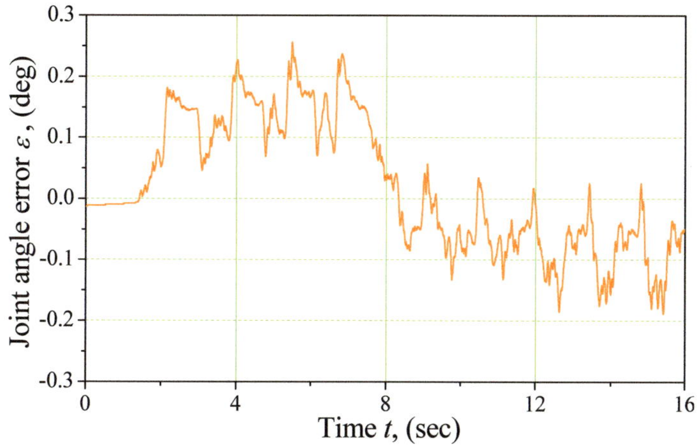

Fig. 23. Joint angle trace error

The experimental results show the successful realization of the joint admittance control. While contacting with human, the robot arm generates compliant soft motion (Fig. 21) according to the sensed torque (Fig. 20). The larger force applied to the robot arm (Fig. 20), the more compliant trajectory is generated by impedance control (Fig. 21).. As we assigned closed to critically damped response of impedance model to disturbance force, output angle ($\Delta\theta_{k+1}$) has ascending-descending exponential trajectory. Admittance control provides small joint angle trace error in borders of -0.2°–0.25° (Fig. 23). The conventionally impedance-controlled robot can realize contacting task only at the tip of the end-effector. By contrast, our approach provides delicate continuous safe interaction of all surface of the robot arm with environment.

5. Conclusion and future work

In Chapter, the stages of the joint torque sensor design are presented. New torque sensors for implementation of virtual backdrivability of robot joint transmissions have been developed. Torque measurement techniques, namely, electrical, electromagnetic, and optical, are discussed in detail. Technical requirements aimed at designing the high-performance torque sensor for humanoid robot arm were formulated. The substantial advantages of the optical technique motivated our choice in its favor. The main novelty of our method is application of the ultra-small size PI as sensitive element to measure relative motion of sensor components. The hub-spoke, ring-shaped, and semi-ring-shaped topologies of the sensor spring member were designed and investigated in order to optimize the mechanical structure of the detector. Hub-spoke structure was proven to be the most suitable solution allowing realization of compact sensor with high resolution. The designed optical torque sensors are characterized by good accuracy, high signal-to-noise ratio, compact sizes, light in weight, easy manufacturing, high signal bandwidth, robustness, low cost, and simple calibration procedure.

In addition to contact force, torque sensor continuously measures the gravity and dynamic load. To extract the value of the contact force from sensor signal, we elaborated algorithm of calculation of torque caused by contact with object.

New whole-sensitive robot arm iSoRA was developed to provide human-like capabilities of contact task performing in a wide variety of environments. Each joint is equipped with optical torque sensor directly connected to the output shaft of harmonic drive. The sizes and appearance of the robot arm were chosen so that the sense of incongruity during interaction with human is avoided. We kept the arm proportions the same as in average height human. The effectiveness of the proposed joint admittance controller ensuring the safety in human-robot interaction was experimentally justified during physical contacting with the entire surface of robot arm body.
Our future research will be focused on elaboration of an approach to estimation of the contact point location, environment stiffness evaluation, and contacting object shape recognition on the basis of knowledge of the applied torques and manipulator geometry.

6. Acknowledgments

The research is supported in part by a Japan Society for the Promotion of Science (JSPS) Postdoctoral Fellowship for Foreign Scholars.

7. References

ATI, Multi-Axis Force/Torque Sensor, ATI Industrial Automation, [Online], Available: http://www.ati-ia.com.

Bicchi, A. & Tonietti, G. (2004). Fast and soft arm tactics: dealing with the safety-performance trade-off in robot arms design and control. *IEEE Robotics and Automation Magazine,* Vol. 11, No. 2, (June 2004) 22-33, ISSN 1070-9932

BL AUTOTEC, BL Sensor, BL AUTOTEC, LTD., [Online], Available: http://www.bl-autotec.co.jp.

Fulmek, P. L.; Wandling, F.; Zdiarsky, W.; Brasseur, G. & Cermak, S. P. (2002). Capacitive sensor for relative angle measurement. *IEEE Transactions on Instrumentation and Measurement,* Vol. 51, No. 6, (December 2002) 1145-1149, ISSN 0018-9456

Golder, I.; Hashimoto, M.; Horiuchi, M. & Ninomiya, T. (2001). Performance of gain-turned harmonic drive torque sensor under load and speed conditions. *IEEE/ASME Transaction on Mechatronics,* Vol. 6, No. 2, (June 2001) 155-160, ISSN 1083-4435

Hashimoto, M.; Kiyosawa, Y. & Paul, R. P. (1993). A torque sensing technique for robots with harmonic drives. *IEEE Transaction on Robotics and Automation,* Vol. 9, No. 1, (February 1993) 108-116, ISSN 1042-296X

Hilton, J. A. (1989). Force and torque converter. U. S. Patent 4 811 608, March 1989.

Hirose S. & Yoneda, K. (1990). Development of optical 6-axial force sensor and its signal calibration considering non-linear interference. *Proceedings of IEEE Int. Conf. Robotics and Automation (ICRA),* pp. 46-53, ISBN 0-8186-9061-5, Cincinnati, May 1990, IEEE Press.

Hirzinger, G.; Albu-Schaffer, A.; Hahnle, M.; Schaefer, I. & Sporer, N. (2001). On a new generation of torque controlled light-weight robots. *Proceedings of IEEE Int. Conf. Robotics and Automation (ICRA),* pp. 3356-3363, ISBN 0-7803-6576-3, Seoul, May 2001, IEEE Press.

Hogan, N. (1985). Impedance control: an approach to manipulation, Part I-III. *ASME Journal of Dynamic Systems, Measurement and Control,* Vol. 107, (March 1985) 1-23, ISSN 022-0434

Horton, S. J. (2004). Displacement and torque sensor. U.S. Patent 6 800 843, October 2004.

Ishino, R.; Saito, T. & Sunahata. M. (1996). Magnetostrictive torque sensor shaft. U.S. Patent 5 491 369, February 1996.

Kaneko, K.; Kanehiro, F.; Kajita, S.; Hirukawa, H.; Kawasaki, T.; Hirata, M.; Akachi, K. & Isozumu, T. (2004). Humanoid robot HRP 2, *Proceedings of IEEE Int. Conf. Robotics and Automation (ICRA),* pp. 1083-1090, ISBN 0-7803-8233-1, New Orleans, 2004, IEEE Press.

Kovacich, J. A.; Kaboord, W.S.; Begale, F.J.; Brzycki, R.R.; Pahl, B. & Hansen, J.E. (2002). Method of manufacturing a piezoelectric torque sensor. U.S. Patent 6 442 812, September 2002.

Lumelsky, V. J. & Cheung, E. (2001). Sensitive skin. *IEEE Sensors Journal,* Vol. 1, No. 1, (June 2001) 41-51, ISSN 1530-437X

Madni, A. M.; Hansen, R. K. & Vuong, J. B. (2004). Differential capacitive torque sensor. U.S. Patent 6 772 646, August 2004.

Minebea, 6-Axial Force Sensor. Minebea Co., Ltd., [Online], Available: http://www.minebea-mcd.com.

Mitsunaga, N.; Miyashita, T.; Ishiguro, H.; Kogure, K. & Hagita, N. (2006). Robovie IV: A communication robot interacting with people daily in an office, *Proceedings of IEEE/RSJ Int. Conf. Intelligent Robots and Systems (IROS)* pp. 5066-5072, ISBN 1-4244-0259-X, Beijing, October 2006, IEEE Press.

Morikawa, S.; Senoo, T.; Namiki, A. & Ishikawa, M. (2007). Real-time collision avoidance using a robot manipulator with light-weight small high-speed vision system. *Proceedings of IEEE Int. Conf. Robotics and Automation (ICRA),* pp. 794-799, ISBN 1-4244-0602-1, Roma, April 2007, IEEE Press.

Nicot, C. (2004). Torque sensor for a turning shaft. U. S. Patent 6 694 828, February 2004.

Okutani, N. & Nakazawa, K. (1993). Torsion angle detection apparatus and torque sensor. U.S. Patent 5 247 839, September 1993.

ROHM, Photointerrupter design guide. *Product catalog of ROHM,* pp. 6-7, 2005.

Sakagami, Y.; Watanabe, R.; Aoyama, C.; Matsunaga, S.; Higaki, N. & Fujimura, K. (2002). The intelligent ASIMO: System overview and integration, *Proceedings of IEEE/RSJ Int. Conf. on Intelligent Robots and Systems (IROS),* pp. 2478-2483, ISBN 0-7803-7399-5, Lausanne, September 2002, IEEE Press.

Sakaki, T. & Iwakane, T. (1992). Impedance control of a manipulator using torque-controlled lightweight actuators. *IEEE Transaction on Industry Applications,* Vol. 28, No. 6, (November 1992) 1399-1405, ISSN 0093-9994

Shinoura, O. (2003). Torque sensor and manufacturing method of the same. U.S. Patent 6 574 853, June 2003.

Takahashi, N.; Tada, M.; Ueda, J.; Matsumoto, Y. & Ogasawara, T. (2003). An optical 6-axis force sensor for brain function analysis using fMRI, *Proceedings of 2nd IEEE Int. Conf. on Sensors,* Vol. 1, pp. 253-258, ISBN 0-7803-8133-5, Toronto, October 2003, IEEE Press.

Tsetserukou, D.; Tadakuma, R.; Kajimoto, H.; Kawakami, N. & Tachi. S. (2007). Towards safe human-robot interaction: joint impedance control of a new teleoperated robot arm. *Proceedings of IEEE Int. Symposium on Robot and Human Interactive Communication (RO-MAN)*, pp. 860-865, ISBN 1-4244-1635-6, Jeju island, August 2007, IEEE Press.

Vischer, D. & Khatib, O. (1995). Design and development of high-performance torque controlled joints. *IEEE Transactions on Robotic and Automation*, Vol. 11, No. 4, (August 1995) 537-544, ISSN 1042-296X

Westbrook, M. H. & Turner, J. D. (1994). *Automotive Sensors*. IOF Publishing, ISBN 0750302933, Philadelphia.

Wu, C. H. & Paul, R. P. (1980). Manipulator compliance based on joint torque control, *Proceedings of IEEE Conf. Decision and Control*, pp. 88-94, ISSN 0191-2216, Albuquerque, December 1980, IEEE Press.

3

CMOS Force Sensor with Scanning Signal Process Circuit for Vertical Probe Card

Jung-Tang Huang, Kuo-Yu Lee and Ming-Chieh Chiu
National Taipei University of Technology
Taiwan

1. Introduction

The constant advancement of semiconductor technology has prompted a further reduction in size and an increase in the density of integrated circuits. In order to meet the various requirements of the industry, higher accuracy, longer fatigue life, and a greater capability to withstand temperature extremes have become important criteria in the design of probe cards (Iscoff, 1994; Gilg, 1997). After a certain period of use, probe cards must be calibrated by a professional machine; several properties need to be verified, such as the probe's maximum current, resistivity (contact resistance), and reaction force. This verification procedure may influence the efficiency of production lines, since it is performed offline. One crucial step in this procedure is monitoring the reaction force exerted by the probes on each other, in order to compute the average reaction force and complete coplanarity to ensure the efficient operation of the probe cards. To expedite this procedure, we designed an array-type CMOS force sensor that is capable of monitoring the status of vertical probe cards online in both die-level and wafer-level applications.
In the past, the fabrication of pressure sensors typically involved an MEMS process with backside etching adopted for its post process (Ghalichechian, 2002; Malhair & Barbier, 2003). However, in recent years, more and more researchers have proposed to combining the standard CMOS process with the MEMS process to manufacture both microsensors and integrated circuits (Yang et al., 2005; Peng et al., 2005; Wang et al., 2006). The combined process also has additional advantages, such as a reduction in the noise and number of pads. Our design also adopted the combined CMOS-MEMS process to fabricate force sensors and their signal conditioning circuits. Moreover, as the conventional post process can barely handle the increasingly smaller pitch between the probes (Wilson, 1999), we etched a cavity on the silicon substrate to deform the membrane in the post process (RLS dry etching).

2. Design principle

The main structure of a piezoresistive pressure sensor is the sensor membrane, which is made from a material that has a piezoresistance effect. As a pressure or force is exerted upon the membrane, the membrane is deformed; the piezoresistance value or resistance of the piezoresistive material changes with the stress; this is referred to as the piezoresistance effect (See Ch.2.1). The resistive materials of the sensor are then connected to a Wheatstone bridge to enhance their sensitivity. Thus, with the help of the Wheatstone bridge, we can obtain the value of ΔV when the resistance values changes. The resistance and ΔV values vary greatly as the pressure increases (See Ch.2.2). The pressure can be measured on the basis of the output value of ΔV once the relation between the pressure and ΔV is established.

2.1 Piezoresistance effect

In 1856, Lord Kelvin discovered the phenomenon of the piezoresistance effect. However, the practicality of the principle was only applied to a strain gauge in 1939, with metal used for the strain gauge material. Strain gauges are now used in many measurement applications, such as for mechanisms, buildings, airplanes, and scales. But, the resistance change rate for a metal strain gauge is very small. As a result, in applications that require the detection of small amounts of strain, metal strain gauges do not provide a sufficient sensitivity or signal/noise ratio. In 1954, Smith first exerted stress in the axial direction on a shaft that had been doped with silicon and germanium, and then measured the change rate of the resistance in the vertical direction (Smith, 1954). The physical relationship between the resistance and stress is first established. The principle and architecture of a piezoresistive pressure sensor is as shown below (Petersen, 1982; Thurston, 1964):

$$\Delta P \Rightarrow \omega \Rightarrow \varepsilon \Rightarrow \sigma \Rightarrow \frac{\Delta R}{R} \Rightarrow \Delta V \tag{1}$$

where ΔP is the change in the pressure; ω, the membrane deformation; ε, strain; σ, stress; ΔR /R, ratio of variation; and ΔV, potential difference. And the change rate of the resistance is:

$$\frac{\Delta R}{R} = \varepsilon(1+2\nu)+\frac{\Delta\rho}{\rho} \tag{2}$$

where v is Poisson's ratio, Δρ/ρ is the change rate of resistivity, (1+2v) is the deformation of the material by the external pressure. Since silicon has a square structure, the relationship between the change rate of resistivity and the stress can be simply shown in the following matrix equation:

$$\frac{1}{\Delta\rho}\begin{pmatrix}\Delta\rho_1\\ \Delta\rho_2\\ \Delta\rho_3\\ \Delta\rho_4\\ \Delta\rho_5\\ \Delta\rho_6\end{pmatrix} = \begin{pmatrix}\pi_{11} & \pi_{12} & \pi_{12} & 0 & 0 & 0\\ \pi_{12} & \pi_{11} & \pi_{12} & 0 & 0 & 0\\ \pi_{12} & \pi_{12} & \pi_{11} & 0 & 0 & 0\\ 0 & 0 & 0 & \pi_{44} & 0 & 0\\ 0 & 0 & 0 & 0 & \pi_{44} & 0\\ 0 & 0 & 0 & 0 & 0 & \pi_{44}\end{pmatrix}\begin{pmatrix}\sigma_1\\ \sigma_2\\ \sigma_3\\ \tau_1\\ \tau_2\\ \tau_3\end{pmatrix} \tag{3}$$

For the function, [π] is the coefficient for the piezoresistance matrix. The material characteristics and coefficient for the piezoresistance matrix can be simply converted by Euler coordinates into the following equation:

$$\pi_T = \pi_{11} + (\pi_{11} - \pi_{12} - \pi_{44})[l_1^2 l_2^2 + m_1^2 m_1^2 + n_1^2 n_2^2] \tag{4}$$

$$\pi_L = \pi_{11} - 2(\pi_{11} - \pi_{12} - \pi_{44})[l_1^2 m_1^2 + m_1^2 n_1^2 + l_1^2 n_1^2] \tag{5}$$

where l1, m1, n1, l2, m2, and n2 are the transverse direction cosine and longitudinal direction cosine, respectively (Kanda, 1982).
The relationship between the change rate of the resistance and the stress can be simply shown in the following equation:

$$\frac{\Delta R}{R} = \frac{\Delta \rho}{\rho} = \pi_L \sigma_L + \pi_T \sigma_T \tag{6}$$

where σ_L and σ_T are the longitudinal stress and transverse stress, and π_L and π_T are the longitudinal piezoresistance coefficient and transverse piezoresistance coefficient, respectively.

2.2 The Wheatstone bridge principle

Piezoresistive pressure sensors use piezoresistive materials embedded in the membrane of the sensor, and adopt the Wheatstone bridge principle, as shown in Fig. 1. Four piezoresistances are placed at the edge of a square membrane (two transverse piezoresistances; two longitudinal piezoresistances). The transverse and longitudinal piezoresistances are influenced by stress when the membrane deforms. The transverse piezoresistances will be widened and reduced, while the other piezoresistances will be lengthened and increased.

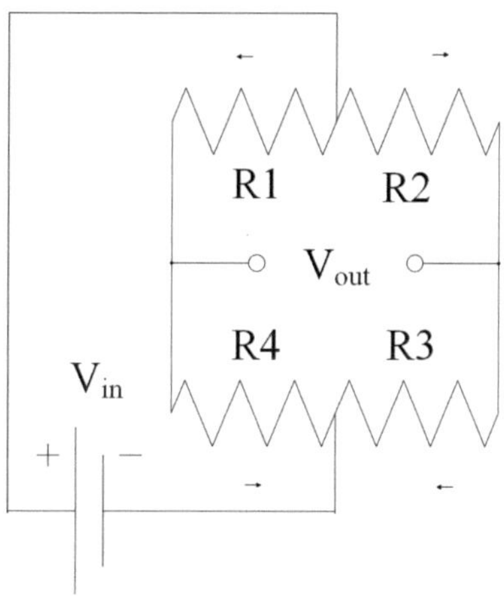

Fig. 1. The circuit structure of Wheatstone bridge.

The relationship between the output and input voltage is shown in the following equation:

$$\frac{V_{OUT}}{V_{IN}} = \frac{R_2}{R_1 + R_2} - \frac{R_4}{R_3 + R_4} \tag{7}$$

When the resistances do not change, the output voltage is zero.

$$R_1 = R_2 = R_3 = R_4 = R,\ V_{OUT} = 0 \tag{8}$$

When pressure is exerted upon the sensor, the resistances will change in an ideal situation:

$$R_2 = R_3 = R + \Delta R\ ;\ \ R_1 = R_4 = R - \Delta R \tag{9}$$

Thus, the relationship between the output voltage and exerted pressure can be established:

$$\frac{V_{out}}{V_{in}} = \frac{\Delta R}{R} = kP \tag{10}$$

Therefore, the situation assumes elasticity:

$$\frac{\Delta R}{R} \propto P \tag{11}$$

As seen in the above equation, we can utilize the change rate of the output voltage to evaluate the change rate of the resistances and thus obtain a pressure measurement.

3. Design and fabrication

First, sensor loading and other properties were simulated. If the properties of the sensor conformed to our requests, a photo mask for the sensor was designed. Then, we adopted the TSMC standard process and the post process (RLS) by CIC. The last step was packaging and measurement.

3.1 Simulation of sensor

There were two parts to the CMOS force sensor simulation: the membrane structure and the piezoresistance locations. The force range of a normal vertical probe card is about 3g ~ 5g, according to a previous reference (Gilg, 1997). Table 1. summarizes the properties of the main materials used in the IC fabrication process for the initial design. The optimal membrane area could be determined based on the probe pitch and membrane depth. The simulation parameter settings are shown in Fig. 2. These include (a) Force: Since the force range of a normal vertical probe card is about 3g ~ 5g, the force range for the simulation was set at 0g ~ 5g, and (b) Boundary condition: The edge of the sensor membrane was firmly fixed. However, since only pressure could be set in the simulation software (CoventorWare), we needed to convert force to pressure (0 MPa ~ 60 MPa) for the simulation.

Material \ Property	Yield stress (GPa)	Young's modulus (GPa)	Poisson ratio
Silicon dioxide	8.4	73	0.17
Silicon	7.0	190	0.23
Polysilicon	2.7	140	0.2
Nitride	14.0	260	0.27
Tungsten	4.0	410	0.28
Aluminum	0.17	70	0.35

Table 1. The property of materials in the TSMC standard process [9]

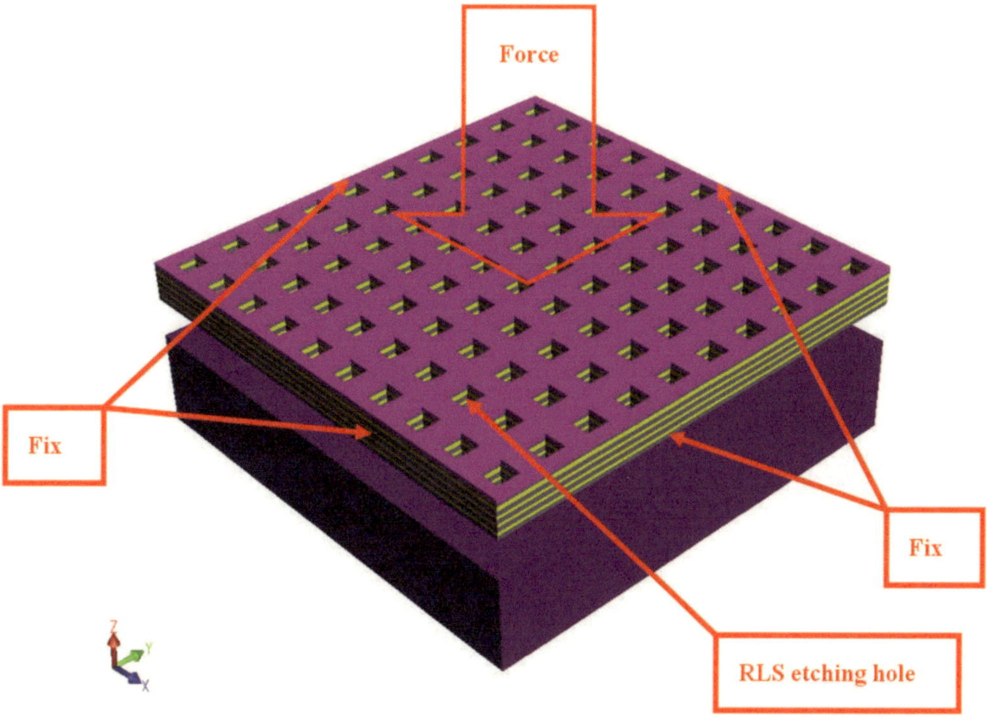

Fig. 2. The setting of boundary condition with the membrane of sensor in our simulation

3.1.1 Simulation of membrane structure

Table 1. summarizes the properties of the main materials used in the IC fabrication process, based on the limits of the TSMC .35 2P4M process. Among these materials, silicon dioxide and metal are the best membrane materials because their Young's modulus and yield stresses are good. In addition, we also considered the probe diameter (40 x 40 μm^2) when assembling a vertical probe card, designing the membrane area to be 100 x 100 μm^2. Fig. 3

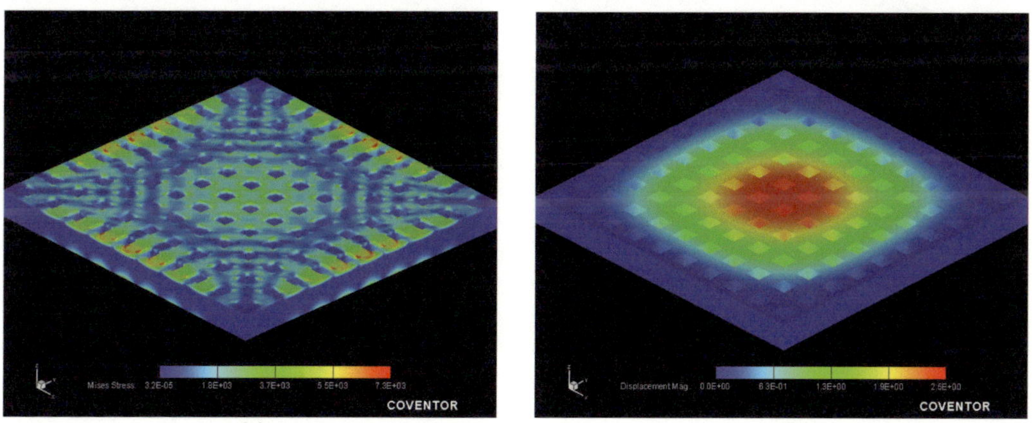

(a) (b)

Fig. 3. The pressure as 65MPa (a) Distribution of the von Mises stress of the COMS force sensor. (b) Distribution of the deformation of the COMS force sensor.

shows a distribution map for the von Mises stress when the membrane was compressed by an external pressure. We were able to resolve the loading range for the force sensor after determining a 100 x 100 μm² area for the membrane. When more metal layers were employed, the structural strength was weaker; when more silicon dioxide layers were employed there was less deformation. For a trade-off between the strength of the membrane structure and the sensitivity of the piezoresistance, we decided to use a membrane structure composed of two metal layers and two silicon dioxide layers. The main membrane material was silicon dioxide, which had a yield stress of 8.4 Gpa, as shown in Table A. According to the simulation results, the von Mises stress over the yield stress when the loading force is 65MPa on the membrane. Therefore, the optimum loading range of the force sensor was 0 to 5 grams (100 x 100 μm²; the safety factor was 2 to 3).

3.1.2 Simulation of piezoresistive location

Based on the simulation results (see Fig. 4) and a previous study (Ghalichechian, 2002), the maximum stress was at the edge of the membrane. Therefore, the piezoresistance was placed at the edge of the membrane to measure the maximum membrane strain. The resistance change rate and sensor sensitivity increased simultaneously. The initial design adopted the traditional type of piezoresistance (rectangular and sheet piezoresistance) for the simulation, and the current density was 14 nA/μm², as shown in Fig. 5. The current was sparse at the corners, so we beveled them (45°). The density of the current then rose to 24 nA/μm², as shown in Fig. 6. The altered design could improve the resistance change rate and raise the sensitivity of the sensor. After deciding on the piezoresistance shape, we moved the piezoresistance to the center of the membrane from its central edge. The maximum current for the transverse piezoresistance was found at 17 μm, as shown in Fig. 7(a). The maximum current for the longitudinal piezoresistance was found at 2, 3, 9 and 18 μm, as shown in Fig. 7(b).

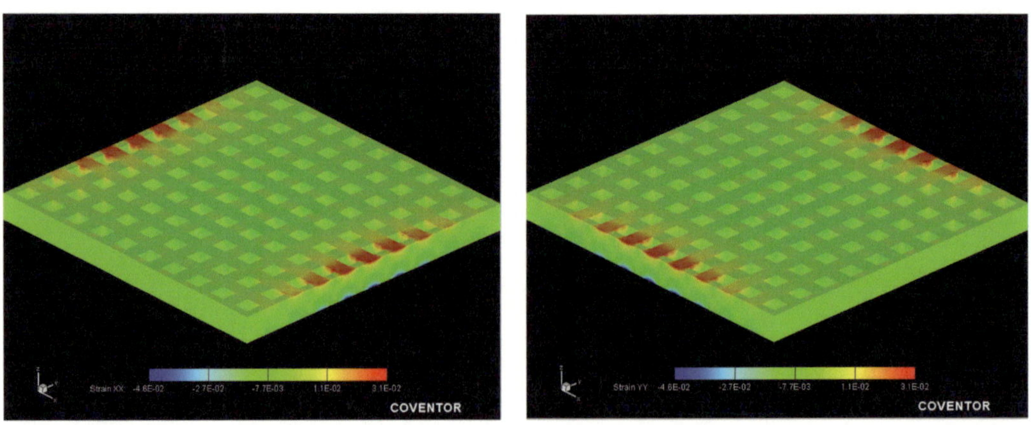

(a) (b)

Fig. 4. Distribution of the von Mises stress of the COMS force sensor (size 100μm²) when the pressure as 65MPa. (a) XX-direction. (b) YY-direction.

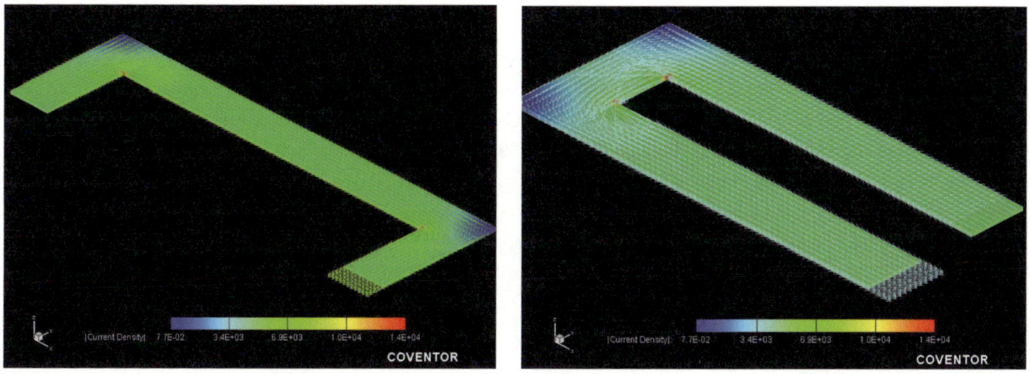

(a) (b)

Fig. 5. (a) The traditional transverse piezoresistance. (b) The traditional longitudinal piezoresistance.

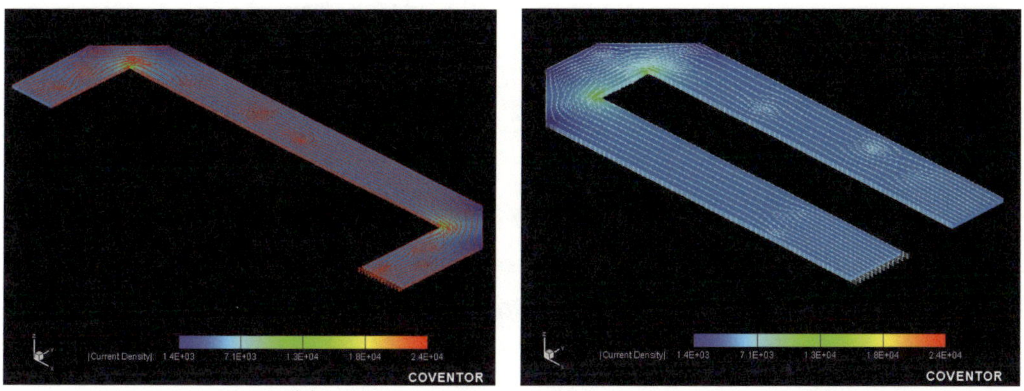

(a) (b)

Fig. 6. (a) The bevel transverse piezoresistance. (b) The bevel longitudinal piezoresistance.

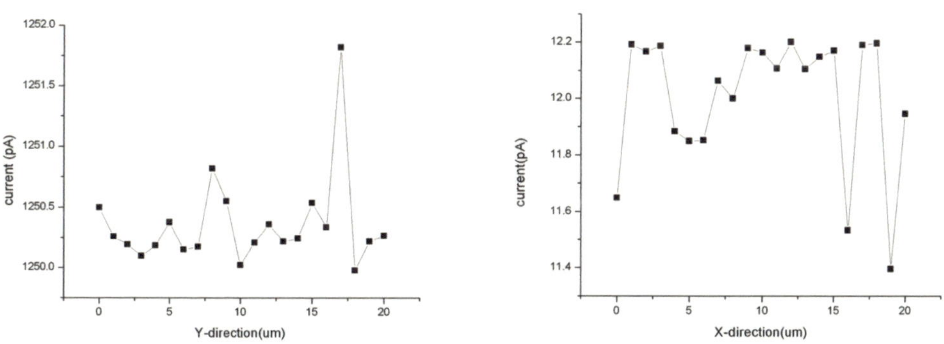

(a) (b)

Fig. 7. (a) The change chart of current when the transverse piezoresistance moves to the center of membrane. (b) The change chart of current when the longitudinal piezoresistance moves to the center of membrane.

3.2 Layout of the chip

Fig. 8 shows a schematic diagram of an array-type CMOS force sensor and signal processing circuit. For a 4x4 array-type CMOS force sensor, it is essential to design an address generator implemented by a counter and several decoders to generate a control signal, in order to measure the output voltage signal of each sensor unit. The specific column bit and row bit from each cycle, working with an analog switch, can send the output voltage signal of each sensor to an on-chip instrumentation amplifier to amplify the output signal for the follow-up signal processing. A single force sensor with an analog switch is shown in Fig. 9.

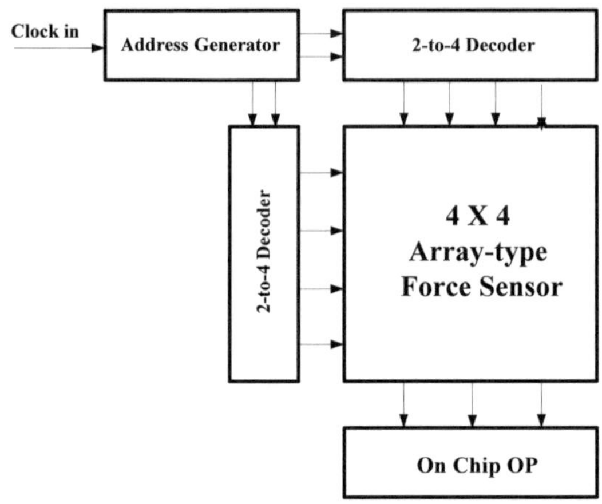

Fig. 8. Schematic diagram of array-type force sensor and signal process system.

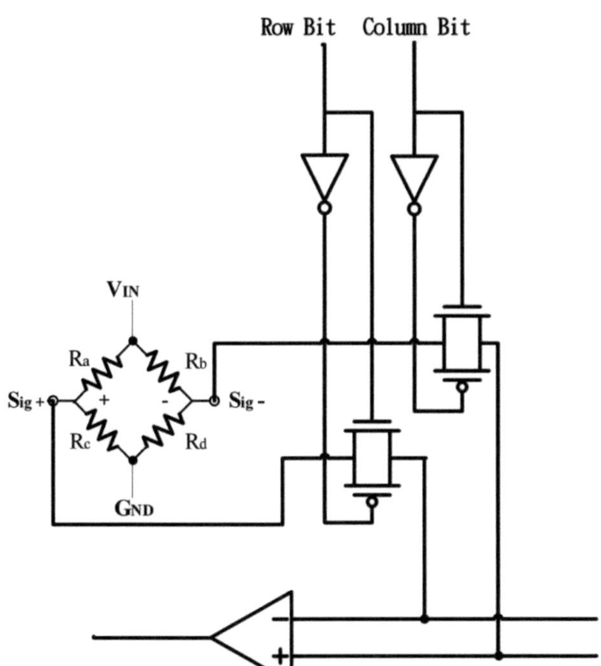

Fig. 9. The single force sensor with analog switch.

The chip had sixteen force sensors, and many pads around the sensors, as shown in Fig. 10. In order to match the probe distances of our vertical probe card, the central distance of each force sensor was designed to be 250 μm. A pad was then opened on the top of each sensor for measuring the probe signal. So, the sensor could simultaneously measure the probe signal and the force sensor signal.

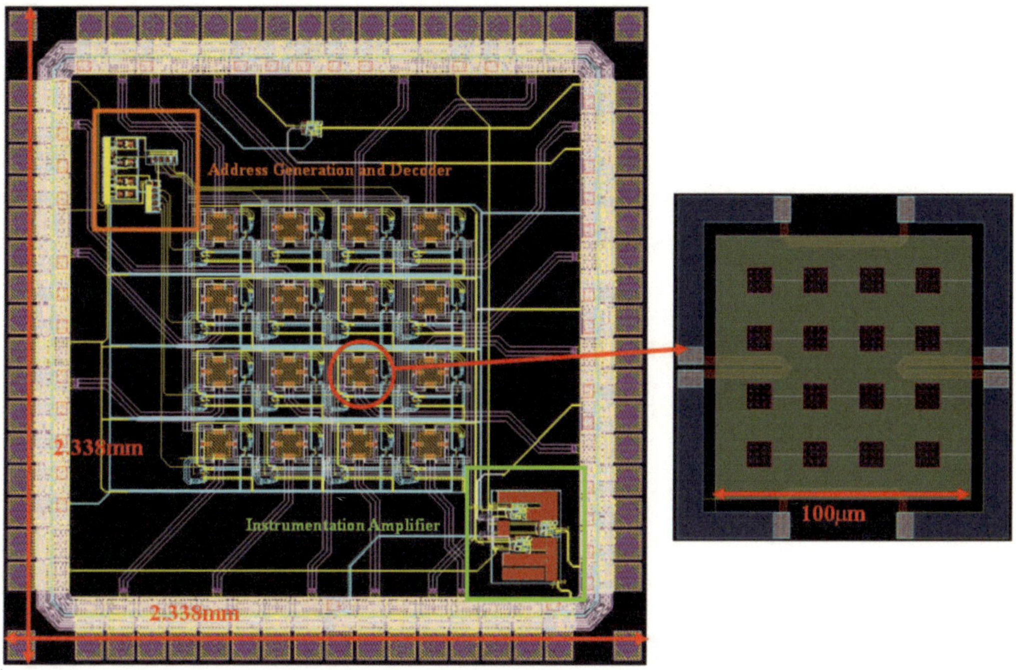

Fig. 10. The layout of array-type CMOS force sensor.

3.3 Process of fabrication

After it was verified that the simulation met the design requirements, the design was drafted and a design rule check (DRC) was performed. The standard TSMC 0.35 μm 2P4M process offered by the Chip Implementation Center (CIC, 2008) was then adopted to fabricate the sensor. Fig. 11(a) shows a cross section of the fabricated chip. Further, we employed a post process (RLS dry etching) supplied by CIC (CIC, 2008). The first step in the process involved the anisotropic etching of silicon nitride and silicon dioxide, as shown in Fig. 11(b). The next step was to etch the silicon substrate by isotropic etching, as shown in Fig. 11(c).

3.4 Piezoresistance

Actually, there are two types of piezoresistance in polycrystalline silicon: p+ and n+, which have well known piezoresistance effects (the gauge factor is 30 (French, 2002; Seto, 1976)). We adopted polycrystalline silicon (poly1 layer n+) in the design of the sensor's piezoresistance. The reason for adopting the poly1 layer is that it is located at the bottommost layer of the membrane. Therefore, the largest thickness can be used for the membrane structure. In addition, the resistivity (ρ) of the poly1 layer is 0.85 mΩ-cm, based on the data offered by CIC.

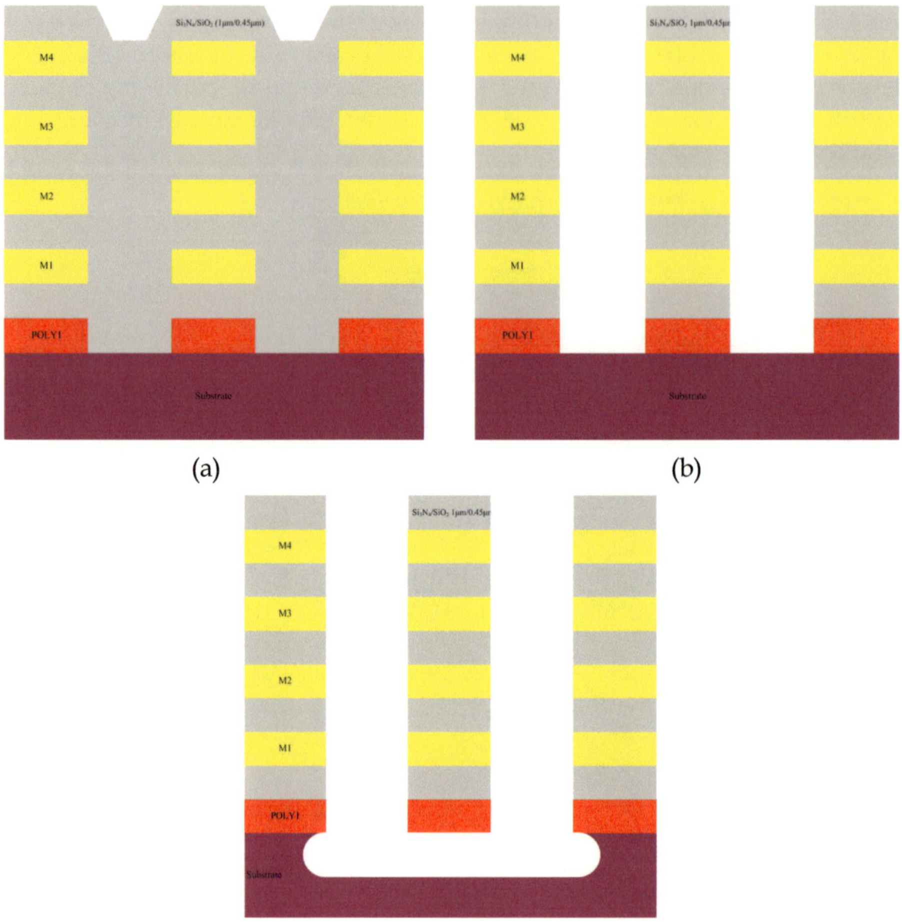

(a) (b)

(c)

Fig. 11. (a) The cross-section with chip. (b) Using anisotropic etching to etch silicon nitride and silicon dioxide. (c) To dig the silicon substrate by isotropic etching.

4. Results and discussion

4.1 Application of a vertical probe card

There are two applications of our force sensor. The first type involves off-line testing. The initial package adopted wire boning, and put the force sensor below the vertical probe card, as shown in Fig. 12(a). When the vertical probe card is off-line, it is made to touch the sensor and the force is measured. All of the data is then compared and the co-planarity of the probe card is adjusted. The second type of application involves on-line testing. Initially, we used micro electroforming and sputtering to form a bump on a pad of the chip, and the sensor and printed circuit board employed flip chip technology, as shown in Fig. 12(b). The probe tip of the probe card touches the sensor membrane. As the probe card is inserted, each probe tip comes into contact with a force sensor and generates a reaction force. The sensor can simultanously transfer this reaction force and the probe current to a testing machine. Thus, we could achieve an array-type force sensor to immediately monitor probes.

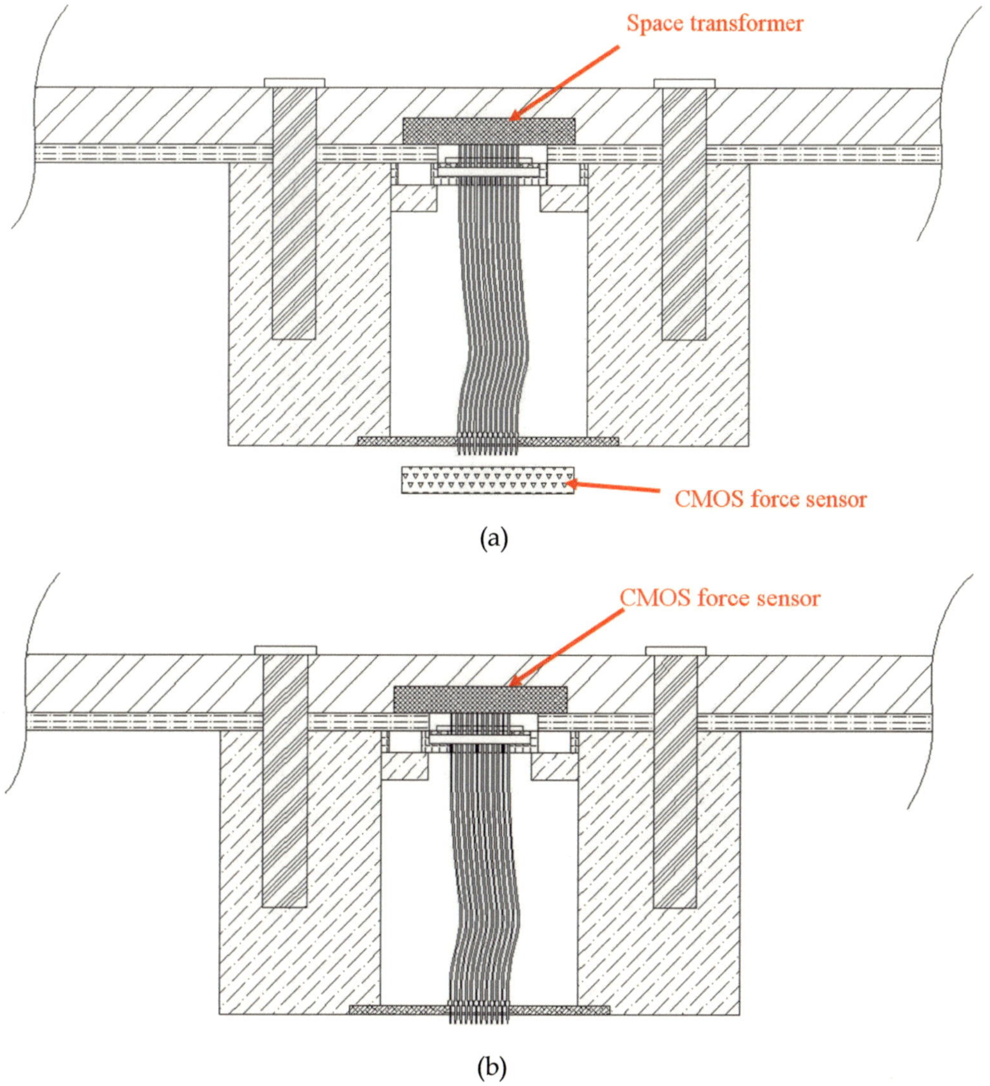

Fig. 12. The cross-sectional view of the vertical probe card with CMOS force sensor (a) off-line (b) on-line

4.2 CMOS force sensor measurement

Fig. 13 illustrates the measurement process for the CMOS force sensor. First, commands were entered into a computer to control a six-axis precision instrument by means of programs and a multi-axis controller. Secondly, an apparatus with a probe was fabricated to compress the sensor chip, and a charge coupled device (CCD) was used to observe whether the probe was able to establish precise contact with the sensor. Finally, for the four resistances of the sensor's Wheatstone bridge, the two input points were connected to a power supply, and both the output points were connected to an oscilloscope.

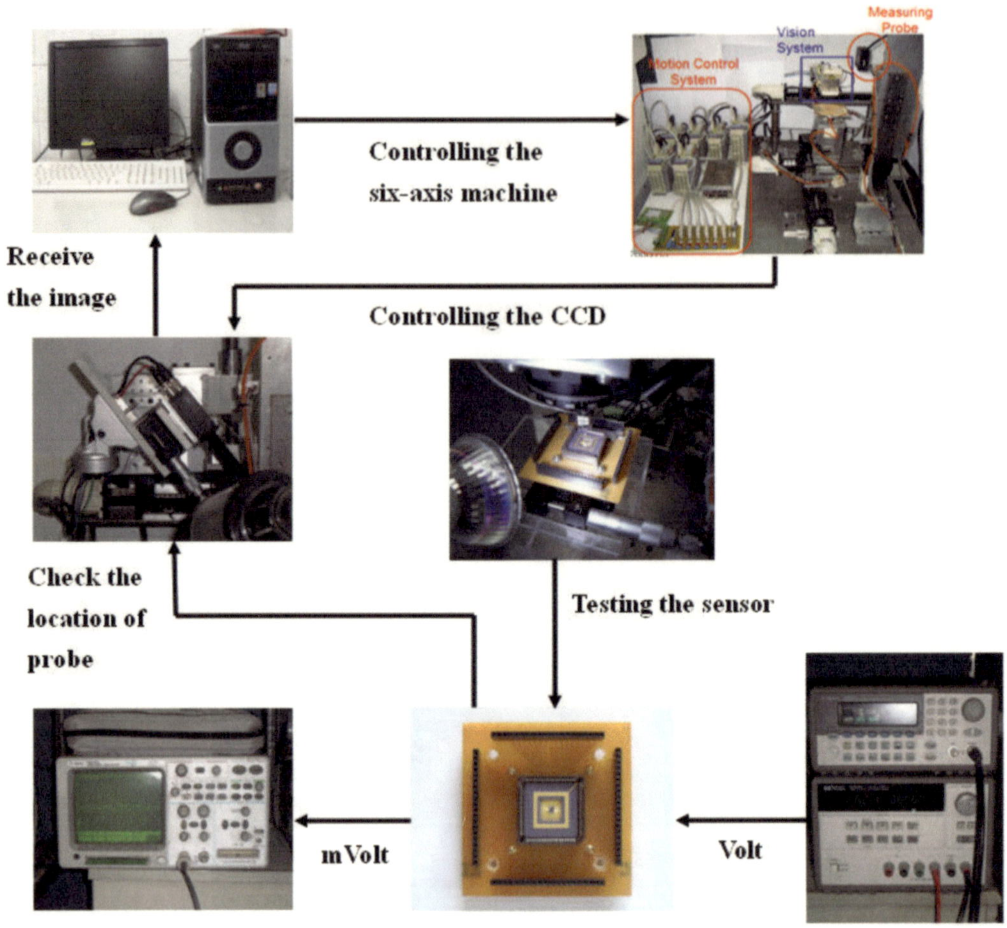

Fig. 13. The process of CMOS forces sensor measurement.

4.2.2 Results of measurement

After cleaning the sensor chip, we observed the surface of the sensor using an optical microscope to check whether the photoresist was clean. Scanning electron microscopy (SEM) and white light interferometers (WLI) were used to take pictures of the sensors, as shown in Fig. 14 and Fig. 15. Then, wire bonding was adopted to package the chip for our experiment.

At first, we simplified a vertical probe card, viewing it as just a probe that touches the sensor in the experiment, as shown in Fig. 16. The probe touches down the electronic scales (see Fig. 17) to plot a diagram of the probe overdrive force, as shown in Fig. 18(a). Second, the probe was pushed against the chip sensor to plot a diagram of the probe overdrive-output signal, as shown in Fig. 18(b). Fig. 19 shows the probe touching the membrane of the sensor in the experiment. We fixed the probe overdrive to get the relationship between the probe's reaction force and the output signal of the sensor. The relationship between the probe's reaction force and probe overdrive is illustrated in Fig. 18(c). The probe's reaction force stays linear as long as the probe overdrive stays within 40μm. The average sensitivity and load-bearing capacity of the tested sensors were 3.114 mV/MPa/V and 0~3 g, respectively.

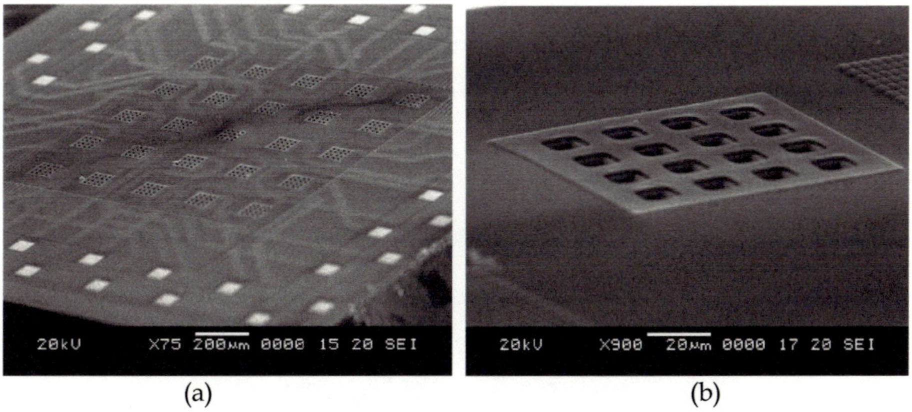

Fig. 14. The photograph of CMOS force sensor chip by SEM (a) The chip of CMOS force sensor (b) The single CMOS force sensor.

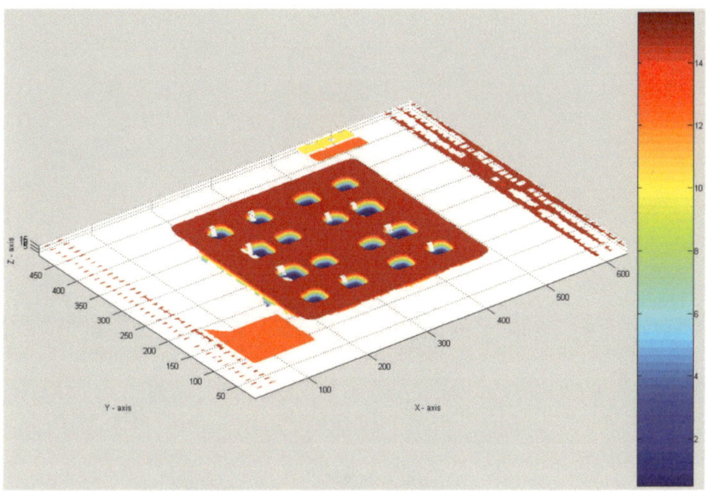

Fig. 15. The photograph of CMOS force sensor by WLI.

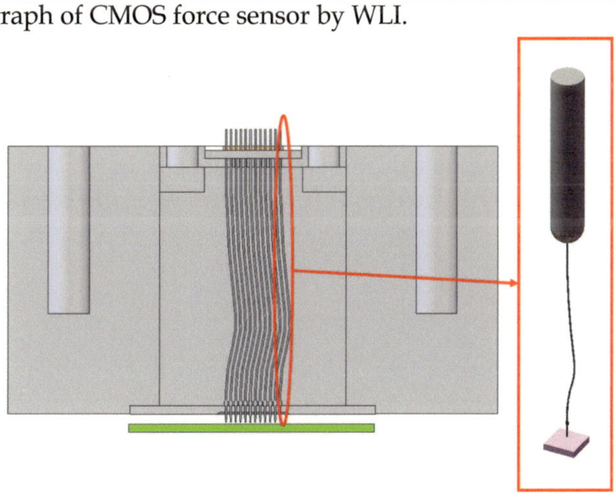

Fig. 16. The vertical probe card simply a probe to touches the sensor in the experiment.

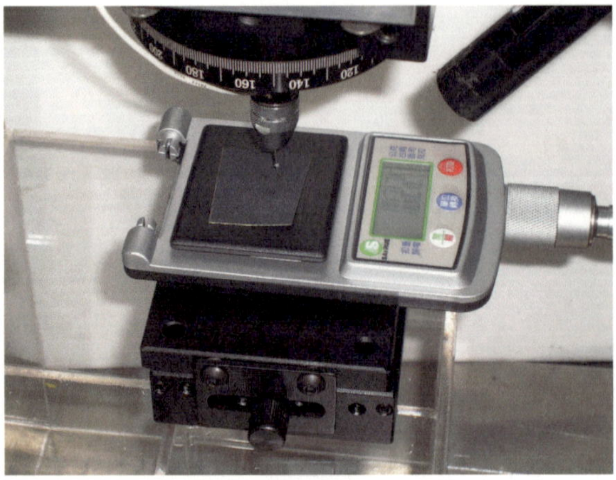

Fig. 17. The probe touches down the electronic scales.

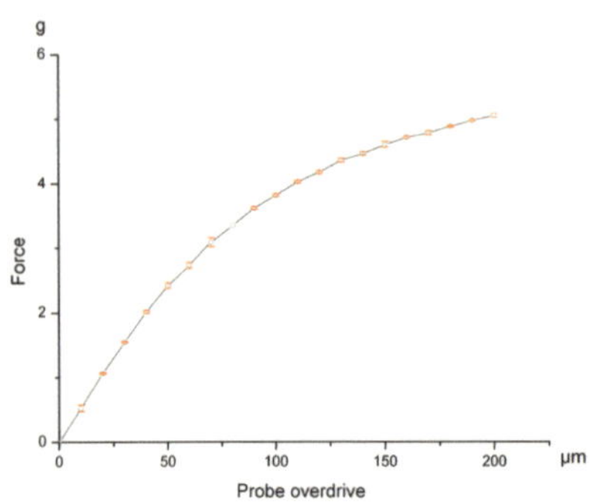

(a)

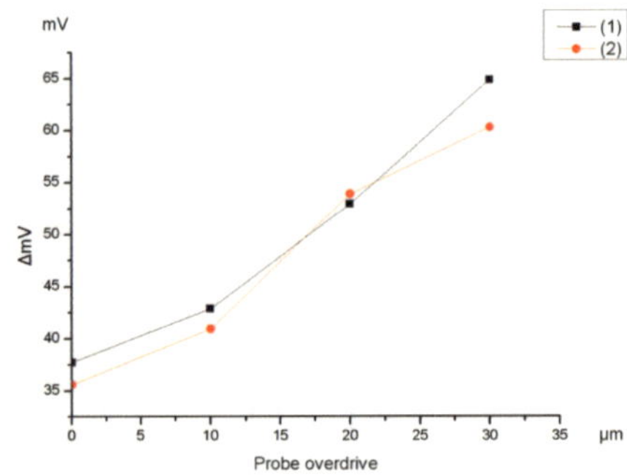

(b)

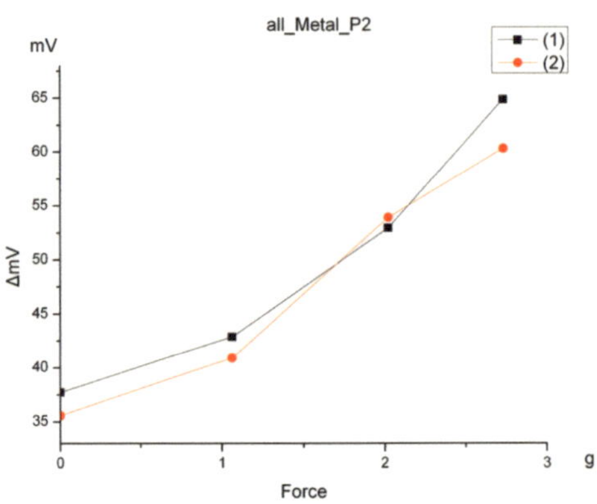

(c)

Fig. 18. (a) The probe reacting force vs. Probe overdrive. (b) The output signal of the sensor vs. Probe overdrive. (c) The output signal of the sensor vs. probe reacting force for the CMOS force sensor that taking two sensors random.

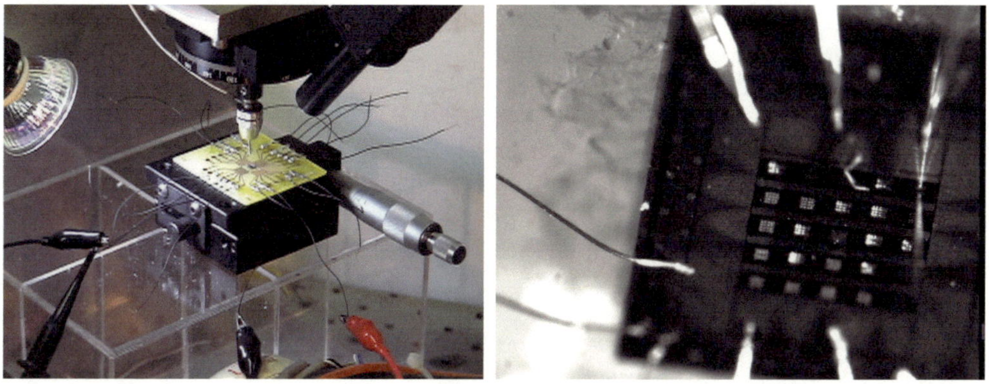

(a) (b)

Fig. 19. The probe tip touches down the membrane of sensor (a) actual experiment (b) CCD camera.

5. Conclusion

Probe cards play an extremely important role in the semiconductor industry. In this study, we designed a CMOS piezoresistive force sensor to be applied to the probe cards. Capable of simultaneously monitoring the probe reacting force and electrical signals, the designed sensor can help operators immediately identify a broken or a deformed probe and recognize that the received signals are erroneous. The repair time and cost of the probes can therefore be reduced. Further, we adopted the TSMC 0.35 μm 2P4M process to fabricate the CMOS force sensor that can be integrated with the circuit. According to the measurement results, the designed sensor reports an average sensitivity of 3.114 mV/MPa/V and a load-bearing capacity ranging from 0 to 3 g.

6. Acknowledgements

The author would like to thank C.C.P. Contact Probes Co., Ltd. for their help. We also thank the Chip Implementation Center of the National Science Council, Taiwan, R.O.C., for supporting the TSMC 0.35um CMOS process. The fund is granted from NSC 94-2622-E-027 -047 -CC3.

7. References

Iscoff, R. (1994). What's in the cards for wafer probing, *Semiconductor. Int.*, June 1994, pp. 76

Gilg, L. (1997). Know good die, *Journal of Electronic Testing: Theory and Applications*, vol. 10, issue 1-2, April 1997, pp. 2019, ISSN:0923-8174

Ghalichechian, N.; Khbeis, M.; Ma, Z.; Moghadam, S. & Tan X. (2002). Piezo-resistor pressure sensor cluster, *ENEE605 Final Project Report, Fall 2002 Department of Electrical and Computer Engineering Group #2*, University of Maryland

Malhair, C. & Barbier, D. (2003). Design of a polysilicon-on-insulator pressure sensor with original polysilicon layout for harsh environment," *Thin Solid Films*, vol. 427, issues 1-2, 3 March 2003, pp. 362-366, ISSN 0040-6090

Yang, L. J.; Lai, C. C.; Dai, C. L. & Chang, P. Z. (2005). A piezoresistive micro pressure sensor fabricated by commercial DPDM CMOS process, *Tamkang Journal of Science and Engineering*, vol. 8, no. 1, 2005, pp. 67-73

Peng, C. T.; Lin, J. C.; Lin C. T. & Chiang, K. N. (2005). Performance and package effect of a novel piezoresistive pressure sensor fabricated by front–side etching technology, *Sensors and Actuators A: Physical*, vol. 119, 2005, pp. 28-37, ISSN 0924-4247

Wang, H. H. & Yang, L. J. (2006). Micro pressure sensors of 50μm size fabricated by a standard CMOS foundry and a novel post process, *MEMS 2006*, pp. 22-26, Turkey, 22-26 January, Istanbul

Wilson, L. (1999). The National Technology Roadmap for Semiconductor, *Semiconductor Industry Association*, San Jose, California, November, 1999

Smith, C. S. (1954). Piezoresistance effect in germanium and silicon, *Physical Review*, vol. 94, issue 1, April 1954, pp. 42-49

Petersen, K. E. (1982). Silicon as a Mechanical Material" *Proceedings of The IEEE.*, vol. 70, no. 5, May 1982, pp. 420-457

Thurston, R. N. (1964). Use of semiconductor transducers in measuring strain, accelerations, and displacements, *Physical Acoustics*, vol. 1, pt. B. New York: Academic Press, 1964, pp. 215-235

Kanda, Y. (1982). A graphical representation of the piezoresistance coefficients in silicon, *IEEE Transactions on Electron Devices*, vol. ED-29, no. 1, January, pp. 64-70

French, P. J. (2002). Polysilicon : a versatile material for microsystems, *Sensors and Actuators A: Physical*, vol. 99, 20 January 2002, pp. 3-12

The CIC CMOS MEMS Design Platform for Heterogeneous Integration, *Chip Implementation Center, CIC*, Taiwan, Document no. CIC-CID-RD-08-01, April 2008

Seto, J. Y. W. (1976). Piezoresistive properties of polycrystalline silicon, *Journal of Applied Physics*, vol. 47, no. 11, November 1976, pp.4780-4783

4

Three-Dimensional Silicon Smart Tactile Imager Using Large Deformation of Swollen Diaphragm with Integrated Piezoresistor Pixel Circuits

Hidekuni Takao and Makoto Ishida
Toyohashi University of Technology
Japan

1. Introduction

Recently, various kinds of tactile sensors have been investigated and reported for tactile applications with robot fingertips. Typical specifications of human fingertips are known as follows; spatial resolution of human fingertip is around 1 mm, time resolution is below 1 msec (1 kHz), and the minimum force resolution is around 1-10 mN. Also, human fingertip can recognize the three-dimensional (3-D) shape of touching object using flexible deformation in the convex shape of fingertip skin. However, it is very difficult to realize all the above requirements/performances in conventional tactile sensors at the same time.

Tactile imager is a spatial distribution type of sensor, which can detect the object contact force and its distribution with an array of force or pressure sensors. In addition, detection ability of 3-D surface shape will be required for object handling. Tactile imagers can be applied to robot applications such as in robots for the assistance of visually handicapped and so on. There are two major trends in the previously reported tactile imagers. One is the polymer-based tactile imager realized by the substrate with organic materials, and the other one is silicon-MEMS type sensors. In polymer-based tactile imagers (Brussel & Belien, 1986; Engel et al., 2003; Shimojo et al., 2004; Someya et al., 2004, Engel et al., 2005), pressure-sensitive conducting rubber has generally been used as a major force sensing element (Brussel & Belien, 1986; Shimojo et al., 2004; Engel et al., 2005). Polymer-based sensors are suitable for wide area tactile sensors since the fabrication cost per unit area is considered to be much lower than that of silicon sensors. Artificial skin mounted on large areas of robot surface is one of the major applications (Someya et al., 2004; Engel et al., 2005). Essential disadvantages of polymer-based sensors are relatively low spatial resolution and upper limitation on the number of pixels due to electronic signal wires. Typical spatial resolution of polymer-based tactile imagers is around 2 ~ 4 mm range, which is not high enough for fingertip tactile sensing applications as mentioned below. Although a tactile imager with a large number of pixels has been reported using organic-FET switching matrix (Someya et al., 2004), it still utilizes conducting rubber sensor elements. Also, the integration density of organic-FET is much lower than the present silicon technology, and its long term reliability in force sensor applications has not yet been demonstrated.

Silicon-MEMS tactile imagers, integrating micro pressure sensor array (Sugiyama et al., 1990) or micro force-sensor array, have been reported earlier (Suzuki et al., 1990(a); (b);

Kobayashi et al., 1990; Souza & Wise, 1997; Mei et al., 1999; Mei et al., 2000; Sato et al, 2003; Charlot et al, 2004). This type of sensors can reduce the number of electronic signal wires by integrated switching matrix fabricated using CMOS technology (Doelle et al, 2004). Also, processing circuits can be integrated for front-end signal processing of the pixel array. As compared to the polymer-based tactile imagers, a higher spatial resolution can be realized using silicon micromachining. 500 dpi spatial resolution has already been reported (Souza & Wise, 1997), and such sensors with high spatial resolutions can be used for fingerprint identification (Sato et al., 2003; Charlot et al., 2004). Most of the silicon tactile imagers are configured as integrated array of individual micromechanical sensor structures. Piezoresistive or capacitive sensors are fabricated in each pixel structure. Since movable stroke of such micro pixels is usually very short (~1μm), it is difficult to realize flexible sensor surface to detect 3-D surface shape of touching object. In order to solve the problem, thick and protective layer of elastomer can be coated on the sensor array. However, such soft materials usually have nonlinearity due to creep and hysteresis in mechanical response. In addition, thick and soft layer works as spatial low-pass filter for the high density pixel array, which degrades the spatial resolution of original sensors. Although silicon-MEMS tactile sensors can realize higher spatial resolution, it is difficult to realize surface flexibility.

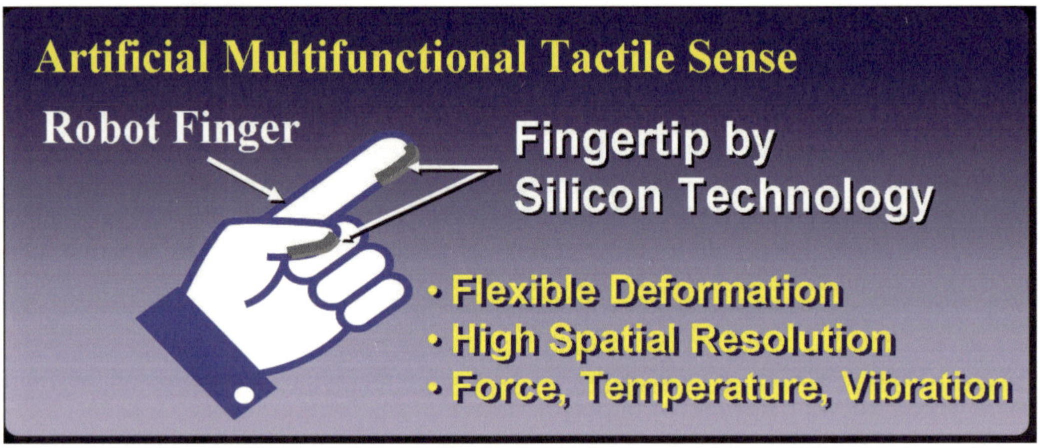

Fig. 1. A future image of tactile imager embedded in robot fingers. They will function as artificial tactile sense of fingertips in human-coexistence type robots.

Considering the tactile sensing in human-coexistence type robots, tactile imagers like human fingertip will be required in near future. Figure 1 shows an image of tactile imager embedded in robot fingers. The embedded imagers will function as artificial tactile sense of fingertips in human-coexistence type robots. In human fingertips, a large number of highly sensitive tactile corpuscles are distributed under skin, and their surface has flexibility for object contact. As explained, it is difficult for previously reported tactile imagers to satisfy the requirements for fingertip applications. In this study, a novel concept of silicon-MEMS tactile imager aimed at fingertip tactile application and the evaluation results of a fabricated device are presented. The final target is multi-functional integrated tactile imager with force, temperature, and vibration sensing elements in sensor arrays as shown in Fig. 1, since silicon technology is very suitable to integrate many kinds of functional sensors/circuits. In this concept, pneumatically swollen single silicon diaphragm integrated with a two-dimensional (2-D) array of strain-sensitive resistors (piezoresistors) is used for tactile sensor array instead of individually separated micromechanical sensor array. This structure has both the large number of pixels and surface flexibility for 3-D object contact (Takao et al., 2004; 2006). The surface shape similar to the diaphragm mechanical stroke can be detected as 3-D information (2-D position in array and depth information). In addition, spatial resolution higher than the polymer-based tactile imagers can be realized. A single tactile sensor with air pressure control has been proposed earlier to detect object hardness (Hasegawa et al., 2003). Air pressure is used to realize elastic surface and mechanical controllability of the surface of sensing area. In this chapter, the new concept, principle, design and experiments are presented in detail.

2. Configuration of tactile imager with silicon-LSI

2.1 A new concept of flexible silicon tactile Imager

In order to realize large mechanical stroke of tactile imager, this device concept uses large deformation of silicon diaphragm. Figures 2 (a) and (b) show schematic diagrams of the silicon tactile-force imager proposed in this study. It consists of three major components; (1) silicon diaphragm with sensing pixel array for contact force imaging, (2) signal processing circuitry integrated with IC technology, and (3) pressure chamber under the sensing diaphragm. 2-D piezoresistor pixel array is integrated on the thin silicon diaphragm. Each

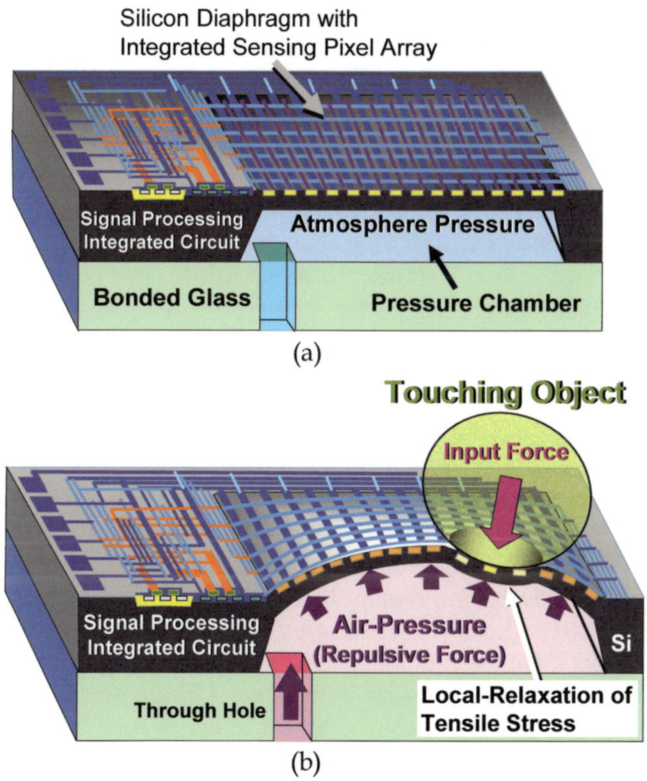

Fig. 2. Schematic diagrams of the tactile imager with skin-like sensing area; (a) Diaphragm backside pressure is equal to atmosphere pressure, (b) Diaphragm is swollen by a pressure, and set in the detection mode of tactile sensing.

strain-sensor pixel is electrically isolated, but, is not isolated mechanically, since all the pixels are formed on a continuous thin diaphragm structure. Switching and signal processing circuits for the sensing pixel array are monolithically integrated around the sensing diaphragm region. Pressurized air is provided to the chamber through the hole in the glass in order to apply the pressure to the diaphragm backside. In Fig. 2 (a), pressure on the diaphragm backside is equal to the atmosphere pressure, and the diaphragm is kept flat. If pressurized air is applied to the diaphragm backside, the diaphragm is deformed and swollen upward like a balloon as shown in Fig. 2 (b). Displacement of the swollen diaphragm depends on the dimensions and applied pressure, and a movable stroke of around 10~200 μm can easily be realized in this approach. Advantages of the tactile imager with pneumatically swollen single diaphragm structure are summarized below.

a. Flexibility of the sensor surface is obtained without any elastomeric materials for high spatial resolution. Convex shape of the swollen surface makes it easier to contact with the sensing target like human's fingertips.
b. Swollen large diaphragm can realize large stroke of surface indentation. 3-D surface shape can be detected by measuring the indentation depth (force) of the swollen diaphragm surface.
c. Pixel pitch of the strain sensor array can be made smaller compared to the polymer-based sensors. In addition, number of pixels in the sensor array can be larger with processing ability of the integrated circuits.
d. Stiffness of the sensing region of diaphragm can be controlled by the backside pressure. This means that characteristics of the sensor can be controlled even after the device packaging is completed (Fig. 2(b)).

2.2 Principle of 3-D Tactile Imaging

In this sensor, contact force image corresponding to 3-D image of the surface shape is detected by reading the stress distribution change on the swollen diaphragm using the 2-D piezoresistor pixel array. Figure 3 shows a cross-sectional view of the detection principle of this tactile imager. The number of piezoresistor pixel array and the pixel pitch can be changed in alternative designs. It mainly depends on the feature size of CMOS fabrication technology used. In the initial state, before the object contact, shown in Fig. 3 (a), tensile membrane stress is distributed with uniform amplitude over the entire piezoresistor array on the diaphragm. Since the swollen silicon diaphragm has a finite thickness, bending stress is generated in addition to the tensile membrane stress on the diaphragm. The surface stress on diaphragm appears according to the principle of superposition of the two components. Tensile membrane stress is caused by the large deformation of diaphragm, and bending stress is caused by the bending moment proportional to the distance from the neutral plane in the diaphragm. If the backside pressure is high, bending stress is negligible as compared to tensile membrane stress (i.e. initial stress on the array can be regarded as uniform value). However, the ratio between the membrane stress and the bending stress becomes only 5 or less depending on the backside pressures in the case of 10 μm diaphragm thickness. In order to cancel out the offset distribution caused by the effect of bending moment, they are once memorized, and subtracted from the output for zero point adjustment. This operation can be performed by software in the measurement system.

Figure 3 (b) shows the sensing mode of contact force of the object. If a hard object touches the surface of the sensing region, swollen diaphragm is deformed at the object contacting points as shown in Fig. 3 (b). Diaphragm region where the object is in contact is pushed downward, and the tensile membrane stress applied initially around the contacting object is eased and reduced by the local deformation causing compressive bending stress around the contacting points. Difference of stress distribution from the initial state corresponds to the signal of the tactile imager, and it can be read out from the 2-D piezoresistor pixel array sequentially. Also, the image corresponds to the depth distribution of the touching object. Thus, the signal component shows peaks at around the tips of contacting object, and the positions and amplitudes of force (i.e. indentation depth) on the diaphragm can be detected as 3-D shape image of the touching object based on this principle.

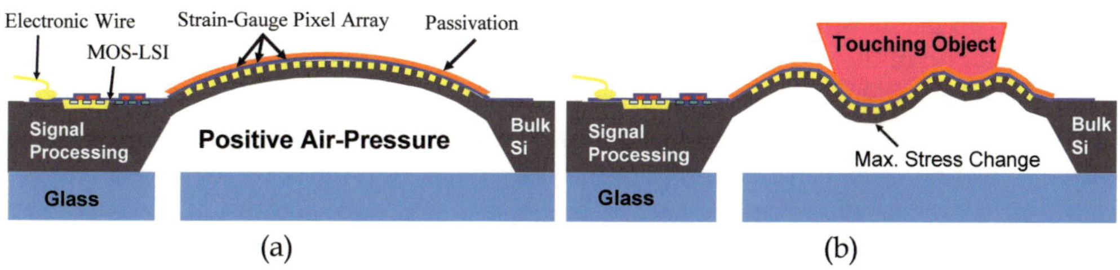

Fig. 3. Detection principle of the tactile imager with surface stress distribution on the diaphragm; (a) Initial state before object touching, (b) 3-D shape detection with deformation.

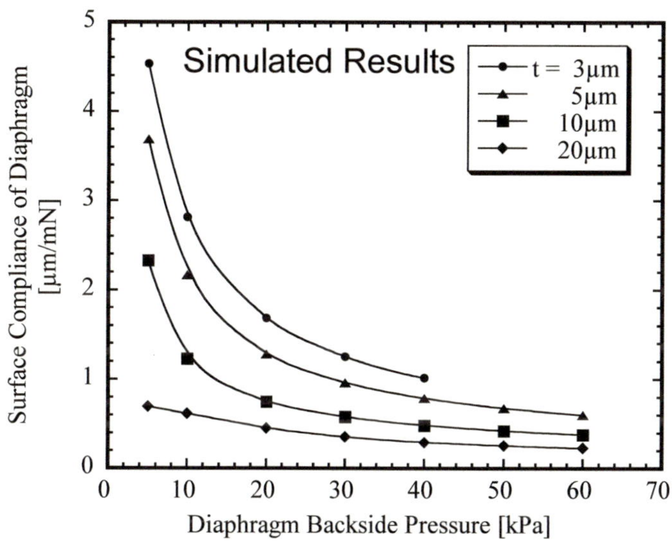

Fig. 4. Simulated surface compliance of pneumatically swollen single-crystal silicon diaphragm for various thicknesses. FEM non-linear analysis was performed for simulations.

Mechanical properties of the sensing diaphragm can be controlled by changing the backside air pressure. For example, compliance of the swollen diaphragm strongly depends on the backside pressure. Finite Element Method (FEM) non-linear analysis was performed to analyze the mechanical property of swollen diaphragm using ANSYS®. Total area of the simulated diaphragm is 3040×3040 μm^2, and the edges are fixed to the silicon substrate like a structure shown in Fig. 2 (a). Figure 4 shows a simulated relationship between the surface compliance of diaphragm and backside air pressure for various diaphragm thicknesses.

Force is applied at a point on the diaphragm surface in the FEM simulation. If the thickness of diaphragm increases, dependence of the surface compliance on the backside pressure becomes small due to its own rigidity as seen in the figure. It is considered from the result that thinner diaphragm is advantageous for controlling the characteristics of tactile imager. Assuming that the thickness of diaphragm is same, higher sensitivity can be obtained with lower backside pressure since the surface stiffness becomes lower and surface stress change will be increased for the same contact force. Conversion factor from the input force into stress change on a pixel is a dominant factor in the force sensitivity of tactile imager. On the other hand, upper limit of the detectable force can be increased by the increased backside pressure. Simulated dependence of the sensitivity and input force range on the pressure are compared with the measured results in a later section.

Spatial resolution of the contact force distribution cannot be determined only by the pitch of piezoresistors. Since the pixels are not mechanically isolated from each other, there is some crosstalk of strain among piezoresistor pixels. If the pixel pitch of piezoresistors is shorter than the effective limit of mechanical crosstalk, spatial resolution of the tactile imager is limited by the crosstalk effect. FEM non-linear analysis was performed to estimate the crosstalk between the multiple force input positions. Figure 5 (a) shows the parameters used in the simulation. Diaphragm size used in the FEM simulation is the same as in the case of Fig. 4 (3040×3040 μm2). Simulation was performed for different distances 'd', of two input forces, varying from 120 μm to 1200 μm. In order to evaluate the spatial resolution, stress change from the initial state (signal component) generated by the two forces is plotted as a function of distance from the center of two forces as shown in Fig. 5 (b). In the simulation result, amplitude of the applied forces is 5 mN, thickness t is 10 μm, and the backside pressure is 30 kPa. This simulation corresponds to the evaluation of two-point discrimination ability of the tactile imager. A parameter of mechanical crosstalk between the two input points is introduced as 'crosstalk ratio' for quantitative evaluation of the spatial resolution. It is determined as a ratio of generated stresses between the input point and the center of the input points when the amplitudes of two input forces are equal. Here, the 'crosstalk length' is determined as the distance at which the crosstalk ratio becomes 0.5. As seen in Fig. 5 (b), the crosstalk ratio becomes approximately 0.5 at 360 μm distance for the boundary condition. If the length d is shorter than 360 μm, crosstalk ratio becomes higher than 0.5. Each peak value at force input point is significantly enhanced by the signal crosstalk, and it is difficult to distinguish the two points of force input. 360 μm is considered as the 'crosstalk length' in this simulation condition. The crosstalk ratio was almost independent of the input force in the simulated range from 0.5 to 15 mN, since it is determined as the ratio of generated stress. On the other hand, the crosstalk ratio has dependence on the backside pressure and diaphragm thickness. If the backside pressure is increased from 30 kPa to 60 kPa, crosstalk ratio is improved by 18.6 % since the deformation around the contact point becomes more local. In this case, the crosstalk length is shortened to below 300 μm. If the thickness of diaphragm is reduced from 10 μm to 5 μm, the crosstalk length of approximately 250 μm can be obtained at 30 kPa backside pressure.

The crosstalk length is a function of both the diaphragm dimensions and the backside pressure. Spatial resolution of the tactile imager is determined either from the crosstalk length or the pixel layout pitch of piezoresistor array.

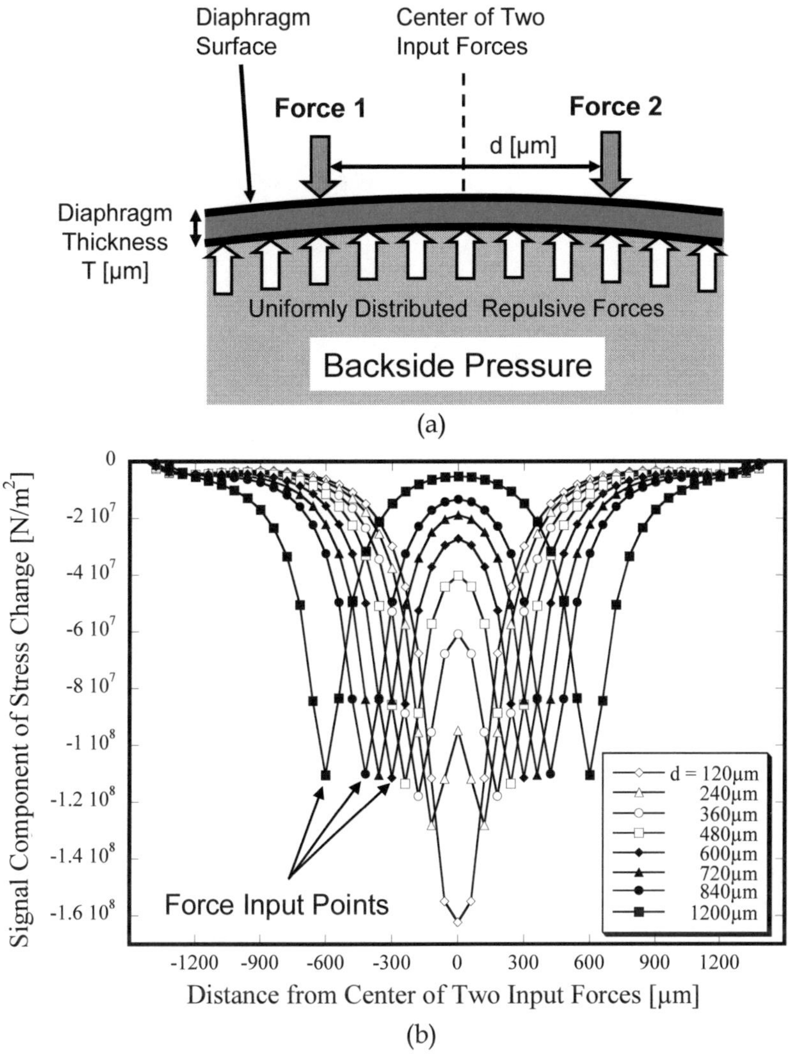

Fig. 5. FEM simulation for spatial resolution analysis; (a) Model parameters in the FEM analysis for the estimation of mechanical crosstalk between two input forces,
(b) Distribution of stress signal component generated by the two input forces for various distances of d. The backside pressure is 30 kPa.

As discussed in this section, the backside pressure of the diaphragm influences both the spatial resolution and the sensitivity for contact input forces. A comparison of FEM results simulated at different pressures is helpful to understand this relationship. Figure 6 (a) and (b) show the simulated stress change (signal component) distributions on a sensing diaphragm of this tactile imager at 23.0kPa and 5.0kPa, respectively. A half model of the sensor structure is used. A load of 8.5mN is applied at the contact point in the figure. In the case of 23.0kPa (Fig. 6 (a)), stress change is distributed locally around the contact point. On the other hand, both the stress level and strained area are increased in the 5.0kPa case as shown in Fig. 6 (b). This means that reduction of the backside pressure results in both improvement of sensitivity (i.e. SNR) and degradation of spatial resolution for an input force applied. Selecting a proper backside pressure adaptively for the device dimensions and expected input force range, the maximum SNR of the tactile imager can be obtained for a required spatial resolution (crosstalk length).

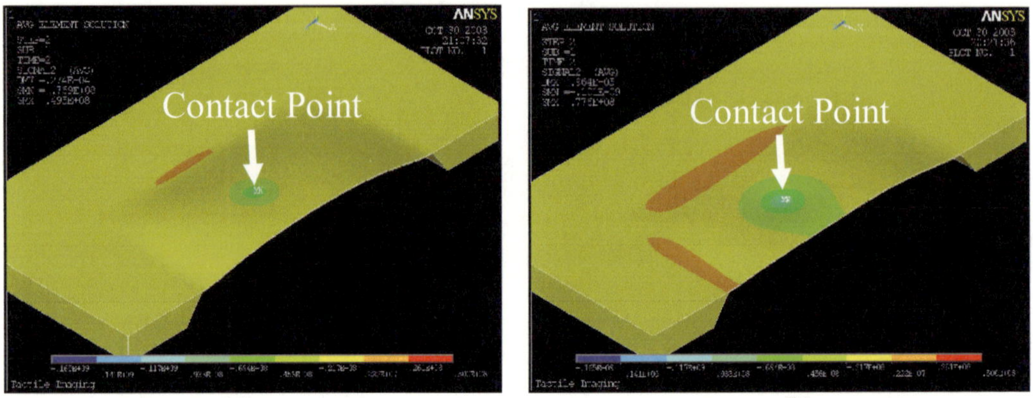

(a) (b)

Fig. 6. Simulated stress change (signal component) distributions on a sensing diaphragm of this tactile imager; (a) Backside pressure is 23.0kPa; (b) Backside pressure is 5.0kPa.

3. Device design and fabrication

3.1 Design of piezoresistor pixel circuit on diaphragm

Signal component of the stress, generated by the object contact, is translated into voltage signal in each pixel circuit with piezoresistor. Figure 7 (a) shows the circuit configuration of each pixel on the diaphragm. A pixel includes n-type diffused piezoresistor for the detection of surface stress (R_{PR}), n^{+}-poly Si reference resistor with very small stress sensitivity (R_{poly}), logic gates for pixel select operation (NAND and NOT), and switch MOSFETs for resistor drive current (M1) and pixel output (M2). Tensile membrane stress generated strongly on the swollen diaphragm is almost isotropic, and the shear component of stress on each pixel is almost zero. Select terminals of line (*X_Sel*) and column (*Y_Sel*) of pixel circuits in the array is driven sequentially in order to read out the distribution of output voltage. If both *X_Sel* and *Y_Sel* in the pixel are pulled up to V_{dd}, switches M1 and M2 are turned on, and drive current for R_{PR} and R_{poly} is provided from the power source through M1. The piezoresistor R_{PR} translates the surface stress level on each pixel into a corresponding resistance value. Voltage of the output line is determined as a partial voltage of R_{PR} and R_{poly} since M2 is turned on in this case. The output voltage of pixel (V_{Pix_Out}) fed to the common amplifier in the following stage is expressed by the next equation (Takao et al, 2006);

$$V_{Pix\_Out} \approx \frac{R_{Poly}\cdot(V_{DD}-V_{SS})}{R_{PR}+R_{Poly}}\cdot\frac{R_{Amp}}{R_{M2}+R_{Amp}} = \frac{R_{Poly}}{R_{PR}+R_{Poly}}\cdot(V_{DD}-V_{SS}) \qquad \text{[V]} \quad (1)$$

where R_{Amp} is equivalent input impedance of the common amplifier (dashed line in the figure), and R_{M1} and R_{M2} are on-channel resistances of M1 and M2, respectively (Takao et al., 2006). In this situation, M1 and M2 are operating in non-saturation region at gate voltage of V_{dd}, and its channel resistance is much lower than the other resistances (i.e. $R_{M1} \ll R_{PR}$, R_{poly}

and R_{M2}<< R_{Amp}). Assuming that only the planar components of surface stress in parallel to the diaphragm surface is dominant, resistance of a piezoresistor can be expressed with a simplified linear expression with the products of piezoresistive coefficient and applied stress. If n-type piezoresistor in the pixel is integrated on Si(100) and oriented to a direction equivalent to <110>, longitudinal and transversal components of piezoresistive coefficient become equal (i.e. $\pi_{long} = \pi_{trans} = \pi_{n<110>}$) (Kanda, 1982). Resistance value of R_{PR} under tensile membrane stress can be expressed as the following expression considering the isotropic distribution of stress on the swollen diaphragm;

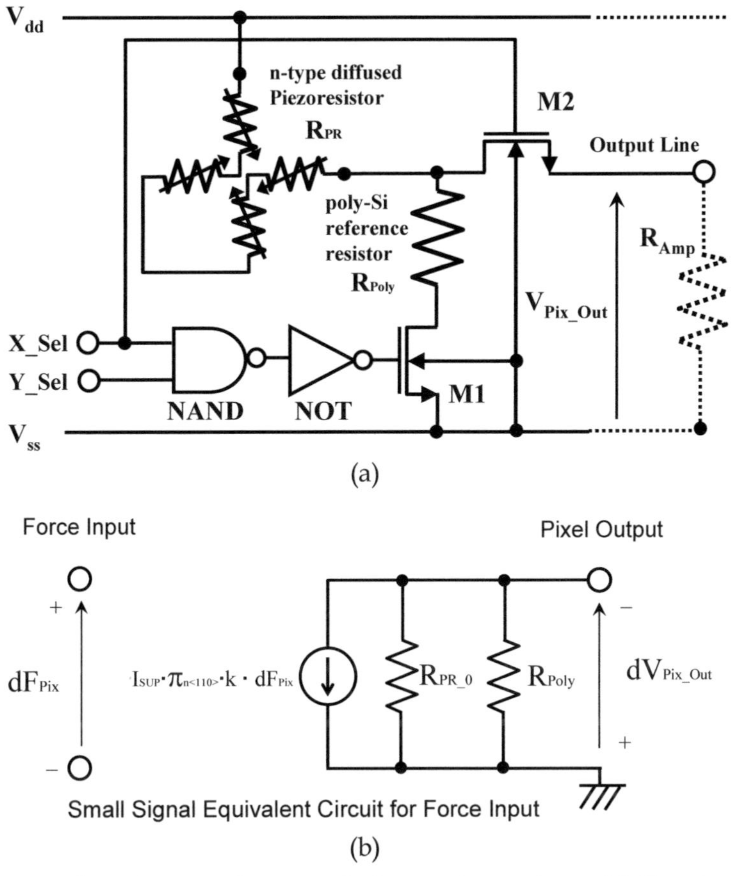

Fig. 7. An example of pixel circuit design including a piezoresistor; (a) Circuit configuration of a unit pixel, (b) Small signal equivalent circuit of the pixel for a force input signal.

$$R_{PR}(\sigma) \approx R_{PR\_0} \cdot (1 + \pi_{long}\sigma_{long} + \pi_{trans}\sigma_{trans}) = R_{PR\_0} \cdot (1 + \pi_{n<110>}\sigma_{EQ}) \quad [\Omega] \quad (2)$$

where R_{PR_0} is the resistance of R_{PR} under zero-stress condition, and σ_{EQ} is an equivalent value of stress corresponding to the sum of the two-axis components of average stress distributed on piezoresistor in a pixel. Typical values for $\pi_{n<110>}$ are around 10^{-10} m^2/N and its polarity is negative. Stress sensitivity of the output voltage in the pixel can be derived by partial differential on σ_{EQ};

$$S_{Pix\_\sigma} = \frac{\partial V_{Pix\_Out}}{\partial \sigma_{EQ}} \approx -\pi_{n<110>} \cdot (V_{DD} - V_{SS}) \cdot \frac{R_{PR\_0} \cdot R_{Poly}}{(R_{PR\_0} + R_{Poly})^2} = -\pi_{n<110>} \cdot I_{SUP} \cdot R_{R_{PR\_0}//R_{Poly}} \quad [V/(N/m^2)] \quad (3)$$

where $R_{RPR\_0//RPoly}$ is the combined resistance of parallel connection of R_{PR_0} and R_{Poly}, and I_{SUP} is the initial supply current to R_{PR} and R_{Poly}. Assuming a linear relationship between average stress change on a pixel and average contact force applied on the pixel area, force sensitivity of the pixel can be expressed as small signal sensitivity in the following expression;

$$S_{Pix\_F} = \frac{dV_{Pix\_Out}}{dF_{Pix}} = \frac{S_{Pix\_\sigma} \cdot d\sigma_{EQ}}{dF_{Pix}} = \frac{S_{Pix\_\sigma} \cdot k \cdot dF_{Pix}}{dF_{Pix}} = -R_{R_{PR\_0}//R_{Poly}} \cdot I_{SUP} \cdot \pi_{n<110>} \cdot k \quad [V/N] \quad (4)$$

where $d\sigma_{EQ} = k \cdot dF_{Pix}$, and k is the conversion factor from the input force (dF_{Pix}) into signal component of stress change on the pixel ($d\sigma_{EQ}$) discussed in the section 2.2. The conversion factor strongly depends on the stiffness of diaphragm surface (i.e. backside pressure). Output signal of a pixel is expressed as a product of S_{Pix_F} and input force.

Figure 7 (b) shows the small signal equivalent circuit of a piezoresistor pixel corresponding to the above relationship between the input force and pixel output. Reading the output signal obtained from sensing pixels in the array, contact force image on the diaphragm can be obtained as two-dimensional distribution. It is easily derived from the simple analysis of the partial differential of pixel sensitivity on R_{PR} that the maximum sensitivity is obtained if the condition $R_{PR} = R_{Poly}$ is satisfied in the real device. Resistance values of R_{PR_0} and R_{Poly} are decided from the power consumption point of view for a fixed power supply voltage and pixel area, while the ratio of resistances should be controlled to bring it close to 1.0 for high sensitivity.

3.2 Layout design of the overall sensor chip

A prototype device with a small scale integrated sensor array was designed and fabricated. The die size of the designed sensor is 5200 × 5200 μm^2, and the sensing diaphragm region with integrated pixel array is 3040×3040 μm^2. 5.0μm technology design rule was used in the pixel circuit design, and the size of pixel layout equivalent to Fig. 7 (a) became 420 × 420 μm^2. R_{PR} and R_{Poly} were designed to be 3 kΩ in the pixel layout. In the prototype device, 6 × 6 sensing pixel array was designed on the sensing diaphragm considering the pixel size (pitch) and the diaphragm size. The sensor array occupies an area of 2520 × 2520 μm^2 on the diaphragm. Tensile membrane stress becomes quite uniform over the array area, and the effect of fixed diaphragm edge, estimated by FEM simulation, is below 10 % in the area. On the peripheral of the diaphragm region, signal conditioning circuits for sensing operation were designed. It includes array scanning logic circuits for sequential readout of pixel array outputs, a reference signal generator for differential readout of each pixel output.

Diaphragm area in the device is same as in the simulated model discussed in section 2.2. Thickness of the sensing diaphragm of the device was decided considering some parameters. A thinner diaphragm is advantageous to a thicker one for controllability of the mechanical property with air pressure as shown in Fig. 4. Also, the thinner the diaphragm, the higher the spatial resolution as discussed in section 2.2. Considering the advantages, designed pixel size of 420 μm, and some difficulties in fabrication process, the thickness was decided to 10 μm. The spatial resolution of the designed device is considered to be approximately equal to the pixel size in this design, since the crosstalk length becomes 360 μm at 30 kPa for 10 μm-thick diaphragm, which is shorter than the pixel pitch, 420 μm. If the backside pressure is reduced to below 18 kPa, spatial resolution will be limited by the increased crosstalk length instead of the pixel pitch.

Calculated displacement of the diaphragm at the center is 31.8μm for 30 kPa, and 48.6 μm for 100 kPa backside pressure. Although, the mechanical stroke is too short to apply it to fingertip tactile sensor application, flexibility of the sensor surface is much higher than the conventional silicon-MEMS tactile imagers. The movable distance can be increased using larger and thinner diaphragm structure.

3.3 Device fabrication from standard LSI wafers

The designed device was fabricated with our silicon IC/MEMS fabrication technology (Takao et al., 1997; 1999; 2001). Figure 8 shows outline of the fabrication process. 1) Starting material of the tactile imager was 2-inch and 300 μm-thick p-type Si (100) wafer with resistivity of 1~3 Ω. 2) 6 × 6 Sensing pixel array including piezoresistors and signal processing circuits were fabricated. 3) After the electrical check of the circuits, surface of the integrated circuit was protected by 6 μm-thick polymer layer (CYTOP®) to prevent physical damage in the following bulk-etching process (Takao et al., 1997). The diaphragm etching pattern was defined by SiO_2 layer using wafer backside aligner. 4) Silicon substrate was etched by 25-wt% TMAH solution at 90 °C for 6.5 hours using the SiO_2 masking layer. In this step, silicon thickness of the sensing diaphragm region was controlled to 10 μm ± 1 μm by etching time control. Variation of thickness in a diaphragm was within 0.5 μm. 5) The protective layer on the surface was removed, and a glass substrate with through holes was bonded to the backside of silicon wafer by adhesive bonding with epoxy glue.

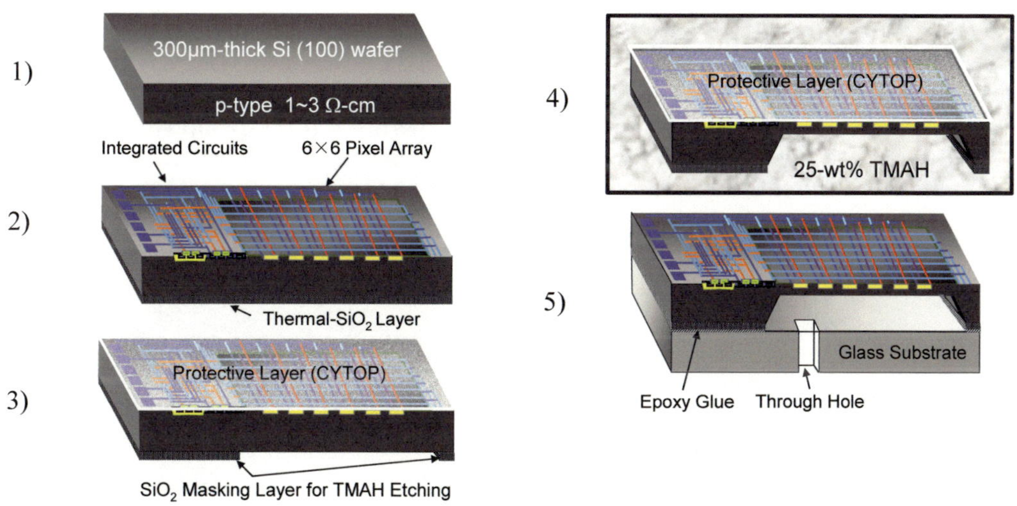

Fig. 8. Fabrication process; 1) Starting material of the tactile imager, 2) Fabrication of 6 × 6 Sensing pixel array including piezoresistors and signal processing circuits, 3) Protection of the circuit surface by a polymer layer (CYTOP®) and definition of the diaphragm etching pattern, 4) Backside-wafer etching with 25-wt% TMAH solution at 90 °C, 5) Removal of the protective layer on the surface and bonding to the glass substrate with epoxy glue layer.

Figure 9 (a) shows a photograph of the fabricated tactile imager with 6 × 6 pixel array integrated on a diaphragm structure. Close-up photograph of a pixel circuit corresponding to Fig. 7 (a) is shown in Fig. 9 (b). In the pixel, n-type piezoresistor device has a box-like shape to detect isotropic stress on the swollen diaphragm effectively, since n-type piezoresistor on Si(100) has bi-axial sensitivity. The areas of the piezoresistor and n^+-poly Si reference resistor are indicated in the figure.

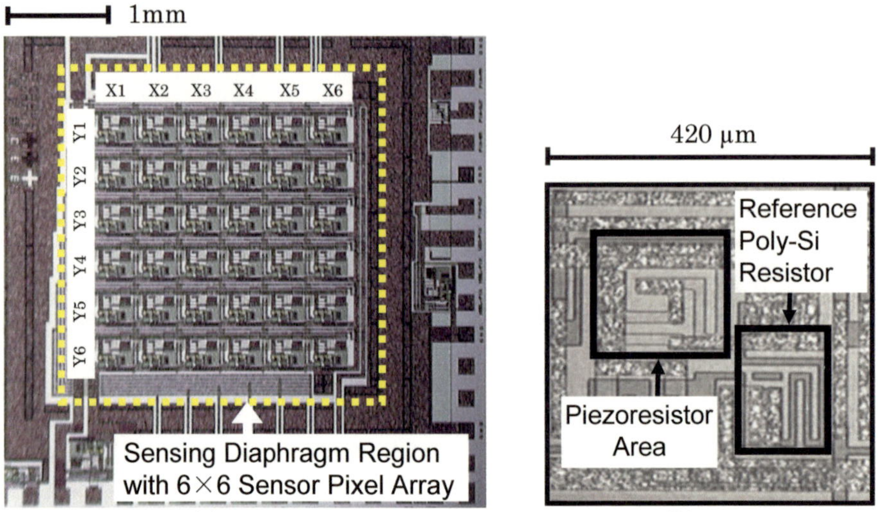

(a) (b)

Fig. 9. Photograph of a fabricated tactile imager with 6 × 6 pixel array integrated on the diaphragm; (a) Overall Chip, (b) Close-up of a pixel circuit in the 6 × 6 pixel array.

4. Characteristics evaluation

4.1 Single point contact force detection

Device characterization was performed with the experimental setup shown in Fig. 10 (a). The fabricated sensor was fixed on a three-axis (X-Y-Z) stage with position resolution of 1 µm in order to control the contact position and input force of measured object on the sensing

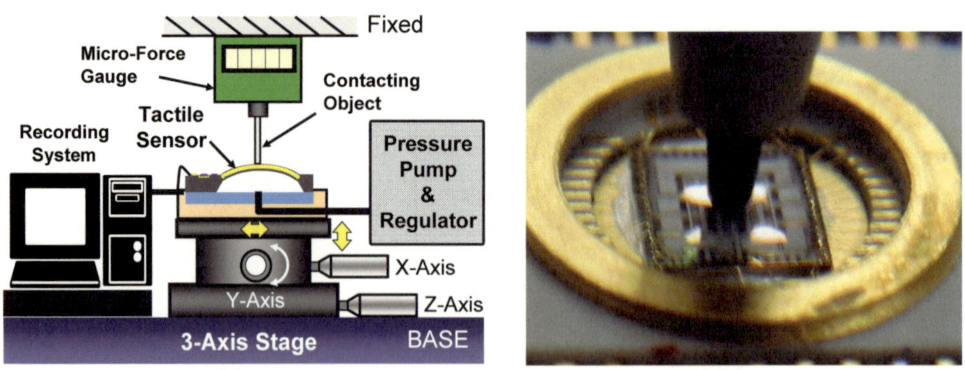

(a) (b)

Fig. 10. Experimental setup; (a) Total system for device characterization; (b) Photograph of a sensing diaphragm deformed by a touching object.

diaphragm. A micro-force gauge with 0.5 mN resolution was used to measure the total contact force of measured object on the sensor surface. Controlled air pressure was applied to the diaphragm backside using a pumping system and a pressure regulator. Signal obtained from the sensor chip was recorded with external measurement equipment and PC. Fig. 10 (b) shows a photograph of the sensing diaphragm deformed by a touching object. Deformation of the diaphragm surface and the indentation can be easily recognized by the reflection of light on the sensing area.

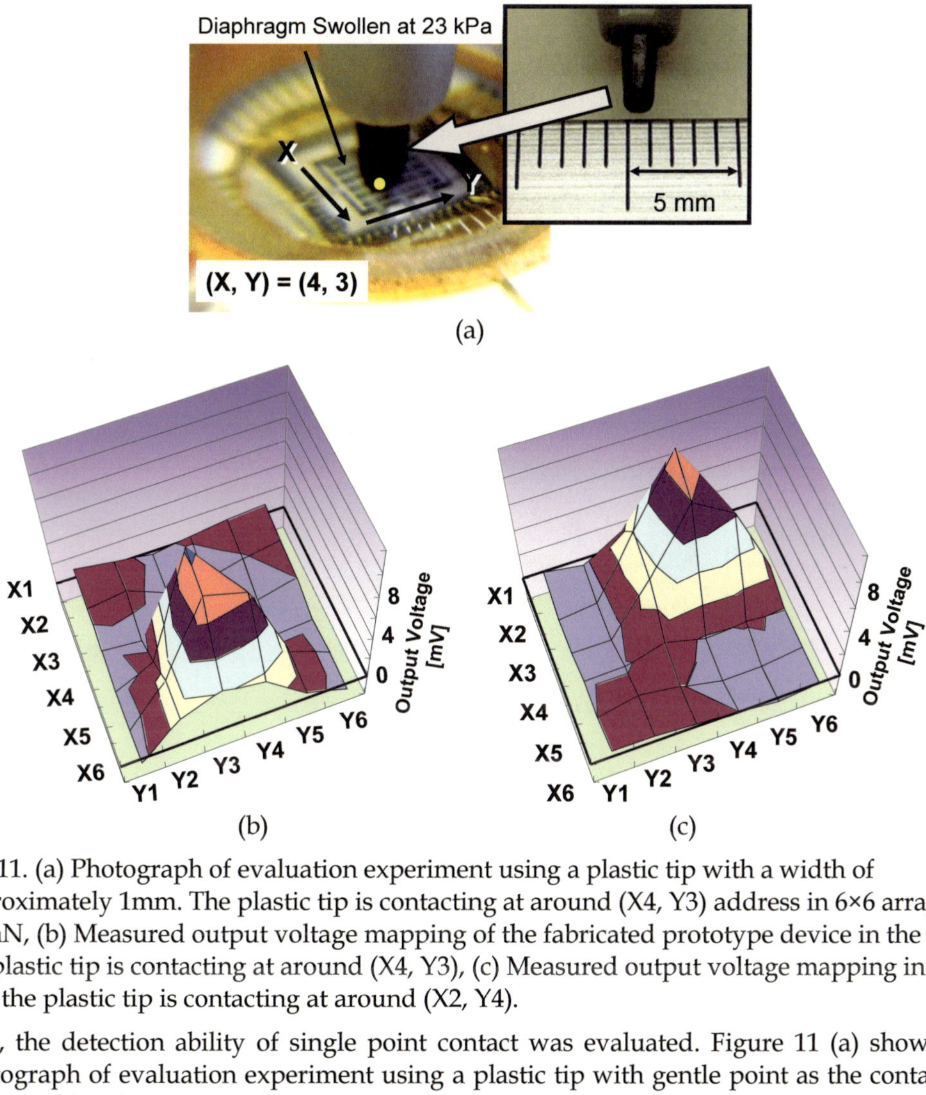

Fig. 11. (a) Photograph of evaluation experiment using a plastic tip with a width of approximately 1mm. The plastic tip is contacting at around (X4, Y3) address in 6×6 array at 29 mN, (b) Measured output voltage mapping of the fabricated prototype device in the case the plastic tip is contacting at around (X4, Y3), (c) Measured output voltage mapping in the case the plastic tip is contacting at around (X2, Y4).

First, the detection ability of single point contact was evaluated. Figure 11 (a) shows the photograph of evaluation experiment using a plastic tip with gentle point as the contacting object. In the photograph, the tip is contacting on a pixel around (X4, Y3) address in 6 × 6 sensor array. A 23.0kPa pressure was applied to the diaphragm backside, and the plastic tip was put on a pixel on the sensing diaphragm with 29mN-force. Movable stroke of the swollen diaphragm from the original surface was approximately 30 μm at the center of the diaphragm. Spatial resolution of the tactile imager is the pixel pitch of 420 μm for 23.0 kPa backside pressure as discussed in the previous section. Considering the movable stroke of the sensor surface, only a small tip area of the plastic tip is contacting on the diaphragm surface. The width of the tip is approximately 1 mm.

Figures 11 (b) and (c) show measured output voltage mapping obtained from the tactile imager in this experiment. In Fig. 11 (b), since the plastic tip is contacting around (X4, Y3), the peak of the pixel output signal appears on the same address in the 3-D plot. In addition, since the spatial resolution of the tactile imager is 420 μm, contact force applied by the gentle point with 1 mm width is detected by the pixels around (X4, Y3). In other words, the gentle shape wider than the spatial resolution is detected by the plural pixels in the sensor array. After the measurement, the plastic tip was moved to another position (X2, Y4), and the output signal obtained from the sensor is as shown in Fig. 11 (c). The peak position of output signal appears at (X2, Y4), and the shape of peak in the output around the contacting point is similar as in the case of (X4, Y3). Amplitude of the output peak is almost same in these two cases.

4.2 Multi-point contact and 3-D surface shape detection

Multi-point detection ability of this tactile imager was also evaluated. Figure 12 (a) shows a photograph of the multi-point contact test arrangement using two sharp fingers of a rubber toy. The fingers are contacting at the points of (X3, Y2) and (X4, Y5) with total load of 29 mN. The right finger contacting at (X4, Y5) is slightly longer than the left finger contacting at

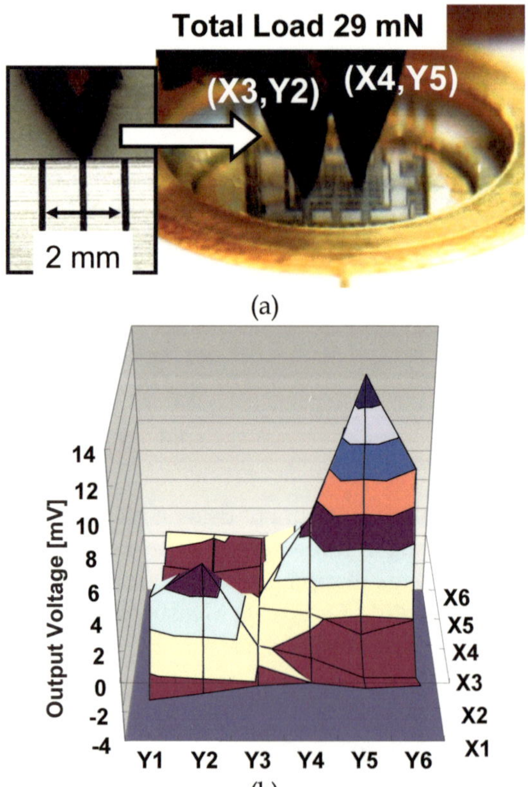

Fig. 12. Multi-point detection using this tactile image sensor; (a) Photograph of multi-point contact test and the sharp point of the rubber finger. Two fingers of rubber toy are contacting at (X3, Y2) and (X4, Y5), respectively, (b) Obtained output distribution with two signal peaks at (X3, Y2) and (X4, Y5), that correspond to the two positions of touch of the two rubber fingers.

(X3, Y2). As seen in the photograph, the shape of the points of the rubber fingers is much sharper than the point of the plastic-tip shown in Fig. 11 (a). The width of the points is approximately below 300 μm under the contacting situation, and it is smaller than the spatial resolution of the tactile imager (i.e. 420 μm in this experiment). Distribution of the output signal obtained from the experiment is shown in Fig. 12 (b). There are two signal peaks at (X3, Y2) and (X4, Y5), and their positions correspond to the two contact points of the two rubber fingers. Signal level of the peak at (X4, Y5) is larger than the peak at (X3, Y2). This corresponds to the fact that the partial contact force of the rubber finger at (X4, Y5) is larger than that of at (X3, Y2). In addition, the shapes of the output peaks around the contacting positions are much sharper than the shapes obtained in Fig. 11 (b) and (c). This fact implies that the shape of peak point in the output distribution correctly reflects the shape of the contacting point within the limit of the spatial resolution, even though some effect of mechanical crosstalk is seen among the pixels. Through the experiments, basic detection ability of multiple contact forces and the spatial resolution of the tactile imager have been demonstrated successfully.

4.3 Ability of characteristic control

Dependence of the contact force sensitivity of each pixel on the backside pressure was evaluated using a sharp tip of rubber finger shown in Fig. 12 (a), since contacting point smaller than the pixel size is necessary to evaluate the pixel sensitivity. Figure 13 (a) shows a measured relationship between pixel force sensitivity of the fabricated sensor and the diaphragm backside pressure. Supply voltage was 5 V, and piezoresistor pixel at (X3, Y4) was used in the measurement. In the figure, sensitivity dependence on the backside pressure, calculated using FEM, is also plotted. Since the FEM results are additionally plotted to have agreement with the measured sensitivity at 20 kPa, only the shape of pressure dependence can be compared in this figure. Dependence of FEM simulation results on the backside pressure are in good agreement with the measured characteristics. Increasing the pressure, force sensitivity decreased due to increase of repulsive force of the swollen diaphragm. The higher the pressure, the lower the pixel force sensitivity. The force sensitivity is backside pressure dependent, whereas measured noise floor of the pixel output is a constant value (approximately 10 $\mu V/Hz^{0.5}$) and independent of the pressure. Hence, a lower backside pressure results in a higher signal to noise ratio (SNR), and is suitable for the detection of small amplitude of the contact force. For example, the minimum detectable input force becomes approximately 1 mN at 2 kHz bandwidth in the case of 60 kPa backside pressure. This result indicates that if the diaphragm backside pressure is reduced to 5 kPa, the minimum detectable input force will be improved to about 400μN with the same signal bandwidth.

Not only the sensitivity but also the maximum input force is dependent on the backside pressure. If the backside pressure is too low for an input force applied to the sensor surface, the diaphragm yields to the over load. Diaphragm surface is swollen to opposite side by a large input force, and pixel output is finally saturated due to the bottoming of the diaphragm deflection. A higher backside pressure is required to detect and support a larger input force on the diaphragm even if the force sensitivity of pixel (i.e. SNR) is degraded. Figure 13 (b) shows the measured relationships between the forced displacement of diaphragm and pixel input force for various backside pressures. Forced displacement calculated with FEM is also plotted in the figure. Even though there are some differences, the results of FEM show a good agreement with the measurement results. Stiffness of the diaphragm surface increases as the backside pressure increases. It is clearly expected from the results that the maximum input force of the tactile imager can be increased by increasing the backside pressure. There is a tradeoff relationship between SNR and the input force range in the tactile imager as seen in the experimental results. The lower the backside pressure, the higher the force resolution and the lower the maximum input force. In

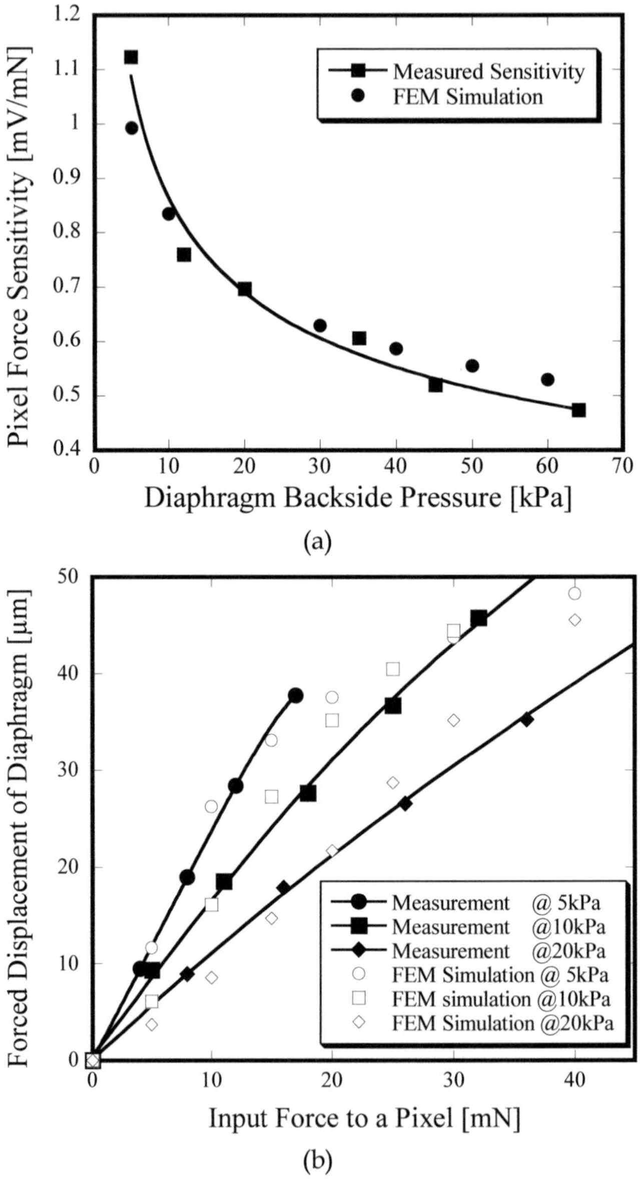

Fig. 13. (a) Measured force sensitivity of a pixel for various diaphragm backside pressures. Simulated sensitivities with FEM non-linear analysis are also plotted, (b) Measured relationship between forced displacement of the diaphragm surface and input force for various backside pressures.

contrast, the higher the backside pressure, the lower the force resolution and the higher the maximum input force. Figure 14 shows an example of over-range input measured with the plastic tip used in Fig. 11. The plastic tip is contacted on (X4,Y5) pixel in the swollen diaphragm at a force of 33mN. The backside pressure is only 4.8kPa which is insufficient for the input. In this case, large area of the diaphragm is largely deformed downward, and precise shape of the tip is not obtained. Once the backside pressure is increased to 23kPa, sharp shape image of the tip is obtained, and contacting position is well recognized similarly with Fig. 11 (b) and (c).

In conclusion, if the backside pressure is adaptively controlled for amplitudes of the input forces, apparent dynamic range of the tactile sensor will be improved. It has been confirmed that the maximum input force range can be controlled from 21 mN to 176 mN by changing the backside pressure from 5 to 64 kPa using the same device (Takao et al., 2006). Even if adaptive control of the pressure is not necessary, sensing characteristics can be chosen or optimized by selecting the sealing pressure suitably for the estimated range of input force in the application.

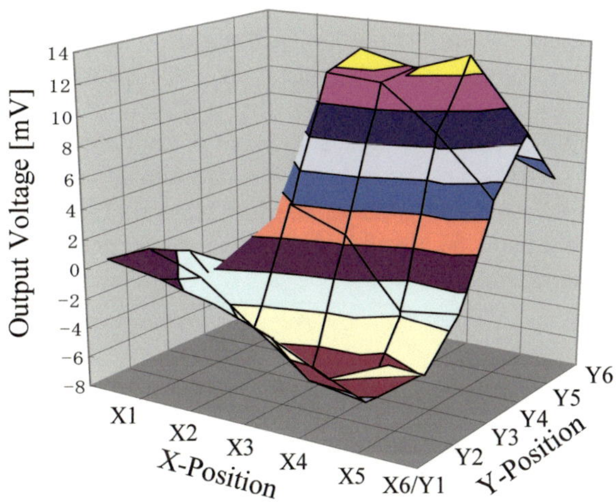

Fig. 14. An example result of over-range input measurement.

5. Conclusions and outlook

In this chapter, a novel concept of silicon-MEMS tactile imager aimed at fingertip tactile applications and evaluation results of a fabricated device have been presented. Array of strain-sensitive sensor pixels was integrated on a pneumatically swollen silicon diaphragm with signal processing circuits in monolithic configuration. Elastic surface of the tactile imager is realized using repulsive force of air pressure applied to the diaphragm backside. Contact force distribution of an object was detected from stress distribution change on the diaphragm. Force range and force sensitivity can be controlled by the pressure even after device packaging step. In this principle, fine pitch of pixels and a large scale sensing array can easily be realized using abilities of silicon CMOS technology. In addition, movable stroke of the sensor surface can be made much longer than individually formed micro mechanical pixels. A designed and fabricated sensor with 3040×3040 μm^2 sensing diaphragm was fabricated integrating with a 6 × 6 array, 420μm-pitch, piezoresistor pixels.

Movable stroke of the swollen diaphragm from the original surface is around 30 μm at 23 kPa pressure at the center. Realized spatial resolution of the fabricated device is approximately 400 μm, which is determined by the relationship between the pixel pitch and mechanical crosstalk among the pixels. The crosstalk depends on both the diaphragm thickness and backside air pressure. Positions of touch and their contact force amplitudes were detected as 2-D distribution of output voltage form the pixel array in multi-point contact test. The maximum input force range can be controlled from 21 mN to 176 mN by changing the backside pressure from 5 to 64 kPa. Through the evaluation of the fabricated device, advantages of the new configuration of tactile imager have been demonstrated successfully.

As introduced in Section 1, multi-functional sensing is one of future directions for tactile imager devices. Recently, our group has presented multi-functional tactile imager with integrated arrays of piezoresistors and temperature sensors for simultaneous detection of force and temperature distribution images (Takao et al., 2005). Also, another group has succeeded to realize simultaneous measurement of stress and temperature using single field-effect transistor structure (Doelle et al., 2006). Realization of multi-functional detection ability is a firm direction of future technology in integrated silicon smart sensor field. The advantage of silicon technology for integration is very promising feature for highly functional tactile imager for advanced tactile sensing applications. On the other hand, once over-ranged strain is applied, silicon is well known as a fragile material even though it has very good mechanical properties and large breaking stress in room temperature. Research to realize robustness of the silicon tactile imager has just started (Takao et al., 2007), and it will be a challenge to change silicon tactile imagers from a research target to practical sensing devices used widely in many fields.

6. References

Brussel, H. V. &Belien, H. (1986). A High Resolution Tactile Sensor for Part Recognition, *Proc. 6th Int. Conf. Robot Vis. Sens. Control*, pp. 49-59.

Engel, J.; Chen, J. & Liu, C. (2003). Development of polyimide flexible tactile sensor skin, *Journal of Micromechanics and Microengineering*, Vol. 13, No. 9, pp. 359-366.

Engel, J.; Chen, J.; Fan, Z. & Liu, C. (2005). Polymer micromachined multimodal tactile sensors, *Sensors and Actuators A*, Vol. A117, No. 1, pp. 50-61.

Shimojo, M.; Makino, R.; Namiki, A.; Ishikawa, M. & Mabuchi, K. (2004). A Tactile Sensor Sheet Using Pressure Conductive Rubber with Electrical Wires Stitched method, *IEEE Sensors journal*, Vol. 4, No. 5, pp. 589-596.

Someya, T.; Sekitani, T.; Iba, S.; Kato, Y.; Kawaguchi, H. & Sakurai, T. (2004). A large-area, flexible pressure sensor matrix with organic field-effect transistors for artificial skin applications, *Proceedings of the National Academy of Sciences of the United States of America*, Vol. 101, Issue 27, pp. 9966-9970.

Sugiyama, S.; Kawahata, K.; Yoneda, M. & Igarashi, I. (1990). Tactile Image Detection Using a 1k-element Silicon Pressure Sensor Array, *Sensors and Actuators A*, Vol. 22, No. 1/3, pp. 397-400.

Suzuki, K.; Najafi, K. & Wise, K. (1990). A 1024 element high-performance silicon tactile imager, *IEEE Transactions on Electron Devices*, Vol. 37, No. 8, pp. 1852-1860.

Suzuki, K.; Najafi, K & Wise, K. (1990). Process alternatives and scaling limits for high-density silicon tactile imagers, *Sensors and Actuators A*, Vol. 23, No. 1/3, pp. 915-918.

Kobayashi, S.; Mitsui, T.; Shoji, S. & M. Esashi (1990). Two-lead Tactile Sensor Array Using Piezoresistive Effect of MOS Transistor, *Tech. Digest of the 9th Sensor Symposium*, pp. 137-140, Tokyo, Japan, June 1990, IEEJ, Tokyo.

Souza, R. & Wise, K. (1997). A very high density bulk-micromachined capacitive tactile imager, *Digest of Tech. Papers of Transducers'97*, Vol. 2, pp. 1473-1477, Chicago USA, June 16-19, 1997.

Mei, T.; Ge, Y.; Chen, Y.; Ni, L.; Liao, W.; Xu, Y. & Li, W. (1999). Design and Fabrication of an Integrated Three-Dimensional Tactile Sensor for Space Robotic Applications, *Proceedings of IEEE MEMS'99*, pp. 130-134, 1999.

Mei, T.; Ge, Y.; Chen, Y.; Ni, L.; Li, W. J. &Chan, M. H. (2000). An integrated MEMS three-dimensional tactile sensor with large force range, *Sensors and Actuators A*, Vol. A80, No. 2, pp. 155-162.

Sato, N.; Machida, K.; Morimura, H.; Shigematsu, S.; Kudou, K.; Yano, M. & Kyuragi, H. (2003). MEMS fingerprint sensor immune to various finger surface conditions, *IEEE Transactions on Electron Devices*, Vol. 50, No. 4, pp. 1109-1116.

Charlot, B.; Galy, N.; Basrour, S. & Courtois, B. (2004). A Sweeping Mode Integrated Fingerprint Sensor with 256 Tactile Microbeams, *Journal of Microelectromechanical Systems*, Vol. 13, No. 4, pp. 636-644.

Doelle, M.; Peters, C.; Gieschke, P.; Ruther, P. & Paul, O. (2004). Two-Dimensional High Density Piezo-FET Stress Sensor Arrays for In-Situ Monitoring of Wire Bonding Processes, *Proceedings of IEEE MEMS2004*, pp. 829-832, Maastricht The Netherlands, Jan. 25-29, 2004.

Doelle, M.; Held, J.; Ruther, P. & Paul, O. (2006). Simultaneous and Independent Measurement of Stress and Temperature Using a Single Field Effect Transistor Based Sensor, *Proceedings of IEEE MEMS2006*, pp. 150-153, Istanbul Turkey, Jan. 22-26, 2006.

Takao, H.; Matsumoto, Y. & Ishida, M. (1997). A Monolithically Integrated Three Axial Accelerometer Using Stress Sensitive CMOS Differential Amplifiers, *Digest of Technical Papers of Transducers '97*, Vol. 2, pp. 1173-1176, Chicago USA, June 16-19, 1997.

Takao, H.; Matsumoto, Y. & Ishida, M. (1999). An Integrated Three-Axis Accelerometer Using CMOS Compatible Stress Sensitive Differential Amplifiers, *IEEE Transactions on Electron Devices*, Vol. 46, No.1, pp. 109-116.

Takao, H.; Fukumoto, H. & Ishida, M. (2001). A CMOS Integrated Three-Axis Accelerometer Fabricated with Commercial Sub-micron CMOS Technology and Bulk-Micromachining, *IEEE Transactions on Electron Devices*, Vol.48, No.9, pp.1961-1968.

Takao, H.; Sawada, K. & Ishida, M. (2004). Silicon Smart Tactile Image Sensor with Pneumatically Swollen Single Diaphragm Structure, *Proceedings of IEEE MEMS2004*, pp.846-849, Maastricht The Netherlands, Jan. 25-29, 2004.

Takao, H., Sawada, K. & Ishida, M. (2005). Multifunctional Smart Tactile-Image Sensor with Integrated Arrays of Strain and Temperature Sensors on Single Air-Pressurized Silicon Diaphragm, *Dig. Tech. Papers of Transducers'05*, pp. 45-48, Seoul Korea, June 5-9, 2005.

Takao, H.; Sawada, K. & Ishida, M. (2006). Monolithic Silicon Smart Tactile Image Sensor with Integrated Strain Sensor Array on Pneumatically Swollen Single Diaphragm Structure, *IEEE Transactions on Electron Devices*, Vol. 53, No. 5, pp. 1250-1259.

Takao, H.; Yawata, M.; Sawada, K. & Ishida, M. (2007). A Robust and Sensitive Silicon-MEMS Tactile-Imager with Scratch Resistant Surface and Over-Range Protection, *Dig. Tech. Papers of Transducers'07*, pp. 1465-1468, Lyon France, June 10-14, 2007.

Hasegawa, Y.; Shimizu, T.; Miyaji, T.; Shikida, M.; Sasaki, H.; Sato, K. & Itogigawa, K. (2003). Hardness Detection Using A Micromachined Active Tactile Sensor, *Digest of Technical Papers of IEEE Transducers'03*, pp. 927-930, Boston USA, June 2003.

Kanda, Y. (1982). A Graphical Representation of the Piezoresistance Coefficients in Silicon, *IEEE Transactions on Electron Devices*, Vol. ED-29, No. 1, pp. 64-70.

5

High-Sensitivity and High-Stiffness Force Sensor Using Strain-Deformation Expansion Mechanism

Yong Yu Takashi Chaen and Showzow Tsujio
Dept. of Mechanical Engineering, Kagoshima University, Kagoshima 890-0065, Japan

1. Introduction

In order to grasp and manipulate an object controllably and dexterously with a multifingered hand of robot, the sensing of fingertip force is required. To these operation, for making the controlability higher and making force sensor be proof against unexpected collisions or weights, not only the sensor's sensibility but also its stiffness are desired as high as possible.

For sensing a force acting on a force sensor, in general, some sensing elastic bodies is equipped in the force sensor. When a force is applied on the sensor, the force will pass through the sensing elastic bodies and make the bodies deform linearly so that it can be measured from the strain-deformations on the elastic bodies. So far, many researches [1]~[7] [9] consider making the sensed force all passed through the sensing elastic bodies for force sensing. To this situation, if we want to make the sensibility higher, the stiffness of the sensing elastic bodies have to be reduced for making their deformations larger. Conversely, if we wan to make the stiffness higher, the sensibility of the sensing elastic bodies will be reduced since their deformations become to small because of the higher stiffness. Thus by using the previous sensing structures, it is hard to realize the force sensing with both of high-stiffness and high-sensitivity.

For instance, about torque sensing, some strain gauges are put on an arm joint shaft to sense the torque deformation on the shaft [1]. In this way, to make the sensitivity higher, reducing the stiffness of joint shaft, for instance, using an elastic shaft as a portion of joint shaft [2], was considered. On the other hand, a joint shaft and an arm link was fixedly connected in series by an elastic body and its elastic deformation was used for joint torque sensing [3]. However this method will reduce the joint stiffness, too. Also, double-cross structure [7], double-membrane structure [6] [8] were proposed for force sensing or acceleration sensing. But these methods will reduce the sensor's stiffness if higher sensibility is required. In general, increasing the sensitivity of joint torque sensing by reducing the joint stiffness, is not desired.

This paper proposes a novel mechanism called *Strain-Deformation Expansion Mechanism* for 3-axis force sensing. By the force sensing mechanism, the small strain-deformation used for force sensing can be expanded while the sensor stiffness will not be reduced but will be heightened. In this paper, the force sensing principle is addressed by analyzing the deformation of the sensing mechanism and the forces acting on the sensor theoretically. Then, the sensitivity of the sensing mechanism and its expansion rate of sensitivity are defined, and a design method for realizing the sensing mechanism with high sensitivity is discussed. Lastly, some experiments are performed to show the basic characteristics and the effectiveness of the proposed force sensing mechanism.

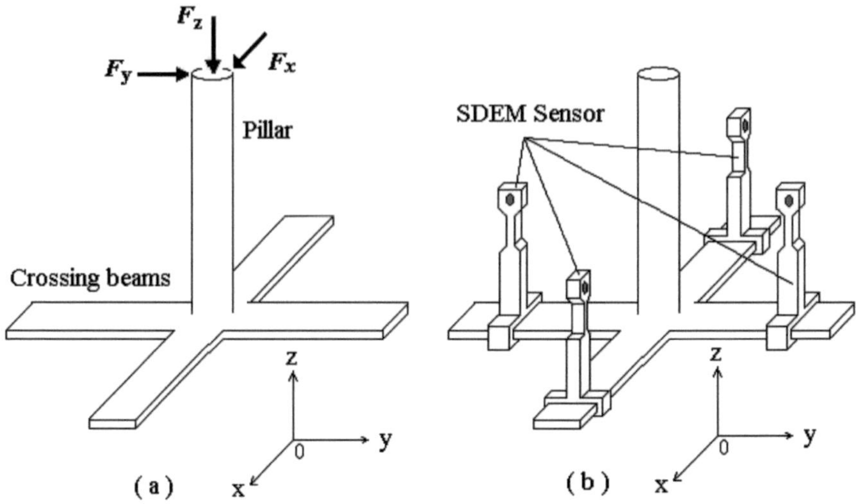

Fig. 1. Proposed force sensing mechanism

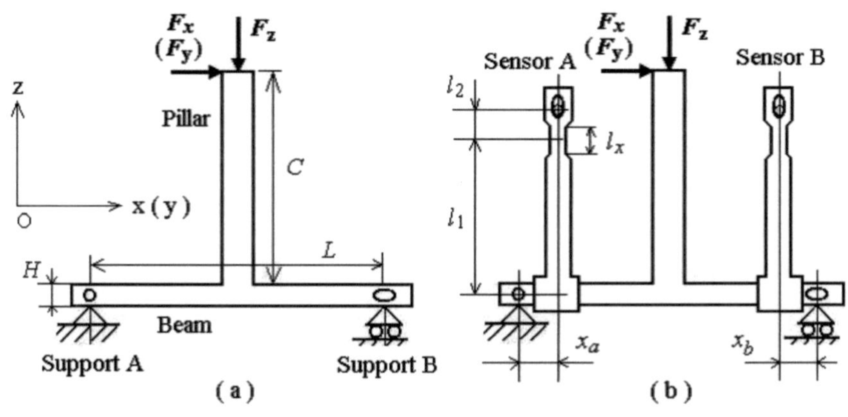

Fig. 2. Side view of proposed force sensing mechanism

2. Force sensing principle

A. Bending deformation on a beam

Fig. 1(a) shows a typical structure of previous 3- axis force sensor, which consists of one pillar and two beams crossing one and another at right angles. 3-axis forces to be sensed will act at the top of the pillar top, and the forces will be sensed by the strain deformations yielding on the crossing beam of the structure. For convenience, let us consider a side view of the sensing structure with height C, where the beam with length L, thickness H and width B is put on a hinge support and a support on rollers (see Fig. 2(a)). If a vertical force or a lateral force is loading on the pillar top, the reactions to the loads at two supports will be upward or downward, so that the beam will be bent and strains will yield on the beam surface. The relation between a strain-deformation ϵ_c at a position x from the support A and the vertical force F_z or the lateral force F_x will be

$$\epsilon_{cz} = -\frac{F_z x}{E_c I_c} \cdot \frac{H}{8}, \tag{1}$$

$$\epsilon_{cx} = \frac{F_x C x}{L E_c I_c} \cdot \frac{H}{2}, \tag{2}$$

where E_c and $I_c(= BH^3/12)$ represent the modulus of longitudinal elasticity (Young's modulus) and the second moment of area of the beam respectively. From the equations, the relation between F_z (or F_x) and ϵ_{cz} (or ϵ_{cx}) is linear and F_z (or F_x) can be measured if ϵ_{cz} (or ϵ_{cx}) is known. Hence, strain gauges can be stuck on the beam surface for measuring the forces form the strain-deformation on the beam.

By this method, however, for making the sensitivity of the sensing structure higher, from eqs. (1) and (2) we can know that I_c (or B and H) of the beam must be reduced, so that the stiffness of the sensing structure will be lower. It is hard to make its sensitivity and stiffness higher in the same time. To this problem, this paper proposes a novel force sensing mechanism, by which the small strain-deformation on the beam can be expanded for the force sensing while the beam stiffness is increased but reduced.

B. Force sensing principle

When a force Φ_z (or Φ_x) acts on the pillar top, the beam will be bent and at a position x there is a bending angle Φ_z (or Φ_x)

$$F_z = \frac{4E_c I_c}{x^2 - L^2/4} \cdot \Phi_z, \tag{3}$$

$$F_x = \frac{2L E_c I_c}{C(x^2 - L^2/12)} \cdot \Phi_x, \tag{4}$$

and the strain at x will vary with Φ_z (or Φ_x). From the above equations, the relation between F_z (or F_x) and Φ_z (or Φ_x) is linear. The novel mechanism in this paper will use the bending angle Φ to realize a high-sensitivity and high-stiffness force sensing.

Fig. 1 to Fig. 3 show the principle of the proposed mechanism, which has four beam-like elastic bodies (sensor hereafter) on two crossing beams for force sensing. For one of two crossing beam, two sensors of the mechanism are fitted on the two side of the beam shown as Fig. 2(b) and Fig. 3. The sensor construct of the mechanism is designed as Fig. 3. We set that the longitudinal direction of the beam-like sensor crosses the beam at right angles, and is parallel with the longitudinal direction of the pillar when no deformation yields. One end of the sensor is fixed on the beam, while another end can displace in the longitudinal and rotational directions of the sensor but on the sensor laterals the end is constrained by the fixtures fixed on the wall around the mechanism.

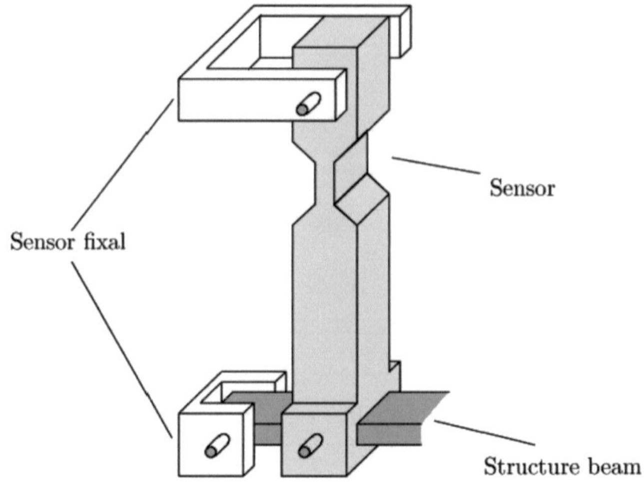

Fig. 3. Proposed force sensor construct

Let l_x denote the length of the narrow portion (whose thickness is h, width is b) of the sensor, l_1 denote the length from the beam connection of the sensor to the middle of l_x, l_2 denote the length from the middle of l_x to the fixtures (see Fig. 4(a)). The portions except the narrow portion are designed enough thickly for checking their deformations. Thus, shown as Fig. 4(a), the sensor's deformation can be focused on the narrow portion when the l_1 portion is displaced rotatively together with the bending angle Φ.

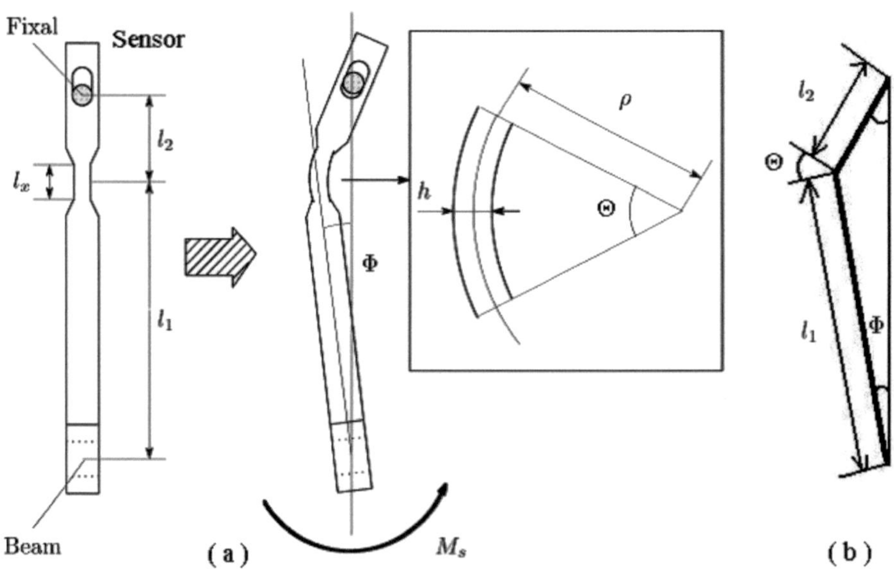

Fig. 4. Force sensor behavior

Fig. 4 gives figures with a side view in Fig. 3, which shows the behavior and characteristic of the proposed sensor. When a vertical force is applied to the pillar top, the force will be delivered to the middle of beam, so that the beam will be bent shown as Fig. 5 and a bending moment will act at the connection between the beam and sensor. From the beam's bending deformation the l_1 portion of sensor will be rotated, if the stiffness of shaft is hard while the narrow portion of sensor is flexible. Then, because the both laterals of the l_2 portion are constrained, bending deformations like Fig. 5 will occur at the deformable l_x portion, and the two sensors on beam's two sides will be bent with opposite rotation. In the same way, if a lateral force is applied at the pillar top, the l_x portion of two sensors will also be bent but thier bent behaviors will be the same rotation (see Fig. 5). By measuring the bending deformations of sensor, it is possible to sense the bending deformation of beam.

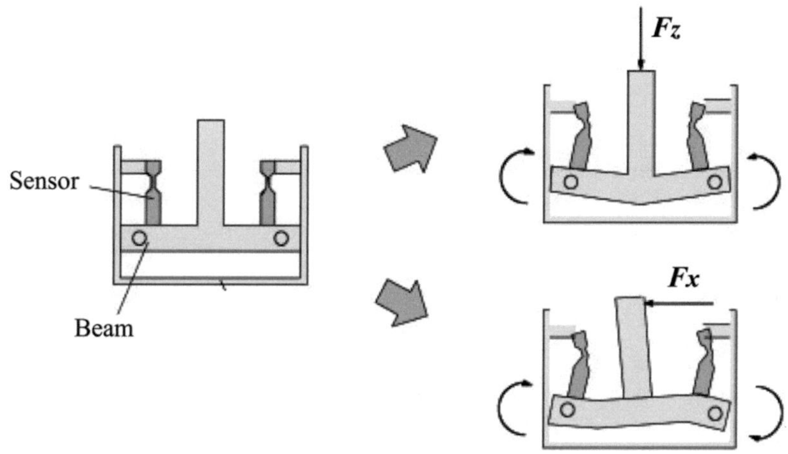

Fig. 5. Behaviors of force sensor

C. Forces acting on beam

When a force is applied on the pillar top, for a beam without the proposed sensing mechanism, two reactions from two beam supports will yield to act against the applied force. For a beam with the proposed sensing mechanism, besides the two support reactions, two reactions at the two sensor's l_2 portions will also yield because of the bending deformation of beam as shown by Fig. 4 and 5. Accordingly, by using the proposed sensing mechanism, part of the applied force will branch off to act on the sensors to yield a bending deformation for force sensing.

In the previous researches [1]~[7], an elastic body for sensing is fitted between a sensor's top and its base in series, so that the whole applied force will pass though the elastic body for yielding a deformation. And for obtaining a higher sensitivity, the beam stiffness should be reduced. In this paper, sensing elastic bodies (sensors) are connected with beams in parallel, so that the applied force will be divided into two parts, one passes though the beams to beam's hinge supports and another passes though the sensors to sensor's fixtures. Thereby, only a part of applied force is required for yielding a sensor deformation. And, considering that the sensors and beams are arranged in parallel, the stiffness of this mechanism can be made higher than that only using the beams.

For the case when the sensors and the beams are arranged in parallel, since the sensors have lower stiffness and the beams have higher stiffness, a bending deformation of beam can make a bending deformation on the sensor. On the other hand, the sensors will give their resistances to the beam bending. Accordingly, the beam stiffness added the sensor stiffness makes the whole stiffness of the mechanism. If we make the sensor stiffness higher, the mechanism stiffness will become higher. For a certain applied force, however, the bending deformation of sensor will become smaller because of its higher stiffness. Conversely, if we make the sensor stiffness lower, its bending deformation will be larger, while the mechanism stiffness will be still higher somewhat than the beam stiffness. Meanwhile, the branched force for bending deformation on sensors will become smaller. Therefore, only by smaller branched forces, we can make a larger bending deformation on sensors for force sensing without reducing the beam stiffness.

3. Analysis of force sensing principle

A. Geometric analysis on sensor deformation

Fig. 4(b) shows a deformed sensor, which is simplified from Fig. 4(a). Since the bending angle of beam Φ is very small, the length of the triangular long-side in Fig. 4(b) is considered as l_1, the length of shortside does l_2. At first, let us consider expressing angle Θ by angle Φ. Noting that only the l_x portion is bent on the deformed sensor, about angles Φ and a, we get

$$l_1 \sin\Phi = l_2 \sin\alpha, \tag{5}$$

$$\alpha = \sin^{-1}\left(\frac{l_1}{l_2}\sin\Phi\right). \tag{6}$$

Thus, Θ can be expressed as

$$\Theta = \alpha + \Phi = \sin^{-1}\left(\frac{l_1}{l_2}\sin\Phi\right) + \Phi. \tag{7}$$

Then, let us consider obtaining the surface strain ε_{l_x} on the bent l_x portion. Shown as Fig. 4(a), let ρ (= l_x/Θ) denote the radius of curvature and Δl denote the deformation of length on its surface. Since the thickness is h and the length of neutral surface is l_x, we have

$$\varepsilon_{l_x} = \frac{\Delta l}{l_x} = \frac{(\rho + h/2)\,\Theta - \rho\Theta}{l_x} = \frac{h}{2l_x}\Theta. \tag{8}$$

According to eq. (5), ε_{l_x} can be expressed by Φ as

$$\varepsilon_{l_x} = \frac{h}{2l_x}\Theta = \frac{h}{2l_x}\left\{\sin^{-1}\left(\frac{l_1}{l_2}\sin\Phi\right) + \Phi\right\}. \tag{9}$$

Hence, from eq. (9) and $\sin\Phi \approx \Phi$, $\cos\Phi \approx 1$ since Φ is very small, there exists

$$\Phi\left(\frac{l_1}{l_2} + \cos\frac{2l_x\varepsilon_{l_x}}{h}\right) = \sin\left(\frac{2l_x\varepsilon_{l_x}}{h}\right). \tag{10}$$

Considering that the strain ε_{l_x} is 10^{-3} orders of magnitude, we have $\sin(2l_x\varepsilon_{l_x}/h) \approx (2l_x\varepsilon_{l_x}/h)$ and $\cos(2l_x\varepsilon_{l_x}/h) \approx 1$ if $(2l_x\varepsilon_{l_x})/h$ is designed small enough. Thereby, ε_{l_x} can be obtained as

$$\varepsilon_{l_x} = \frac{(l_1 + l_2)h}{2l_x l_2} \cdot \Phi. \tag{11}$$

Accordingly, the strain ε_{l_x} on the bent l_x portion has a linear relation with the bending angle of beam Φ.

B. Sensor deformation and forces

About the bent l_x portion of a sensor shown as Fig. 4, the relation between an bending moment acted on the sensor M_s and the radius of curvature ρ can be described as

$$\frac{1}{\rho} = \frac{M_s}{E_s I_s}, \tag{12}$$

where, E_s denotes the modulus of longitudinal elasticity (Young's modulus) of the sensor's material, I_s denotes second moment of area, which is $I_s = bh^3/12$ when the cross section of the l_x portion is rectangular. From eqs. (6), (12) and $l_x = \rho\Theta$, we have

$$\varepsilon_{l_x} = \frac{h}{2\rho} = \frac{6}{E_s b h^2} \cdot M_s. \tag{13}$$

Thus, if we make the thickness h and width b on l_x portion smaller, in the same material, a certain strain ε_{l_x} is able to be realized with a smaller M_s. By eq. (13), the moment M_s for realizing a strain ε_{l_x} can be written as

$$M_s = \frac{E_s b h^2}{6} \cdot \varepsilon_{l_x}. \tag{14}$$

For a beam with the proposed sensing mechanism, the relations among an applied force F_z (or F_x), the sensor's bending moment M_{sa} and M_{sb}, and the beam's bending angle Φ_a and Φ_b at the connecting positions x_a and x_b, can be written as follows, that is,

$$\Phi_a = \frac{F_z(x_a^2 - L^2/4)}{4E_c I_c} - \frac{M_{sa}(x_a - L/2)}{E_c I_c}, \tag{15}$$

$$\Phi_b = -\frac{F_z[x_b^2 - L^2/4]}{4E_c I_c} + \frac{M_{sb}(x_b - L/2)}{E_c I_c}, \tag{16}$$

for a vertical force F_z;

$$\Phi_a = \frac{F_x C(x_a^2 - L^2/12)}{2LE_c I_c} + \frac{M_{sa}(2x_a^2 - x_a L + L^2/6)}{LE_c I_c}, \tag{17}$$

$$\Phi_b = \frac{F_x C(x_b^2 - L^2/12)}{2LE_c I_c} + \frac{M_{sb}(2x_b^2 - x_b L + L^2/6)}{LE_c I_c}, \tag{18}$$

for a lateral force F_x. From eqs. (11), (14) and the above equations, we have

$$F_z = \frac{4E_cI_c}{x_a^2 - L^2/4}\left[\frac{(l_1+l_2)h}{2l_xl_2} + \frac{E_sbh^2(x_a - L/2)}{6E_cI_c}\right]\varepsilon_{zal_x} \triangleq B_a\varepsilon_{zal_x}, \quad (19)$$

$$F_z = \frac{4E_cI_c}{x_b^2 - L^2/4}\left[-\frac{(l_1+l_2)h}{2l_xl_2} + \frac{E_sbh^2(x_b - L/2)}{6E_cI_c}\right]\varepsilon_{zbl_x} \triangleq B_b\varepsilon_{zbl_x}, \quad (20)$$

$$F_x = \frac{2LE_cI_c}{C(x_a^2 - L^2/12)}\left[\frac{(l_1+l_2)h}{2l_xl_2} - \frac{E_sbh^2(2x_a^2 - x_aL + L^2/6)}{6LE_cI_c}\right]\varepsilon_{xal_x} \triangleq A_a\varepsilon_{xal_x}, \quad (21)$$

$$F_x = \frac{2LE_cI_c}{C(x_b^2 - L^2/12)}\left[\frac{(l_1+l_2)h}{2l_xl_2} - \frac{E_sbh^2(2x_b^2 - x_bL + L^2/6)}{6LE_cI_c}\right]\varepsilon_{xbl_x} \triangleq A_b\varepsilon_{xbl_x}, \quad (22)$$

where ε_{zal_x} and ε_{zbl_x} are the strains on the sensors at x_a and x_b corresponding F_z, ε_{xal_x} and ε_{xbl_x} are those corresponding F_x.
For an applied force whose direction is between F_z and F_x, the corresponding strains ε_{al_x} and ε_{bl_x} on the two sensor of a beam will be

$$\varepsilon_{al_x} = \varepsilon_{zal_x} + \varepsilon_{xal_x}, \quad (23)$$

$$\varepsilon_{bl_x} = -\varepsilon_{zbl_x} + \varepsilon_{xbl_x}. \quad (24)$$

And since $|\varepsilon_{zal_x}| = |\varepsilon_{zbl_x}|$ and $|\varepsilon_{xal_x}| = |\varepsilon_{xbl_x}|$ for the proposed mechanism, we have

$$\varepsilon_{zal_x} = \varepsilon_{zbl_x} = (\varepsilon_{al_x} - \varepsilon_{bl_x})/2, \quad (25)$$

$$\varepsilon_{xal_x} = \varepsilon_{xbl_x} = (\varepsilon_{al_x} + \varepsilon_{bl_x})/2. \quad (26)$$

Therefore, an arbitrary force $\boldsymbol{F}_{xz}$ between F_x and F_z can be obtained by

$$\boldsymbol{F}_{xz} = \frac{1}{4}\begin{bmatrix} (A_a + A_b) & (A_a + A_b) \\ (B_a + B_b) & -(B_a + B_b) \end{bmatrix}\begin{bmatrix} \varepsilon_{al_x} \\ \varepsilon_{bl_x} \end{bmatrix}. \quad (27)$$

In the same way, we can obtain the relation between an arbitrary 3-axis force $\boldsymbol{F}$ and the strains $\varepsilon_{al_x}^X$, $\varepsilon_{bl_x}^X$, $\varepsilon_{al_x}^Y$, $\varepsilon_{bl_x}^Y$ on the 4 sensors of 2 crossing beams as following

$$\boldsymbol{F} = \frac{1}{8}\begin{bmatrix} 2(A_a^X + A_b^X) & 2(A_a^X + A_b^X) \\ 0 & 0 \\ (B_a^X + B_b^X) & -(B_a^X + B_b^X) \end{bmatrix}$$

$$\begin{matrix} 0 & 0 \\ 2(A_a^Y + A_b^Y) & 2(A_a^Y + A_b^Y) \\ (B_a^Y + B_b^Y) & -(B_a^Y + B_b^Y) \end{matrix}\Bigg] \begin{bmatrix} \varepsilon_{al_x}^X \\ \varepsilon_{bl_x}^X \\ \varepsilon_{al_x}^Y \\ \varepsilon_{bl_x}^Y \end{bmatrix} \tag{28}$$

where $*^X$ and $*^Y$ denote the beams respectively along X-axis and Y -axis. Thereby, we can know that there exist linear relations between an applied 3-axis force and the proposed sensor's strains.

According to eq. (28), the sensitivity and stiffness of the proposed sensing mechanism can be regulated within a certain extent by the designing of the sensor's dimensions and material.

C. Sensor's sensitivity and expansion rate of sensitivity

This subsection discusses the expansion rate of sensitivity of the proposed sensing mechanism for a certain applied force. For this purpose, the strain on the proposed sensor will be compared with the strain on the beam which was employed for force sensing in previous methods. In this paper, the sensitivity of joint torque sensor is defined as the magnitude of the strain for sensing for every unit force.

At first, about the previous method which directl sticks gauges on the crossing beams, the strains on beam ε_{cz} and ε_{cx} corresponding to forces F_z and F_x are represented as eqs. (1) and (2) respectively. Accordingly, their sensitivity can be expressed as

$$\frac{\epsilon_{cz}}{F_z} = -\frac{xH}{8E_cI_c}, \tag{29}$$

$$\frac{\epsilon_{cx}}{F_x} = -\frac{CxH}{2LE_cI_c}. \tag{30}$$

On the other hand, the sensitivity of the proposed sensor can be written as

$$\frac{\varepsilon_{l_x}}{F} = \frac{\varepsilon_{l_x}}{\varepsilon_c} \cdot \frac{\varepsilon_c}{F}. \tag{31}$$

From eqs. (19), (21) and (29), (30), $\varepsilon_{l_x}/\varepsilon_c$ will be

$$\frac{\varepsilon_{l_x}}{\varepsilon_{cz}} = \left[\frac{(l_1 + l_2)h}{2l_x l_2} + \frac{E_s bh^2(x - L/2)}{6E_cI_c}\right] \frac{L^2/4 - x^2}{xH}, \tag{32}$$

$$\frac{\epsilon_{l_x}}{\epsilon_{cx}} = \left[\frac{(l_1 + l_2)h}{2l_x l_2} - \frac{E_s bh^2(2x^2 - xL + L^2/6)}{6LE_cI_c}\right] \frac{x^2 - L^2/12}{xH}. \tag{33}$$

When $\varepsilon_{l_x}/\varepsilon_c$ is larger than 1, sensitivity ε_{l_x}/F will be larger than ε_c/F . In this paper, $\varepsilon_{l_x}/\varepsilon_c$ is referred to as *Expansion Rate of Sensitivity* relative to a strain on beams.

As the above discussion, by means of the proposed joint torque sensor, the applied force can be sensed by "expanding" the strain-deformation on beams. The sensor will be referred to as *SDEM (Strain- Deformation Expansion Mechanism) Force Sensor* hereafter.

4. Basic characteristics and implementation of SDEM force sensor

A. Experimental SDEM force sensor

Overview of a SDEM force sensor with size ϕ 18 [mm]× 18 [mm] is shown in Fig. 6. The dimensions about the proposed sensor is shown in Table I. The materials used for sensors and beams are duralumin (A2014-T4) and austenitic stainless steel (SUS303) respectively. According to eqs. (32) and (33), the expansion rate of sensitivity of the sensor are 3.071 in x and y directions and 9.309 in z direction theoretically.

L	10.0×10^{-3} [m]	l_1	9.0×10^{-3} [m]
C	15.3×10^{-3} [m]	l_2	3.0×10^{-3} [m]
H	1.5×10^{-3} [m]	l_x	3.5×10^{-3} [m]
B	3.0×10^{-3} [m]	h	1.0×10^{-3} [m]
E_c	20.6×10^{10} [Pa]	b	2.0×10^{-3} [m]
E_s	73.0×10^{9} [Pa]	x	0.5×10^{-3} [m]

Table I. Dimensions of experimental force sensor

B. Evaluation of static characteristics

A series of forces in x (y) and z directions from −20.0 [N] to 20.0 [N], and forces continuously and slowly changing in x (y) direction from 0 up to 6.2 [N] then down to −6.2 [N] and up to 0, in z direction from 0 up to 4.0 [N] then down to −4.0 [N] and up to 0 were respectively applied on the SDEM force sensor to confirm the linearity and hysteresis characteristics.

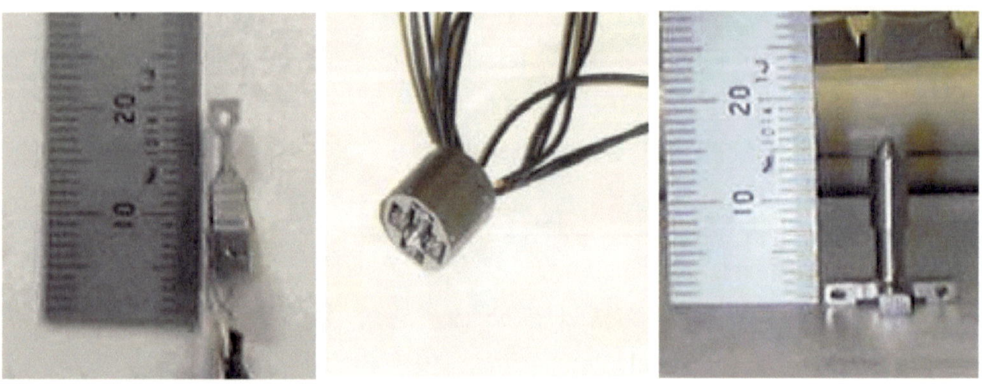

Fig. 6. Experimental SDEM 3-axis force sensor

Fig. 7 shows the experiment results on the linearity of the sensor respectively in x (y) and z directions, which plots the applied force values in abscissa and the SDEM sensor outputs in ordinate. According to the errors which are less than ±0.004 [Nm], the degree of linearity (100×error / measured range) is ±0.95 [%].

Based on the data from the SEDM force sensor and a Force Gauge (FGC-5N, made by Nihon Densan), Fig. 8 gives the graphs plotting the hysteresis characteristics, which plots the Force

Gauge outputs in abscissa and the SDEM sensor outputs in ordinate. The hysteresis differences in x (y) direction are less than 0.75 [N], those in z direction are less than 0.70 [N].

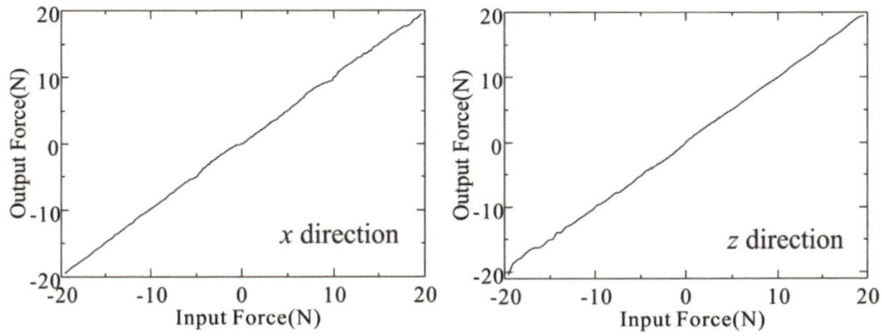

Fig. 7. Linearity between applied forces and sensor outputs

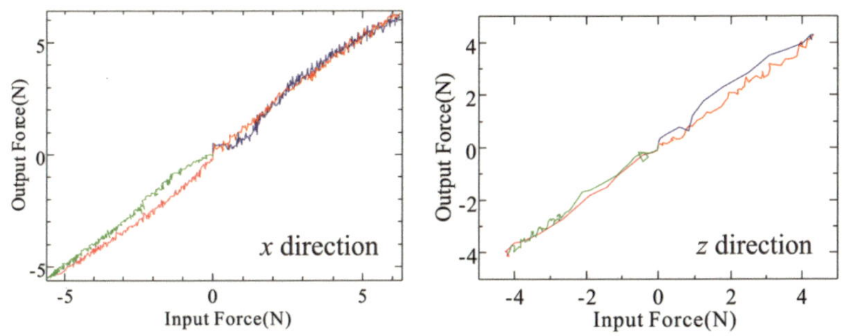

Fig. 8. Hysteresis characteristics

C. Evaluation of dynamic characteristics

The step response of the SDEM force sensor was examined. In the experiments of step response, the applied forces in x (y) and z directions were varied respectively from 0.5 [kgf], 1.0 [kgf], 2.0 [kgf] to 0 instantaneously. The results are shown by Fig. 9, where almost no overshoot or time lag appears.

D. Experimental verifications on expansion rate of sensitivity

In order to verify the expansion rate of sensitivity of the SDEM sensor experimentally, we use a previous crossing beam sensor without the proposed sensing mechanism. In the experiments, 3 forces of 0.5 [kgf], 1.0 [kgf] and 2.0 [kgf] were applied respectively in x (y) and z directions, and the SDEM sensor outputs and the outputs from the previous sensor were accumulated respectively. The 3 ratios of the two kind of outputs are plotted in Fig. 10.

According to the experimental results, the expansion rate of sensitivity of the SDEM force sensor is 3.195 about x (y) direction, 9.429 about z direction on the average. On the other hand, the expansion rate calculated theoretically is 3.079 about x (y) direction and 9.309 about z direction. The theoretical value and the experimental value are almost equal.

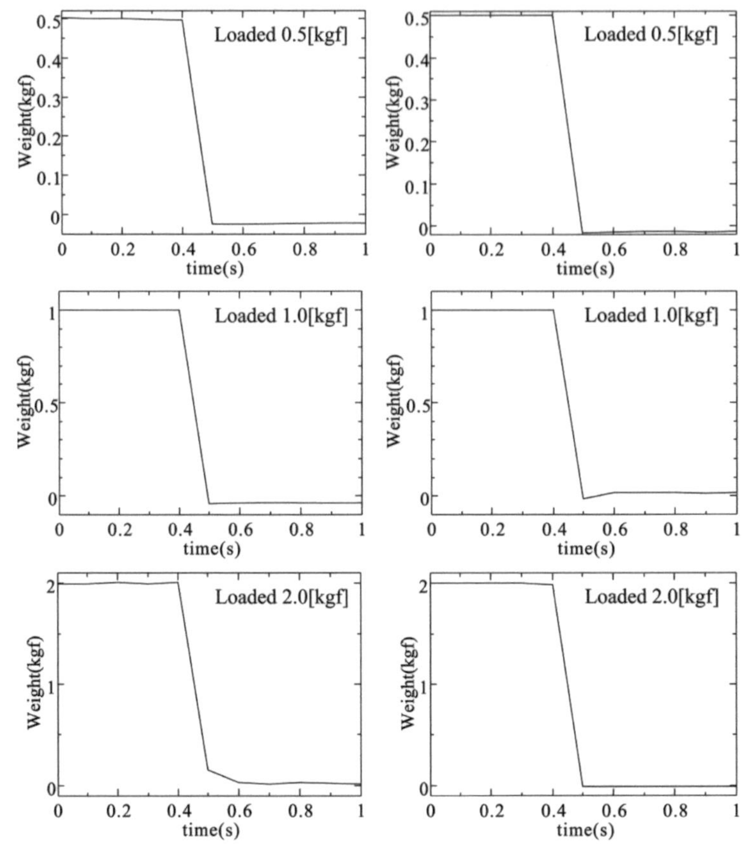

(a) In x direction (b) In z direction

Fig. 9. Step responses

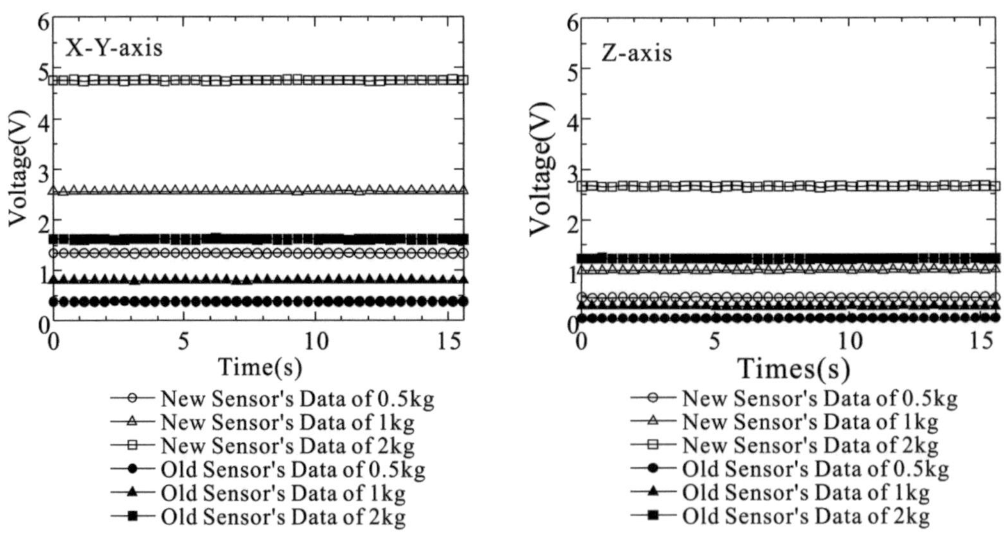

Fig. 10. Expansion rate of sensitivity

5. Conclusion

For dexterously performing object grasping and manipulation with multifingered hand of robot, sensing the fingertip forces with high-sensitivity and highstiffness is desired. In general, from previous sensing structures, if the stiffness of a sensor is made be high, its sensitivity will be reduced, so that It is hard to realize the force sensing with both of high-sensitivity and high-stiffness. This paper proposes a novel mechanism called *Strain-Deformation Expansion Mechanism* for 3-axis force sensing. By the force sensing mechanism, the small strain-deformation used for force sensing can be expanded while the sensor stiffness will not be reduced but will be heightened. In this paper, the force sensing principle was addressed by analyzing the deformation of the sensing mechanism and the forces acting on the sensor theoretically. Then, the sensitivity of the sensing mechanism and its expansion rate of sensitivity were defined, and a design method for realizing the sensing mechanism with high sensitivity was discussed. Lastly, some experiments with robot finger were performed to show the basic characteristics and the effectiveness of the proposed force sensing mechanism.

The proposed force sensing mechanism can be also applied to other cases besides robot and the like, for force sensing with high sensitivity and high stiffness.

6. References

L. E. Pfeffer, O. Khatib and J. Hake, "Joint Torque Sensory Feedback in Control of a PUMA Manipulator," *IEEE Trans. on Robotics and Automation,* Vol.5, No.4, pp.537-544, 1989.

H. Asada *et al.*, "Joint Torque Measurement of a Direct-Drive Arm," *Proc. of IEEE Int. Conf. on Decision and Control,* pp.1332, 1984.

D. Vischer and O. Khatib, "Design and Development of High-Performance Torque-Controlled Joints," *IEEE Trans. on Robotics and Automation,* Vol.11, No.4, pp.537-544, 1995.

Y. F. Zhang and Y. Fan, "Robot Force Sensor Interaction with Environments," *IEEE Trans. on Robotics and Automation,* Vol.7, No.1, pp.156-164, 1991.

U. Uchiyama, E. Bayo and E. Palma-Villalon, "A Mathematical Approach to the Optimal Structural Design of a Robot Force Sensor," *Proc. of USA-JAPAN Symposium on Flexible Automation,* Vol.1, pp.539-546, 1988.

M. Kaneko and T. Nishihara, "Basic Study of Six-Axis Force Sensor Design Based on Combination Theory," *Journal of the Robotics Society of Japan,* Vol.11, No.8, pp.1261-1271, 1993. (*in Japanese*)

K. Nagai, Y. Ito, M. Yazaki, K Higuchi and S. Abe, "Development of a Small Six-Component Force/Torque Sensor Based on the Double-Cross Structure," *Journal of the Robotics Society of Japan,* Vol.22, No.3, pp.361-368, 2004. (*in Japanese*)

Y. Yu, T. Ishitsuka and S. Tsujio, "Torque Sensing of Finger Joint Using Strain-Deformation Expansion Mechanism," *Proc. of IEEE Int. Conf. on Robotics and Automation,* Vol.2, pp.1850-1856, 2003.

M. Meng, Z. Wu, Y. Yu, Y. Ge and Y. Ge, "Design and Characterization of a Six-axis Accelerometer," *Proc. of IEEE Int. Conf. on Robotics and Automation,* Vol.3, pp.2367-2372, 2005.

Y. Yu, T. Arima and S. Tsujio, "Estimation of Object Inertia Parameters on Robot Pushing Operation," *Proc. of IEEE Int. Conf. on Robotics and Automation,* Vol.2, pp.1669-1674, 2005.

6

High-Precision Three-Axis Force Sensor for Five-Fingered Haptic Interface

Takahiro Endo[1], Haruhisa Kawasaki[1], Kazumi Kouketsu[2]
and Tetsuya Mouri[1]
[1]Gifu University
[2]Tec Gihan Co. LTD
Japan

1. Introduction

Haptic interfaces that present force and tactile feeling have been utilized in the areas of telemanipulation (Ivanisevi & Lumelsky, 2000; Elhajj et al., 2001), interaction with micro/nano scale phenomena (Ando et al., 2001; Marliere et al., 2004), medical training and evaluation (Langrana et al., 1994; Basdogan et al., 2001), and so on. Haptic interfaces are key devices in constructing virtual reality environments. In contrast with single-point haptic interfaces, multi-fingered haptic interfaces hold promise for the above-mentioned applications and should dramatically increase the believability of the haptic experience (Magnenat-Thalmann & Bonanni, 2006). From these points of view, several multi-fingered haptic interfaces (Kawasaki & Hayashi, 1993; Ueda & Maeno, 2004; Walairacht et al., 2001; Bouzit et al., 2002; Adachi et al., 2002; Yoshikawa & Nagara, 2000; Immersion Corporation) have been developed.

A haptic interface consisting of an arm and fingertips (Adachi et al., 2002; Yoshikawa & Nagara, 2000; Immersion Corporation) can be used in a wide space. However, most of them consist of a hand and arm exoskeleton system. With this system, it is hard to represent the weight of virtual objects through the fingertips because the hand mechanism is mounted on the back of a human hand. Fixing the haptic interface to the hand binds the hand and creates an oppressive sensation in the operator. Moreover, the operator is subject to a strong sense of unease when the system performs abnormally. The haptic interface must be safe, function in a wide space, and represent not only the force at the contact points but also the weight of virtual objects. In addition, it should not cause an oppressive feeling when it attached to humans and should not represent its own weight. In order to solve these problems, we have developed a multi-fingered haptic interface robot, which is placed opposite to the human hand: HIRO (Kawasaki et al., 2003) and HIRO II$^+$ (Kawasaki & Mouri, 2007). However, HIRO and HIRO II$^+$ have high reduction mechanisms at all finger joints, an arrangement that ensures the compactness of the mechanism, but requires force sensors at the haptic fingertips.

Most haptic devices have a low reduction ratio, which permits impedance control without the use of a force sensor. However, this requires a large mechanism and entails difficulty in construction. Therefore, HIRO and HIRO II$^+$ have high reduction mechanisms and require a force sensor at each haptic fingertip. And, to accomplish high-precision force presentation to the operator, they require high-precision force sensors. From this point of view, we have developed high-precision three-axis force sensors and a compact sensor amplifier circuit with 15 channels. The developed force sensor uses strain gauges, and its diameter and length are 14 [mm] and 27 [mm], respectively. The size of the force sensor is small enough to install at the haptic fingertip. The force sensor signals are inputted to an interface FPGA circuit through a sensor amplifier circuit with a 24-bit A/D converter, which is mounted on the back side of the haptic hand and communicates to a main control PC with LAN. Therefore, high-precision force control was achieved while the number of wires in the control system was minimized.

This paper presents the design and specifications of a newly developed three-axis force sensor for the five-fingered haptic interface robot HIRO II$^+$. The paper is organized as follows: In the next section, the mechanical design of HIRO II$^+$ used here is summarized, and the newly developed control system for HIRO II$^+$ is presented. In section 3, the design and specifications of the developed force sensor are presented, and then we consider the experiments of HIRO II$^+$ with the developed force sensor and control system in section 4. The experimental results in free space and constraint space demonstrate the high potential of the five-fingered haptic interface robot equipped with the developed three-axis force sensors. Finally, section 5 presents our conclusions.

2. Five-fingered haptic interface

The authors have developed a five-fingered haptic interface, HIRO II$^+$, consisting of a robot arm and a five-fingered haptic hand as shown in Fig. 1. HIRO II$^+$ can present force and tactile feeling at the five fingertips of the human hand. First we introduce HIRO II$^+$ briefly in subsection 2.1, and then we present the control system with the newly developed force sensor's amplifier circuit and interface FPGA circuit in subsection 2.2.

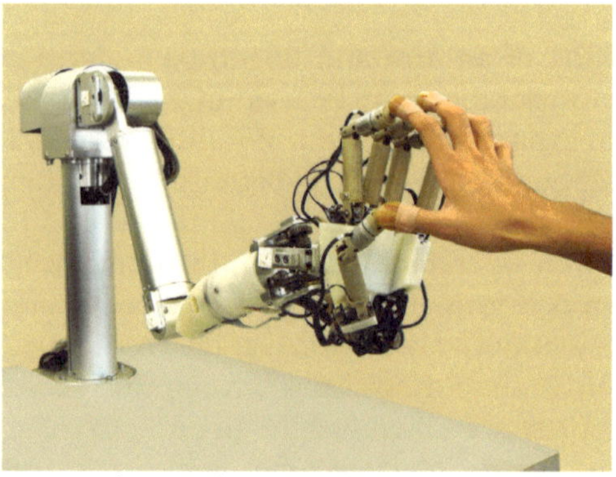

Fig. 1. Five-fingered Haptic Interface Robot: HIRO II$^+$.

2.1 Mechanical design

Fig. 1 shows the five-fingered haptic interface HIRO II$^+$ where it is coupled to the five fingers of an operator's hand. The haptic interface consists of an interface arm, a haptic hand with five haptic fingers, and a controller. When the operator moves his/her hand, the haptic interface follows the motion of the operator's fingertips and presents the sensation of force.

The operator feels only a small constriction because the only coupling between the human hand and the haptic interface occurs through the fingertips of the operator. The features of HIRO II+ are the following: (a) HIRO II+ can present force at the human five fingertips, (b) HIRO II+ can represent not only the force at the contact points but also the weight of virtual objects, (c) HIRO II+ should not cause an oppressive feeling when it is attached to humans and does not represent its own weight.

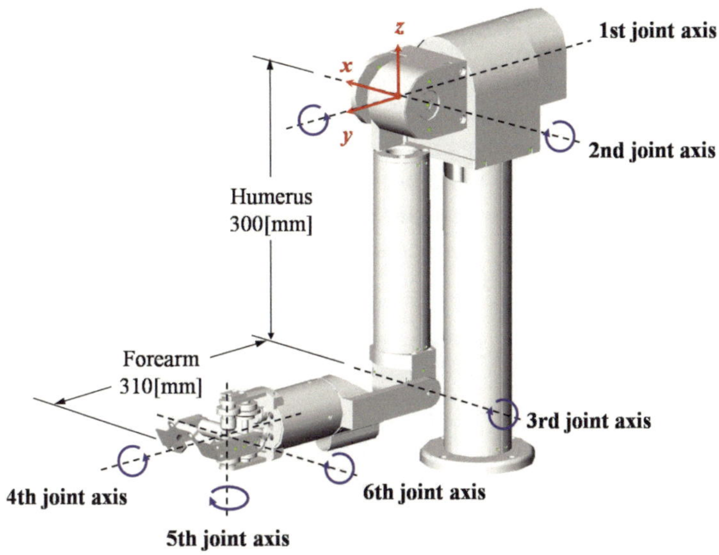

Fig. 2. Interface Arm design.

Degree of freedom	6	[dof]
Output force	45	[N]
Output moment	2.6	[Nm] (max)
Translational velocity	0.4	[m/s] (max)
Rotational velocity	1.4	[rad/s] (max)
Weight	6.9	[kgf]

Table 1. Specifications of the interface arm.

The interface arm is designed to be as close as possible to the human arm in geometry and motion ability, as shown in Fig. 2. The lengths of the upper arm and the forearm are 0.3 and 0.31 [m], respectively. The arm joints are actuated by AC servomotors equipped with rotary encoders and gear transmissions. The interface arm has 2 DOF at the shoulder joint, 1 DOF at the elbow joint, and 3 DOF at the wrist joint. The interface arm therefore has 6 joints allowing 6 DOF. Virtual work using the haptic interface can comfortably take place on the work space of a desktop. Table 1 shows the specifications of the interface arm.
The haptic hand starts from the wrist but does not include it, and ends at the fingertips. The hand base and five haptic fingers from the haptic hand are configured as shown in Fig. 3. The haptic fingers are designed to be similar to the human fingers in geometry and motion ability. Table 2 shows the specifications of the haptic hand. Each finger has 3 joints, allowing

3 DOF. The first joint, relative to the hand base, allows abduction/adduction. The second joint and the third joint allow flexion/extension. All joints are driven by DC servomotors with gear transmissions and rotary encoders. Another important issue in haptic finger design is the installation of the force sensors. In order to read the finger loading forces, a 6-axis force sensor (NANO sensor made by BL AUTOTEC, LTD.) in the second link of each finger is installed, of which 3-axis outputs, namely x-, y-, and z-elements, are used. The resolution of the force sensors F_x, F_y, and F_z, are 32, 32, and 98 [mN], respectively. Here, F_x, F_y, and F_z, are the x-, y-, and z-element output of the force sensor, respectively. To manipulate the haptic interface, the operator wears a finger holder on his/her fingertips as shown in Fig. 4. The finger holder has a sphere which, attached to the permanent magnet at the force sensor tip, forms a passive spherical joint. The role of the passive spherical joint attached by permanent magnet is to adjust the differences between the human and haptic fingers orientations and to ensure that the operator can remove his/her fingers from the haptic interface if it malfunctions. The suction force created by the permanent magnet is 5 [N].

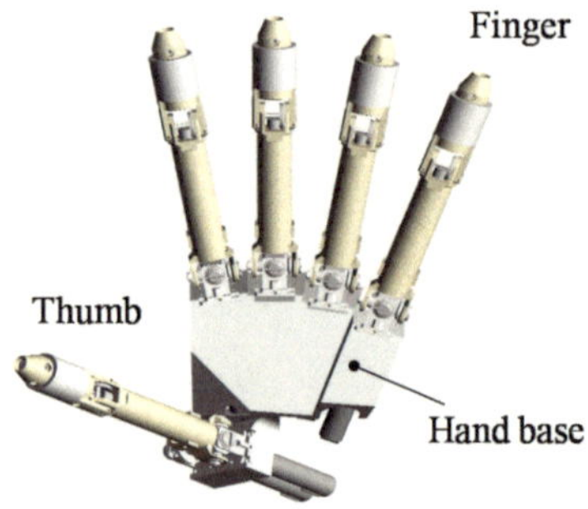

Fig. 3. Haptic hand design.

Hand	Number of fingers	5
	Degree of freedom	15 [dof]
	Weight	0.73 [kgf]
Finger	Degree of freedom	3 [dof]
	Output force	3.5 [N](max)
	Velocity	0.23 [m/s] (max)
	Weight	0.13 [kgf]

Table 2. Specifications of the haptic hand.

Fig. 4. Finger holder.

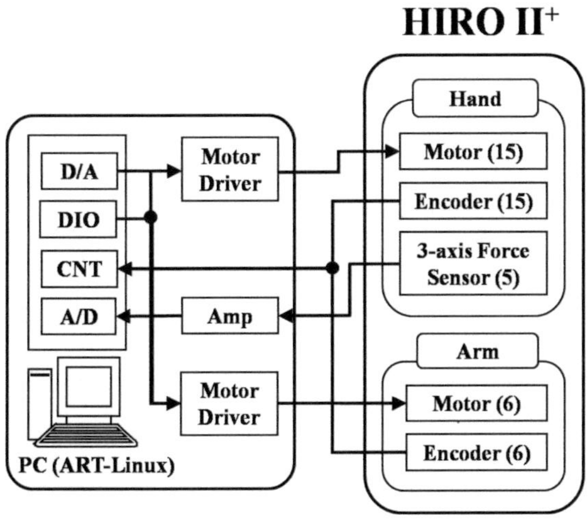

Fig. 5. Control system of HIRO II+.

The control system of HIRO II+ is shown in Fig. 5. The control system consists of a PC, a 12-bit digital-to-analogue converter (D/A), a 16-bit up/down counter (CNT), a 12-bit analogue-to-digital converter (A/D), digital input and output (DIO), a hand motor analogue power amplifier and an arm motor PWM driver. The real-time operating system ART-Linux is adopted to guarantee 1[ms] sampling time of the control.

2.2 Developed control system for haptic hand

Most haptic devices have a low reduction ratio, which permits impedance control without the use of a force sensor. However, this requires a large mechanism and entails difficulty in construction. The fact that the haptic finger of HIRO II+ uses a DC servomotor with a high reduction ratio ensures the compactness of the mechanism, but requires force sensors at the haptic fingertips. In particular, in order to accomplish high-precision force presentation to the operator, we need high-precision force sensors. On the other hand, in order to present the force at the five human fingertips, 15 DC servomotors and 5 force sensors are installed in the haptic hand of HIRO II+. Hence, the communication cable between the PC and haptic hand consists of 32 wires for the 15 encoders, 30 wires for the 15 DC servomotors, and 45 wires for the 5 force sensors, for a total wire count of 107. These wires greatly obstruct smooth movement of the haptic interface. Further, the amplifiers of the force sensors are stored under the haptic interface. Although the size of each force sensor amplifier is not large (each is 11×13×16 mm), many amplifiers are needed, and the total size becomes large. In order to solve these problems, we have developed high-precision force sensors and a compact sensor amplifier circuit with 15 channels. The force sensor signals are inputted to an interface FPGA circuit through a 24-bit A/D converter, which is mounted on the back side of the haptic hand and communicates to the main control PC with LAN. This leads to high-precision force control and reducing the number of wires of the control system. Next, we introduce the newly developed haptic hand control system, which consists of an interface FPGA circuit and a compact sensor amplifier circuit.

Fig. 6 shows the newly developed haptic hand control system made up of an interface FPGA circuit and a compact sensor amplifier circuit. The size of both circuits is 70×70 [mm], and we can install these circuits on the back side of the haptic hand. The sensor amplifier circuit has a 24-bit A/D and 12-bit D/A converter. The five force sensors at the haptic fingertips are connected to the sensor amplifier circuit, and the output signals of the amplifier circuit are inputted to the interface FPGA circuit. The interface FPGA circuit and control PC are connected by Ethernet, so we could reduce the number of wires between the haptic hand and the control PC from 107 to 72: namely, 8 wires for Ethernet, 32 wires for the 15 encoders, 30 wires for the 15 DC servomotors and 2 wires for the power supply to the FPGA circuit. Furthermore, the sensor amplifier circuit has 15 channels. Hence the five force sensors can be covered by this one circuit. In the former system, we needed five amplifiers for the five force sensors. Although the size of each amplifier is not large, when five amplifiers are needed the total size becomes large: 190×105×150 [mm]. Therefore, by developing the sensor amplifier circuit, we could also miniaturize the total size of the haptic interface system.

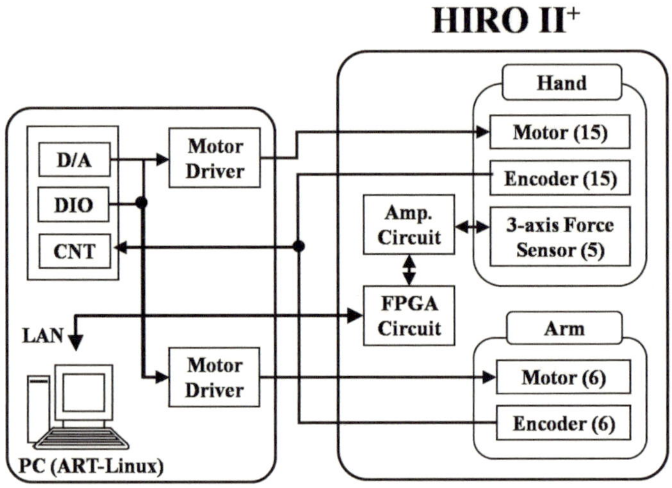

Fig. 6. Developed Control System.

Fig. 7. Interface FPGA circuit.

Fig. 7 shows the interface FPGA circuit, which has the following functions:

- Reading analogue voltages of the five force sensors
- Offset adjustment of voltages of the five force sensors
- Communication between the FPGA circuit and the control PC by UDP/IP

Table 3 shows the interface FPGA circuit specifications. In the FPGA chip, the digital signals of the force sensors from A/D are read, and FPGA sends these signals to the control PC each time the PC sends the commands. Moreover, when the PC sends the offset adjustment signals of the force sensors, FPGA sends digital signals corresponding to the offset adjustment signals to the D/A in the sensor amplifier circuit. As a result, the reading of the analogue voltages of the 5 force sensors and the offset adjustment of the voltages of the 5 force sensors are accomplished. The interface FPGA circuit is connected with the PC by Ethernet. The communication is fulfilled by UDP/IP, and is a command-and-response type communication. For the communication, we established several commands (Table 4).

Size	70×70×1.6 [mm]	
FPGA	Type	Xilinx Corp., Spartan3E(XCS500E)
	Clock frequency	100 [MHz]
Ethernet	Communication	100Base-TX
	Baud rate	100 [Mbps]
	protocol	UDP
Communication to sensor circuit	Serial communication specification	SPI (Serial Peripheral Interface)
	Communication clock frequency	2 [MHz]

Table 3. Specifications of the interface FPGA circuit.

Commands	Function
HiroHand_open_connection	Open connection between PC and FPGA by UDP/IP
HiroHand_close_connection	Close connection
HiroHand_force_offset	Offset adjustment of the force sensor
HiroHand_get_force	Measurement fingertips forces

Table 4. Examples of communication commands.

Fig. 8 shows the newly developed sensor amplifier circuit for the force sensors, and Table 5 shows the interface FPGA circuit specifications. The size of the circuits is 70×70×5 [mm], and the number of channels is 15. We installed a differential amplifier and low-pass filter in the amplifier output of the sensor amplifier circuit, and we gave 24-bit high resolution to the A/D converter in the sensor amplifier circuit. Thus, we consider that high-precision force

Fig. 8. Sensor amplifier circuit.

control, namely the reduction of the operational drag, is achieved. Further, we installed a 12 -bit D/A converter in the circuit because of the offset adjustment of the force sensors. The interface FPGA circuit and the sensor amplifier circuit are mounted on the back side of the haptic hand as shown in Fig. 9.

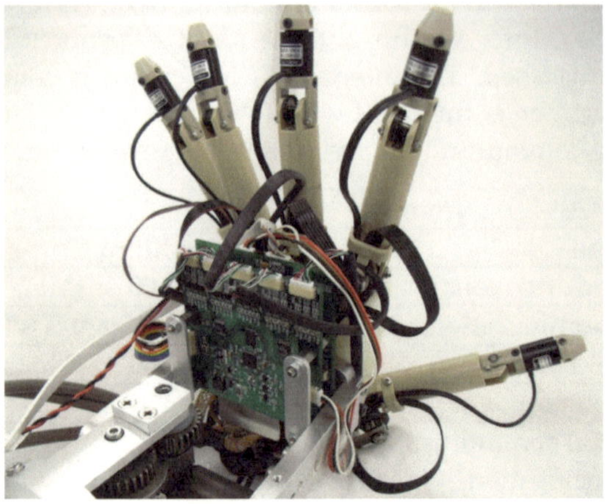

Fig. 9. Developed circuits installed in HIRO II+.

Channel number		15 (3ch×5)
bridge voltage		2.048 [v](all-channel common)
Amplifier output	differential amplifier	0~1.25~2.5 [V] with respect to the outputs of the force sensor -500~0~500 [με]
	low-pass filter	6 [dB/oct], Fc = 100 [Hz]
A/D converter	Type	ADS1258 (Texas Instruments Inc)
	resolution	24 [bit]
	channel	16 (we used 15ch)
D/A converter	Type	AD5328 (Analog Devices Inc)
	Adjustment	By using 0~1.25~2.5[V], -500~0~500[με] can be adjustment.
	Resolution	12 [bit]
	channel	16 (we used 15ch: 8ch×2)
Power supply		DC 5[V](0.5[A]) and DC3.3[V](0.1[A])
Size		W70×D70×H5 [mm]

Table 5. Specifications of the sensor amplifier circuit.

3. Three-axis force sensor

3.1 Design concept and specifications

The purpose of the developed force sensor is to measure the three-axis forces, F_x, F_y, and F_z, at the haptic fingertip. Unlike most haptic devices that have a low reduction ratio, HIRO II+ uses a DC servomotor with a high reduction ratio. A high-precision force sensor is required to accomplish high-precision force presentation. The force sensor is installed in the second link of each haptic finger and is connected to the developed sensor amplifier circuit, introduced in subsection 2.2. The size of the force sensor has to be small enough to install at the haptic fingertip. We set the following development goals for the force sensor:

- Low load $F_x = F_y = F_z$ =10 [mN] can be measured;
- The outside diameter is 14 [mm]; and
- The interference between each axis force in the sensor element is minimal.

Fig. 10 shows the newly developed three-axis force sensor, and Fig. 11 shows its dimensional outline drawing. By using the strain gauge in the pressure port of the sensor element, the force sensor measures the force from the deformation of the metal beam. The force sensor with strain gauge has the following advantages:

- The sensor has better linearity;
- The sensor has a small effect of temperature change; and
- The force sensor is suitable for the measurement of dynamic phenomena because the amount of the displacement is extremely small.

In order to accomplish the development goals, we need a sensor element that can measure the three axes forces with accuracy in a tight space.

Fig. 10 Three-axis force sensor.

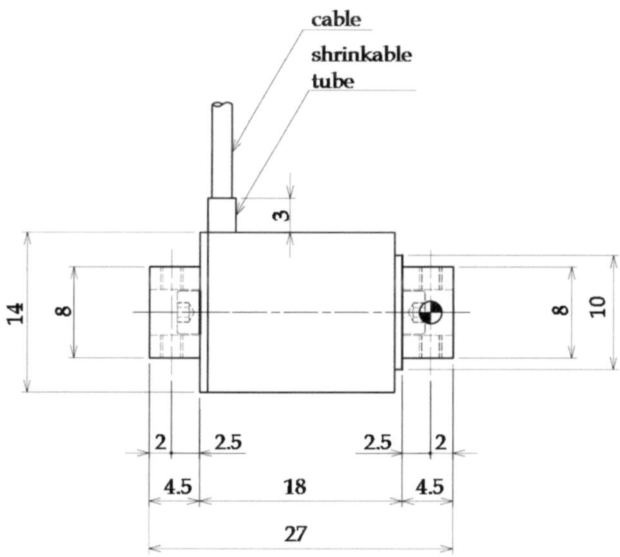

Fig. 11 The dimensional drawing of the force sensor.

3.2 Sensor structure

Fig. 12 shows the sensor structure of the developed three-axis force sensor. For the acting axis force, the strain gauges are installed at the surface of the spot where the stress of the sensor element reaches a maximum, and the force is detected as the change in electrical resistance. The total number of strain gauges is 16: x1~x4, y1~y4, and z1~z8. For the forces in the X- and Y-directions, the bending strains of x1~x4 and y1~y4 are detected, respectively. Further, the shear strains of z1~z8 are detected for the force in the Z-direction. Fig. 13 shows the circuit of the strain gauges. For each pair of strain gauges which detect the corresponding axis force, a Wheatstone bridge is applied. In order to simplify the wire system, the power supply to the bridge circuits is common, and we set up 8: 2 wires for the input (power supply to bridge) and 6 wires for the output (F_x-output, F_y-output, and F_z-output). The output of the circuit is expressed as the strain [$\times 10^{-6}$], and the force is measured by calibrating the relation between the load and the strain.

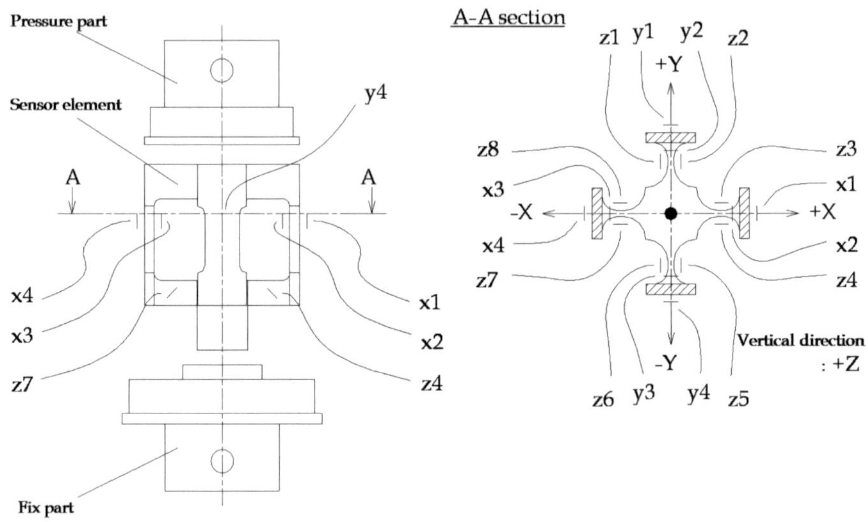

Fig. 12 Force sensor structure.

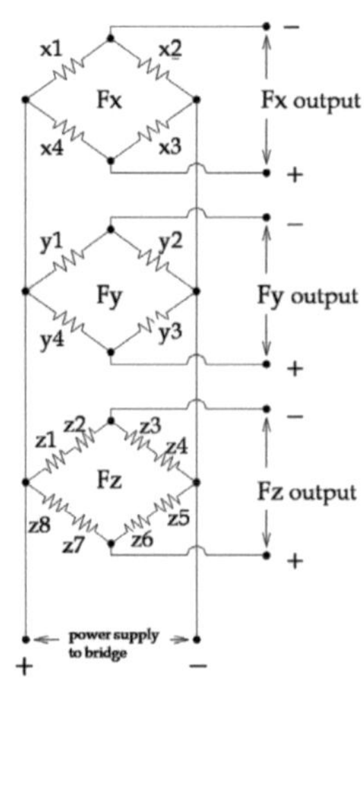

Fig. 13 Bridge circuit.

3.3 Basic characteristics

Fig. 14 shows the three-axis forces of the force sensor. From the calibration tests, we could obtain the relationship between the load and the strain in each axis direction as shown in Figs. 15, 16, and 17. The basic characteristics of the developed three-axis force sensor are the following: The rated load is 10 [N]. For the rated load, the rated outputs of the X-direction, Y-direction, and Z-direction are $\varepsilon x=1300\times10^{-6}$, $\varepsilon y=1300\times10^{-6}$, and $\varepsilon z=350\times10^{-6}$, respectively, where ε^* is the strain of the *-direction. Nonlinearity and hysteresis, which reflect the performance of the force sensor, are about 0.3%RO (Rated Output). Therefore, it can be said that the output strain is linear with respect to the load in each axis direction. Further, the interference output between each axis force also has linearity. Hence, the acting force can be derived by correcting the mutual interference. Table 6 shows the specifications of the developed three-axis force sensor.

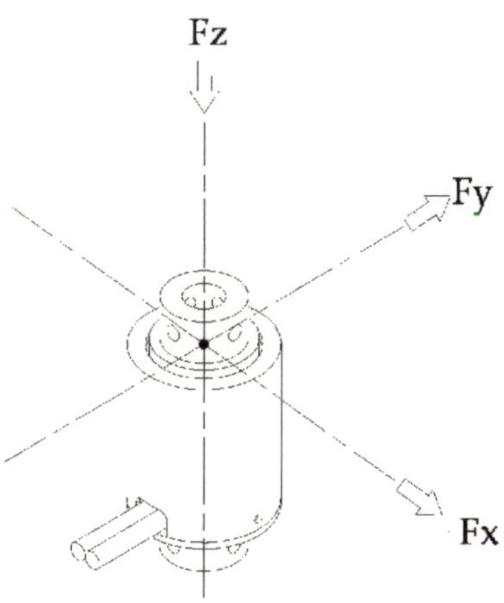

Fig. 14. Three axis forces.

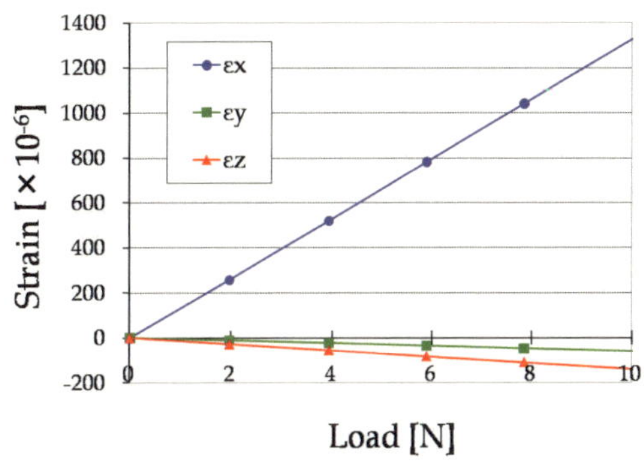

Fig. 15. Strain output w.r.t. the load in the +X direction.

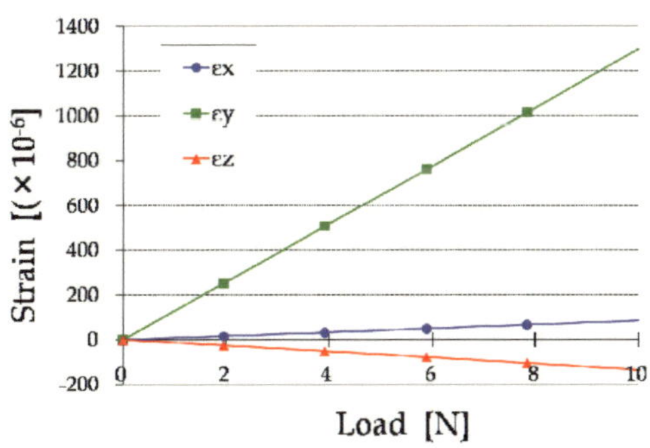

Fig. 16. Strain output w.r.t. the load in the +Y direction.

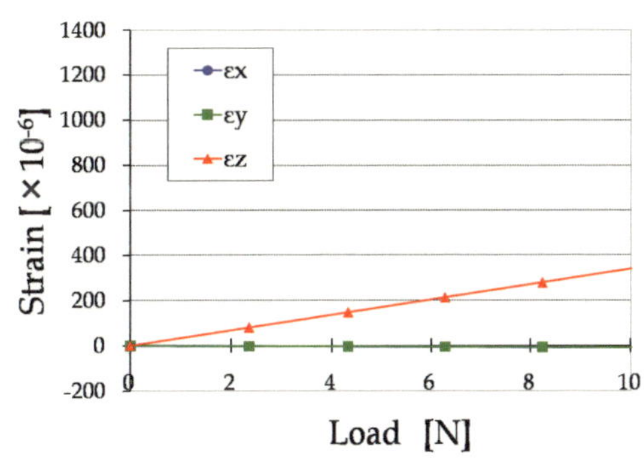

Fig. 17. Strain output w.r.t. the load in the +Z direction.

Rated capacity		10	[N]
Rated output	X-, Y-direction	±0.5	[mV/V] (±1000×10-6 strain)
	Z-direction	±0.15	[mV/V] (±300×10-6 strain)
Safe overload		200	[%]
Nonlinearity		≤ |1.0|	[%RO]
Hysteresis		≤ |1.0|	[%RO]
Repeatability		≤ 1.0	[%RO]
Safe temperature range		10~70	[°C]
Temperature effect on zero	X-, Y-direction	±0.05	[%RO/°C]
	Z-direction	±0.10	[%RO/°C]
Temperature effect on output	X-, Y-direction	±0.05	[%RO/°C]
	Z-direction	±0.10	[%RO/°C]
Input-output resistance	input	48±5%	[Ω]
	X-, Y-output	120±5%	[Ω]
	Z-output	240±5%	[Ω]
Weight		14.7	[g]

Table 6. Specifications of the developed force sensor.

4. Experiment of haptic interface using three-axis force sensor

In order to evaluate the developed three-axis force sensor, including the compact sensor amplifier circuit and the interface FPGA circuit, we carried out two kinds of experiments. One was the manipulation of the haptic interface in free space, and the other was the manipulation of the haptic interface in constraint space. In each experiment, we compared the experimental results of HIRO II+ with the developed force sensor and HIRO II+ with the former force sensor. Figs. 5 and 6 show the control system of the HIRO II+ with the developed force sensor and HIRO II+ with the former force sensor, respectively. As the control law of the haptic interface, we used the following redundant force control: The control of the haptic hand is given by

$$\tau(t) = J^T \left\{ K_1 F_e(t) + K_2 \int_0^t F_e(s)ds + F_d \right\} - K_3 \dot{q}_f(t) \tag{1}$$

where $\tau \in R^{15}$ is a joint torque vector of each finger of the haptic hand, $J \in R^{15\times15}$ is a kinematic Jacobian, $F = [F_1^T, F_2^T, \cdots, F_5^T]^T \in R^{15}$ is a force vector whose sub vector is the force vector at the fingertip, $F_d = [F_{d1}^T, F_{d2}^T, \cdots, F_{d5}^T]^T \in R^{15}$ is the desired force, $F_e = F_d - F$, and $q_f = [q_1^T, q_2^T, \cdots, q_5^T]^T \in R^{15}$ is a joint angle vector of the haptic finger. Further, $K_i \in R^+$ is the feedback gain matrix. On the other hand, the control of the interface arm is given by

$$\tau_a(t) = K_{a1}\tau_e(t) + K_{a2} \int_0^t \tau_e(s)ds - K_{a3}\dot{q}_f(t) + J_a^T \begin{pmatrix} \sum_{i=1}^{k} F_{di} \\ \sum_{i=1}^{k} (p_i - p_{hb}) \times F_{di} \end{pmatrix} \tag{2}$$

where $K_{ai} \in R^+$ is the feedback gain matrix, $q_{fa} = [q_{a1}, q_{a2}, \cdots, q_{a6}]^T$ is the joint angle vector, $J_a \in R^{6\times6}$ is a kinematic Jacobian, $p_i \in R^3$ is the *i*-th fingertip position vector, $p_{hb} \in R^3$ is the tip of the interface arm and

$$\tau_e(t) = J_a^T \begin{pmatrix} \sum_{i=1}^{5}(F_{di} - F_i) \\ \sum_{i=1}^{5}(p_i - p_{hb}) \times (F_{di} - F_i) \end{pmatrix}$$

is an equivalent joint torque to the offset forces $(F_{di} - F_i)$ at the five fingertips. Details of the control law have been shown in (Kawasaki & Mouri, 2007). On the other hand, the control PC uses a real-time OS (ART-Linux) to guarantee 1 [ms] sampling time, for the experiments.

4.1 Manipulation in free space

Fig. 18 shows the responses of HIRO II$^+$ with the developed force sensor in free space, and Fig. 19 shows the responses of HIRO II$^+$ with the former force sensor in free space. In each figure, (a) shows the responses of the five fingertip positions and (b) shows the responses of the five fingertip forces. Note that we considered the following force error for the *i*-th fingertip's force error: $F_e^i = \sqrt{F_x^2 + F_y^2 + F_z^2}$. In both cases, the five fingers of the operator

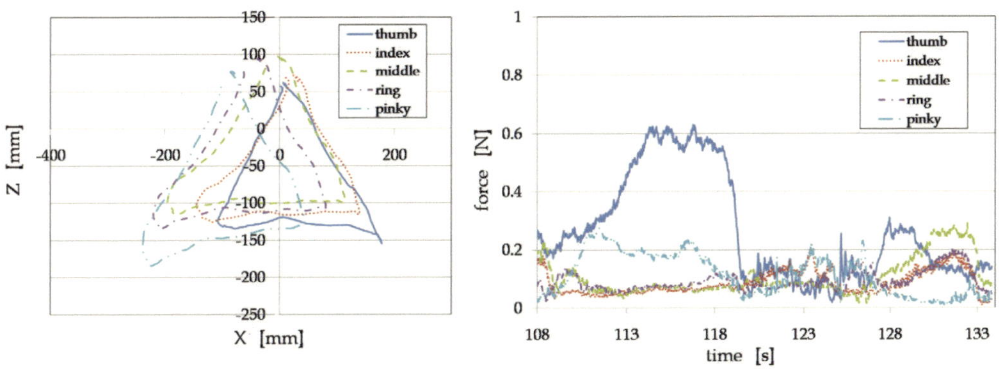

(a) fingertip positions (b) fingertip forces

Fig. 18. Responses of HIRO II$^+$ with developed force sensor in free space.

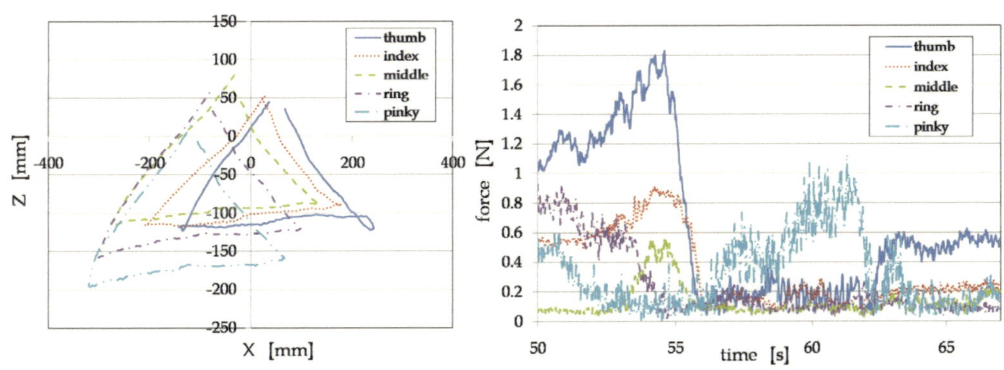

(a) fingertip positions (b) fingertip forces

Fig. 19. Responses of HIRO II$^+$ with former force sensor in free space.

are connected to the haptic interface, and the operator moved his five fingertips like drawing triangle. In the experiment, we considered the manipulation of the haptic interface in free space. Hence, the desired forces at the five fingertips are set to zero. But, the responses of the fingertip forces have a slight force error. In order to consider the force error quantitatively, we considered the average force error. The average force errors of HIRO II+ with the developed force sensor and HIRO II+ with the former force sensor were 0.13 [N] and 0.33 [N], respectively. The responses of HIRO II+ with the developed force sensor in free space were apparently improved. However, in contrast with the finger joints' backlash and the friction, those of the arm joints are large, and we have considered that a large part of the above average force error 0.13 [N] was caused by the interface arm.

4.2 Manipulation in constrained space

Next, we considered the manipulation of the haptic interface in constraint space. In the experiment, the five fingers of the operator are connected to the haptic interface, and the operator moved his five fingertips like drawing triangle while grasping the cube in the VR space as shown in Fig. 20.

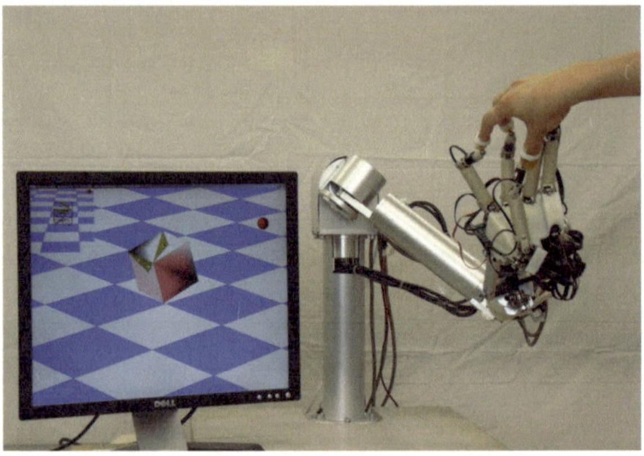

Fig. 20. Experimental environment.

Fig. 21 shows the responses of HIRO II+ with the developed force sensor in the constraint space, and Fig. 22 shows the responses of HIRO II+ with the former force sensor in the constraint space. In each figure, (a) shows the responses of the five fingertip positions and

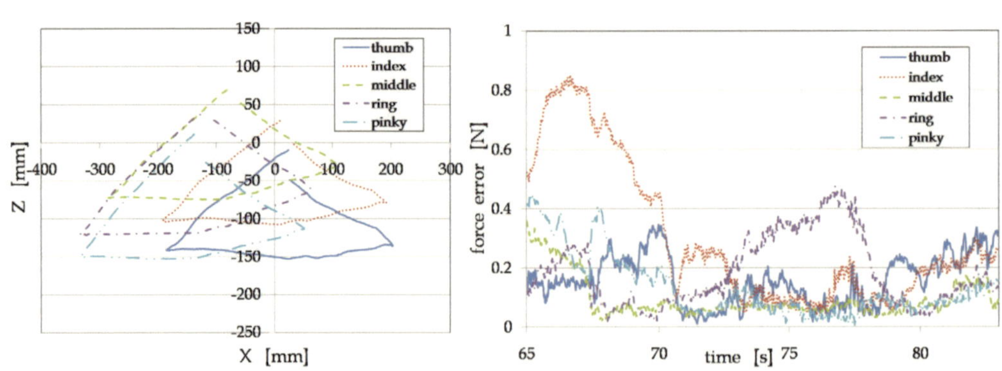

(a) fingertip positions (b) fingertip force errors

Fig. 21. Responses of HIRO II+ with developed force sensor in constraint space.

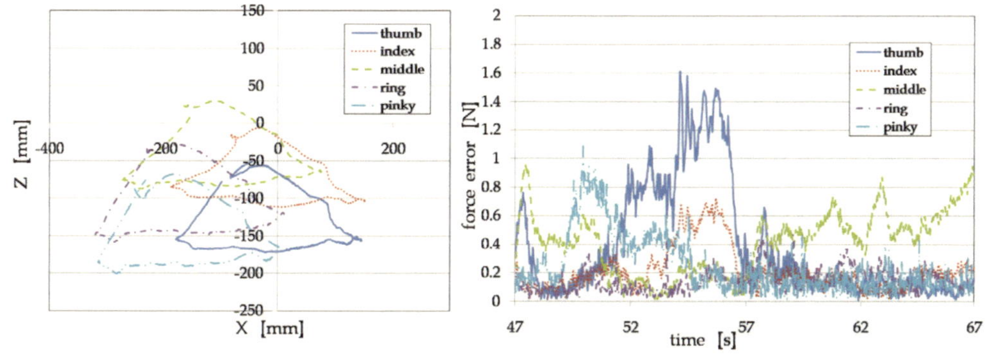

(a) fingertip positions (b) fingertip force errors

Fig. 22. Responses of HIRO II+ with former force sensor in constraint space.

(b) shows the responses of the five fingertip force errors. The following equation was used for the *i*-th fingertip's force error: $F_e^i = \sqrt{(F_x - F_x^d)^2 + (F_y - F_y^d)^2 + (F_z - F_z^d)^2}$. The desired force at each finger was set to be $F_i^d = Kx_i + D\dot{x}_i$, where x_i is the penetration depth of the *i*-th finger into the cube, K (=600[N/m]) is the stiffness of the cube, and D (=1.0×10-4[Ns/m]) is the damping coeeficient of the cube. The average force errors of HIRO II+ with the developed force sensor and HIRO II+ with the former force sensor were 0.17 [N] and 0.28 [N], respectively. The responses of HIRO II+ with the developed force sensor in the constraint space were apparently improved. However the responses of the fingertip forces have a slight force error.

We have considered that the resolution of the force sensors, characteristics of the motor amplifiers, adjustments of the feedback gain in the control law, and the backlash and friction in the mechanism are related to the existing slight force error. In this study, we mainly considered the resolution of the force sensor. In order to develop a better haptic interface, we must clarify the cause and plan to investigate these problems as challenges for the future.

5. Conclusion

In this paper, we have developed a small size high-precision three-axis force sensor, a compact sensor amplifier circuit with 24-bit high resolution, and an interface FPGA circuit. These circuits are mounted on the back of the haptic hand, and the output signals of the force sensors are input to the interface FPGA circuit, which communicate to the main control PC with LAN. Therefore, we could reduce the number of wires in the control system. Furthermore, we carried out an experimental test and confirmed the improvement of the haptic interface. This experiment demonstrates the high potential of the five-fingered haptic interface robot HIRO II+ with the developed three-axis force sensors. The next problem to be tackled is to investigate the cause of the force error and based on those results, to improve HIRO II+ further. This paper was supported by the Ministry of Internal Affairs and Communications Strategic Information and Communications R&D Promotion Programme (SCOPE) and by a JSPS Grant-in-Aid for Scientific Research (B) (19360190).

distributed logic array. Based on questionnaires of robot engineers, he summarized such specifications of tactile sensors as a 10 x 10 array size, 1 - 10 ms response time, a wide dynamic range of 1:1000, linearity, and a skin-like surface.

So far, various tactile sensors have been developed on the basis of several physical phenomena, such as the piezoresistance effect (Ohka, M. et al., 1994; Takeuchi et al., 1994), the capacitance variation (Novak, J. L., 1989), the piezoelectric effect (Dario et al., 1984; Bicchi et al., 1989; Howe & Cutkosky, 1993; Yamada & Cutkosky, 1994), the resistance variation of conductive rubber (Raibert & Tanner, 1982; Shimojo & Ishikawa, 1990), magnetic variation (Hackwood et al., 1983), and the optical effect(Mott et al., 1984; Tanie et al., 1986; Nicholls, 1990; Maekawa, 1992; Borovac et al., 1996). However, even today, multirole tactile sensors have not been developed since every tactile sensor has both merits and demerits. For example, although semiconductor type tactile sensors utilizing the piezoresistance effect possess good linearity and can internally incorporate data processing functions with IC technology, they are fragile. Such fragility is fatal since tactile sensing intrinsically requires contact between the finger surface and an object. While conductive rubber-type tactile sensors possess such excellent characteristics as large detective area and skin-like soft surface, they emit spontaneous noise generated by the chattering of carbon particles distributed in the rubber medium and also have hysteresis.

Although optical effect type tactile sensors (hereafter, optical tactile sensor) possess several defects as well, they can be easily produced using simple equipment and Charge Coupled Device (CCD) cameras and image data processing, which are mature techniques. Thus we developed an optical three-axis tactile sensor based on the principle of an optical uniaxial tactile sensor comprised of an optical waveguide plate, made of transparent acrylic and illuminated along its edge by a light source (Mott et al., 1984; Tanie et al., 1986; Nicholls, 1990; Maekawa, 1992). As shown in Fig. 1, the light directed into the plate remains within it due to the total internal reflection generated because the plate is surrounded by air, which has a lower refractive index than the plate. A rubber sheet featuring an array of conical feelers is placed on the plate to maintain array surface contact. If an object contacts the back of the rubber sheet and produces contact pressure, the feelers collapse, and at the points where these feelers collapse, light is diffusely reflected out of the reverse surface of the plate. The distribution of contact pressure is calculated from the bright areas viewed from the plate's reverse surface.

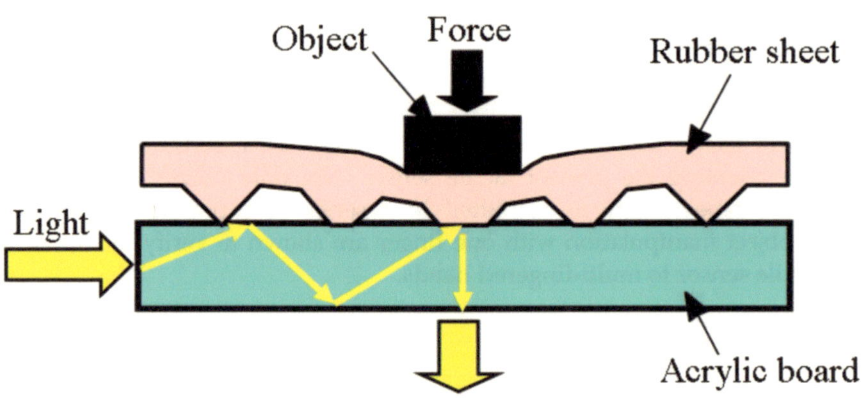

Fig. 1. Conventional optical tactile sensor

Improving the uniaxial tactile sensor to design a three-axis tactile sensor is possible. A new sensing element for it has been developed based on previous investigations (Ohka et al., 1995; Ohka et al., 2004a; Ohka et al., 2004b; Ohka et al., 2005c); it has a columnar feeler and four conical feelers that maintain contact with the acrylic surface while the tip of the columnar feeler touches an object. Normal and shearing forces applied to the columnar feeler tip are calculated from the area-sum and area-difference for the contact areas, respectively. An optical three-axis tactile sensor for micro robots was developed on the basis of simplified structure with a miniaturization advantage (Ohka et al., 2005b). Using simple rubber sheets having only a conical feeler array and an image processing technique, normal and shearing forces are calculated from the integrated grayscale value and the movement of its centroid, respectively. However, these three-axis tactile sensors can approach an object surface within limited direction due to their flat surfaces.

3. Sensing principle of optical three-axis tactile sensor

3.1 Structure of the present optical tactile sensor

The flat surface type tactile sensor described in the previous section is improved to be a hemispherical tactile sensor, which we intend to mount on the fingertips of a multi-fingered hand, as shown in Fig. 2 (Ohka et al., 2005c; Ohka et al., 2006). The multi-fingered robotic hand for general-purpose use in robotics that includes links, fingertips equipped with the three-axis tactile sensor, and microactuators (YR-KA01-A000, Yasukawa). Each microactuator consists of an AC servo motor, a harmonic drive, and an incremental encoder and is particularly developed for multi-fingered hand applications.

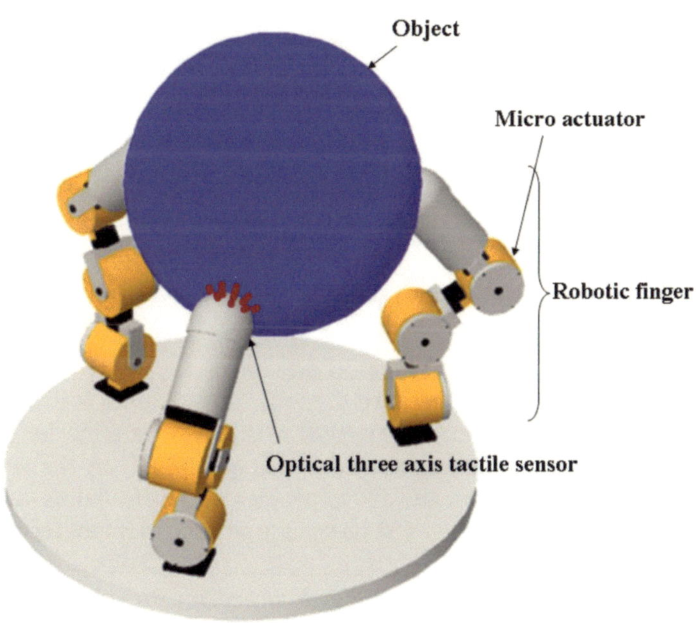

Fig. 2. Rendering of a three fingered hand equipped with optical three-axis tactile sensors

Since the tactile sensor essentially requires a lens system, it is difficult to make it be thinner, so it should be designed as an integral type of a fingertip and a hemispherical three-axis tactile sensor, as shown in Fig. 3. It is composed of a fiberscope, an acrylic dome, a fixing dome, optical fibers and 41 sensing elements. The sensing element comprised of a columnar feeler and eight conical feelers is adopted, because the element showed wide measuring range and good linearity in a previous paper (Ohka et al., 2005c). The sensing elements are made of silicone rubber (KE119, Shinetsu Co.) and are designed to maintain contact with the conical feelers and the acrylic board and to make the columnar feelers touch an object. Each columnar feeler features a flange to fit it into a counter bore portion in the fixing dome to protect the columnar feeler from horizontal displacement caused by shearing force.
Figure 4 shows a schematic view of the present tactile processing system to explain the sensing principle. The light emitted from the light source is directed into the acrylic dome through the optical fiber. Contact phenomena are observed as image data, acquired by the CCD camera through a bore scope guide and the fiberscope, and transmitted to the computer to calculate the three-axis force distribution.

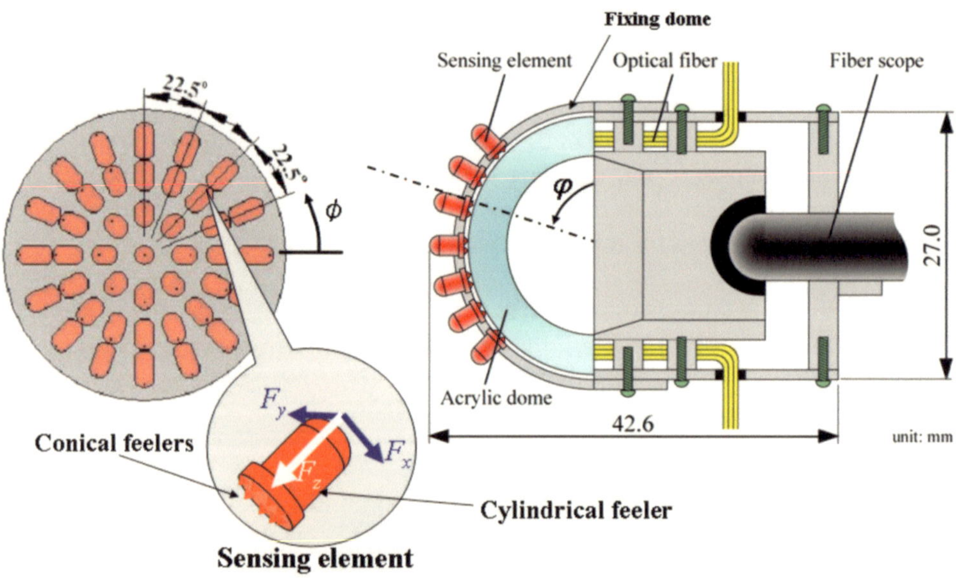

Fig. 3. Design of a fingertip including optical three-axis tactile sensor

3.2 Expressions for sensing element located on vertex

Dome brightness is inhomogeneous because the edge of the dome is illuminated and light converges on its parietal region. Since the optical axis coincides with the center line of the vertex, the apparent image of the contact area changes based on the sensing element's latitude. Although we must consider the above problems to formulate a series of equations for the three components of force, the most basic case sensing element located on the vertex will be considered first.

Coordinate O-*xyz* is adopted, as shown in Fig. 5. Based on previous studies (Ohka et al., 2005c), since grayscale value $g(x, y)$ obtained from the image data is proportional to pressure $p(x, y)$ caused by contact between the acrylic dome and the conical feeler, normal force is calculated from integrated grayscale value G. Additionally, shearing force is proportional to the centroid displacement of the grayscale value. Therefore, the F_x, F_y, and F_z values are calculated using integrated grayscale value G and the horizontal displacement of the centroid of grayscale distribution $\mathbf{u} = u_x\mathbf{i} + u_y\mathbf{j}$ as follows:

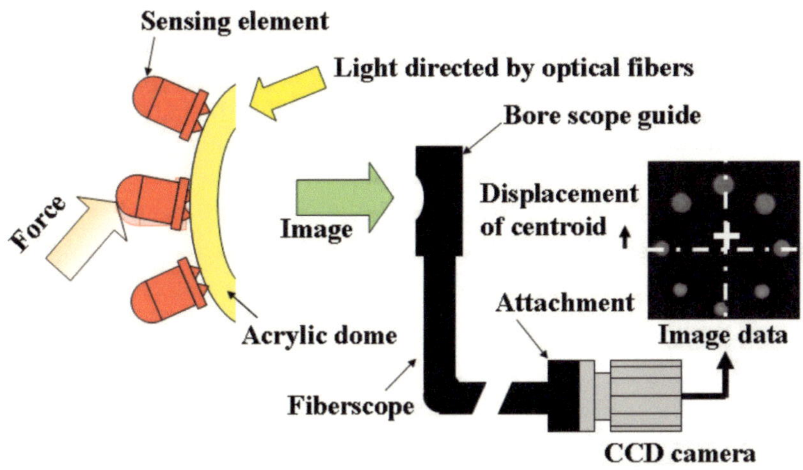

Fig. 4. Principle of three-axis tactile sensor system

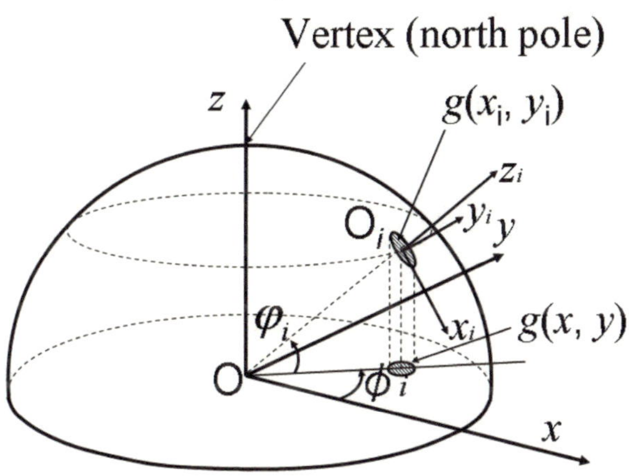

Fig. 5. Relationship between spherical and Cartesian coordinates

$$F_x = f_x(u_x), \tag{1}$$

$$F_y = f_y(u_y), \tag{2}$$

$$F_z = -g(G) \tag{3}$$

where **i** and **j** are the orthogonal base vectors of the x- and y-axes of a Cartesian coordinate, respectively, and $f_x(x)$, $f_y(x)$, and $g(x)$ are approximate curves estimated in calibration experiments.

3.3 Expressions for sensing elements other than those located on vertex

For sensing elements other than those located on the vertex, each local coordinate O_i-$x_i y_i z_i$ is attached to the root of the element, where suffix *i* denotes element number. Each z_i-axis is aligned with the center line of the element and its direction is along the normal direction of the acrylic dome. The z_i-axis in local coordinate O_i-$x_i y_i z_i$ is taken along the center line of sensing element *i* so that its origin is located on the crossing point of the center line and the acrylic dome's surface and its direction coincides with the normal direction of the acrylic dome. If the vertex is likened to the North Pole, the directions of the x_i- and y_i-axes are north to south and west to east, respectively. Since the optical axis direction of the CCD camera coincides with the direction of the *z*-axis, information of every tactile element is obtained as an image projected into the O-*xy* plane. The obtained image data $g(x, y)$ should be transformed into modified image $g(x_i, y_i)$, which is assumed to be taken in the negative direction of the z_i-axis attached to each sensing element. The transform expression is derived from the coordinate transformation of the spherical coordinate to the Cartesian coordinate as follows:

$$g(x_i, y_i) = g(x, y) / \sin\varphi_i . \tag{4}$$

Centroid displacements included in Eqs (1) and (2) $u_x(x,y)$ and $u_y(x,y)$ should be transformed into $u_x(x_i, y_i)$ and $u_y(x_i, y_i)$ as well. In the same way as Eq. (4), the transform expression is derived from the coordinate transformation of the spherical coordinate to the Cartesian coordinate as follows:

$$u_x(x_i, y_i) = \frac{u_x(x,y)\cos\phi_i + u_y(x,y)\sin\phi_i}{\sin\varphi_i}, \tag{5}$$

$$u_y(x_i, y_i) = u_x(x,y)\sin\phi_i + u_y(x,y)\cos\phi_i . \tag{6}$$

4. Basic sensing characteristics

4.1 Experimental apparatus

We developed a loading machine shown in Fig. 6 that includes an *x*-stage, a *z*-stage, rotary stages, and a force gauge (FGC-0.2B, NIDEC-SIMPO Co.) to detect the sensing characteristics of normal and shearing forces. The force gauge has a probe to measure force and can detect force ranging from 0 to 2 N with a resolution of 0.001 N. The positioning precisions of the *y*-, the *z*-, and rotary stages are 0.001 mm, 0.1 mm, and 0.1° , respectively.

Output of the present tactile sensor is processed by the data processing system shown in Fig. 6. The system is composed of an image processing board (Himawari PCI/S, Library, Co.) and a computer. Image data acquired by the image processing board are processed by software made in-house. The image data acquired by the CCD camera are divided into 41 sub-regions, as shown in Fig. 7. The dividing procedure, digital filtering, integrated grayscale value and centroid displacement are processed on the image processing board. Since the image warps due to projection from a hemispherical surface, as shown in Fig. 7, software installed on the computer modifies the obtained data. The motorized stage and the force gauge are controlled by the software.

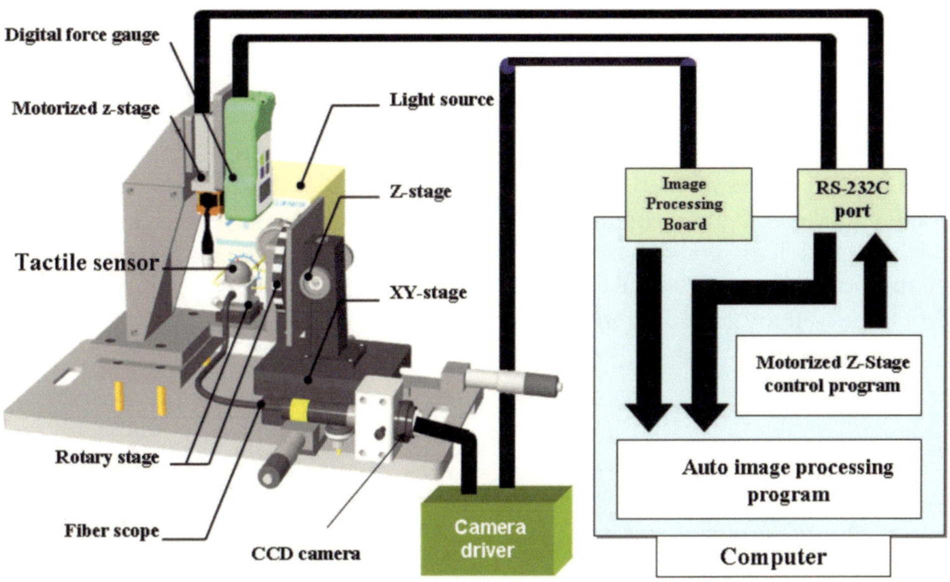

Fig. 6. Loading machine and tactile data processing system

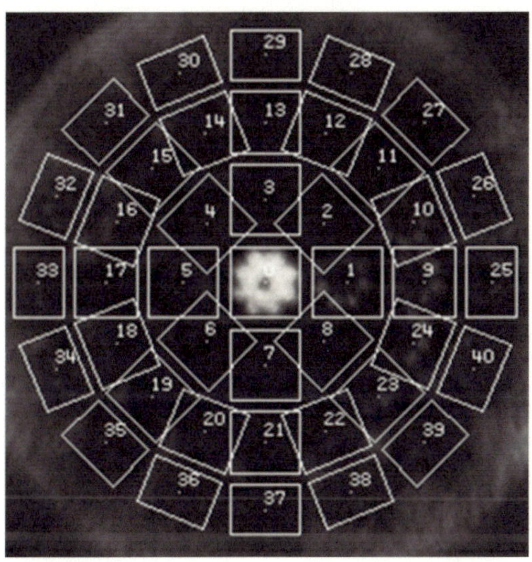

Fig. 7. Address of sensing elements

4.2 Sensing normal force

Because the present tactile sensor can detect not only normal force but also shearing force, we must confirm the sensing capability of both forces. In normal-force testing, by applying a normal force to the tip of a sensing element using the *z*-stage after rotating the attitude of the tactile sensor, it is easy to test the specified sensing element using the rotary stage. Since the rotary stage's center of rotation coincides with the center of the present tactile sensor's hemispherical dome, testing any sensing element aligned along the hemisphere's meridian is easy.

To evaluate the sensing characteristics of sensing elements distributed on the hemispherical dome, we need to measure the variation within the integrated grayscale values generated by the sensor elements. Figure 8 shows examples of variation in the integrated grayscale value caused by increases in the normal force for sensing element No. 0, 1, 5, 9, 17, 25, and 33. In these experiments, normal force is applied to a tip of each tactile element. As the figure indicates, the gradient of the relationship between the integrated grayscale value and applied force increases with an increase in φ; that is, sensitivity depends upon the latitude on the hemisphere. Dome brightness is inhomogeneous because the edge of the dome is illuminated and light converges on its parietal region. Brightness is represented as a function of latitude φ, and since sensitivity is uniquely determined by latitude, it is easy to modify the sensitivity according to φ.

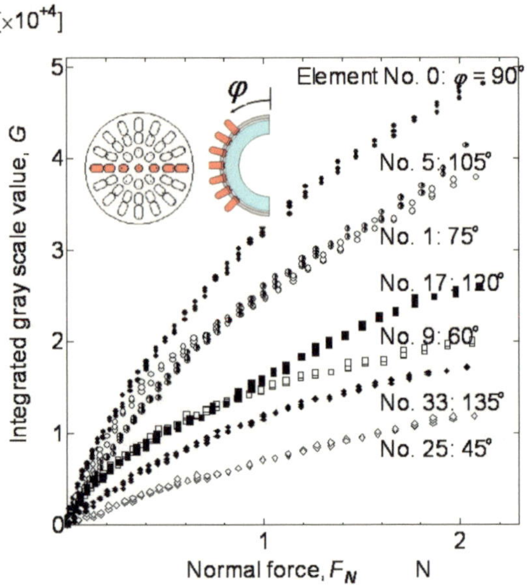

Fig. 8. Relationship between applied force and grayscale value

However, sensing elements located at the same latitude show different sensing characteristics. For example, the sensitivities of element No. 9 and 17 should coincide since they have identical latitude; however, as Fig. 8 clearly indicates, they do not. The difference reflects the inhomogeneous brightness of the acrylic dome. Therefore, we need to obtain the sensitivity of every sensing element.

When generating the shearing-force component, both the rotary and x-stages are controlled to specify the force direction and sensing element. First, the rotary stage is operated to give force direction θ. The x-stage is then adjusted to the applied tilted force at the tip of the specified sensing element. The inserted figure in Fig. 9 shows that the sensing element located on the parietal region can be assigned based on the procedure described above. After that, a force is loaded onto the tip of the sensing element using the z-stage. Regarding the manner of loading, since the force direction does not coincide with the axis of the sensing element, slippage between the probe and the tip of the sensing element occurs. To eliminate this problem, a spherical concave portion is formed on the probe surface to mate the concave portion with the hemispherical tip of the tactile element. Normal force F_N and shearing force F_S applied to the sensing elements are calculated using the following formulas, when force F is applied to the tip of the tactile element:

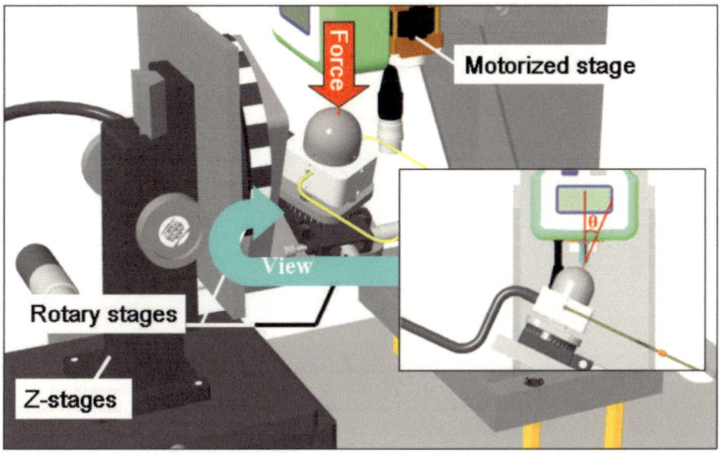

Fig. 9. Generation of shearing force component

$$F_N = F\cos\theta\,, \tag{7}$$

$$F_S = F\sin\theta\,. \tag{8}$$

To show that under combined loading condition normal force component was independently obtained with Eq. (3), we applied inclined force to the tip of the tactile element to examine the relationship between the normal component of applied force obtained by Eq. (7) and integrated grayscale value. Figure 10 displays the relationship for

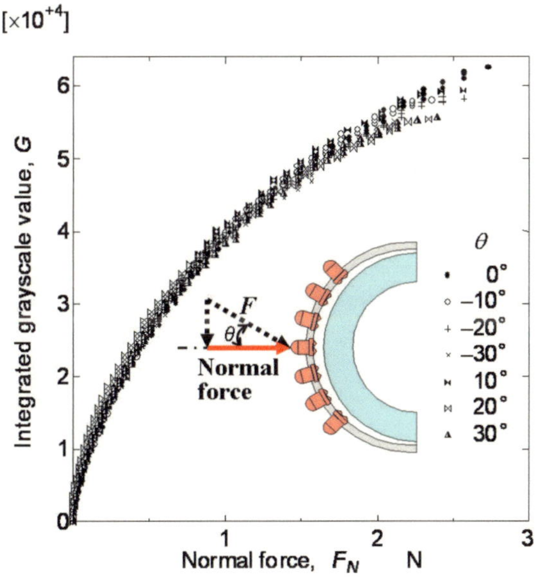

Fig. 10. Relationship between integrated grayscale value and applied normal force at several inclinations

No. 0. Even if the inclination is varied from -30 to 30°, the relationship coincides within a deviation of 3.7%. Therefore the relationship between the normal component of applied force and the integrated grayscale value is independent of inclination θ.

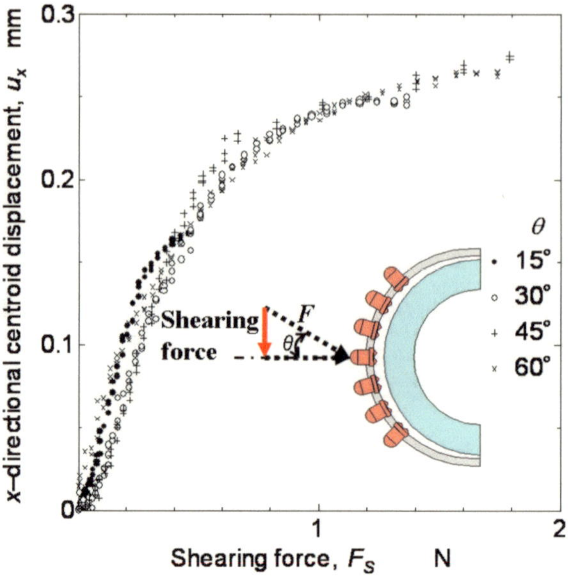

Fig. 11. Relationship between centroid displacement and applied shearing force

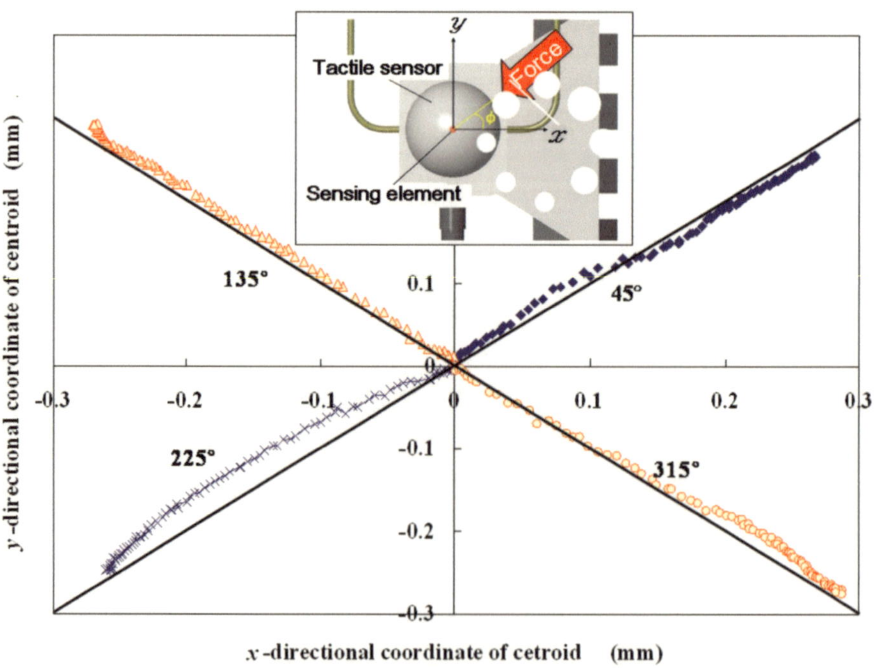

Fig. 12. Precision of two-dimensional shearing force detection

4.2 Sensing shearing force

When force is applied to the tip of the sensing element located in the parietal region under several θ s, the relationships between the displacement of the centroid and the shearing-force component calculated by Eq. (8) are obtained, as shown in Fig. 11. Although the inclination of the applied force is varied in a range from 15 to 60°, the curves converge into a single one. Therefore, the applied shearing force is obtained independently from centroid displacement.

When the tactile element accepts directional forces of 45, 135, 225, and 315°, centroid trajectories are shown in Fig. 12 to examine shearing force detection under various directions except for the *x*- and *y*-directions. If the desired trajectories shown in Fig. 12 are compared to the experimental results, they almost trace identical desired trajectories. The present tactile sensor can detect various detections of applied force.

4.3 Repeatability

Figure 13 shows repeatability of relationship between centroid displacement and applied shearing force. The relationship between the integrated grayscale value and applied force has high repeatability. Experimental results from 1,000 repetitions on No. 8 are superimposed in Fig. 13 and show that all the curves coincide. The deviation among them is within 2%.

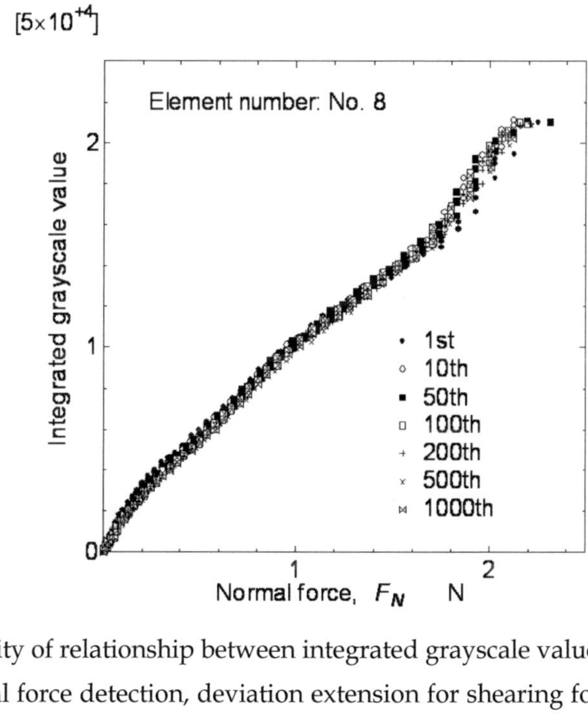

Fig. 13. Repeatability of relationship between integrated grayscale value and applied force

Contrary to normal force detection, deviation extension for shearing force is higher than for normal force, as shown in Fig. 14. The relationship between centroid movement and shearing force depends on the number of times applied force is repeated. However, if we compare the 1st to the 100th cycles, the 200th to the 500th cycles, and the 750th to the 1,000th cycles, the difference between cycles decreases with an increase of repetitions. It appears that centroid displacement is more easily disturbed and displaced by a slight change in loading direction than the integrated grayscale value, even if the probe tip has a concave portion mated with the tip of the sensing element. Increasing the number of repetitions reduces the disturbance and stabilizes the state.

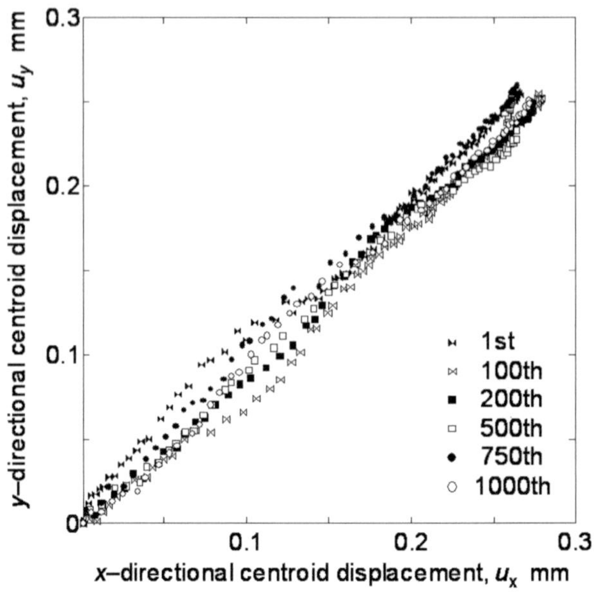

Fig. 14. Repeatability of shearing force detection

5. Robotic finger equipped with the three-axis tactile sensor

5.1 Robotic finger driven by dual computer system

A robotic finger equipped with the three-axis tactile sensor is developed to verify the present tactile sensor as shown in Fig. 15. Experiments using the finger are performed also as the first step of a series of studies on a dexterous multi-fingered hand. Since the present tactile sensor is based on image processing required heavy calculation time, a computer is used to only process image data acquired by the CCD camera. Toward a dexterous robotic hand equipped with the present tactile sensors, we develop tentatively a system of a robotic finger shown in Fig. 15, which possesses two computers; one is for tactile information processing; the other is for controlling the robotic finger; these computers are connected with a local area network.

After tactile data are obtained by the aforementioned process, they are sent to the computer for robotic finger control through the local area network. In the computer, signals applied to joint motors of the robotic finger are calculated to make the fingertip follow a trajectory according to an algorithm of Resolved Motion Rate Control (Whitney, 1969; Muir & Neuman, 1990). The signals are sent to the motor driver through the digital I/O board to drive the micro actuators.

In order to verify the robotic hand, experiments are carried out on basic motions such as surface scanning and object manipulation, which are performed in object recognition and stable grasping. Results of scanning test and object manipulation using are described in the following section.

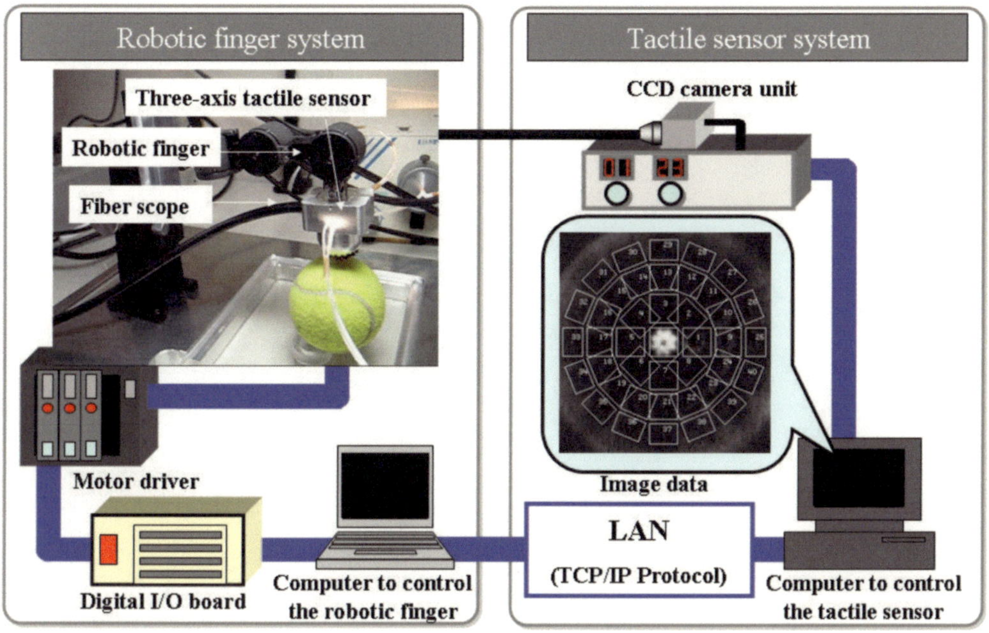

Fig. 15. Overview of the present system comprised of robotic hand and tactile data processing systems

5.2 Scanning test on flat surfaces

In scanning on a flat surface, sensing element located on the vertex of the tactile sensor is made to perpendicularly contact with a flat table by adjusting angles θ_2 and θ_3 in Fig. 16. After that, a z-stage equipped with a robotic finger is adjusted to obtain appropriate contact force (0.1 N). Precision abrasive paper (produced by Sumitomo 3M) is mounted on the table. In this experiment, three kinds of abrasive paper, 1, 30, and 60 μm, were adopted as specimens. To examine the dependence of friction coefficient on scanning speed, we chose three speeds: 1.4, 6.2, and 25 mm/sec.

Variations in normal force, shearing force, and friction coefficient obtained during scanning are shown in Fig. 17 to examine the dynamic characteristics of the tactile sensor for the case of 1 μm and 6.2 mm/sec. Shearing force starts at zero because it is not applied at zero speed. After the start, it increases abruptly to reach a constant value. Normal force almost shows a constant value. The coefficient of friction almost shows a constant value except near the origin. The mean value of the friction coefficient is 0.39. Friction coefficients for 30 and 60 μm abrasive paper are 0.40 and 0.53, respectively.

Next, variation in friction coefficient against variation in scanning velocity is shown in Fig. 17. In this experiment, 1 μm abrasive paper is adopted as a specimen and eight trials are performed for each scanning speed. As shown in Fig. 18, variation in friction coefficient decreases slightly with an increase of scanning velocity. Since cutting resistance decreases with an increase of cutting speed in grinding theory, it is assumed that this cutting effect will arise in this experiment.

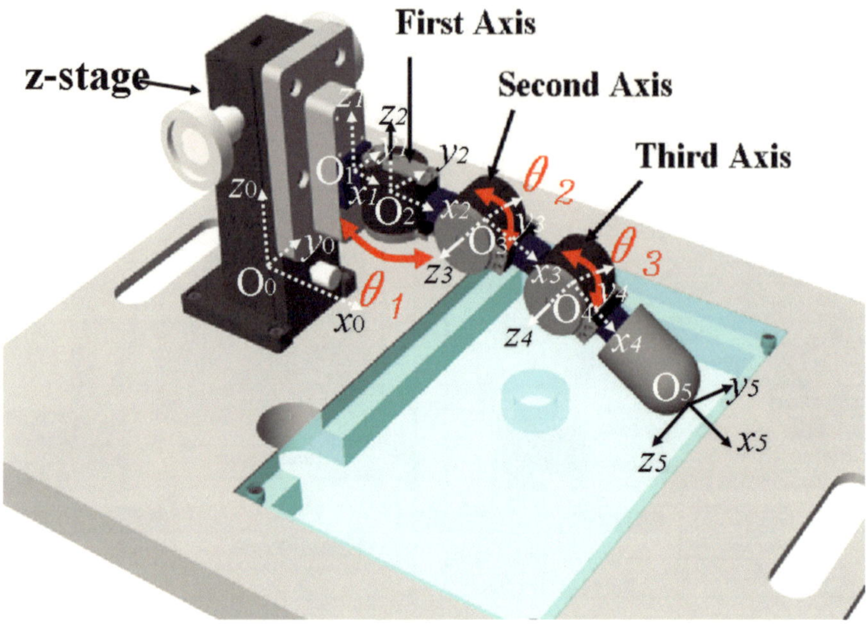

Fig. 16. Robotic finger equipped with a three-axis tactile sensor

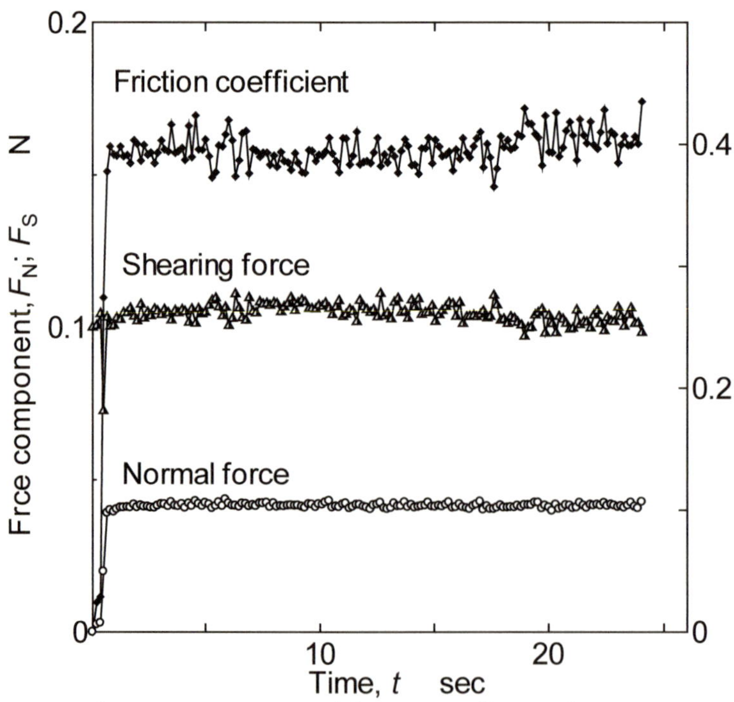

Fig. 17. Variation in force components and friction coefficient during scanning test (abrasive paper of 1 m and velocity of 6.2 mm/sec)

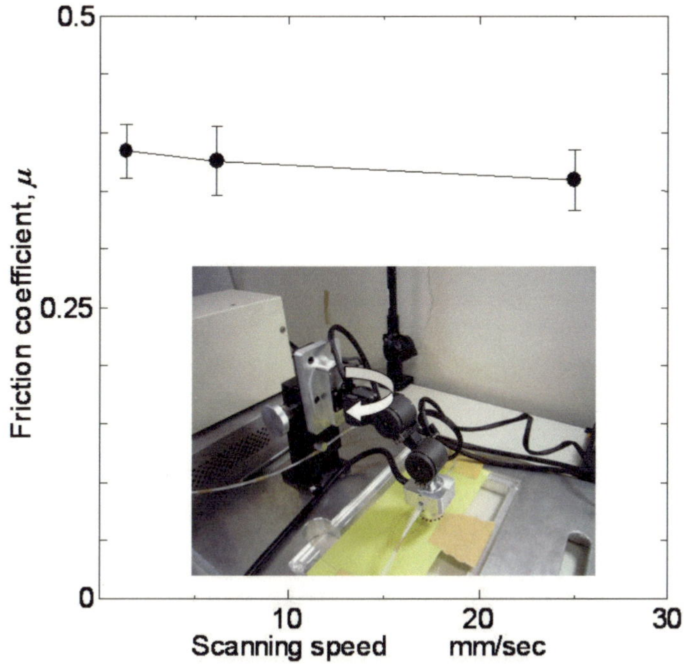

Fig. 18. Dependence of friction coefficient on scanning velocity

6.3 Object manipulation test

If slippage occurs between a finger and an object, a robotic hand cannot manipulate the object without any control based on acquired slippage information. Object manipulation tests check its capability to acquire slippage information using one-finger manipulation.

The robotic finger moves the parallelepiped object, which is put on the acrylic table (Fig. 19). Since the object is only put on the table, it is moved based on finger movement. During this manipulation, if the time derivative of the shearing force caused on the tactile sensor exceeds a specified threshold, slippage is assumed, and the finger moves slightly downward to increase compressive force. Since the sensing element is made of silicone rubber, friction between the finger and the object can be increased without considerably increasing friction between the object and the table. In this experiment, the robotic finger moves along a rectangular trajectory, and the object's movement is measured by a position sensitive detector (PSD; PS1100, Toyonaka Kenkyusyo, Co.).

Figure 20 shows the trajectory and attitude of the manipulated object. To see the attitude easier, the parallelepiped object is shown as 1/10 size in Fig. 20. As shown in Fig. 20, the object moves along the desired trajectory with considerable deviation. To analyze slippage phenomenon, variations in normal force and shearing force derivatives are shown in Fig. 21. In this experiment, since sensor elements No. 0 and 7 emit rather large signal compared to elements No. 1, 3 and 5, which touch the surface, their variations are shown in Fig. 21. Where elements such as element No. 0, 1, 3 and 5 are shown in Fig. 18. To show the relationship between the representative points of Figs. 20 and 21, corresponding points are shown in both figures as identical characters.

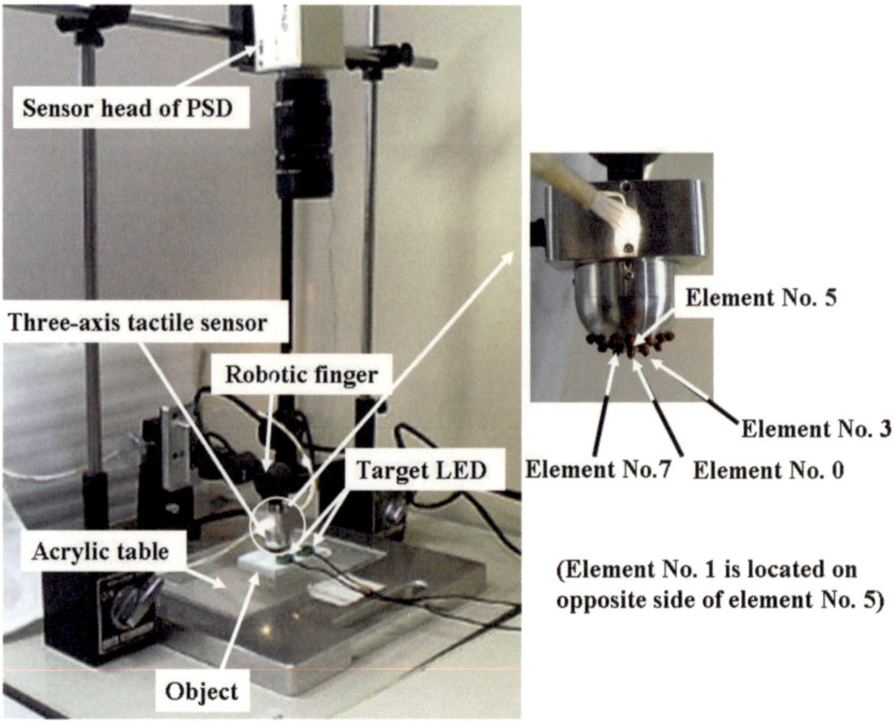

Fig. 19. Set up for object manipulations

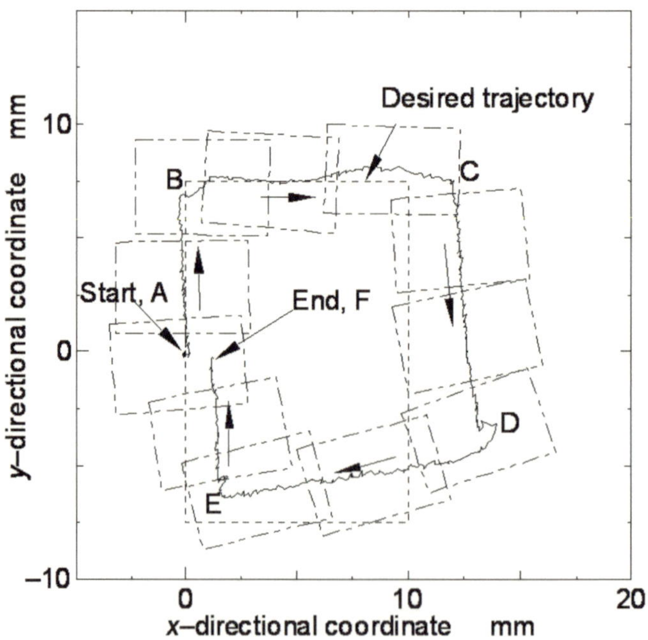

Fig. 20. Manipulation of parallelepiped object

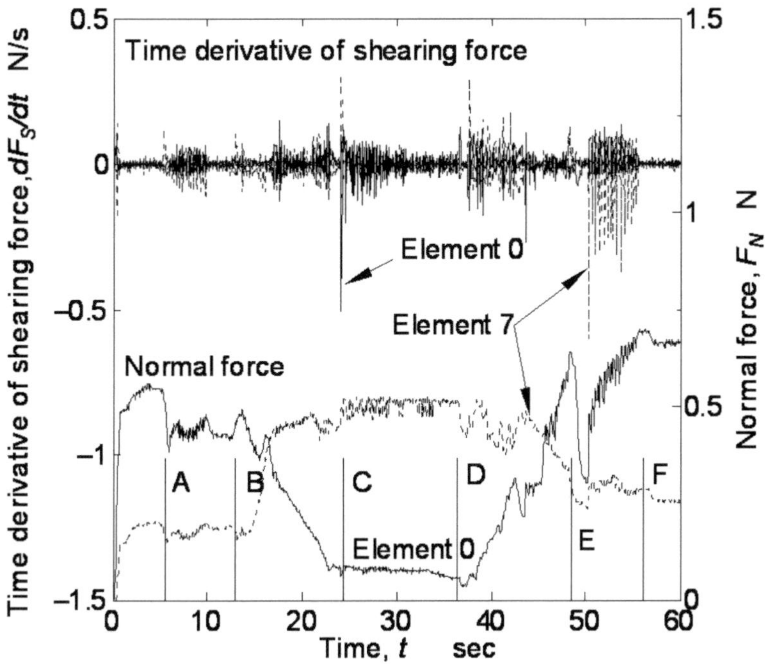

Fig. 21. Time derivative of shearing force in object manipulation

In this experiment, since finger moving starts when the compressive force exceeds 0.5 N at point A, normal force abruptly decreases just after point A in terms of the inclination of the sensing element. Since the present robotic finger only possesses three degrees of freedom and cannot control its inclination, the contact point is changed. Consequently, just after point B's normal force of element No. 0 decreases, the normal force of element No. 7 increases.

Next, we examine the time derivative of the shearing force in Fig. 21. If the derivative vibration is examined on the segments, the derivative on AB is smaller than on others. This result means that slippage on others is more considerable than on AB. Consequently, deviation after point C becomes considerable in terms of the slippage.

Through the above experimental results, the present robotic finger possesses sensing ability for acquiring the friction coefficients of the object surface and the slippage phenomenon, which are useful bits of information for a multi-fingered hand.

7. Conclusion

A new three-axis tactile sensor to be mounted on multi-fingered hands is developed based on the principle of an optical waveguide-type tactile sensor comprised of an acrylic hemispherical dome, a light source, an array of rubber sensing elements, and a CCD camera. The sensing element of the present tactile sensor includes one columnar feeler and eight conical feelers. A three-axis force applied to the tip of the sensing element is detected by the contact areas of the conical feelers, which maintain contact with the acrylic dome. Normal and shearing forces are calculated from integration and centroid displacement of the grayscale value derived from the conical feeler's contacts.

A series of experiments is conducted using a y-z stage, rotational stages, and a force gauge to evaluate the present tactile sensor. Although the relationship between the integrated grayscale value and normal force depended on the sensor's latitude on the hemispherical surface, it was easy to modify sensitivity based on the latitude. Sensitivity to normal and shearing forces was approximated with bi-linear curves. The results revealed that the relationship between the integrated grayscale value and normal force converges into a single curve despite the inclination of the applied force. This was also true for the relationship between centroid displacement and shearing force. Therefore, applied normal and shearing forces can be obtained independently from integrated grayscale values and centroid displacement, respectively. Also, the results for the present sensor had enough repeatability to confirm that the sensor is sufficiently sensitive to both normal and shearing forces.

Finally, the three-axis tactile sensor was mounted on a robotic finger of three degrees of freedom to evaluate the tactile sensor for dexterous hands. The robotic hand touched and scanned flat specimens to evaluate its friction coefficient. Subsequently, it manipulated a parallelepiped case put on a table by sliding it. Since the present robotic hand can perform the abovementioned tasks with appropriate precision, it is applicable to the dexterous hand in subsequent studies.

8. References

Bicchi, A.; Salisbury, J. K. & Dario, P. (1989), Augmentation of Grasp Robustness Using Intrinsic Tactile Sensing, Proc. of 1989 IEEE Int. Conf. On Robotics and Automation, 303-307.

Borovac, B.; Nagy, L. & Sabli, M. (1996), Contact Tasks Realization by sensing Contact Forces, Theory and Practice of Robots and Manipulators, Proc. of 11th CISM-IFToNN Symposium, Springer Wien New York, 381-388.

Dario, P., Rossi, D.D., Domenci, C., and Francesconi, R.(1984), Ferroelectric Polymer Tactile Sensors with Anthopomorphic Features, Proc. 1984 IEEE Int. Conf. On Robotics and Automation, pp. 332-340.

Hackwood, S.; Beni, G.; Hornak, L. A.; Wolfe, R. & Nelson, T. J.(1983), Torque-Sensitive Tactile Array for Robotics, Int. J. Robotics Research, Vol. 2-2, 46-50.

Hakozaki, M. & Shinoda, H. (2002), Digital Tactile Sensing Elements Communicating Through Conductive Skin Layers, Proc. of 2002 IEEE Int. Conf. On Robotics and Automation, 3813-3817.

Harmon, L. D. (1982), Automated Tactile Sensing, The International Journal of Robotic Research, Vol. 1, No.2, 3-32.

Hosoda, K.; Tada, Y. & Asada, M. (2006), Anthropomorphic Robotic Soft Fingertip with Randomly Distributed Receptors, Robotic and Autonomous Systems, vol. 54, 104-109.

Howe, R. D. & Cutkosky, M. R. (1993), Dynamic Tactile Sensing: Perception of Fine Surface Features with Stress Rate Sensing, IEEE Trans on Robotics and Automation, Vol. 9, No. 2, 140-151.

Maekawa, H.; Tanie, K.; Komoriya, K.; Kaneko M.; Horiguchi, C. & Sugawara, T. (1992), Development of a Finger-shaped Tactile Sensor and Its Evaluation by Active Touch, in Proc. of the 1992 IEEE Int. Conf. on Robotics and Automation, 1327-1334.

Mott, H.; Lee, M. H. & Nicholls, H. R. (1984), An Experimental Very-High-Resolution Tactile Sensor Array, in Proc. 4th Int. Conf. On Robot Vision and Sensory Control, 241-250.

Muir, P. & Neuman, C. P. (1990), Resolved Motion Rate and Resolved Acceleration Servo-Control of Wheeled Mobile Robots, Proceedings of the 1990 IEEE International Conference on Robotics and Automation, Vol. 2, 1133-1140.

Nicholls, H. R. & Lee, M. H. (1989), A Survey of Robot Tactile Sensing Technology, Int. J. Robotics Res., Vol. 8-3, 3-30.

Nicholls, H. R. (1990), Tactile Sensing Using an Optical Transduction Method, Traditional and Non-traditional Robot Sensors (Edited by T. C. Henderson), Springer-Verlag, 83-99.

Novak, J. L.(1989), Initial Design and Analysis of a Capacitive Sensor for Shear and Normal Force Measurement, Proc. of 1989 IEEE Int. Conf. On Robotic and Automation, 137-145.

Ohka, M. et al. (1994), Tactile Expert System Using a Parallel-fingered Hand Fitted with Three-axis Tactile Sensors, JSME Int. J., Series C, Vol. 37, No. 1, 138-146.

Ohka, M. ,Mitsuya; Y., Takeuchi, S.; Ishihara, H. & Kamekawa, O. (1995), A Three-axis Optical Tactile Sensor (FEM Contact Analyses and Sensing Experiments Using a Large-sized Tactile Sensor), in Proc. of the 1995 IEEE Int. Conf. on Robotics and Automation, 817-824.

Ohka, M.; Mitsuya, Y.; Matsunaga, Y. & Takeuchi, S. (2004), Sensing Characteristics of an Optical Three-axis Tactile Sensor Under Combined Loading, Robotica, vol. 22, pp. 213-221.

Ohka, M.; Kawahara, T.; Kobayashi, H. & Mitsuya, Y. (2004), A Basic Study on Optical Three-axis Tactile Sensor, Eight Inter. Conf. on Manufacturing and Management, 1047-1052.

Ohka, M.; Kawamura, T.; Itahashi, T.; Miyaoka, T. & Mitsuya, Y. (2005), A Tactile Recognition System Mimicking Human Mechanism for Recognizing Surface Roughness, JSME International Journal, Series C. Vol. 48, No. 2, 278-285.

Ohka, M.; Mitsuya, Y.; Higashioka, I. & Kabeshita, H. (2005), An Experimental Optical Three-axis Tactile Sensor for Micro-robots, Robotica, vol. 23, 457-465.

Ohka, M.; Kobayashi, H. & Mitsuya, Y. (2005), Sensing Characteristics of an Optical Three-axis Tactile Sensor Mounted on a Multi-fingered Robotic Hand, IEEE/RSJ Inter. Conf. on Intelligent Robots and Systems, 1959-1964.

Ohka, M.; Kobayashi, H.; Takata, J. & Mitsuya, Y. (2006), Sensing Precision of an Optical Three-axis Tactile Sensor for a Robotic Finger, Proc. of ROMAN 2006-The 15th IEEE Inter. Sympo. on Robot and Human Interactive Communication, 214-219.

Raibert, H.M. & Tanner, J.E.(1982). Design and Implementation of a VSLI Tactile Sensing Computer, Int. J. Robotics Res., Vol. 1-3, 3-18.

Shimojo, M. & Ishikawa, M.; Thin and Flexible Position Sensor, J. Robotics and Mechatronics, Vol.2, No.1, pp.38-41, 1990.

Takeuchi, S.; Ohka, M. & Mitsuya, Y. (1994), Tactile Recognition Using Fuzzy Production Rules and Fuzzy Relations for Processing Image Data from Three-dimensional Tactile Sensors Mounted on a Robot Hand, Proc. of the Asian Control Conf., Vol. 3, 631-634.

Tanie, K.; Komoriya, K.; Kaneko M.; Tachi, S. & Fujiwara, A. (1986), A High-Resolution Tactile Sensor Array, Robot Sensors Vol. 2: Tactile and Non-Vision, Kempston, UK: IFS (Pubs), 189-198.

Whitney, D. E. (1969), Resolved Motion Rate Control of Manipulators and Human Prostheses, IEEE Transaction on Man-Machine Systems, Vol. 10-2, 47-53.

Yamada, Y. & Cutkosky, R. (1994), Tactile Sensor with 3-Axis Force and Vibration Sensing Function and Its Application to Detect Rotational Slip, Proc. of 1994 IEEE Int. Conf. On Robotics and Automation, 3550-3557.

8

Measurement Principles of Optical Three-Axis Tactile Sensor and its Application to Robotic Fingers System

Hanafiah Yussof[1], Jumpei Takata[2] and Masahiro Ohka[3]
[1,3]Graduate School of Information Science, Nagoya University
[2]Olympus Corporation
[1]Faculty of Mechanical Engineering, Universiti Teknologi MARA
[1,2,3]Japan
[1]Malaysia

1. Introduction

A tactile sensor is a device that can measure a given property of an object or contact event through physical contact between the sensor and the object. Traditionally, tactile sensors have been developed using measurements of strain produced in sensing materials that are detected using physical quantities such as electric resistance and capacity, magnetic intensity, voltage and light intensity (Nicholls, 1990). Research on tactile sensor is basically motivated by the tactile sensing system of the human skin. In humans, the skin's structure provides a mechanism to simultaneously sense static and dynamic pressure with extremely high accuracy. Meanwhile in robotics, several tactile sensing principles are commonly used nowadays, such as capacitive, piezoelectrical, inductive, piezoresistive, and optoelectrical sensors (Schmidt et al., 2006, Lee & Nicholls, 1999).

In our research lab, with the purpose to establish object manipulation ability in robotic fingers, we developed a hemispherical shaped optical three-axis tactile sensor capable of acquiring normal and shearing forces to mount on the fingertips of robot fingers. This tactile sensor uses an optical waveguide transduction method and applies image processing techniques. Such a sensing principle is expected to provide better sensing accuracy to realize contact phenomena by acquiring the three axial directions of the forces, so that normal and shearing forces can be measured simultaneously. This tactile sensor is designed in a hemispherical dome shape that consists of an array of sensing elements. This shape is to mimics the structure of human fingertips for easy compliance with various shapes of objects. For miniaturization of the tactile sensor, measurement devices are placed outside the sensor. The small size of the sensor makes it easy for installation at robotic fingers.

The optical three-axis tactile sensor developed in this research is designed in hemispherical shape, and the sensing elements are distributed in 41-sub region. Due to this structure, the acquired images by CCD camera, except for sensing element at the sensor tip area, are not the actual image of contact pressure at the sensing elements. Therefore, to compensate with the sensor structure, it is necessary to conduct coordinate transformation calculations for each sensing element except for the element at the sensor tip area. In this chapter, we present calculations to define coordinate transformation of the sensing elements on the sensor's hemispherical shape dome. Meanwhile, in the tactile sensor controller, since the image was warped due to projection from a hemispherical surface, image processing software Cosmos32 installed in the computer modifies the warped image data based on the coordinate transformation formulations, and calculates the integrated gray-scale value and displacement of gray-scale distribution to obtain the applied three-axis forces at the sensing element. Finally, we conduct experiment to evaluate the performance of the optical three-axis tactile sensor system using 3-dofs robotic fingers. We analyze the performance of tactile sensing feedback on the robot finger system to define optimum grasp pressure on the object surface. Figure 1 shows the robotic fingers mounted with the optical three-axis tactile sensors on each of its fingertips.

Fig. 1. Robotic fingers with optical three-axis tactile sensor.

2. Motivations and current state-of-the-art survey

The sense of touch is one of the five main sensing modalities in humans besides sight, sound, smell, and taste. It will play an important role in robotic paradigms toward effective manipulation and collaboration with humans in built-for-human environments. Research on tactile sensor is basically motivated by tactile sensing system of human skin. In daily life, humans regularly apply tactile sensing to support motions and perform tasks. However, in a developmental robot, tactile sensors are especially appropriate sensing devices that have too often been neglected in favor of vision-based approaches.

To date, while much research has developed visual and auditory sensors, comparatively little progress has been made on sensors that translate the sense of touch. This apparent neglect reflects the complexity of tactile sensing itself, because tactile sensing through the skin is not a simple transduction of one physical property into electronic signals. Furthermore tactile sensing is difficult to imitate, unlike sight and sound, which are well-defined physical quantities. In addition, the fact that a tactile signal is distributed over a much wider area and lacks such localized sensory organs as eyes and ears complicates the developmental of artificial sensory devices.

Nonetheless, realizing that the development of intelligent tactile sensors will help advance the evolution of human and robots working together in real life is encouraging (Ohmura et al., 2006, Kuniyoshi et al., 2004, Natale & Torres-Jara, 2006, Ohmura & Kuniyoshi, 2007). Indeed, researchers have recently agreed that a tactile sensor system is an essential sensory device to support the robot control system, particularly for object manipulation tasks (Omata et al., 2004, Kerpa et al., 2003, Lee & Nicholls, 1999). This agreement reflects the tactile sensor's capability to simultaneously sense normal force, shearing force, and slippage, thus offering exciting possibilities for determining object shape, texture, and property.

To date, several basic sensing principles are commonly in use in tactile sensor, such as capacitive sensor, piezoelectrical sensor, inductive sensor, optoelectrical sensor and piezoresistive sensor (Lee & Nicholls, 1999). In this research, with the aim of establishing object manipulation ability in real humanoid robot, we have developed an optical three-axis tactile sensor using optical waveguide transduction method, applying image processing techniques. This type of sensing principle is comparatively provides better sensing accuracy to detect contact phenomena from acquisition of three axial directions of forces, thus normal force and shearing force can be measured simultaneously (Ohka et al., 2004, Hanafiah et al. 2006, Hanafiah et al. 2007). The proposed three-axis tactile sensor has high potential compared to ordinal tactile sensor for fitting to a dextrose robotic arm to perform robot manipulation tasks.

3. Hardware structure

The optical three-axis tactile sensor developed in this research is designed in a hemispherical dome shape that consists of an array of sensing elements. This shape is to mimics the structure of human fingertips for easy compliance with various shapes of objects. For miniaturization of the tactile sensor, measurement devices are placed outside the sensor. The small size of the sensor makes it easy for installation at the robotic fingers.

The hardware novelty is shown in Fig. 2. It consists of an acrylic hemispherical dome, an array of 41 pieces of sensing elements made from silicon rubber, a light source, an optical fiber-scope, and a CCD camera. The optical fiber-scope is connected to the CCD camera to acquire image of sensing elements touching acrylic dome inside the tactile sensor. At this moment, light emitted from the light source is directed toward the edge of the hemispherical acrylic dome through optical fibers. A total of 24 pieces optical fibers are used; 12 pieces each at left and right side of the sensor, transmitting halogen light from the light source. The light directed into the acrylic dome remains within it due to total internal reflection generated, since the acrylic dome is surrounded by air having a lower reflection index than the acrylic dome. This make the acquired image by the CCD camera become clear even the sensor is hemispherical shape.

Meanwhile, the silicone rubber sensing element is comprised of one columnar feeler and eight conical feelers which remain in contact with the acrylic surface while the tip of the columnar feeler touches an object. The sensing elements are arranged on the hemispherical acrylic dome in a concentric configuration with 41 sub-regions. Such orientation is expected to provide good indication of contact pressure during performing object manipulation.

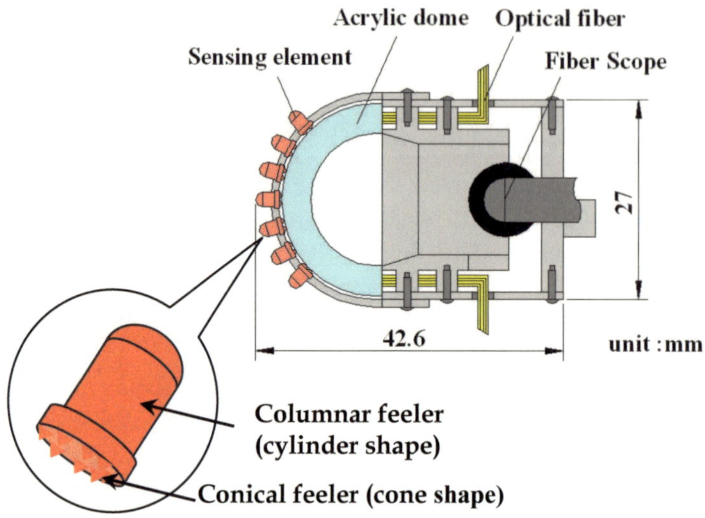

Fig. 2. Structure of hemispherical dome shaped optical three-axis tactile sensor.

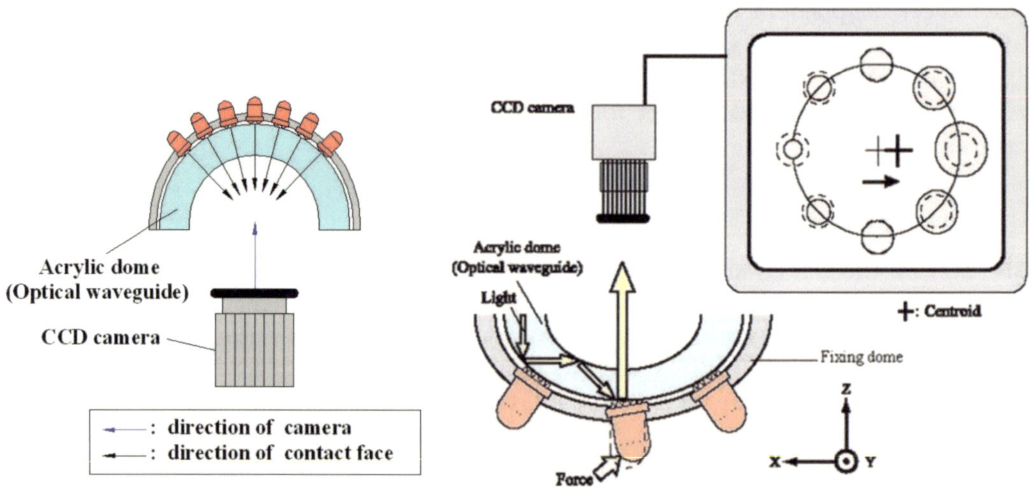

Fig. 3. Sensing principle of optical three-axis tactile sensor system.

4. Sensing principles

The optical three-axis tactile sensor is based on the principle of an optical waveguide-type tactile sensor. Figure 3 shows the sensing principle of the optical three-axis tactile sensor system. The light emitted from the light source is directed towards the edge of the hemispherical acrylic dome through optical fibers. When an object contacts the columnar feelers, resulting in contact pressure, the feelers collapse. At the points where the conical feelers collapse, light is diffusely reflected out of the reverse surface of the acrylic surface because the rubber has a higher reflective index. Contact phenomena consisting of bright spots caused by the collapse of the feelers are observed as image data, which are retrieved by the optical fiber-scope connected to the CCD camera and transmitted to the computer.

Figure 4 shows the real image data of contact phenomenon inside the tactile sensor acquired by the CCD camera where some bright area resulted from contact pressure can be observed at the sensing elements. Referring to Fig. 3, for normal force detection, when load is applied vertically to sensing elements, the conical feelers will collapse on acrylic dome surface. At this moment, the image retrieved by CCD camera shows brightness change at the area where conical feelers are collapse. The normal force is calculated based on the brightness of this area. At this moment, centroid point of the sensing element is remaining unchanged. Meanwhile, in shearing force detection, when tangential force is applied to the sensing element, the sensing element is collapse according to the applied load direction. In the same time, the centroid point of the sensing element also shifted. Therefore, the shearing force can be calculated based on horizontal displacement of this centroid point.

The system conception of the optical three-axis tactile sensor system is shown in Fig. 5. As explained previously, when contact pressure is applied on the tactile sensor elements, a bright spot areas are appeared inside the tactile sensor which then captured as image data by a CCD camera. The image data retrieved by the CCD camera are delivered to PC via PCI bus of image processing board Himawari PCI/S. Then the image data are saved in an internal buffer area that created inside the PC internal memory space.

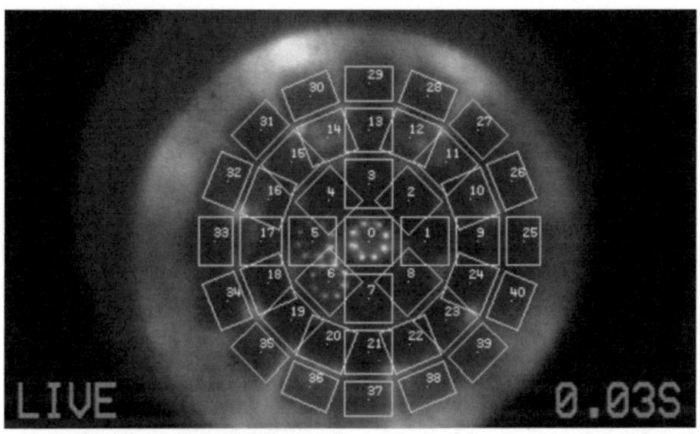

Contact phenomenon and 41 sub-regions of sensor elements inside the optical three-axis tactile sensor

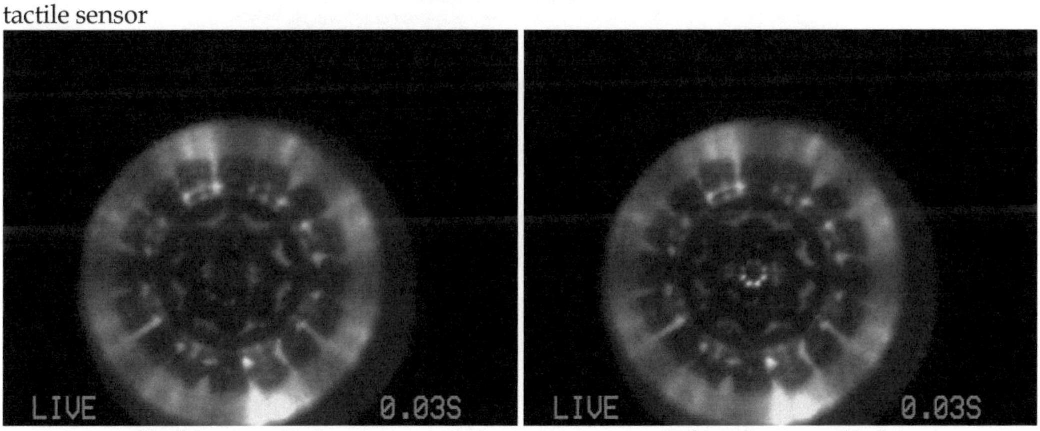

No force applied When force applied

Fig. 4. CCD camera images of contact phenomenon in the hemispherical shaped optical three-axis tactile sensor.

The image capture cycle was initially fixed at 1/30 seconds. Sensing program inside the PC is using Visual C++ and we utilized image analysis software Cosmos32 to analyze and measure the image data. During measurement process, the dividing procedure, digital filtering, integrated gray-scale value and centroid displacement are controlled on the PC using sensing program which created based on the software Cosmos32 functions.
Figure 6 shows system conception diagram of the sensing program which consists of image analysis module and connection module. This sensing program embedded a user interface and a tactile information structure which designed for both modules to share the tactile sensing data. The image analysis module uploads the image data from internal buffer and performs image analysis and measurement to define forces that applied to the sensing elements and also centroid point displacement. The upload cycle was fixed by the operator so that it not less than the image captured cycle.

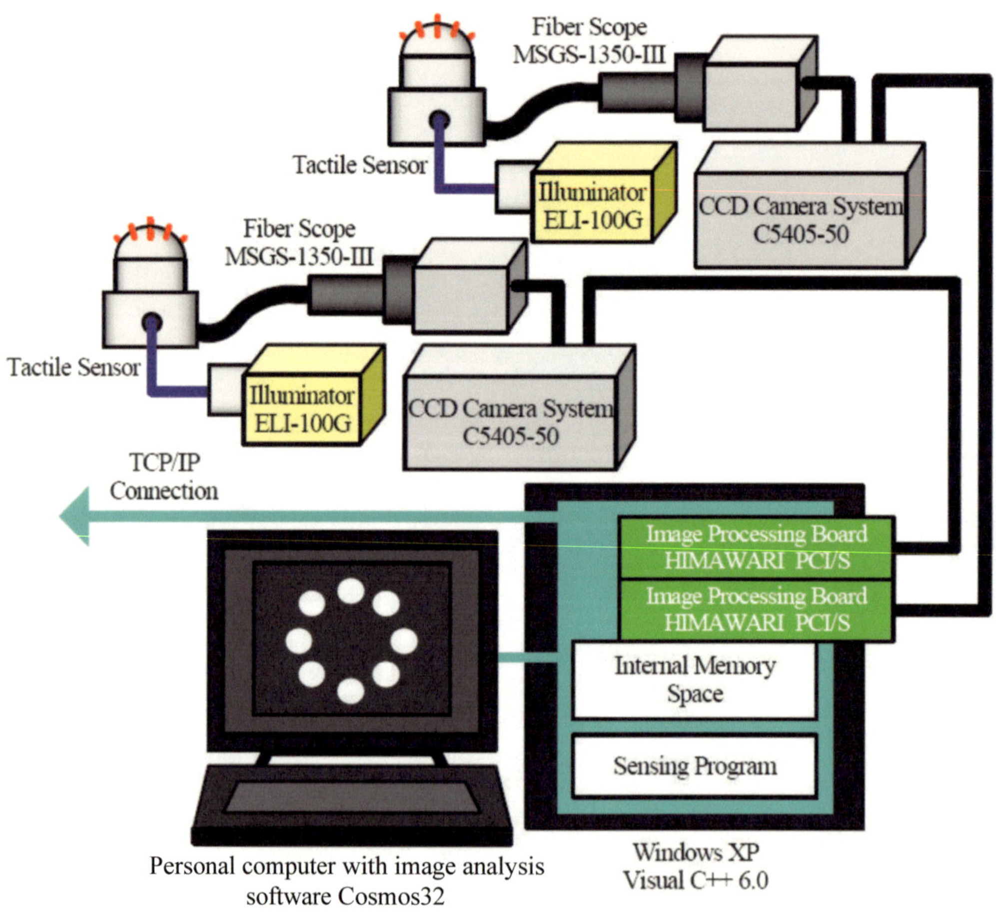

Fig. 5. System conception diagram of tactile sensor system.

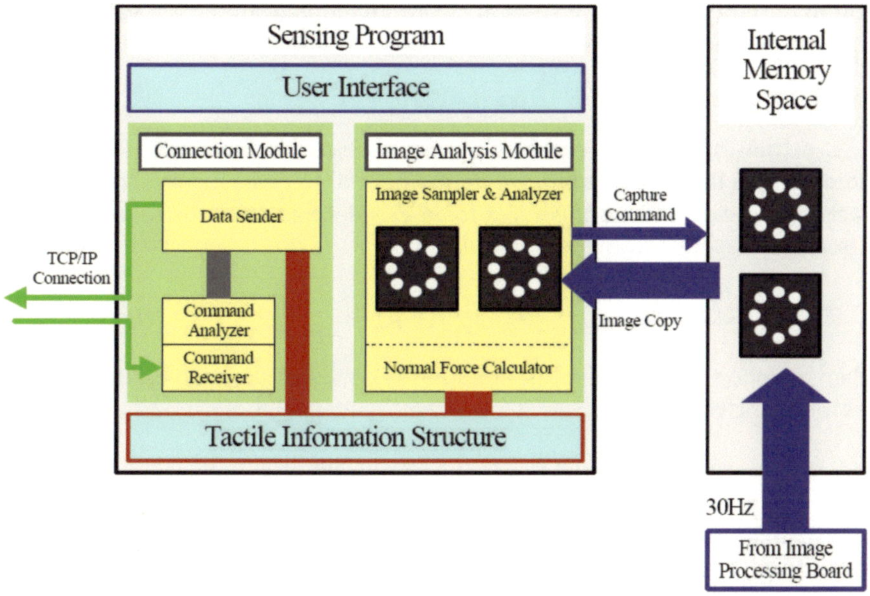

Fig. 6. System conception of sensing program.

5. Measurement principles

In measurement process, the normal force of the F_x, F_y and F_z values are calculated using integrated gray-scale value G, while shearing force is based on horizontal centre point displacement. The displacement of gray-scale distribution $\boldsymbol{u}$ is defined in (1), where $\boldsymbol{i}$ and $\boldsymbol{j}$ are the orthogonal base vectors of the x- and y-axes of a Cartesian coordinate, respectively. This equation is based on calibration experiments, and material functions are identified with piecewise approximate curves (Ohka et al. 2006, Takata, 2005). Consequently, each force component is defined in (2).

$$\boldsymbol{u} = u_x\boldsymbol{i} + u_y\boldsymbol{j} \quad (1)$$

$$F_x = f(u_x),\ F_y = f(u_y),\ F_z = g(G) \quad (2)$$

The optical three-axis tactile sensor developed in this research is designed in hemispherical shape, and the sensing elements are distributed in 41-sub region. Due to this structure, the acquired images by CCD camera, except for sensing element at the sensor tip area, are not the actual image of contact pressure at the sensing elements. Therefore, to compensate with the sensor structure, it is necessary to conduct coordinate transformation calculations for each sensing element except for the element at the sensor tip area.

5.1 Calculation of contact pressure and centroid position

As explained in previous section, when force is applied to columnar feeler of the tactile sensor element, the conical feeler collapsed on the acryl dome surface which created a bright spot retrieved by the CCD camera. To measure the applied forces, we measure the integrated gray-scale value of the bright area and also centroid point displacement.

According to current research (Ohka et al, 2006), integrated gray-scale value of the contact area $g(x,y)$ is proportional with contact force $p(x,y)$, as shown in (3).

$$p(x,y)= C_v \Delta g(x,y) \tag{3}$$

Here, C_v is a transformation coefficient, $\Delta g(x,y)$ is increment of the integrated gray-scale value. According to this relationship, we define normal force from gray-scale distribution of the contact area. Measurement of contact force P is defined from the following integration, where S is size of the gray-scale measurement area.

$$P = \int_S p(x,y)dS \tag{4}$$

Here, when equation (3) is applied to equation (4), we can define measurement of contact force P as following equation (5).

$$P = C_v \int_S \Delta g(x,y)dS \tag{5}$$

Next, we calculate the centroid position. The centroid position is measured according to center point of the bright spots area, which equal to center position of integrated gray-scale measurement area. When tangential force is applied to the sensor element, contact area of conical feeler at the sensor element with acryl surface is shifted horizontally. To define shearing force, we measure the horizontal centroid point displacement at x and y axes. At first, by applying the increment of integrated gray-scale value $\Delta g(x,y)$, the centroid positions at xy-axes which described at as x_G and y_G are define within the measurement area of integrated gray-scale value as shown in (6) and (7), respectively.

$$x_G = \frac{\int_S \Delta g(x,y)xdS}{\int_S \Delta g(x,y)ydS} \tag{6}$$

$$y_G = \frac{\int_S \Delta g(x,y)ydS}{\int_S \Delta g(x,y)xdS} \tag{7}$$

According to the above equations, displacement of centroid point at x y-axes in time t are defined as following equations.

$$dx_G^{(t)} = x_G^{(t)} - x_G^{(t-1)} \tag{8}$$

$$dy_G^{(t)} = y_G^{(t)} - y_G^{(t-1)} \tag{9}$$

5.2 Coordinate transformation on hemispherical shaped dome

The optical three-axis tactile sensor developed in this research is designed in hemispherical shape, and the sensing elements are distributed in 41-sub region as shown in Fig. 4. Due to this structure, the acquired images by CCD camera, except for sensing element at the sensor tip area #0, are not the actual image of contact pressure at the sensing elements. Therefore, to compensate with the sensor structure, it is necessary to conduct coordinate transformation calculation for each sensing element except for element #0.

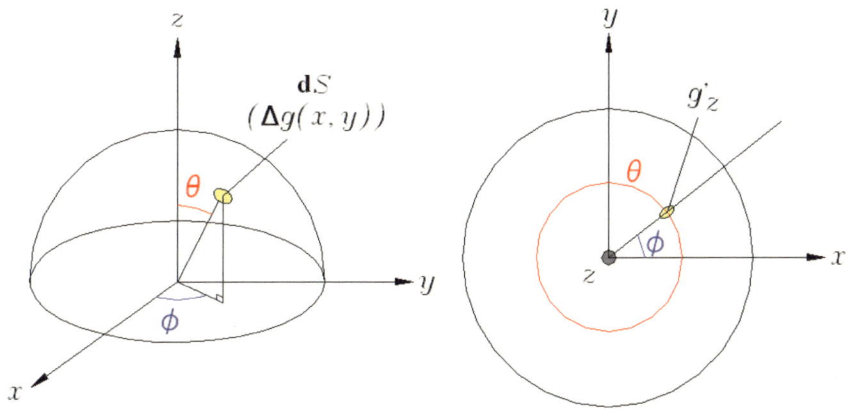

Fig. 7. Diagram of hemispherical dome indicates measurement area of integrated gray-scale value for coordinate transformation calculation.

Referring to Fig. 7, where *dS* is a measurement area of integrated gray-scale value, coordinate transformation is described in equation (10). Here (g_x, g_y, g_z) is integrated gray-scale value before coordinate transformation, meanwhile (g'_x, g'_y, g'_z) is after the transformation.

$$\begin{bmatrix} g'_x \\ g'_y \\ g'_z \end{bmatrix} = \begin{bmatrix} \cos\theta\cos\phi & -\sin\phi & \sin\theta\cos\phi \\ \cos\theta\sin\phi & \cos\phi & \sin\theta\sin\phi \\ -\sin\theta & 0 & \cos\theta \end{bmatrix} \begin{bmatrix} g_x \\ g_y \\ g_z \end{bmatrix} \tag{10}$$

If *dS* is put on the sensor tip area, equation (10) is become like equation (11), where the increment of the integrated gray-scale value can be described as $\Delta g(x,y) = g_z(x,y)$.

$$\begin{bmatrix} g'_x \\ g'_y \\ g'_z \end{bmatrix} = \begin{bmatrix} \cos\theta\cos\phi & -\sin\phi & \sin\theta\cos\phi \\ \cos\theta\sin\phi & \cos\phi & \sin\theta\sin\phi \\ -\sin\theta & 0 & \cos\theta \end{bmatrix} \begin{bmatrix} 0 \\ 0 \\ g_z \end{bmatrix} \tag{11}$$

Here, equation (11) is solved as following equations:

$$g'_x = g_z \sin\theta\cos\phi \tag{12}$$

$$g'_y = g_z \sin\theta\sin\phi \tag{13}$$

$$g'_z = g_z \cos\theta \tag{14}$$

Hence, when measurement area of integrated gray-scale value is put on the sensor tip area, the integrated gray-scale value after coordinate transformation is coincide with $g_z(x,y)$ as shown in the following equation.

$$g'_x \sqrt{g'_x + g'_y + g'_z} = g_z \sqrt{\sin^2\theta\cos^2\phi + \sin^2\theta\sin^2\phi + \cos^2\theta}$$
$$= g_z \sqrt{\sin^2\theta + \cos^2\theta} \quad (15)$$
$$= g_z$$

Here, from (14), the increment of the integrated gray-scale value is define as follows:

$$\Delta g(x,y) = \frac{g_z}{\cos\theta} \quad (16)$$

From equation (16), the coordinate transformation for measurement of contact force P from equation (5) is described as follows:

$$P = C_v \int_S \Delta g(x,y) dS$$
$$= C_v \int_S \frac{g'_z}{\cos\theta} dS \quad (17)$$

Consequently, the captured image by CCD camera for measurement area dS' on the hemispherical dome is corrected to measurement area dS by defining integrated gray-scale value g'_z using equation (17).

6. Application in robotic fingers system

The hemispherical shaped optical three-axis tactile sensors are mounted on fingertips of robotic finger. The robotic finger system is comprised of two articulated fingers, each of which has 3-dofs with micro-actuators that are used in each joint. This system is comprised of two main controllers: finger controller and tactile sensor controller. Each of these controllers is connected to each other using TCP/IP protocols via the internet.

6.1 Finger controller

The control system architecture of the robot finger controller, which is based on tactile sensing, is shown in Fig. 8. This controller is comprised of three modules: connection module, thinking routines, and hand/finger control module. It is connected with tactile sensor controller by the connection module using TCP/IP protocols (Takata, 2005, Hanafiah et al., 2008). The most important considerations in controlling finger motions during performing object manipulation tasks are: what kind of information are acquired from the tactile sensor, how to translate and utilize this information, and how to send command to robot finger so that velocity of the finger motion can be control properly. These processes are performed inside the thinking routines module.

As shown in Fig. 8, inside the thinking routines module, there is thinking routine chooser consists of pin status analyzer and velocity generator. Moreover, there is motion information structure which connecting to both pin status analyzer and velocity generator. The pin status analyzer module is functioned to receive information from the tactile sensor about sensing elements condition, and use this information to decide suitable motion mode. Then it sends to the connection module a list of sensing elements that acquire tactile sensing information. Meanwhile, the velocity generator module is functioned to decide finger velocity based on finger information structure and motion information structure. The motion information structure consists of initial velocity, motion flag mode, etc, which is used to control finger movement. Meanwhile, finger information structure provides connection all modules so that they can share data of finger orientation, joint angle and tactile sensing data from each sensor elements.

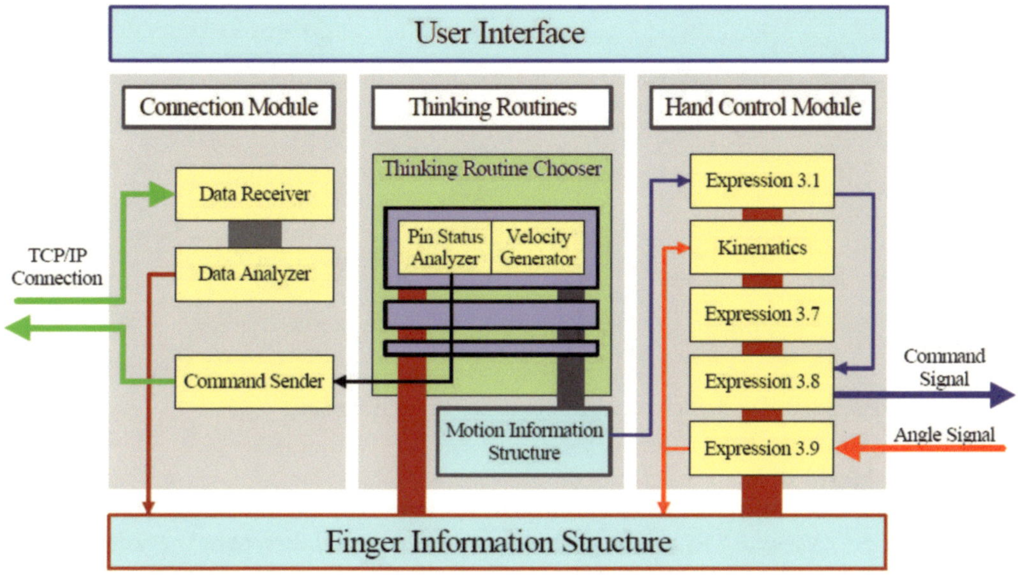

Fig. 8. Control system structure of robotic fingers.

User Interface was designed for the operator to provide commands to the finger control system. Finger control module controls the finger motion by calculating joints velocity and angle. In fact, this module can move finger without using sensing feedback. Thinking routines module receives tactile sensing data from tactile sensor and uses it to calculate fingertip velocity. In addition, to obtain low force interactions of the fingers during exploring object surface without causing damage, rotation velocity at each joint is defined precisely based on joint angle obtained in kinematics calculations, whereby force-position controls are performed.

6.2 Sensor controller

Figure 9 shows layout of tactile sensor controller. In the tactile sensor controller, based on image data captured by CCD camera, an image processing board Himawari PCI/S (Library Corp.) function as PCI bus picks up the image and sends it to internal buffer created inside the PC main memory. Sampling time for this process is 1/30 seconds. We use PC with Windows XP OS installed with Microsoft Visual C++. The image data are then sent to image analysis module applying Cosmos32 software which controls the dividing procedure, digital filtering, calculation of integrated gray-scale value and centroid displacement.
Since the image warps due to projection from a hemispherical surface as shown in Fig. 2, the software Cosmos32 with auto image analysis program installed in the computer modifies

the warped image data and calculates G, u_x and u_y to obtain the three-axis force applied to the tip of the sensing element using equation (2). These control schemes enable the finger controller to perform force-position control to adjust grasp pressure of the two fingers.

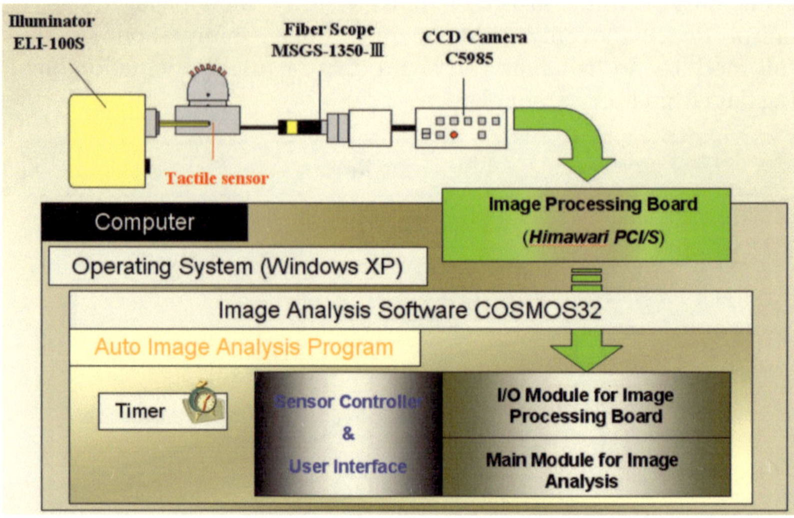

Fig. 9. Control system structure of optical three-axis tactile sensor.

6.3 Control algorithm

To further understand about data communication process in the finger controller, we present a simple case study where finger touches an object and then avoid/evade the object by moving the finger to reverse direction. At first, the finger moving velocity to search for object is defined as V_0. Next, we fix thresholds of normal force F_1 and F_2. During searching process, when any of sensor elements touch an object, and if the detected normal force F_n is exceeding normal force threshold F_1 [N], the finger will stop moving. Meanwhile, if the detected normal force F_n is exceeding threshold F_2 [N], the finger will move towards reverse direction of the sensing element that detects the highest force. At this moment the reverse velocity is defined as $|V_{re}|$. The parameters values of V_0, F_1, F_2 and $|V_{re}|$ are saved inside the motion information structure. The thresholds F_1 and F_2 are also delivered to sensor controller. When finger start moving, command to request status of each sensor elements are delivered to the sensor system according to control sampling phase of the finger system. Detail of data communication process for the pin status analyzer is shown in Fig. 10.

The processes at the pin status analyzer are explained as follows:

1. When sensor system received request command pin status analyzer, it will feedback status flag of each requested sensing element condition.
2. Connection module received the feedback data and then sends this data to the pin status analyzer, as well keeps it inside the finger information structure.
3. Pin status analyzer will then reset the finger motion ("STOP" and "EVADE") inside the motion information structure.
4. If the pin status analyzer received data flag that exceeds F_1 or F_2, or both of them, it will list up the concerned sensor elements. The pin status analyzer will rise up "STOP" flag if any listed sensor element is exceeded F_1, meanwhile it will rise-up "EVADE" flag if any listed sensor element is exceeded F_2. Then it sent the lists of data to the connection module.

5. The connection module will create a command to request normal force data of related sensor elements, and send the request to sensor system.
6. When sensor system received this request command, it will feedback normal force data of the requested sensor element to connection module at the finger controller.
7. Connection module received the feedback normal force data and then sends it to finger information structure. Based on this data, the velocity generator module will decide the velocity of the finger.

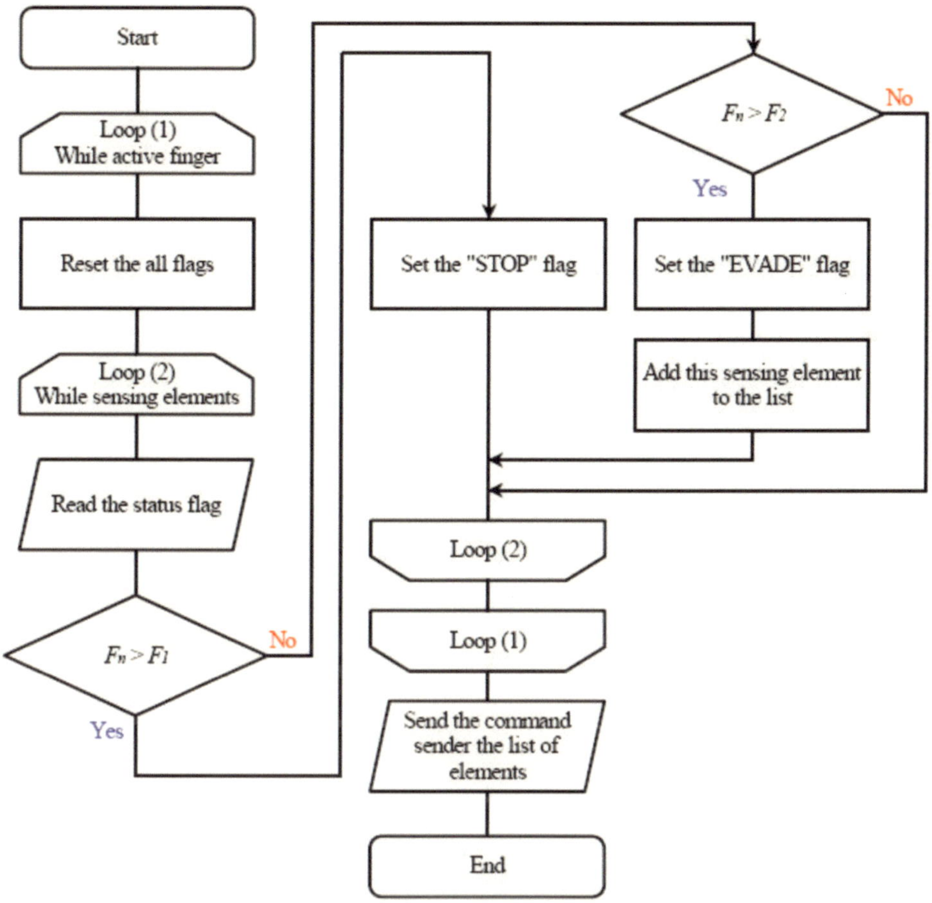

Fig. 10. Example of flowchart at the pin status analyzer for case study.

Figure 11 shows flowchart at the motion generator for case study. The processes are explained as follows:

1. If no flags "STOP" or "EVADE" rise-up, finger will move according to initial velocity V_0.
2. If "STOP" flag is rise-up, finger velocity becomes 0.
3. If "EVADE" flag is rise-up, the finger will move towards reverse direction of the sensing element that detects the highest normal force value. To decide the finger velocity, when finger evading velocity is described as $V_r = (V_{rx}, V_{ry}, V_{rz})$, the direction cosine in the frame of workspace ($\alpha_{Gk}, \beta_{Gk}, \gamma_{Gk}$) is calculated as equation (18).

$$V_r = -|V_{re}| \begin{bmatrix} \alpha_{Gk} \\ \beta_{Gk} \\ \gamma_{Gk} \end{bmatrix} \tag{18}$$

Here, basically this generation of velocity is sent to hand/finger control module to solve the joint rotation velocity at the finger derived by kinematics-based Resolved Motion Rate Control (RMRC), which commonly known as an algorithm for solving path-tracking problem in robotic control (Umetani & Yoshida, 1989). Therefore, controls of the finger based on tactile sensing information are conducted.

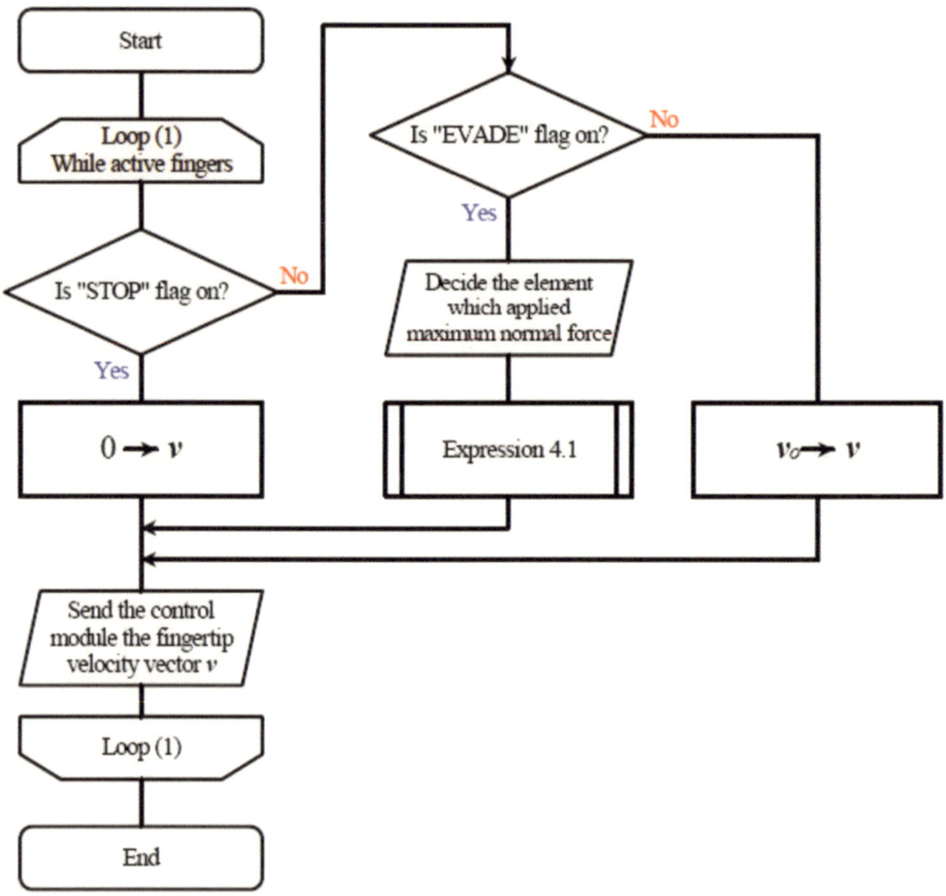

Fig. 11. Example of flowchart at the motion generator for case study.

7. Performance evaluation

We conduct object manipulation experiment using the robotic fingers system. We used a wood block as an object. In this experiment, at first the two fingers grasp the object to define optimum gripping pressure. At this moment, the grasp pressure is controlled by parameters of normal force thresholds. Then both fingers lift up the object to z-axis direction while maintaining the optimum grasp pressure. During this motion, both normal pressure and slippage are concerned. Therefore the finger controller utilized parameters of normal force and centroid change thresholds. Here, when shearing force exceeds the centroid change threshold, the finger's velocity for reinforcing the grasping pressure is calculated using equation (19), whereby vector velocity of the finger $v+\Delta v$ is defined by finger control module in the finger controller.

$$\Delta v = \left| v_p \right| \begin{bmatrix} \alpha_{Gk} \\ \beta_{Gk} \\ \gamma_{Gk} \end{bmatrix} \tag{19}$$

Table 1 shows control parameters in the robotic fingers system. These parameters value was determined from calibration experiments conducted using soft and hard objects (Hanafiah et al. 2008). Figure 12 shows photographs of the robot arm performing object manipulation with wood block. In this experiment, both fingers move along x-axis direction to grasp the wood block. When optimum grasping pressure is defined and the robot recognized the hardness of the object, both fingers lift up the wood block along y-axis, and then move forward along z-axis. Figure 13 shows relation between normal force and fingertip movement at x-, y- and z-axes for left finger. Accordingly, Fig. 14 shows relation between amount of x-directional centroid change and fingertip movement at x-, y-, and z-axes for left finger.

From these graphs, we can observe that the tactile sensor managed to detect normal and shearing forces applied to the sensing elements during manipulating object. Finger controller used this information to control grasp pressure against the object. The fingers movement stopped when the detected normal force reached to threshold F_1. Meanwhile, threshold of centroid change is used to control re-push velocity of the fingers when slippage occurred during grasping.

Category		Parameter
Sampling interval	Sensor	100 ms
	Finger	25 ms
Threshold of normal force	F_1	0.5 N
	F_2	1.8 N
Threshold of centroid change	dr	0.004 mm
Velocity of re-push	v_p	2 mm/s
Progress time	Δt	0.1 s

Table 1. Control parameters in robotic fingers system.

This experimental result shows that the fingers managed to grasp the objects within optimum grasp pressure, lift it to upwards direction, and then performed some movements manipulating the objects. The experimental results also revealed that the robotic fingers system mounted with the hemispherical shaped optical three-axis tactile sensor managed to grasp the object within optimum grasp pressure without causing damage to the object and the sensor elements. In addition, the formulations applied in this system enabled precise control of the fingertips from determination of joint rotation angles and velocity.

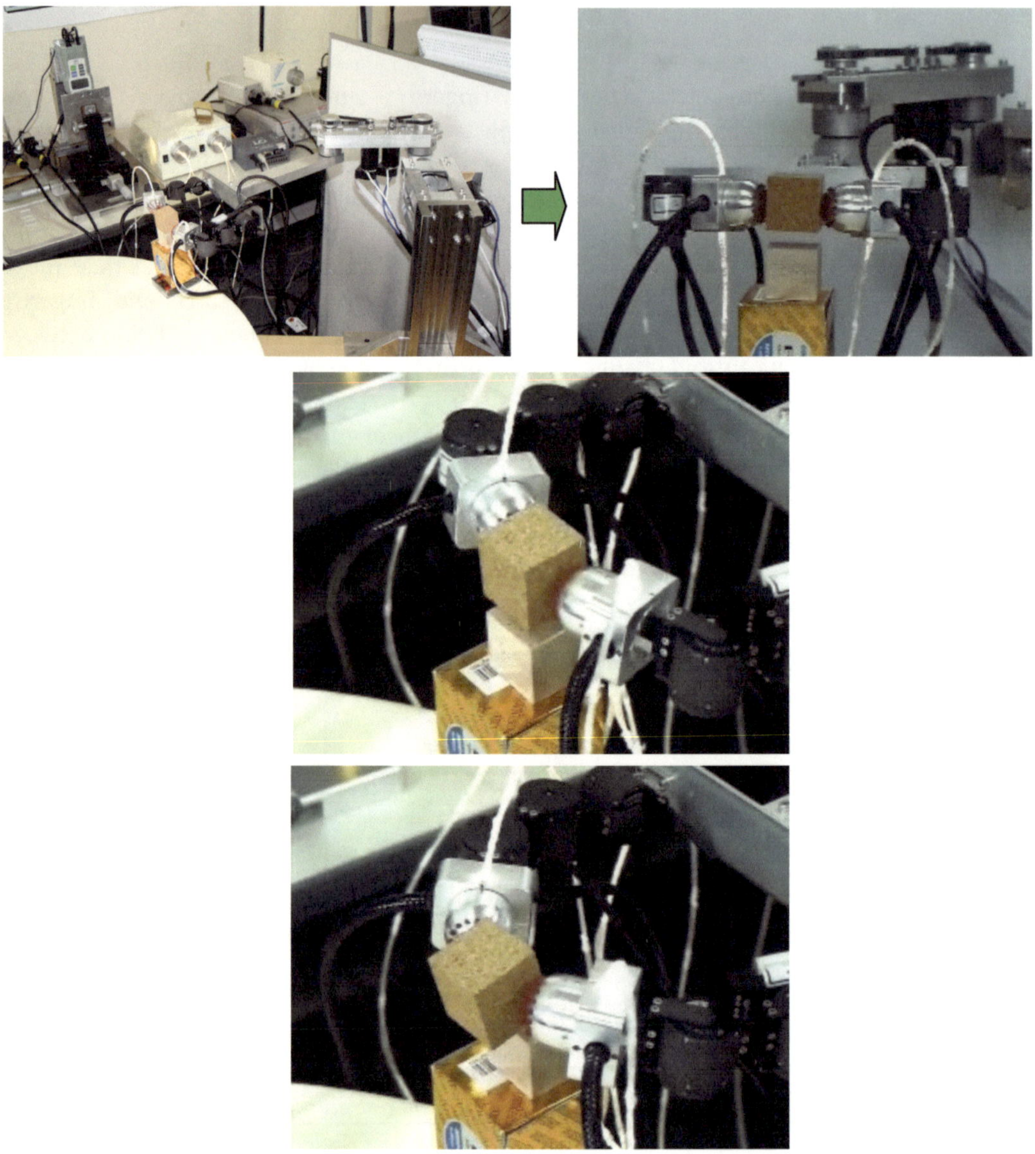

Fig. 12. Experiments of two robotic fingers manipulate wood block.

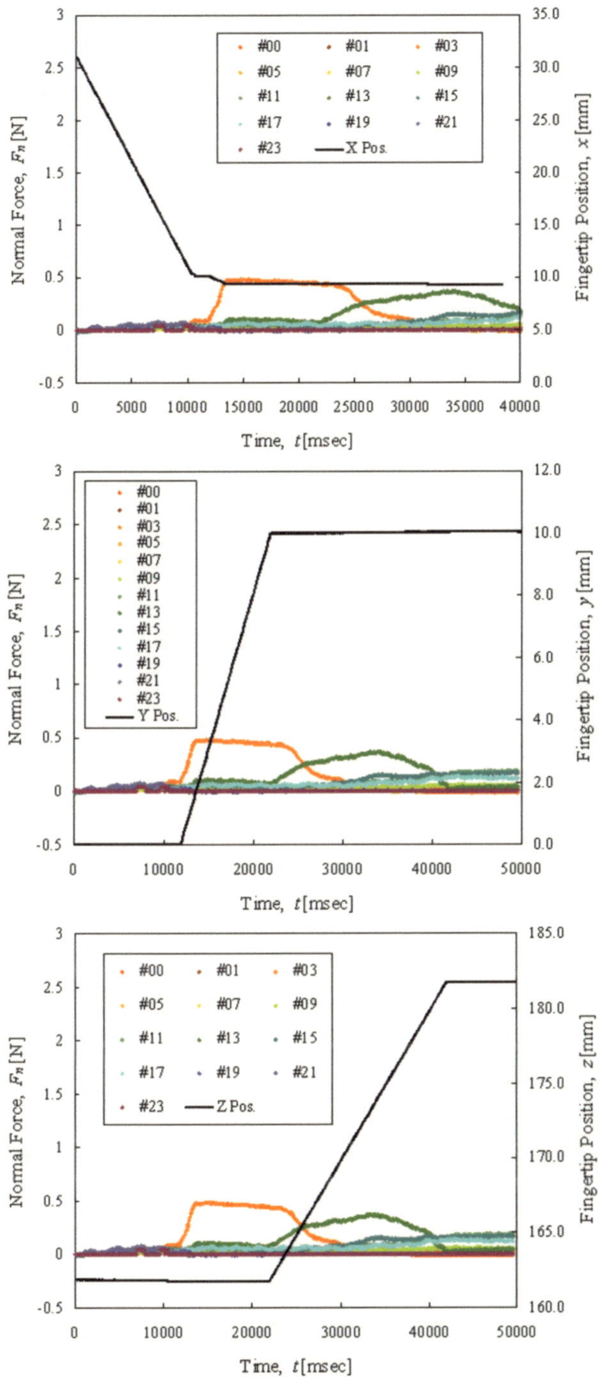

Fig. 13. Experiments of two robotic fingers manipulate wood block; data for left finger: *(Top)* Relation between normal force and fingertip movement at x-axis. *(Middle)* Relation between normal force and fingertip movement at y-axis. *(Bottom)* Relation between normal force and fingertip movement at z-axis.

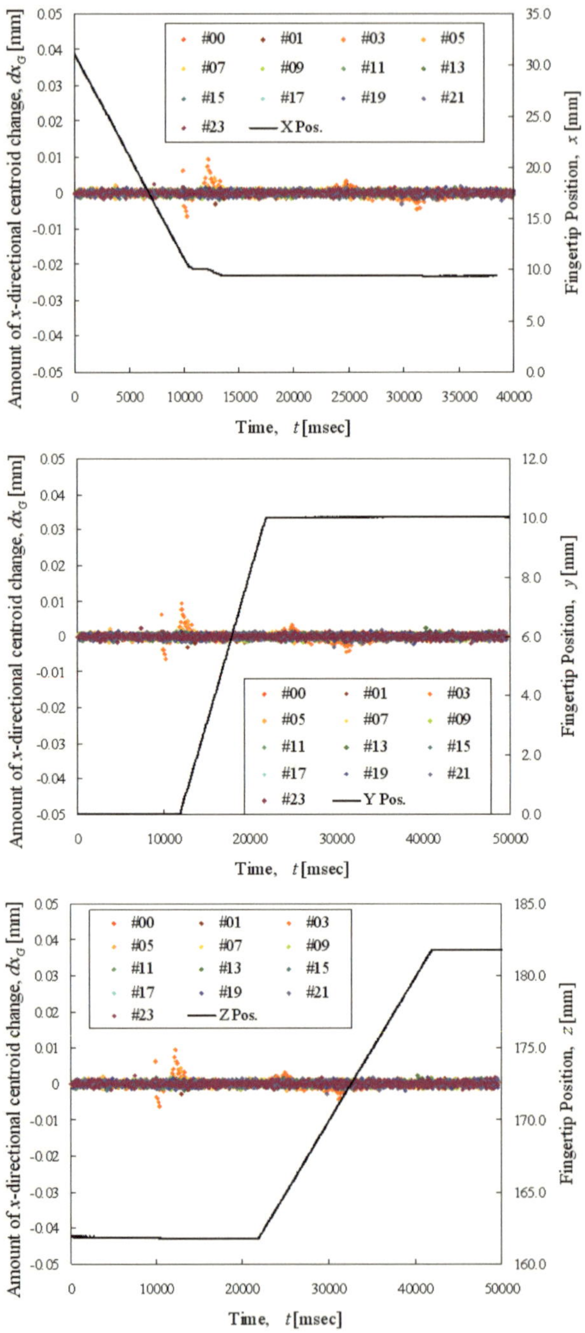

Fig. 14. Experiments of two robotic fingers manipulate wood block; data for left finger: *(Top)* Relation between amount of x-directional centroid change and fingertip movement at *x*-axis. *(Middle)* Relation between amount of x-directional centroid change and fingertip movement at *y*-axis. *(Bottom)* Relation between amount of x-directional centroid change and fingertip movement at *z*-axis.

8. Conclusion

In this research we developed the original hemispherical shaped optical three-axis tactile sensor system to mount on robotic fingers. The tactile sensor is capable of acquiring normal and shearing forces, which are the most important sensing elements in object manipulation tasks. This tactile sensor is designed in a hemispherical dome shape that consists of an array of sensing elements. This shape is to mimics the structure of human fingertips for easy compliance with various shapes of objects. This tactile sensor uses an optical waveguide transduction method and applies image processing techniques. Such a sensing principle is expected to provide better sensing accuracy to realize contact phenomena by acquiring the three axial directions of the forces, so that normal and shearing forces can be measured simultaneously.

In this chapter, we have presented force detection and measurement principles of the tactile sensor for normal and shearing forces. The normal force is calculated based on brightness changes of visual image taken by CCD camera. To define the applied force, we measure the integrated gray-scale value of the bright area. Meanwhile, shearing force is calculated by measuring centroid point displacement of the bright area retrieved by the CCD camera.

The optical three-axis tactile sensor developed in this research is designed in hemispherical shape, and the sensing elements are distributed in 41-sub region. Due to this structure, the acquired images by CCD camera, except for sensing element at the sensor tip area, are not the actual image of contact pressure at the sensing elements. Therefore, to compensate with the sensor structure, we conducted calculations to define coordinate transformation of the sensing elements on the sensor's hemispherical shape dome.

The optical three-axis tactile sensors are mounted on fingertips of two robotic fingers to perform object handling tasks. We have developed a control system consist of finger and sensor controllers. A control algorithm was designed in the robotic fingers system which applying normal and shearing forces obtained by the tactile sensors to control the movements of fingers during grasping tasks. The performance of this system was evaluated in object handling experiment using a wood block as an object. Experimental results shows that the robotic fingers managed to grasp the object within optimum grasp pressure and performed handling tasks without causing damage to the object or the sensor elements. The optical three-axis tactile sensor revealed good performance to use in robotic finger system.

In this chapter, we have shown that the robotic finger system equipped with the hemispherical shaped tactile sensors is suited for refining grips on object surface. Furthermore, the applied control algorithms are capable of preventing the probability of damage to the sensors and the object during robust grasping tasks. It is anticipated that using this novel optical three-axis tactile sensor in robotic grippers, with further improvement on hardware structure and image processing technique, will help advance the evolution of real-time object manipulation based on tactile sensing in robotic systems.

9. Acknowledgement

A part of this study was supported by fiscal 2006 Grant-in-Aid for Scientific Research in Exploratory Research from the Japan Ministry of Education, Culture, Sports, Science and Technology (Grant no. 18656079), and Grant-in-Research from the Japan Society for the Promotion of Science (JSPS) under Postdoctoral Fellowship for Foreign Researcher program 2008-2010.

8. References

Hanafiah, Y., Ohka, M., Kobayashi, H., Takata, J., Yamano M. & Nasu Y. (2006). Contribution to the development of contact interaction-based humanoid robot navigation system: Application of an optical three-axis tactile sensor, *3rd Int. Conf. on Autonomous Robot and Agents (ICARA06)*, pp. 63-68

Hanafiah Y., Ohka, M., Suzuki, H., Morisawa N. & Takata, J. (2007). Sensing performance of an optical three-axis tactile sensor system with application in multi-fingered humanoid robot arm, *World Congress on Engineering & Computer Science (WCECS07), Int. Conf. on Intelligent Automation and Robotics (ICIAR07)*, pp. 504-509

Hanafiah Y., Ohka, M., Takata, J., Nasu Y. & Yamano, M. (2008). Low force control scheme for object hardness distinction in robot manipulation based on tactile sensing, *IEEE International Conference on Robotics and Automation (ICRA2008)*, pp. 3443-3448

Kerpa, O., Weiss, K. & Worn, H. (2003). Development of a flexible tactile sensor system for a humanoid robot, *Intl. Conf. on Intelligent Robots and Systems IROS2003*, CDR

Kuniyoshi, Y., Ohmura, Y. & Terada, K. (2004). Embodied basis of invariant features in execution and perception of whole-body dynamic actions - knacks and focuses of roll-and-rise motion, *Journal Robotics and Autonomous Systems*, Vol. 48, pp. 189-201

Lee, M. H. & Nicholls, H. R. (1999). Tactile sensing for mechatronics - a state of the art survey, *Journal Mechatronics*, Vol. 9, pp. 1-31

Natale, L. & Torres-Jara, E. (2006). A sensitive approach to grasping, *Proceeding 6th International Conference on Epigenetic Robotics*, CDR

Nicholls, H. R. (1990). Tactile sensing using an optical transduction method, *Traditional and Non-Traditional Robot Sensors (Edited by T. C. Henderson)*, Springer-Verlag, pp. 83-99

Ohka, M., Mitsuya, Y. & Matsunaga Y. (2004). Sensing characteristics of an optical three-axis tactile sensor under combined loading, *Robotica*, vol.22, pp. 213-221

Ohka, M., Kobayashi, H. & Mitsuya, Y. (2006). Sensing precision of an optical three-axis tactile sensor for a robotic finger, *15th IEEE International Symposium on Robot and Human Interaction Communication (RO-MAN2006)*, pp. 220-225

Ohmura, Y., Kuniyoshi, Y. & Nagakubo, A. (2006). Conformable and scalable tactile sensor skin for curved surfaces, *Proceeding International Conference on Robotics and Automation (ICRA2006)*, pp. 1348-1353

Ohmura, Y. & Kuniyoshi, Y. (2007). Humanoid robot which can lift a 30kg box by whole body contact and tactile feedback, *2007 IEEE/RSJ International Conference on Intelligent Robots and Systems (IROS2007)*, pp. 1136-1141

Omata, S., Murayama, Y. & Constantinou, C.E. (2004). Real time robotic tactile sensor system for determination of the physical properties of biomaterials, *Journal of Sensors and Actuators A*, Vol. 112, pp. 278-285

Schmidt, P. A., Mael, E. & Wurtz, R. P. (2006). A sensor for dynamic tactile information with applications in human-robot interaction and object exploration, *Journal Robotics & Autonomous Systems*, Vol. 54, Issue 12, Dec. 2006, pp. 1005-1014

Takata, J. (2006). Object recognition and manipulation using a robotic hand equipped with optical three-axis tactile sensors, *Master Thesis*, Department of Electro-Mechanical Engineering, Graduate School of Engineering, Nagoya University, Japan

Umetani Y. & Yoshida K. (1989). Resolved Motion Rate Control of Space Manipulators with Generalized Jacobian Matrix, *IEEE Transactions on Robotics and Automation*, Vol. 5(3), pp. 303-314

Three Dimensional Capacitive Force Sensor for Tactile Applications

Jose Gerardo Rocha and Senentxu Lanceros-Mendez
University of Minho,
Portugal

1. Introduction

Tactile sensing is one of the most common ways of interacting with the environment. So common that it is most of the time unnoticed unless unpleasant sensations are involved (e.g. burning). The human being is completely covered, though the skin, by tactile sensors providing information about pressure and temperature variations translated into pleasant or unpleasant sensations, warnings messages and so on.

The tactile resolution of the human skin can be as precise as tenths of millimeters in the fingertips up to some centimeters in other parts of the body.

One interesting issue of the tactile sensing of the human skin is that it is not only responsive to normal forces, but also to shear, which is particularly important by graving delicate object and sensing roughness, among others. We refer two examples:

- People with serious motor limitations have, in most of the cases, a decrease of sensitivity in the areas of the body in contact with the supporting surfaces (e.g. beds or wheel chairs). Their limited motor capacity does not allow them to regularly change position autonomously, as a healthy person would do unconsciously. Consequently, insufficient sanguineous irrigation occurs as a result of pressure exceeding the tissue capillary pressure for a long time, depriving tissues of oxygen and essential nutrients, leading to ischemia and hypoxia, which then causes the development of pressure ulcers. These ulcers can be developed either due to a constant pressure exerted normally to the skin or due to a shear pressure also in the skin. One way to reduce the probability of pressure related ulcer formation is to monitor the three axes force that people with serious motor limitations exert on the supporting surfaces. In order to do that, an array of three axes force sensors must be placed along a bed sheet, for example.
- Gripping and manipulation of objects by robots require robust and reliable feedback of forces. For example, when a manipulator holds a fragile object, on the one hand, the force normal to the object must be as low as possible, and on the other hand, with a low normal force, the falling risk increases. One solution consists in measuring the shear forces and appropriately feedback them to electronics in order to adjust the necessary normal force. In order to do this a three axis soft flexible force sensor system seems to be more appropriate for delicate objects.

A great challenge in the development of artifacts such as robot fingers and portable terminals is to create tactile information processing systems that closely resemble those of humans, both in terms of flexibility, durability and versatility. Such artifacts will allow application areas for tactile sensors such as food processing, bio-medical, entertainment, future domestic and service industries and robotics in medicine for minimally invasive and micro-surgeries.

A major requirement of a tactile sensor is to have the ability of measuring the magnitude and location of an exerted force over the sensor. Further, well-designed tactile sensors must be thin and flexible, similarly to sensitive skin.

Further, capacitive sensors are used in a wide variety of measurement and control systems, such as liquid-level gauges, pressure meters, accelerometers, precision positioners, and as described in this chapter, as force sensors. In these applications, the capacitances to be measured often range from tenths to tens of picofarads. In particular, there are a large variety of applications for capacitive sensors acting as three axis force sensors.

This chapter is organized as follows: Section 2 describes different types of tactile sensors, giving examples of the ones using electromagnetic, piezoelectric, resistive and piezoresistive working principles. In section 3, a complete description of a three dimensional tactile sensor based on capacitive elements is performed. Section 4 describes some readout electronic techniques that could be applied to the sensor of section 3, namely switched-capacitor, capacitance to pulse-width, capacitance to duty-cycle and capacitance to frequency techniques. Section 5 is devoted to a complete description of a capacitance to voltage conversion technique that shows very good results when associated to the sensor of section 3. Finally, section 6 is dedicated to the conclusions.

2. Overview about tactile sensors

According to their working principle, the tactile sensors can be classified as electromagnetic, piezoelectric, resistive, piezoresistive, capacitive, etc. In the following subsections and in section 3, some examples of sensors based on some of these principles are described.

2.1 Electromagnetic tactile sensors

Among all types of tactile sensors using the electromagnetic principle, the fiber Bragg grating sensors (FBG) are the ones showing very promising results.

FBG sensors are fabricated by taking a given length of a single mode optical fiber, subjecting it to high pressure hydrogenation to increase its photosensivity, and then focusing an intense interference pattern of ultra violet light onto the core of the fiber using a laser and tipically a phase mask [1]. The grating consists on an axially periodically varying refractive index along the core of the fiber. The effect of the grating is to convert the fiber into a high wavelength-selective mirror. If broadband light is directed through the fiber, a narrow band of light will be reflected. Its center, or Bragg wavelength, is given by:

$$\lambda_B = 2n\Lambda. \quad (1)$$

where n is the effective refractive index of the fiber core mode and Λ is the period of the grating. When the fiber is subjected to an increase in strain or temperature, both n and Λ are affected and the Bragg wavelength increases [2].

Sensor systems involving such gratings typically work by injecting light from a spectrally broadband source into the fiber, with the result that the grating reflects a narrow spectral component at the Bragg wavelength or, in transmission mode, with this component missing from the observed spectrum. Fig. 1 illustrates this process [3].

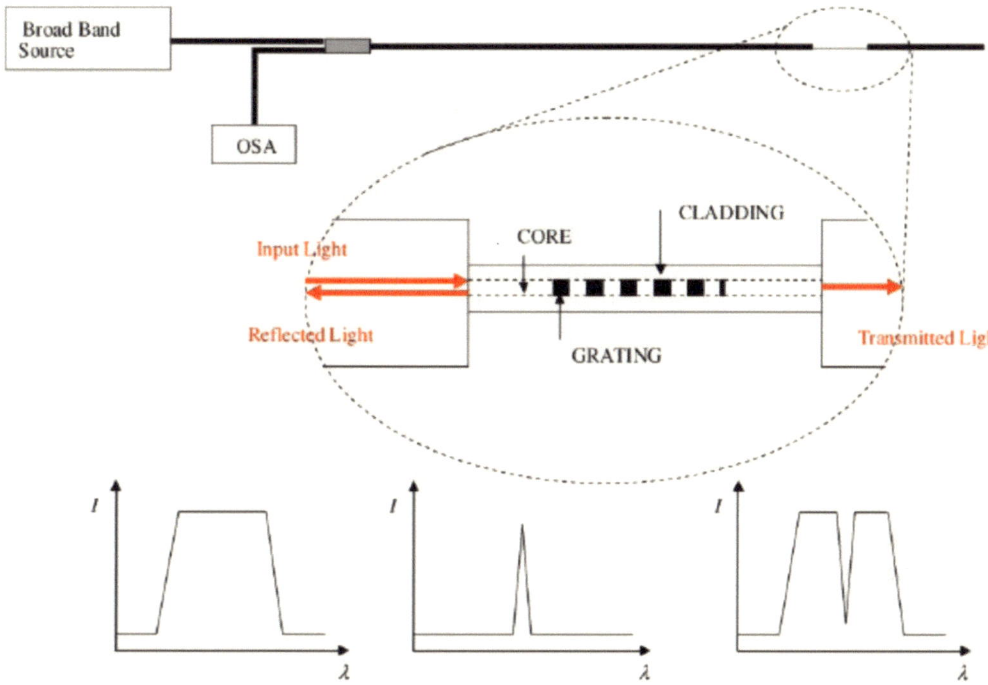

Fig. 1. Fiber Bragg Grating sensor encoding operation [3].

2.2 Piezoelectric tactile sensors

Piezoelectric tactile sensors are based on materials that show the piezoelectric effect, that is, when a pressure change is applied to them, it is possible to measure an electrical potential difference between their surfaces. As the potential difference is proportional to the variation of pressure instead to the pressure itself, some techniques are necessary to build piezoelectric tactile sensors. Tanaka et. al., presented a sensor system based on a PVDF piezoelectric polymer film and a soft rubber [4]. The surface of the sensor is heated through a temperature controller mechanism. The sensor is attached on the tip of a robot finger and the base of the finger is mounted on a linear slider. Using this sensor system, the active sensing was designed resembling human motions for tactile perception in mind. Two kinds of active sensing were introduced to obtain the information on feelings of vibration and warmth, as follows:

- For obtaining the information on warmth, the sensor is moved and contacted with the object through the trajectory control of the contact pressure.
- For obtaining the information on feelings of vibration, the sensor is slid over the object at a constant speed.

The geometry of the sensor is shown in Fig. 2. The sensor is attached on the tip of a robot finger. The base of the sensor is composed by an acrylic block and a vulcanized rubber. Around the base, a copper tape overlaid with Nichrome wire, a PVDF film, and a protective layer of acetate film are stacked in sequence. A thermistor is attached on the side of the sensor. The Nichrome wire and the thermistor are used for heating the sensor and measuring the sensor's temperature, respectively. The robot finger is composed by a piezoelectric bimorph strip and an aluminum beam. The piezoelectric bimorph strip is used to press the sensor on an object. A strain gauge is attached on the base of the aluminum beam for the measurement of the contact force applied by the finger.

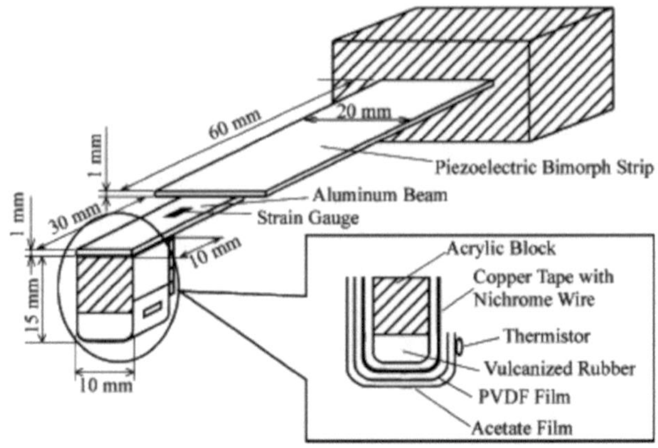

Fig. 2. Geometry of a piezoelectric tactile sensor. The sensor is attached on the tip of the robot finger driven by the piezoelectric bimorph strip. The sensory receptor is the PVDF film[4].

Fig. 3 shows the schematic illustration of the sensing process.

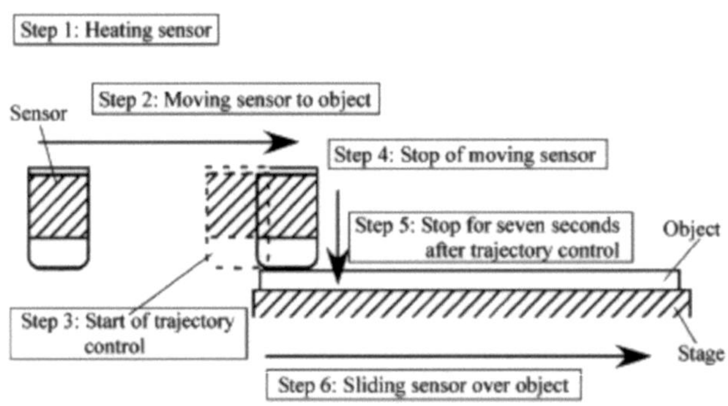

Fig. 3. Schematic illustration of sensing process. Two kinds of active sensing are introduced. First, the sensor is heated. After the sensor's temperature is constant, the experiment through contact is performed (steps 2to 5) in order to obtain the information on warmth. Then, the experiment through slide is performed (step 6) in order to obtain the information on feelings of vibration [4].

First, the sensor is heated until its temperature is constant. Then, the experiment through contact is performed (steps 2 to 5) in order to obtain the information on warmth. Then, the experiment through slide is performed (step 6) in order to obtain the information on feelings of vibration

Another technique that employs piezoelectric elements for tactile sensing is the one based on resonant vibrating sensors [5, 6]. The principle of the piezoelectric vibration-type sensor can be explained with the help of Fig. 4 [6]. This sensor has two electrodes: one for driving and the other for measuring. The driving electrode is supplied with an alternating current, and the sensor resonates due to the inverse piezoelectric effect. Due to this, a voltage is generated by the piezoelectric effect in the measuring electrode. An external force applied to the sensor causes a change in the mechanical impedance between the piezoelectric elements. This change can be measured at the measuring electrode.

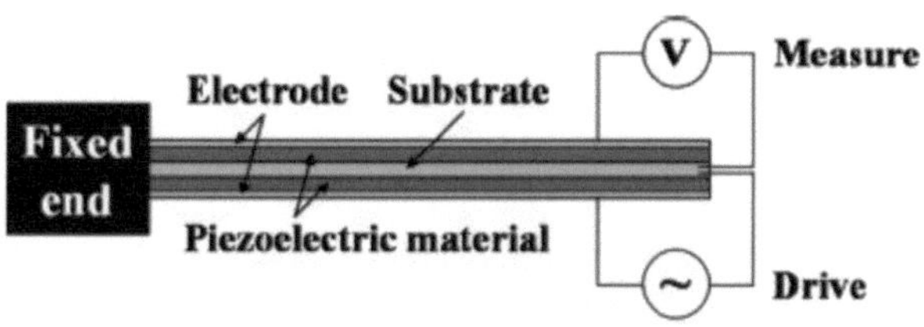

Fig. 4. Piezoelectric vibration-type sensor. The system is composed by a substrate, piezoelectric materials, and electrodes [6].

2.3 Resistive tactile sensors

Resistive tactile sensors usually use strain gauge as sensing elements. Hwang et. al. describe an approach that uses the deformation of a polymer substrate where the strain gauge elements are fabricated. Fig. 5 shows a schematic view of the proposed tactile sensor. The strain sensitive elements, i.e., the strain gauges, were embedded into a ductile polymer substrate. A thin film of metal is used for fabricating the strain gauges. Both the polymer and the thin film allow the sensor to be flexible. A bump structure is placed on top of the sensor surface to facilitate load distribution. When an external force is applied to the device, the thin film structure and the polymer suffers a deformation that induces strain and therefore changes in the electrical resistance of the strain gauge [7].

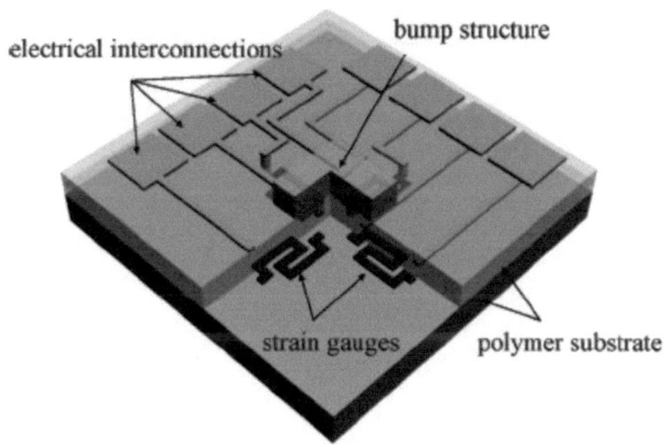

Fig. 5. Schematic view of the proposed tactile sensor. The strain-sensitive elements, i.e., the strain gauges, are embedded in a ductile polymer substrate that is used to measure the strains in the polymer substrate. A bump structure, used for load distribution, is placed on top of the sensor surface [7].

The normal and shear load sensing principles of the proposed sensor are shown in Fig. 6. Strain gauges are embedded at the center of a ductile polymer substrate. When a normal load is applied on the surface of the sensor, the substrate deforms. This deformation induces equal strains on both strain gauges, as shown in Fig. 6(a). When a shear load is applied, as shown in Fig. 6(b), one strain gauge experiences tensile strains, while the other experiences compressive strains. This difference results in different measurable voltage drops across each strain gauge, which allows the shear load to be determined. An unknown load can be found by superposition of these two cases. From these sensing principles, the tactile-sensor unit cell consists of four strain gauges for shear load detection in the x- and y-directions.

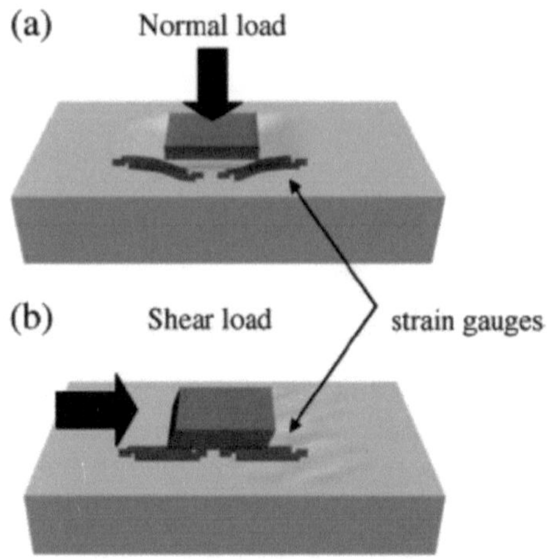

Fig. 6. Normal and shear load sensing principles. (a) In the case of a normal load, both strain gauges are subjected to tensile stress. (b) In the case of a shear load, one strain gauge is subjected to tensile stress, and the other is subjected to compressive stress[7].

2.4 Piezoresistive tactile sensors

The piezoresistive tactile sensors are based on piezoresistive elements, that is, elements based on materials that show changes in their electrical resistance proportional to the applied pressure. Contrarily to the piezoelectric elements, these ones can act as tactile sensors even in the presence of static pressures. The material most effectively used to build piezoresistive elements is silicon. As this material is also the basis of the microelectronic circuits, it is a common procedure to place the sensors as well as the readout electronics into the same device [8].

Balke et. al. [9] presented a sensor system consisting on a cantilever structure with integrated probing tip with variable heights and shapes. The signal of the deflection is processed by a piezoresistive Wheatstone bridge configuration into an electrical signal.

The sensor is fabricated in a standard silicon micromachining process (Fig. 7). The process started with {100}-n-silicon wafer with a thickness of 350 μm. The wafer was scratched and broken into samples of 26 x 26 mm^2. The basic steps are lithography with a positive S1818-photoresist, oxidation of the samples at 1100°C, wet chemical etching in aquaeous potassium hydroxide and tetramethylammonium hydroxide solutions and boron diffusion. Some of the standard processes can be modified or substituted by other processes.

There are however piezoresistive tactile sensors that do not use silicon. Wisitsoraat et. al. present a tactile sensor using indium tin oxide (ITO) as piezoresistive element [10]. The MEMS tactile sensor structure (Fig. 8) consists of a multi-layer AlN/Al/Cr square shuttle plate (200 μm× 200 μm) with four arms on which four piezoresistors of ITO are placed. In addition, contacts of each piezoresistor are coated with Cr/Au. When an external force is applied, the piezoresistors changes electrical resistance due to induced strain. The gold layer is used as the electrode and bonding pad of the piezoresistor. The AlN layer acts as an insulator providing isolation between the Cr/Al membrane and the piezoresistor. The Cr layer is used as an adhesive film and the thick Al layer is used to increase the flexibility and robustness of the membrane structure.

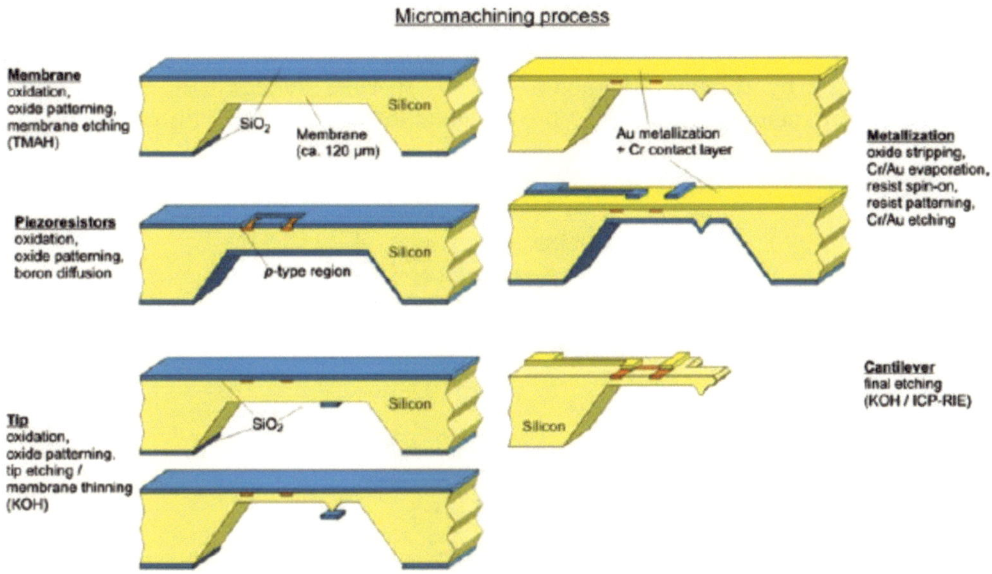

Fig. 7. Representation of the five main stages of a standard silicon micromachining process for the fabrication of the piezoresistive tactile sensor [9].

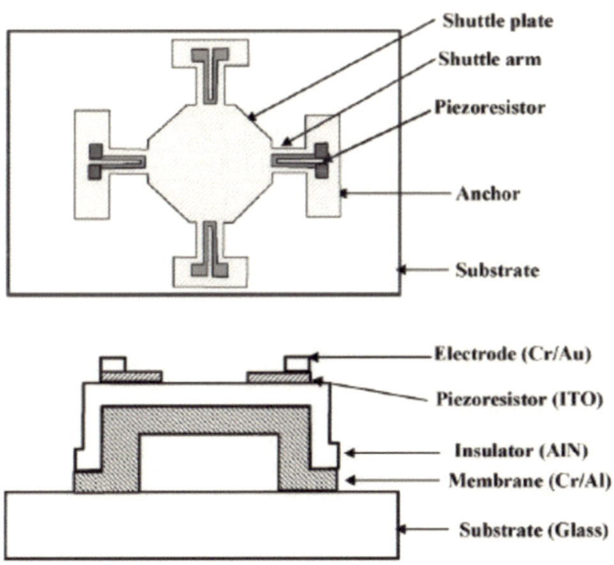

Fig. 8. Structure of a ITO-based piezoresistive tactile sensor [10].

3. Theoretical description of a three dimensional capacitive force sensor

Among the capacitive tactile sensors, it is possible to find recent articles describing epoxy-based [11] or foam-based [12] ones. In the following a general a three dimensional, 3D, capacitive tactile sensor based on a flexible dielectrics will be described in detail.

Fig. 9 shows a schematic diagram of the sensor. It consists on a flexible dielectric material such as rubber or foam in which a square of conductive material (such as aluminum) is placed at its top and four squares of conductive material are placed at its bottom. The thickness of the dielectric material is t, the dimension of the conductive square side is D and the distance between the two conductive squares of the bottom is d. The conductive material can be deposited by different methods, such as thermal evaporation, physical vapor depositions, etc, depending on the dielectric used and the dimensions of the capacitors.

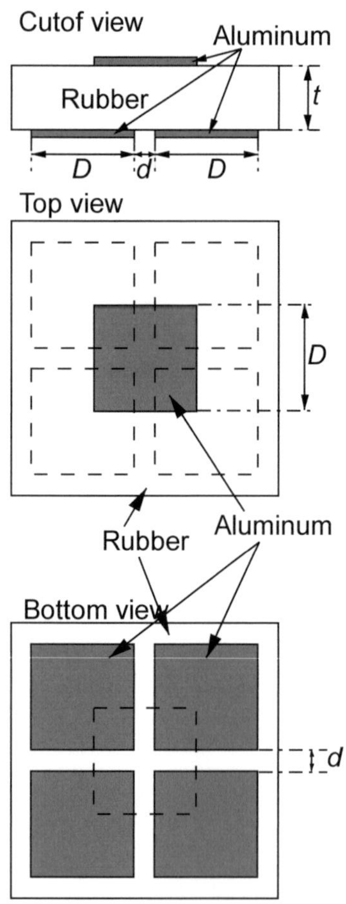

Fig. 9. Schematic diagram of the force sensor consisting on a flexible dielectric material (e.g. rubber) in which a square of aluminum is placed at its top and four squares of aluminum are placed at the bottom.

The arrangement forms four capacitors, C_1, C_2, C_3 and C_4, as shown in Fig. 10. In the following sections, the x, y and z displacement values are calculated for forces applied along the xx, yy and zz axes either independently or simultaneously.

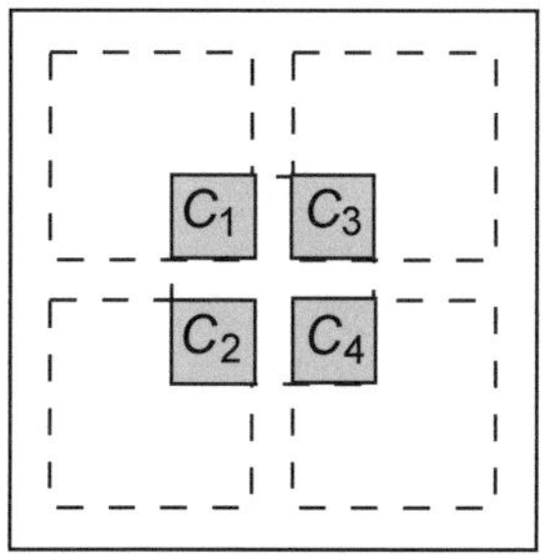

Fig. 10. Representation of the four capacitors formed by the conductive material arrangement.

3.1 Sensor with no applied pressure

At the steady state and with no applied pressure, the capacitance of the capacitors is given by:

$$C_1 = C_2 = C_3 = C_4 = \frac{\varepsilon_r A}{t}, \tag{2}$$

where ε_r is the dielectric constant of the flexible dielectric material and A is the area of the parallel plate capacitor, given by:

$$A = \left(\frac{D-d}{2}\right)^2. \tag{3}$$

Therefore, the capacitances are given by:

$$C_1 = C_2 = C_3 = C_4 = \frac{\varepsilon_r (D-d)^2}{4t}, \tag{4}$$

3.2 Sensor with a pressure applied along the zz axis

The sensor with a force applied in the *zz* axis is represented in Fig. 11.

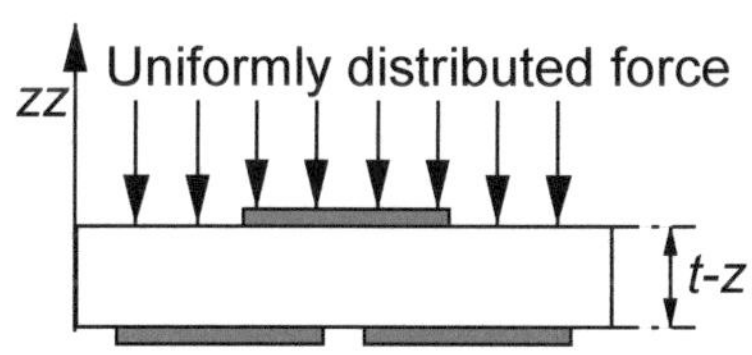

Fig. 11. Sensor with a force applied along the zz axis.

In this case, the values of the capacitances are given by:

$$C_1 = C_2 = C_3 = C_4 = \frac{\varepsilon_r (D-d)^2}{4(t-z)}, \tag{5}$$

where z is the displacement along the *zz* axis. The minus sign in front of z is due to the fact that the applied pressure will decrease the thickness of the dielectric. Therefore, the displacement, z, along the *zz* axis can be calculated by solving equation 5 in order to z:

$$z = -\frac{(d-D)^2 \varepsilon_r}{4C_1} + t = -\frac{(d-D)^2 \varepsilon_r}{4C_2} + t = -\frac{(d-D)^2 \varepsilon_r}{4C_3} + t = -\frac{(d-D)^2 \varepsilon_r}{4C_4} + t \tag{6}$$

Equation 6 shows four different forms to calculate the z displacement value. The final value is obtained through a simple average procedure, reducing the resultant error.

3.3 Sensor with a pressure applied along the xx axis or the yy axis

The force sensor with a pressure applied along the xx axis is shown in Fig. 12.

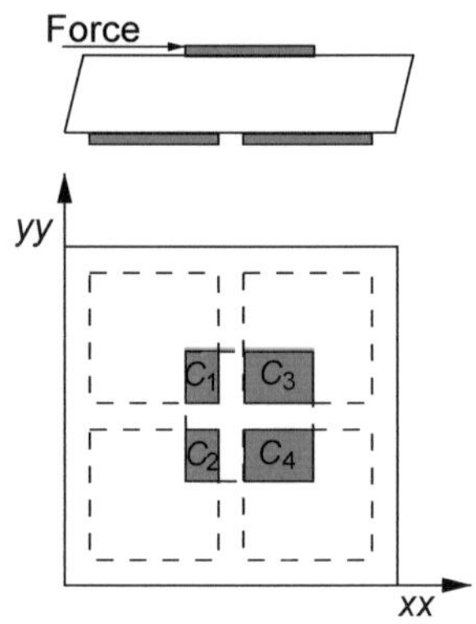

Fig. 4. Force sensor with a pressure applied along the xx axis.

In this case, the values of the capacitances are given by:

$$C_1 = C_2 = \frac{\varepsilon_r (d-D)(d-D+2x)}{4t}$$
$$C_3 = C_4 = \frac{\varepsilon_r (d-D)(d-D-2x)}{4t} \tag{7}$$

where x is the displacement of the conductive square of the top along the xx axis. Solving equation 7 in order to x, it is obtained:

$$x = \frac{1}{2}\left(-d + D + \frac{4C_1 t}{\varepsilon_r (d-D)}\right) = \frac{1}{2}\left(-d + D + \frac{4C_2 t}{\varepsilon_r (d-D)}\right) = \frac{1}{2}\left(-d - D - \frac{4C_3 t}{\varepsilon_r (d-D)}\right) = \frac{1}{2}\left(-d - D - \frac{4C_4 t}{\varepsilon_r (d-D)}\right) \tag{8}$$

In a similar way, the values of the capacitances expressing a displacement y along the yy axis are given by:

$$\begin{aligned} C_1 = C_3 &= \frac{\varepsilon_r (d-D)(d-D+2y)}{4t} \\ C_2 = C_4 &= \frac{\varepsilon_r (d-D)(d-D-2y)}{4t} \end{aligned} \tag{9}$$

In this case, the value of y can be obtained from:

$$\begin{aligned} y &= \frac{1}{2}\left(d - D - \frac{4C_1 t}{\varepsilon_r (d-D)}\right) \\ &= \frac{1}{2}\left(-d + D + \frac{4C_2 t}{\varepsilon_r (d-D)}\right) \\ &= \frac{1}{2}\left(d - D - \frac{4C_3 t}{\varepsilon_r (d-D)}\right) \\ &= \frac{1}{2}\left(-d + D + \frac{4C_4 t}{\varepsilon_r (d-D)}\right) \end{aligned} \tag{10}$$

Both in the x and y displacements, the thickness of the dielectric, t, decreases, so in precision applications, this change must be taken into account. The calculation can be performed as follows: considering Fig. 13, the horizontal displacement hd is given by:

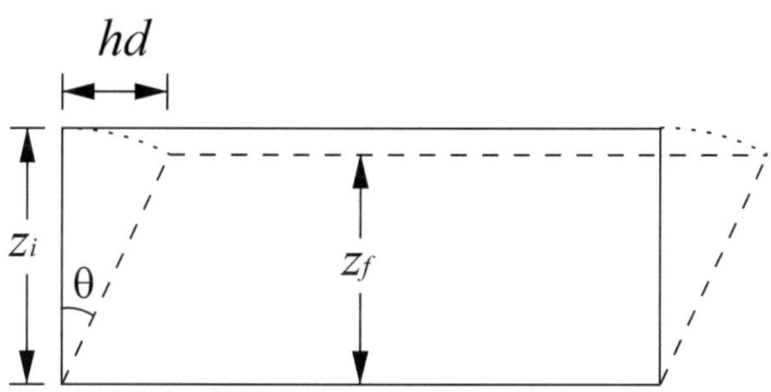

Fig. 13. Decrease of the thickness of the dielectric caused by forces applied horizontally.

$$hd = \sqrt{x^2 + y^2} \tag{11}$$

The angle θ can be calculated from the horizontal displacement:

$$\theta = \arcsin \frac{hd}{z_i} \tag{12}$$

where z_i is the initial value of z. Finally, the final value of z (z_f) is given by:

$$z_f = z_i \cos\theta \tag{13}$$

In the more general case of forces applied along the 3 axes a specific formalism, described below (section 3.4), must be taken into account.

3.4 Sensor with a pressure applied along the three axes

In the case of pressures applied along the three axes, the conductive material on the top of the sensor will be moved along the *xx*, *yy* and *zz* axes, and the values of the capacitances are:

$$
\begin{aligned}
C_1 &= \frac{\varepsilon_r (d-D+2x)(d-D-2y)}{4(t-z)} \\
C_2 &= \frac{\varepsilon_r (d-D+2x)(d-D+2y)}{4(t-z)} \\
C_3 &= \frac{\varepsilon_r (d-D-2x)(d-D-2y)}{4(t-z)} \\
C_4 &= \frac{\varepsilon_r (d-D-2x)(d-D+2y)}{4t}
\end{aligned} \qquad (14)
$$

In this case, four equations and three unknowns are obtained, meaning that there are several solutions for the unknowns. The values of the *x*, *y* and *z* displacements are:

$$
\begin{aligned}
x &= \frac{(C_1 - C_3)(d-D)}{2(C_1 + C_3)} \\
&= \frac{(C_2 - C_4)(d-D)}{2(C_2 + C_4)}
\end{aligned} \qquad (15)
$$

$$
\begin{aligned}
y &= \frac{(C_1 - C_2)(d-D)}{2(C_1 + C_2)} \\
&= \frac{(C_3 - C_4)(d-D)}{2(C_3 + C_4)}
\end{aligned} \qquad (16)
$$

$$
\begin{aligned}
z &= -\frac{\varepsilon_r C_1 (d-D)^2}{(C_1 + C_2)(C_1 + C_3)} + t \\
&= -\frac{\varepsilon_r C_2 (d-D)^2}{(C_1 + C_2)(C_2 + C_4)} + t \\
&= -\frac{\varepsilon_r C_3 (d-D)^2}{(C_1 + C_3)(C_3 + C_4)} + t \\
&= -\frac{\varepsilon_r C_4 (d-D)^2}{(C_2 + C_4)(C_3 + C_4)} + t
\end{aligned} \qquad (17)
$$

In this case, there are two different forms to calculate the *x* and *y* displacements and four different forms to calculate the *z* value. The final value of each displacement can be obtained through a simple average procedure, reducing the resultant errors, as described for the previous cases.

Finally, the applied force along the three axes can be calculated by multiplying the displacement by the elasticity constant of the dielectric medium.

4. Overview about readout electronic circuits for capacitive sensors

Regarding readout circuits for capacitive sensors, there are several configurations that can be used and are found in literature, such as:

- switched-capacitor capacitance to voltage converter [13, 14],
- capacitance to pulse-width modulator [1, 16],
- capacitance to duty-cycle converter [17],
- capacitance to frequency converter [18] and complete signal processing units [19].

4.1 Switched-capacitor capacitance to voltage converter

The first circuit for capacitance measurements described here is a switched-capacitor capacitance to voltage converter, which can be applied to differential capacitive sensors. The functional block schematic of the proposed signal conditioner is shown in Fig. 14.

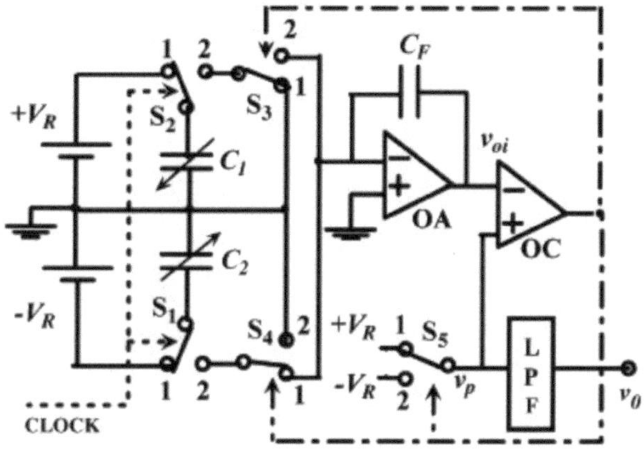

Fig. 14. Block diagram of the switched-capacitor signal-conditioning circuit [14].

Switches S_1, S_2, S_3, S_4, and S_5 are all single-pole double-throw analog switches. Switches S_1 and S_2 are controlled by a high-frequency clock having 50% duty-cycle and switch the capacitances C_1 and C_2 of the sensor between the charge and discharge positions. Switches S_3 and S_4 are connected such that only one of the capacitors (C_1 or C_2) charges the feedback capacitance C_F at a given time. The circuit is essentially a relaxation oscillator with the operational amplifier (OA) working as a switched capacitor integrator, while switch S_5, in conjunction with comparator OC, provides the necessary hysteresis and positive feedback required for oscillation [14].

Let us assume that initially, the integrator output is positive and increasing. The comparator output is high, and hence, switches S_3, S_4, and S_5 will be at position 1. S_1 and S_2 will toggle between positions 1 and 2 for every clock cycle. When S_1 and S_2 are in position 1, the C_1 voltage will be $+V_R$, and the C_2 voltage will be $-V_R$. When the clock becomes low, switches S_1 and S_2 will go to position 2, and the charge in the capacitor C_2 will be transferred to the integrator capacitor C_F, while the charge in C1 is discharged to ground. Hence, the charge in the capacitor C_F will increase for every clock cycle, and hence, the voltage across C_F will also increase with a positive polarity on the terminal of C_F that is connected to the output of the OA. The output of the integrator will now ramp in the positive direction with a step of $V_R C_2/C_F$ for every clock cycle. This condition prevails until the integrator voltage v_{oi} reaches $+V_R$. As soon as $v_{oi} = +V_R$, the comparator output will toggle, and switches S_3, S_4, and S_5 will be changed to position 2. Now, during every clock cycle, the charge in the capacitor C_1 will be transferred to the integrator capacitor C_F, and the charge in C_2 is discharged to ground. Since C_1 is charged from $+V_R$, the charge transferred to the feedback capacitor C_F will now be in the opposite direction compared to the earlier case, and hence, the output voltage of the integrator will ramp down with a step size of $V_R C_1/C_F$ for every clock. The integrator voltage v_{oi} will ramp down until it reaches $-V_R$. As soon as v_{oi} becomes $-V_R$, the comparator output will toggle again, and switches S_3, S_4 and S_5 will go back to position 1. Now, the circuit will operate as explained initially, thus leading to sustained oscillation.

4.2 Capacitance to pulse-width modulator

The capacitance-to-pulse-width modulator (PWM) circuit is schematically shown in Fig. 15, where, for the sake of simplicity, the sensor C_x is represented without parasitic elements, while C_R and C_I are constant capacitances [16]. The circuit includes three Miller integrators based on the operational transconductance amplifiers (OTAs) INT1, INT2 and INT3, and an OTA with two in-phase matched output ports, indicated by OTA2 in Fig. 15.

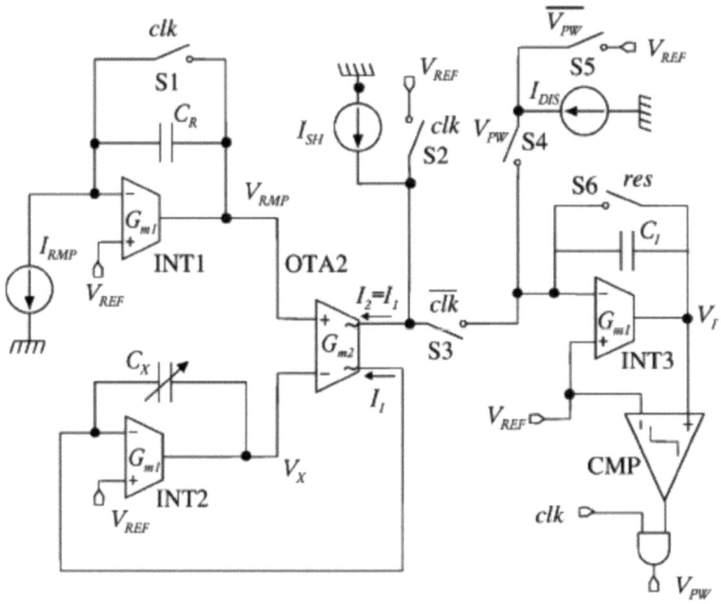

Fig. 15. Simplified schematic view of the capacitance to pulse duration converter [16].

The control signals of switches S1 to S6 are indicated by the labels in the figure. V_{PW} is the output pulse-duration-modulated signal, synchronous to the clock, which, in turn, is a 50% duty-cycle square wave with a period of T_{ck}. In normal operation the res signal is high, so that S_6 is open. INT1 integrates the current during the clock lower level, producing a ramp, and it is discharged to the reference voltage during the clock higher level. V_{REF} works as the zero level for all the three integrators. The integrator INT2 is closed in a feedback loop through OTA2 forcing INT2 output voltage, V_X, to track the INT1 output V_{RMP}. For this reason, the increments of V_X and V_{RMP} across the whole clock lower level are the same, that is,

$$\frac{1}{C_X} = \int_0^{T_{CK}/2} I_1 dt = \frac{1}{C_R} = \int_0^{T_{CK}/2} I_{RMP} dt = \frac{T_{CK} I_{RMP}}{2C_R} \tag{18}$$

Notice that, due to the AND gate, the voltage is forced to be low during the clock lower level, regardless of the comparator (CMP) output. Therefore, the integrator INT3 integrates the difference between the current I_2, which is a replica of I_1, and the constant current I_{SH}

(integration phase). The INT3 initial output voltage (V_I) must be equal to V_{REF}. Setting I_{SH} smaller than the maximum value of I_2, V_I gets higher than V_{REF} during the integration phase and the CMP output turns high. As the clock switches to the high level, INT3 is disconnected from OTA2 by S3, V_{PW} turns high and, consequently, C_I starts to be discharged (de-integration phase) by the constant current I_{DIS}, connected to INT3 by S4. This phase ends when V_{PW} is set low by CMP, i.e., when the INT3 output voltage crosses again the starting voltage V_{REF}. This occurs when the whole charge stored during the integration phase is released. Since INT3 output voltage is held at V_{REF} until the clock is raised again, the value V_{REF} is actually the starting point of the next integration phases, as supposed in the discussion above. The output pulse duration coincides with the duration of the de-integration phase:

$$\int_0^{T_{CK}/2} \left(I_2 - I_{SH}\right) dt = \int_0^{T_{PW}} I_{DIS} dt = T_{PW} I_{DIS} \tag{19}$$

Combing (18) and (19) and considering that I_{DIS} is a constant current, we get the expression of the pulse duration:

$$T_{PW} = \frac{T_{Ck}}{2} k_1 \left(\frac{C_X - k_2 C_R}{C_R} \right) \tag{20}$$

where $k_1 = I_{RMP}/I_{DIS}$ and $k_2 = I_{SH}/I_{RMP}$.

4.3 Capacitance to duty-cycle converter

A capacitance to duty-cycle (CDC) conversion circuit is shown in Fig. 16.

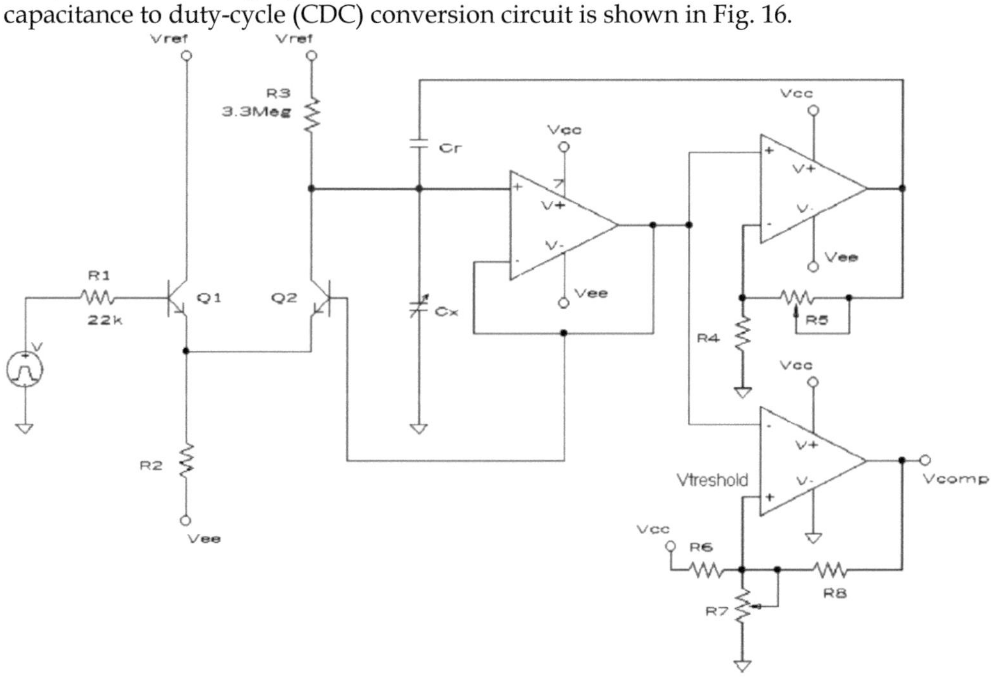

Fig. 16. Capacitance to Duty-Cycle conversion circuit[17].

The CDC converter consists of an RC circuit, a switch, a voltage follower, a feedback amplifier and an output comparator. From the response of an RC circuit, it is well known that the time T at which the voltage across the capacitor reaches a given value $V_{threshhold}$ depends linearly on the capacitance

$$T = RC \ln \frac{V_{CC}}{V_{CC} - V_{threshold}} \tag{21}$$

To achieve as large a bandwidth as possible, feedback has been incorporated in the circuit, as shown in Fig. 17.

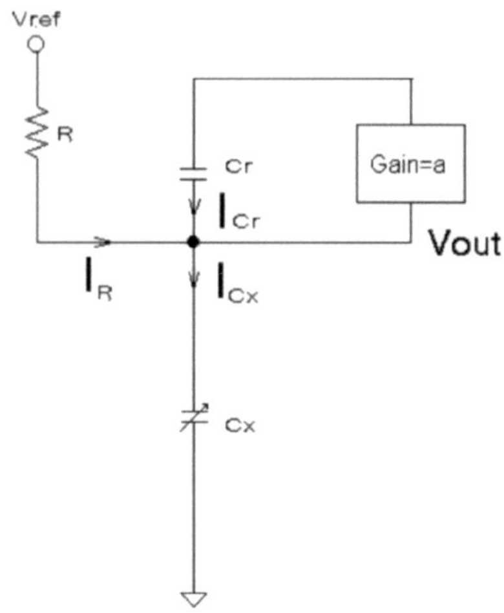

Fig. 17. Feedback loop in the CDC converter.

The feedback path consists on an amplifier and a reference capacitance C_r. For the circuit in Fig. 17, we have:

$$I_R + I_{Cr} = I_{Cx} \tag{22}$$

$$\frac{V_{ref} - V_{out}}{R} + C_r \frac{d}{dt}\left(aV_{out} - V_{out}\right) = C_x \frac{dV_{out}}{dt} \tag{23}$$

From equation 23, it can be shown that:

$$T = R\left[C_x - (a-1)C_r\right] \ln \frac{V_{ref}}{V_{ref} - V_{threshold}} \tag{24}$$

Without feedback, the time T at which the voltage reaches $V_{threshold}$ depends solely on C_x. In this case, the minimum value of CDC (T_{min}) occurs when $C_x=C_{xmin}$, and the maximum value of T (T_{max}) occurs when $C_x=C_{xmax}$. In many capacitive sensing elements, the difference between the maximum and minimum values of C_x is a small fraction of its nominal value (C_{xnom}). Hence, without feedback, the nominal value T_{nom}, would be much larger than the dynamic range of T. Furthermore, the time duration for one CDC conversion must be at least T_{min}; therefore, without feedback, the conversion time will have a redundant baseline T_{nom} value resulting in reduced bandwidth of the CDC. Employing the feedback path shown, the conversion time depends linearly on the difference between the variable capacitance C_x and the reference capacitance C_r. This significantly decreases the fixed component of the rise time, and leads to a significant increase of the bandwidth.

4.4 Capacitance to frequency converter

The schematic circuit of the first-order relaxation oscillator is shown in Fig. 18, which is the core of the capacitance to frequency converter [18].

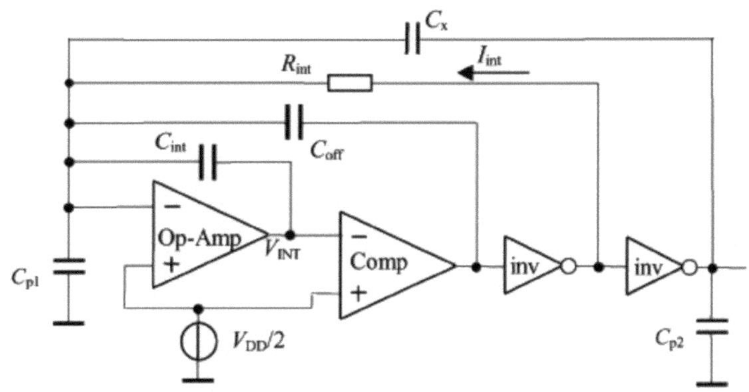

Fig. 18. First-order relaxation oscillator [18].

The oscillator is implemented with an operational amplifier (op-amp), a comparator (comp), two digital inverters, the capacitances C_{off} and C_{int}, and a controlled current source I_{int}, whose value depends on the resistor R_{int}. The capacitor C_x represents the capacitance of the sensor to be measured and C_{p1} and C_{p2} model the parasitic capacitances to ground due to, for example, the connecting cables. When the voltage at the output of the op-amp crosses the threshold level $V_{DD}/2$, the output of the comparator and the inverters switch, and the charge is transferred from capacitors C_x and C_{off} to C_{int}, generating a step voltage at the op-amp output. Next, the current I_{int} removes the charge stored at C_{int} until the op-amp output reaches the threshold level again. The period (T) of the oscillator output signal is given by:

$$T = 4R_{\text{int}}\left(C_{off} + C_x\right). \tag{24}$$

5. Readout electronic circuit for the three dimensional capacitive sensor

The electronic interfaces whose output signals are period, frequency or duty-cycle modulated are the most attractive as their output can be read directly by a digital circuit such as a microcontroller. The main drawback is that the readout speed usually is slow and that, for some architectures, it is capacitance-dependent.

In order to measure the impedance with precision, some care with the parasitics must be taken into account both at the project and at the connection levels. Specifically, the effects of the series and shunt impedances of the wires that connect the sensing elements to the readout electronics have to be taken into account. There are, however, some techniques that reduce these effects.

A good technique to reduce the wire effects consist in the two port measurement. Fig. 19 (a) shows the case of unknown impedance Z_x with series parasitic impedances Z_1 to Z_4. In this case, the impedances Z_1 to Z_4 do not affect the measurement of the Z_x value: Z_1 and Z_2 are in series with a current source, therefore the current in Z_x is not influenced by them; Z_3 and Z_4 are in series with a voltmeter, therefore their current (and their voltage drop) will be null. Fig. 19 (b) shows the case of the shunt parasitic impedances. The impedances Z_1 and Z_2 do not influence the measurement of Z_x: Z_1 is in parallel with the voltage source V, therefore its current does not influence the measured value; Z_2 is in parallel with the ammeter (which has very low impedance), therefore its current will be null. In a real situation, both the non infinite impedance of the current source and of the voltmeter, for the circuit of Fig. 19 (a), and the non null impedance of the voltage source and of the ammeter, for the circuit of Fig. 19 (b) influence the final result.

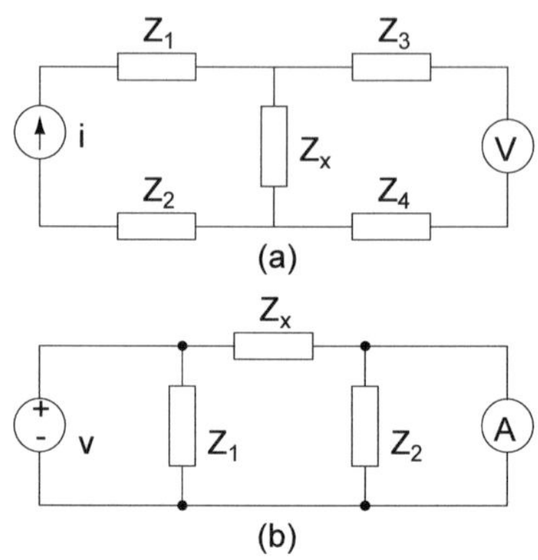

Fig. 19. Two port measurement techniques: (a) with series parasitic impedances; (b) with shunt parasitic impedances.

Based on the two port measurement technique, it is possible to obtain the capacitance value of small capacitors (C_x of Fig. 20) in a precise way, even in the presence of the parasitic capacitances of the wires (C_{P1} and C_{P2}) several orders of magnitude superior than C_x [20, 21].

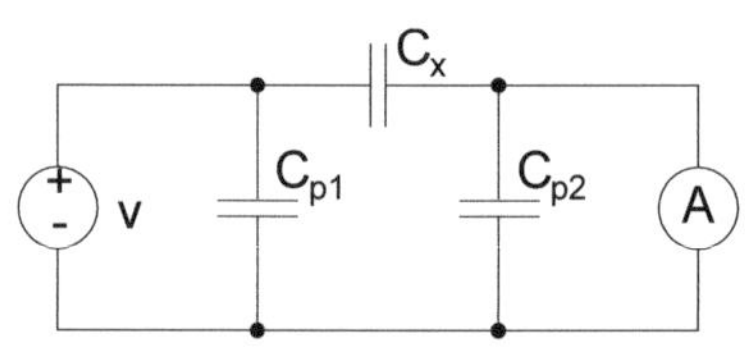

Fig. 20. Two port network for capacitance reading.

In this circuit, the current i measured by the ammeter is given by:

$$i = v\omega C \quad (25)$$

where v is the voltage of the source V and $\omega = 2\pi f$. The parasitic capacitances C_{P1} and C_{P2} do not influence the measurement: C_{P1} is in parallel with the voltage source, therefore its current is not measured by the ammeter; C_{P2} is in parallel with the ammeter, i. e., a very low impedance device, therefore its current is close to zero. Notice that the current i measured by the ammeter is directly proportional to the capacitance C of the capacitor, as it is indicated by equation 25. The proportionality constant is $v\omega$.

Fig. 21 shows a block diagram of the capacitance to voltage converter circuit. The voltage source is implemented by an oscillator and the ammeter is implemented by a very low input impedance current to voltage converter. In order to output a voltage signal proportional to the capacitance, the current to voltage converter must be followed by a full wave rectifier and a low pass filter.

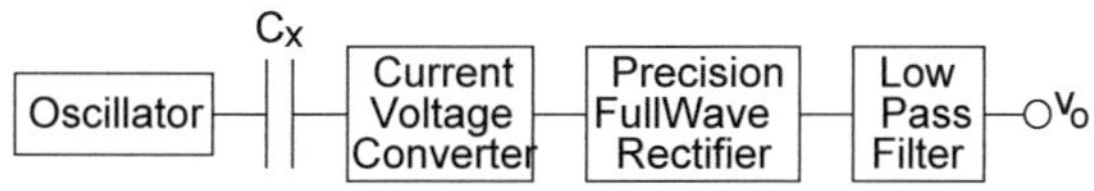

Fig. 21. Block diagram of the circuit.

Fig. 22 shows the schematic diagram of the circuit where the functional blocks of Fig. 21 are implemented by simple configurations.

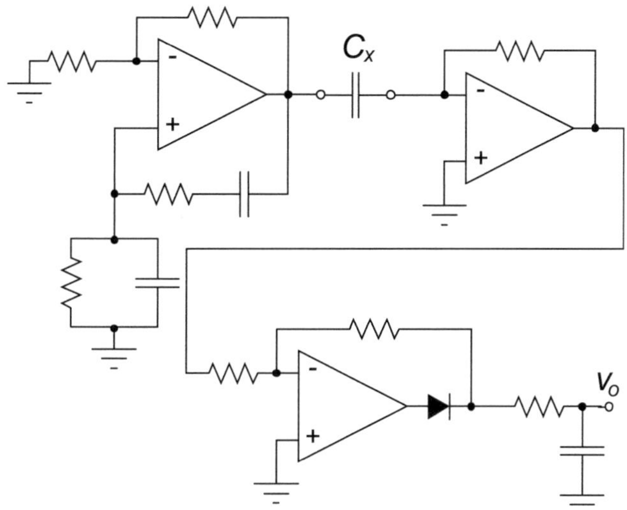

Fig. 22. Schematic diagram of the readout circuit.

6. Conclusions

A three dimensional capacitive force sensor for tactile applications has been presented. In order to better understand its capabilities, the different technologies developed in order to achieve tactile sensors have been reviewed. In this way, electromagnetic, piezoelectric, resistive and piezoresistive based tactile sensors have been discussed. Finally, the three dimensional capacitive force sensor has been presented. In a first stage, the theoretical description of the sensor has been developed through the simulation of the sensor response to forces applied along different directions. Finally, the readout electronics for capacitive sensors in general and for the three dimensional capacitive sensor has been shown.

As a conclusion, capacitive sensors allow an easy and low-cost fabrication of a flexible and precise three-dimensional force sensor suitable for tactile applications.

7. References

I. Bennion, J. A. R. Williams, L. Ahang, K. Sugden, and N. J. Doran; "UV-written in-fibre Bragg gratings," Opt. Quantum Electron., Vol. 28, 1996, pp. 93-135.

B. M. Cowie, D. J. Webb, B. Tam, P. Slack and P. N. Brett, "Fibre Bragg grating sensors for distributive tactile sensing," Meas. Sci. Technol., vol. 18, 2007, pp. 138-146.

J. S. Heo, J. H. Chung, J. J. Lee, "Tactile sensor arrays using fiber Bragg grating sensors," Sensors and Actuators A, vol. 126, 2006, pp. 312-327.

Y. Tanaka, M. Tanakaand S. Chonan, "Development of a sensor system for collecting tactile information," Microsyst Technol, vol. 13, 2007, pp. 1005-1013.

T. Hemsel, R. Stroop, D. Oliva Uribe, J. Wallaschek, "Resonant vibrating sensors for tactile tissue differentiation," Journal of Sound and Vibration, vol. 308, 2007, pp. 441-446.

K. Motoo, F. Arai and T. Fukuda, "Piezoelectric Vibration-Type Tactile Sensor Using Elasticity and Viscosity Change of Structure" IEEE Sensors Journal, Vol. 7, 2007, pp.1044-1051.

E. S. Hwang, J. Seo and Y. J. Kim, "A Polymer-Based Flexible Tactile Sensor for Both Normal and Shear Load Detections and Its Application for Robotics," Journal of Microelectromechanical Systems, Vol. 16, 2007, pp. 556-563.

M. Ádám, T. Mohácsy, P. Jónás, C. Dücso, E. Vázsonyi, I. Bársony, "CMOS integrated tactile sensor array by porous Si bulk micromachining," Sensors and Actuators A 142, 2008, 192-195.

M. Balke, E. Peiner and L. Doering, "A new micromachined sensor system for tactile measurements of high aspect ratio microstructures," Microsyst Technol Vol. 14, 2008, pp. 543-549.

A. Wisitsoraat, V. Patthanasetakul, T. Lomas, A. Tuantranont, "Low cost thin film based piezoresistive MEMS tactile sensor," Sensors and Actuators A 139, 2007, 17-22.

R. Matsuzaki and A. Todoroki; "Wireless flexible capacitive sensor based on ultra-flexible epoxy resin for strain measurement of automobile tires," Sensors and Actuators A 140 (2007) 32–42.

C. Metzgera, E. Fleisch, J. Meyer, M. Dansachmüller, I. Graz, M. Kaltenbrunner, C. Keplinger, R. Schwödiauer and Siegfried Bauerb; "Flexible-foam-based capacitive sensor arrays for object detection at low cost," Applied Physics Letters 92, 013506 2008.

W. Bracke, P. Merken, R. Puers, C. Van Hoof; "Design methods and algorithms for configurable capacitive sensor interfaces," Sensors and Actuators A 125 (2005) 25–33.

B. George and V. J. Kumar, "Switched Capacitor Signal Conditioning for Differential Capacitive Sensors," IEEE Transactions on Instrumentation and Measurement, Vol. 56, No. 3, 2007 pp. 913-917.

P. Bruschi, D. Navarrini, G. Barillaro and A. Gola; "A Precise Capacitance-to-Pulse Width Converter for Integrated Sensors," Analog Integrated Circuits and Signal Processing, 42, 2005, 185–189.

P. Bruschi, N. Nizza, and M. Piotto, "A Current-Mode, Dual Slope, Integrated Capacitance-to-Pulse Duration Converter," IEEE J. of Solid-State Circuits, Vol. 42, 2007, pp. 1884-1891.

Z. Ignjatovic, and M. F. Bocko; "An Interface Circuit for Measuring Capacitance Changes Based Upon Capacitance-to-Duty Cycle (CDC) Converter," IEEE Sensors Journal, Vol. 5, No. 3, June 2005, 403-410.

M. Gasulla, X. Li and G. C. M. Meijer; "The Noise Performance of a High-Speed Capacitive-Sensor Interface Based on a Relaxation Oscillator and a Fast Counter," IEEE Transactions on Instrumentation and Measurement, Vol. 54, No. 5, October 2005, 1934-1940.

A. S. Hou and S. X. Su; "Design of a capacitive-sensor signal processing system with high accuracy and short conversion time," Sensors and Actuators A 119 (2005) 113–119.

A. Srivastava, "Digital CMOS interface circuit for current and capacitance sensing," Proceedings of SPIE, Vol. 5389, 2004, pp. 1-12.

F. M. van der Goes and G. C. Meijer, "A universal transducer interface for capacitive and resistive sensor elements," Analog Integrated Circuits and Signal Processing, 14, 1997, pp. 249-260.

10

Study on Dynamic Characteristics of Six-axis Wrist Force/torque Sensor

Ke-Jun Xu
Institute of Automation, Hefei University of Technology, Hefei 230009, PRC

1. Introduction

It is necessary to detect complex force and torque components of three-dimensional space in research and manufacture, therefore the multi-axis force and torque sensors are developed rapidly. For examples, the six-axis wrist force/torque sensor for robot assembling and automated polishing [1,2], the multi-component strain gauge balance for wind tunnel testing to determine loads on models [3], the multi-axis measuring force platform for biomedical research and development, and so on [4]. Now the dynamic performances of multi-axis force and torque sensors are demanded strictly because these sensors are used in robot operation, dynamic testing and process control. However, the dynamic performances of multi-axis force sensor are not satisfied with the needs of applications. First, due to the small damped ratio and low natural frequency of sensor, the dynamic response of sensor is slow, and the time to reach steady state is long. Secondly, there are dynamic couples among various directions of multi-axis force and torque sensor because the elastic body of sensor is an integer structure and the interaction of various directions can not be avoided completely. Thirdly, the multi-axis force sensor possesses the nonlinear dynamic characteristics under some conditions, which will affect the measurement accuracy. It is necessary to study and solve these key problems. Therefore, this chapter focuses on the six-axis wrist force sensor, and introduces the dynamic calibration experiments, dynamic modeling, dynamic compensation and dynamic decoupling. Finally, it presents some researches on the nonlinear dynamic characteristics of the six-axis wrist force sensor.

2. Dynamic calibration experiments

Generally speaking, there are three kinds of dynamic calibration methods, that is, the frequency response method, the impact response method and the step response method. As for the wrist force sensor, it is difficult to generate sine wave to make an excitation because the natural frequency of some channels may be larger than 800 Hz and there are six channels. Therefore we carried out the impulse response experiment and step response experiment.

In the impulse response experimental, the impact force was generated by a hammer. A piezoelectric sensor was installed in the hammer to reflect the impulse force. When the impact force was applied to the wrist force sensor, both the eight-channel output of the wrist force sensor and the piezoelectric sensor output was collected by a data acquisition equipment at the same time. It is difficult to control the striking point and angle of the hammer and the strength of the force in the experiments, and it is impossible to perform the torque calibration.

In order to test the dynamic performance of the wrist force sensor accurately and strictly, we carried out the dynamic calibration experiments based on the step response method. The wrist force sensor was mounted horizontally in the center of a dynamic calibration test bench designed by the Institute of Intelligent Machines, Chinese Academy of Sciences. A round steel plate was fixed on the sensor through screws. In the center of the plate a bar was installed vertically, and in the top end of the bar there was a hook, which could be used to hang weights through a string so as to apply the force and torque. At four sides of the plate, four short bars were also mounted symmetrically, and there was a groove in the top end of each short bar. At four sides of the bench four supports were fixed, and fixed pulleys were mounted on the supports. A string was winded through the fixed pulleys of the supports, the hook of the bar and the grooves of the short bars. A weight hung on one end of the string, and the other end of the string was connected with a one-dimensional force sensor. When the string was cut with scissors fast, the weight suddenly fell to the ground, which generated a negative step force or torque for the wrist force sensor. This force or torque was measured and transformed into electrical signals by the one-dimensional force sensor. Fig. 2-1 (a), (b) and (c) showed the calibrations in different directions.

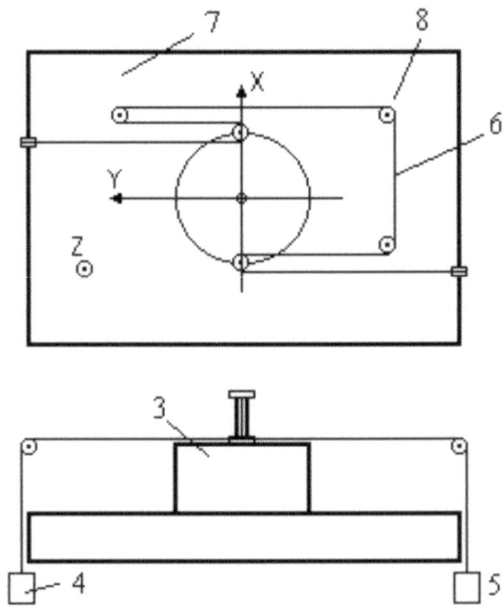

(a) Channel M_z

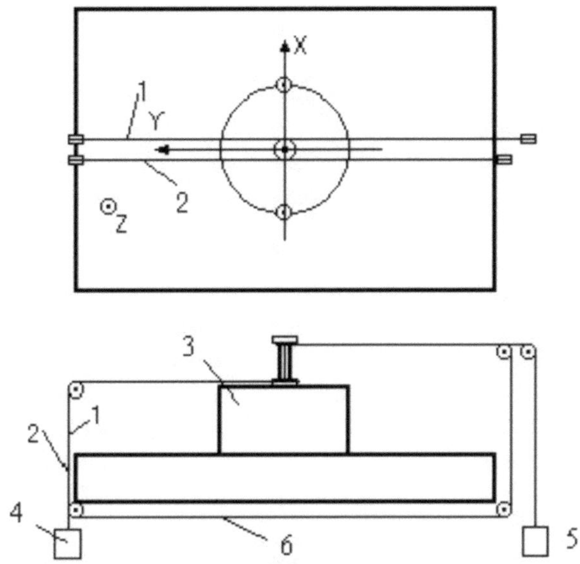

(b) Channel M_x

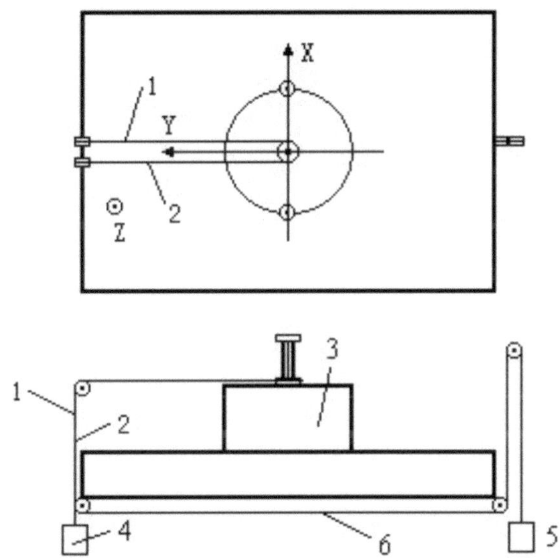

(c) Channel F_y

Fig. 2-1 Calibration schematic diagram (1) and (2) strip, (3) wrist force sensor, (4) one-dimensional force sensor, (5) weight, (6) cutting place, (7) calibration test table, (8) pulleys.

The elastic body of the wrist force sensor was a floating cross - beam, and shown in Fig. 2-2. Thirty-two pieces of strain gauges were adhered to the beam, forming eight electrical bridges. The outputs of the bridges were amplified and sent into the DSP-based real time dynamic compensation system developed by us. The signals of eight channels, which were acquired and decoupled by this system, are transformed into the signals of six channels, expressing F_x, F_y, F_z, M_x, M_y and M_z that were three force components and three torque components in the three dimensional space. At the same time, the system also conducted the dynamic compensation for the output signals of six channels so as to enhance the dynamic response of the wrist force sensor. The programs of real time data acquisition, decoupling and compensation are stored in an EPROM of the system. They could also be downloaded via the serial communication port by a PC. In the dynamic calibration, the exciting signal measured by a one-dimensional force sensor and the outputs of six channels of the wrist force sensor were collected by a MR30C recorder. At the same time, the exciting signal and the outputs of main channels of the wrist force sensor were captured by a COM7101A digital oscilloscope, and then stored in HP PC via a GP-IB bus.

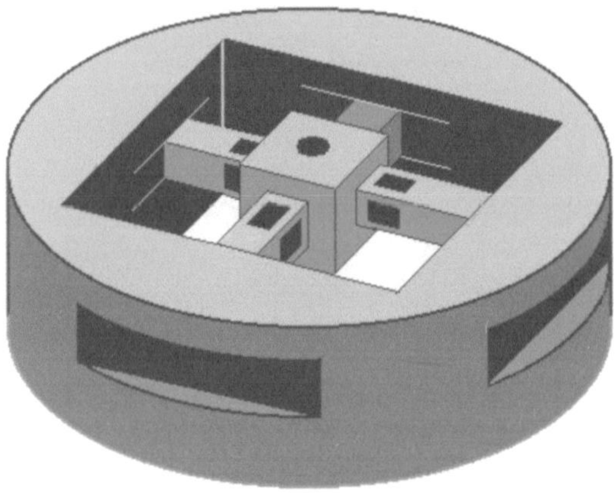

Fig. 2-2 Structure of elastic body

The calibration procedure was as follows.

1. Adjusting the zero point of the amplifiers of the wrist force sensor.
2. Applying the force or torque to the wrist force sensor through hanging weights, and measuring the static output voltages of eight channels of the wrist force sensor and one of six channels of the dynamic correcting system.
3. Cutting the string on which the weights were hung, and generating the step form excitation force or torque. Then measuring and recording the dynamic exciting signal and response signals of the wrist force sensor. The work of measuring and recording was done three times for each channel of the wrist force sensor.

3. Dynamic modeling

In order to describe accurately the dynamic characteristics of the wrist force sensor, its dynamic models are built with various methods, such as the system identification, the Walsh transformation, the time series analysis and the artificial neural network according to the step response experimental data.

3.1 Preprocessing

Before modeling, the experimental data must be pretreated.

1) Construction of exciting signal

In the dynamic calibration experiments, not suitable one-dimensional force sensor (that is to say, the sensor has no ideal frequency characteristic) could be used to test the practical step-form exciting signal, i.e. the release process of the weight. Although the string was cut off fast, the weight suddenly released, which was equal to the negative step force theoretically. In the actual situation, however, the string was elastic and goes through the pulleys, so the string was subjected to the frictional force and its speed of release slowed down. Therefore, the step signal was not an ideal step wave. If the step wave was regarded as the exciting signal, it was not satisfied with the actual conditions, and the results of modeling were poor. So we have to construct the exciting signal according to the step response because the step response of the wrist force sensor is caused by the exciting signal.

a. It is supposed that the sensor is a two-order system.
b. The starting points of both the exciting signal and step response are the same.
c. In many cases, the amplitude of the exciting signal is equal to the steady status value of the step response.
d. The rising part of the exciting signal is constructed so as to make the difference between the step response and the exciting signal be an attenuated oscillation curve.

2) Data filtered

In the measurement and transforming processes of dynamic signals, the experimental data may be polluted by noise. In order to reduce the influence of noise, a digital low pass filter is designed with the cut-off frequencies being adjusted from $200\,Hz$ to $1000\,Hz$, and the step responses are filtered.

3.2 Modeling method

The model parameters of the wrist force sensor are estimated by the least squares method [7]. As the oscillation attenuation of the sensor's dynamic response is slow, when the model response is compared with the actual response, it is we found out that the fitting precision is not high. At this time, the model parameters in the discrete domain are transformed into the continuous domain, and the natural frequency ω_n (or f_n) and damp ratio ξ are solved out. Then ξ is modified and transformed back to the discrete domain. The model response is obtained once again, and the modeling accuracy is improved.

3.3 Modeling results

The modeling results of six channels of a wrist force sensor (No. 1) are shown in Fig. 3-1 (a)~(f). In the figures, curve 1 expresses the actual response, and curve 2 expresses the model output. The performance indexes in the frequency domain of one sensor are shown in Table 3-1. In the dynamic experiments, we hung the weight with different kinds of strings, such as nylon string, badminton racket string and tennis racket string, and the strings were cut off at different positions, such as the center of above or under calibration test bench.

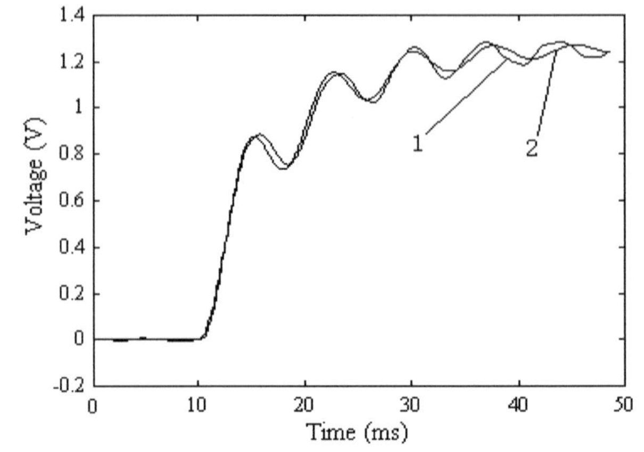

(a) Channel F_x

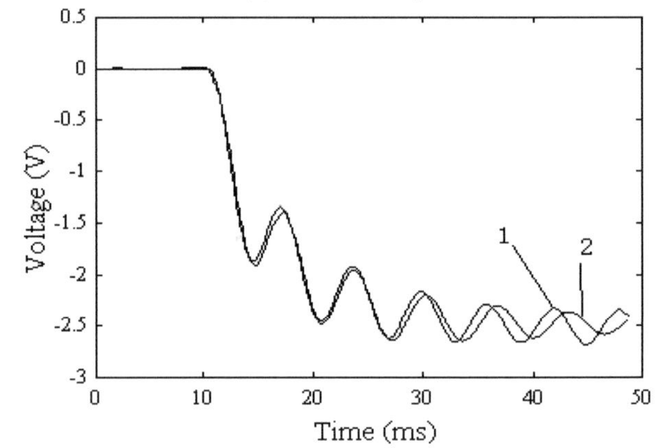

(b) Channel F_y

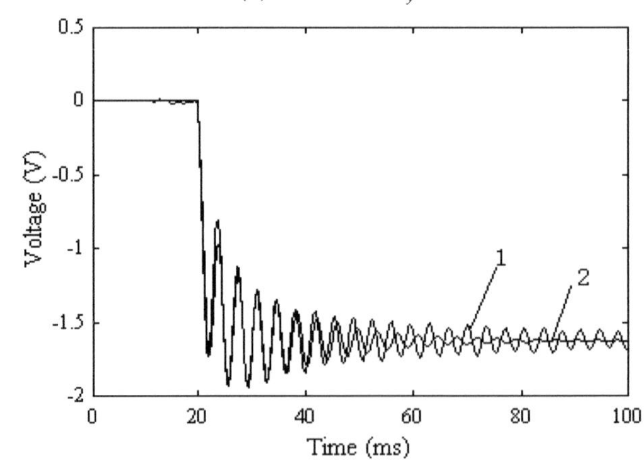

(c) Channel F_z

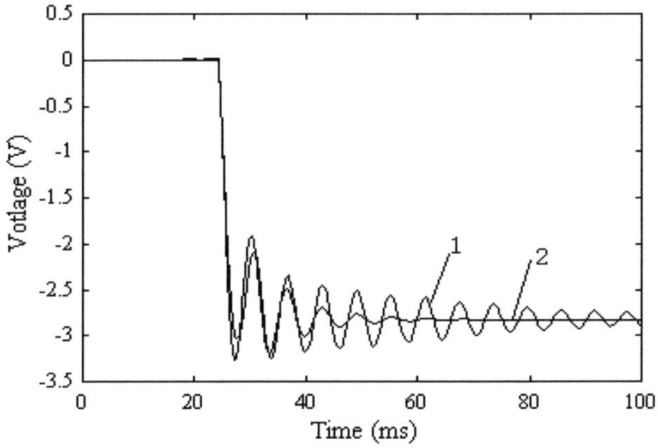

(d) Channel M_x

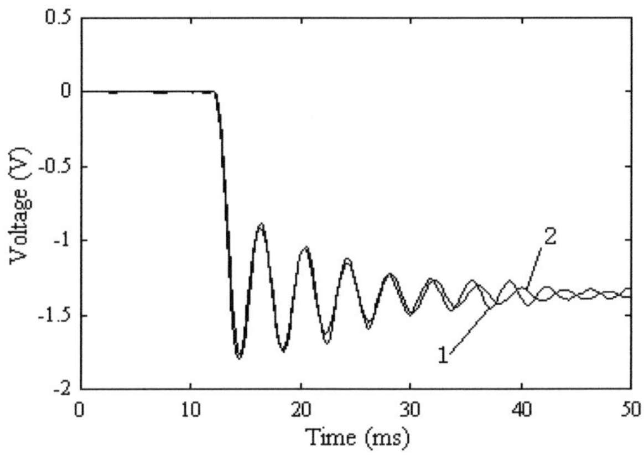

(e) Channel M_y

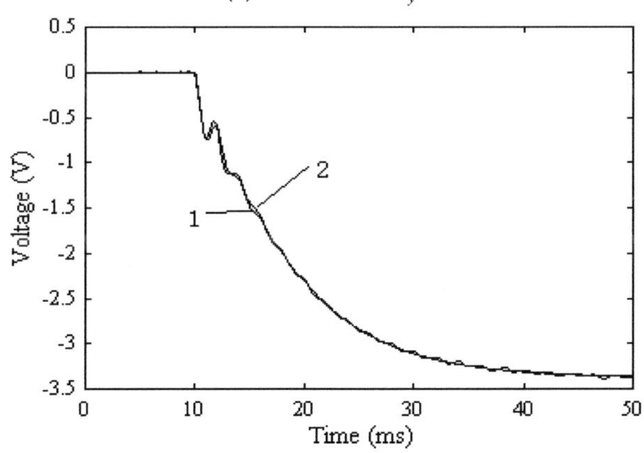

(f) Channel M_z

Fig. 3-1 Modeling results (1) Actual response, (2) model response.

	ξ	f_n	Remarks
F_x	0.105127	130.095841	Cutting on side
F_y	0.059806	150.582778	Cutting on side
F_z	0.0251005	274.554810	Cutting in middle
	0.0392930	266.282844	Cutting on side
M_x	0.162782	148.175903	Cutting on side
M_y	0.066125	127.218132	Cutting on side
M_z	0.267158	524.732768	Cutting in middle
	0.107061	483.605486	Cutting on side

Table 3-1 Performance indexes of sensor in frequency domain

4. Dynamic compensation

The experimental results showed that the step response time of the wrist force sensor varied from 20 ms to 100 ms (error being within ± 10 %), which was not satisfying with some applications, so the dynamic compensation methods and system should be studied and developed to speed the sensor dynamic responses. Firstly, the dynamic compensation system should be designed using the system identification method, pole-zero configuration or artificial neural network. Secondly, the dynamic compensation system should be implemented with DSP chip, which includes six compensating devices.

4.1 Design principle of dynamic compensation device

A designing method of the dynamic compensation device was proposed based on FLANN (Function link artificial neural network) by us. FLANN has the excellent ability to approximate many functions [8]. Patra et al. utilized FLANN to estimate and correct the static nonlinearly of pressure sensor, but FLANN has not been used for studying the dynamic characteristics of sensors at that time [9,10]. We constructed the structure of the neural network, made it be of time delay and present the dynamic characteristic. It is supposed that $u(k)$ is the input signal of sensor, i.e. the measured signal, that $y(k)$ is the output signal, and that $u'(k)$ is the output signal of the dynamic compensating device attached to the sensor. In order to shorten the time of dynamic response and make the measured signal reappear accurately, $u'(k)$should be approximate to $u(k)$ as closely as possible. The dynamic inverse model of the sensor is established using the method of FLANN off-line, and the inverse model is used as the model of the dynamic compensation device. In designing, both $u(k)$ and $y(k)$ are used as the input of the network, and $u(k)$ is regarded as the reference output signal of the compensation device. A schematic diagram of training is shown in Fig. 4-1. Assuming m=2 and n=3, where m refers to the delay step number of $u(k)$ and n refers to the delay step number of $y(k)$, so as to perform real time processing and make the realization be easy, the training equation of the network are written as

$$u'(k) = \begin{aligned} &W_0(k)y(k) + W_1(k)y(k-1) + W_2(k)y(k-2) + W_3(k)u(k-1) \\ &+W_4(k)u(k-2) \end{aligned} \tag{4.1}$$

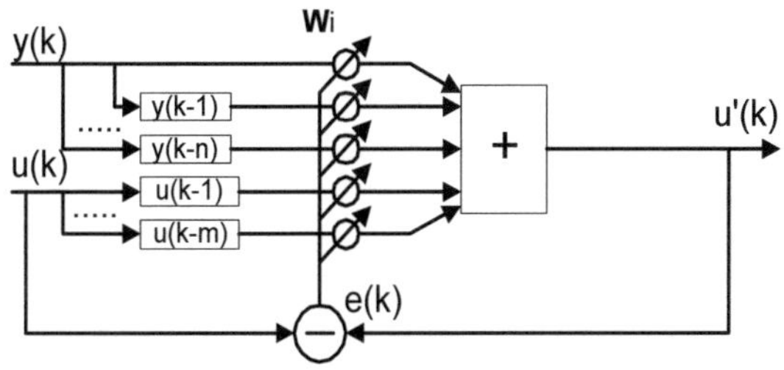

Fig. 4-1 Design schematic of dynamic compensating device by FLANN

Where W_i are the weights of the network, i.e. the model coefficients of the dynamic compensation device, k is the point number of data.
The error is

$$e(k) = u(k) - u^{'}(k) \tag{4.2}$$

The weight updating equations are given by

$$W_n(k+1) = W_n(k) + \alpha e(k) y(k-n) \quad (n=0,1,2) \tag{4.3}$$

$$W_{m+2}(k+1) = W_{m+2}(k) + \alpha e(k) u(k-m-2) \quad (m=1,2) \tag{4.4}$$

where the learning constant α governs the stability and the rate of convergence. If the value of α is too small, the speed of convergence is slow. If the value of α is too large, the result may diverge. Generally speaking, the value of α varies from 0 to 1. The simulation results show that it is suitable to set α about 0.1 for our case. After training of many times, when the average mean square error attains a minimum value, the obtained weights are the coefficients of the compensation device.
At the beginning of the on-line compensation, we suppose $u^{'}(k) = y(k)$, where k =0,1,2, and $u(k)$ is replaced by the output feedback $u^{'}(k)$. The equation of dynamic compensating is

$$u^{'}(k) = W_0 y(k) + W_1 y(k-1) + W_2 y(k-2) + W_3 u^{'}(k-1) + W_4 u^{'}(k-2) \; (k \geq 3) \tag{4.5}$$

It should be noted that the designing equations mentioned above are used for one channel of the wrist force sensor, and the equations for other channels are on the analogy of the above equations.

4.2 Design procedure of dynamic compensation device

1. An ideal equivalent measurement system including the sensor and the dynamic compensation device is constructed by adjustment the damp ratio and natural frequency.
2. The exciting signal (constructed or practical) is inputted, and the dynamic response of the equivalent measurement system is obtained.
3. Based on the dynamic responses of both the wrist force sensor and the equivalent measurement system, a dynamic compensation device is designed.

4. The dynamic response of the wrist force sensor is corrected.
5. In the light of the effects of compensation, the dynamic compensation devices are improved until the requirement is satisfied.

4.3 Dynamic compensation system

1. Realization of the dynamic compensation system

This dynamic compensation system consists of six dynamic compensating devices for six directions of the wrist force sensor, and the data acquisition, decoupling, dynamic compensating and output can be performed with the system.

Fig. 4-2 shows the hardware block of the dynamic compensation system. This system mainly includes an ADSP-2181 EZ-KIT Lite, an analog input part, an output part and the logic control circuit. The analog input part consists of eight sampling and holding circuits (S/H), a multiplexer (MUX), an amplifier (AMP) and an analog-to-digital converter (A/D). The output part contains six digital to analog converters (D/A) and six RC filters. The logic control circuit mainly consists of a decoder. The ADSP-2181 EZ-KIT Lite board is a minimal implementation system of an ADSP-2181 processor designed by ADI Corporation, and mainly includes an ADSP-2181, an EPROM and a serial communication port. The outputs of eight channels of the wrist force sensor are connected to the inputs of eight S/Hs. Whether the sampling mode is switched to the holding mode is controlled by ADSP-2181 based on the sampling frequency. Eight channel signals are switched and connected sequentially by the MUX, amplified by the AMP, and sent to the A/D. A busy pin of the A/D is connected to a programmable input/output pin. The ADSP-2181 determines the reading time according to the status of the busy pin.

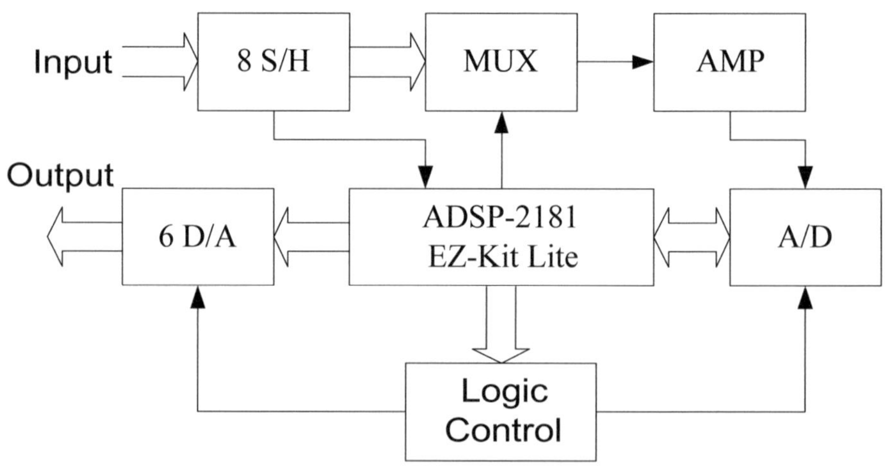

Fig. 4-2 Schematic block digram of dynamic compensating system

After eight channel signals are acquired by the ADSP-2181 at the same time, they are decoupled statically to be six channel signals, i.e. F_x, F_y, F_z, M_x, M_y and M_z. Then the six channel signals are compensated dynamically, and output by six D/As. Under the program control of the ADSP-2181, the logic control circuit determines the chip selection of A/D and D/A.

The software design of the system includes a data acquisition module, a data processing module and a result output module. The sampling interval of the system is determined by the interrupt of the timer, and is 50 μs or 250 μs so as to sample enough data in the sensor's dynamic response process. When the power is applied to the system, it starts initialization, and then enters the state of waiting for interruption. When the timer generates an interruption, the system begins a circle of data acquisition, processing and output. In one sampling period, the eight channel signals of the same time are acquired, decoupled statically to become six channel signals, then compensated dynamically, and output. The system runs the program continuously in this way.

2. Experimental results of the dynamic compensation system

The dynamic compensation system was connected to the wrist force sensor, and the dynamic experiments of step response were conducted to verify the effectiveness of dynamic compensation. The dynamic compensation results of six channels of a wrist force sensor (No. 3) are shown in Fig. 4-3 (a)~(f). In figures, curve 1 was the dynamic response of the wrist force sensor, and curve 2 is the output of the dynamic compensation system. The compensation coefficients of six channels were shown in Table II. The experimental results indicate that the adjusting time (within ±10 error of steady status) of dynamic response of the wrist force sensor is less than 5 ms, i. e. the adjusting time is reduced to less than 25%, and the dynamic performance indexes is greatly improved via dynamic compensation.

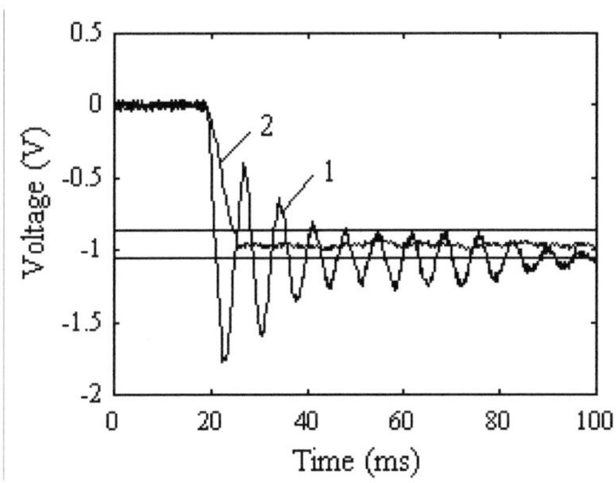

(a) Channel F_x

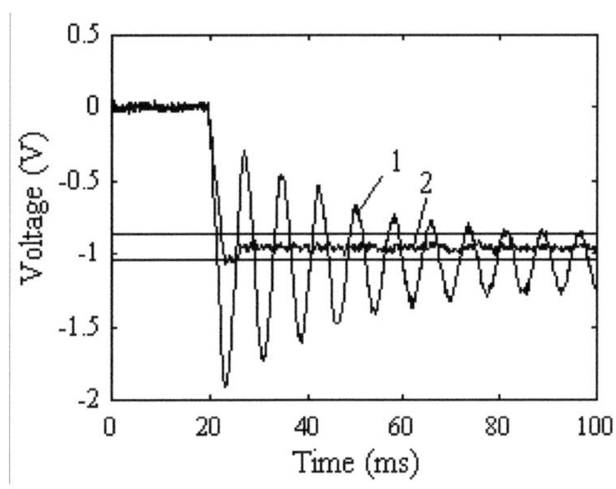

(b) Channel F_y

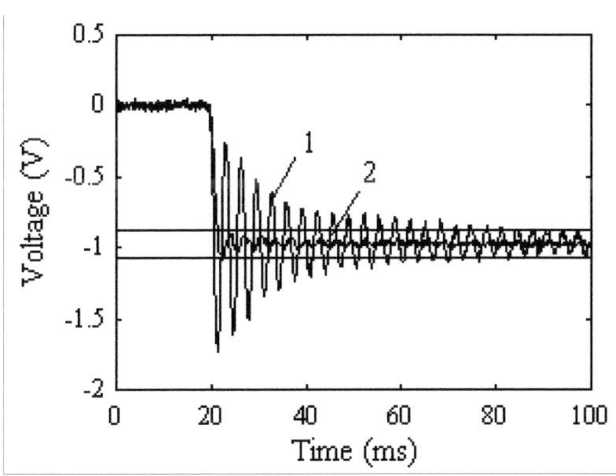

(c) Channel F_z

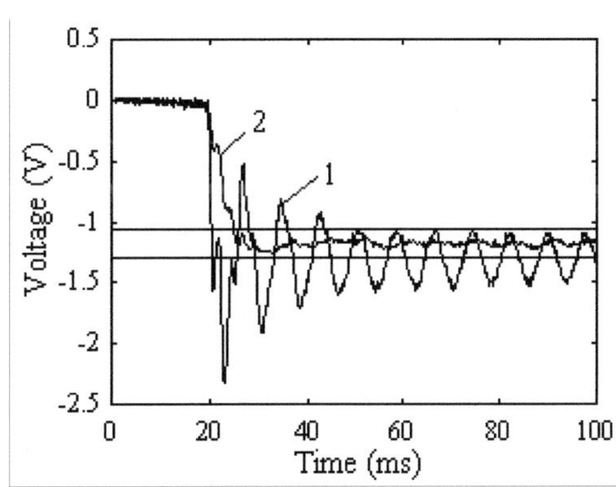

(d) Channel M_x

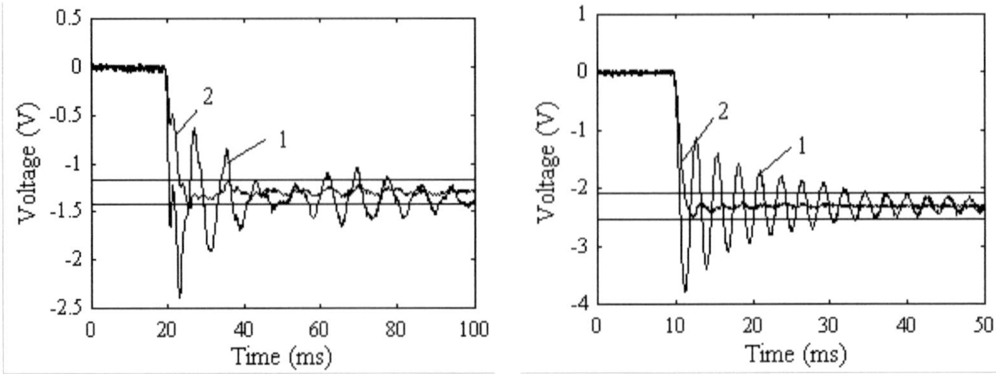

(e) Channel M_y (f) Channel M_z

Fig. 4-3 Experiment results (1) dynamic response of sensor, (2) output of dynamic compensating system.

	F_x	F_y	F_z	M_x	M_y	M_z
W_0	0.264231	0.959694	0.777428	0.251256	0.359567	0.920652
W_1	-0.525321	-1.915169	-1.541528	-0.4987790	-0.713234	-1.818204
W_2	0.261612	0.957071	0.772072	0.248014	0.354212	0.909576
W_3	1.966720	1.956619	1.871834	1.969605	1.970583	1.846556
W_4	-0.967239	-0.958209	-0.879801	-0.970085	-0.971131	-0.858568

Table 4-1 Coefficients of dynamic compensating system

5. Dynamic decoupling-compensation

There are dynamic couples among various channels of multi-axis force and torque sensors because the elastic body of the sensor is an integer structure and the interaction of various channels cannot be avoided completely. In addition, due to their small damped ratio and low natural frequency, the sensors dynamic response is slow, and the time to reach steady state is long. Both dynamic coupling and slow dynamic response are two main factors affecting the dynamic performances of sensors. We proposed the dynamic compensating and decoupling methods of multi-force sensors, constructed four types of dynamic decoupling and compensating networks, gave the design procedures and determines the order and parameters of the networks. The parameters of the networks are determined using the method based on FLANN. The dynamic decoupling and compensating results of a wrist force sensor have proved the methods to be correct and effective.

5.1 Structures of dynamic decoupling and compensating networks

The different places of compensating part result in different structure of dynamic decoupling and compensating network. In general, the compensating part is not put in front of decoupling part; otherwise it will make the design of decoupling part complex. The structure in which decoupling is done first and then compensation is carried out is called a serial decoupling and compensating network .The structure in which decoupling and compensation are completed at the same time is called a parallel network. Taking two dimensional force sensor as an example, the structures of various networks are shown in Fig.5-1 (a)~ (b).

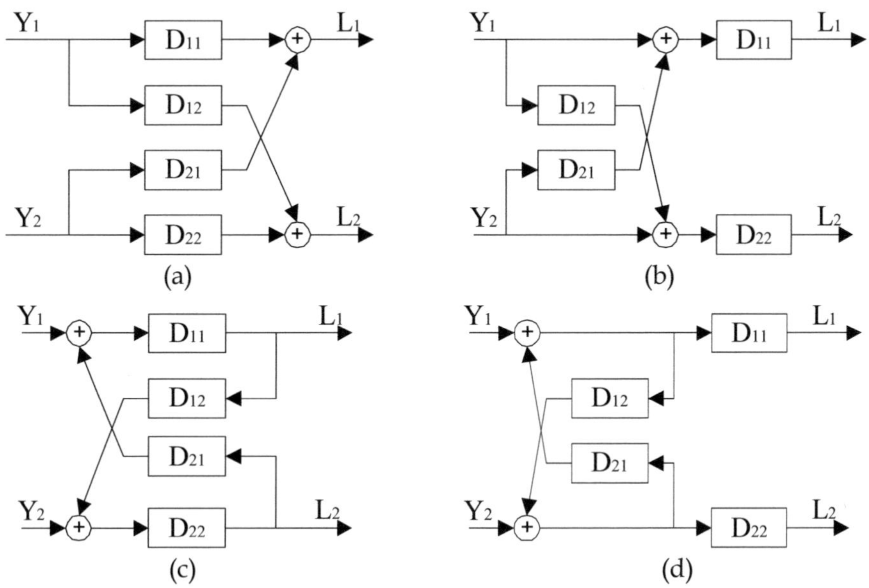

Fig. 5-1 Four kinds of network structures

In Fig. 5-1, (a) expresses the P parallel decoupling and compensating network (PPDCN), (b) is the P serial decoupling and compensating network (VPDCN), (c) describes the V Parallel Decoupling and Compensation Network (VPDCN),(d) is the V Serial Decoupling and Compensation Network (VSDCN). In figures, Y_i are the outputs of sensor, and L_i are the outputs of decoupling and compensating network.

5.2 Designs of dynamic decoupling and compensating networks

1. Design of PSDCN

The design procedure of PSDCN includes two steps. At first the decoupling part is designed, and then the compensating part is done. The design goal of decoupling part is to make the elements of non-main diagonal line in the matrix which is product of sensor transfer function matrix and decoupling matrix be zero. The design goal of compensating part is to make the compensating matrix equal to inverse of product matrix obtained by multiplying the sensor transfer function matrix and the decoupling matrix. For n-dimensional sensor, the decoupling matrix D_{psd} and compensating matrix D_{psc} in PSDCN respectively are given by equation (5.1).

$$D_{psd} = \begin{bmatrix} 1 & D_{12} & \dots & D_{1n} \\ D_{21} & 1 & \dots & D_{2n} \\ \dots\dots & & & \\ D_{n1} & D_{n2} & \dots & 1 \end{bmatrix} \qquad D_{psc} = \begin{bmatrix} D_{11} & 1 & \dots & 1 \\ 1 & D_{22} & \dots & 1 \\ \dots\dots & & & \\ 1 & 1 & \dots & D_{nn} \end{bmatrix} \tag{5.1}$$

To decouple completely, we must have

$$G \cdot D_{psd} \cdot D_{psc} = I \tag{5.2}$$

Therefore $$D_{psd} \cdot D_{psc} = \begin{bmatrix} D_{11} & D_{12}D_{22} & \dots & D_{1n}D_{nn} \\ D_{21}D_{11} & D_{22} & \dots & D_{2n}D_{nn} \\ \dots\dots & & & \\ D_{n1}D_{11} & D_{n2}D_{22} & \dots & D_{nn} \end{bmatrix} = \frac{1}{|G|}G^{*} \tag{5.3}$$

Where, $|G|$ is the determinant matrix of sensor transfer function G, and G^{*} is the companion matrix of G.
To make the corresponding elements in equation (5.3) equal, the elements of D_{psd} and D_{psc} are resolved as follows.

$$D_{ii} = \frac{G_{ii}^{*}}{|G|}, \quad D_{ij} = \frac{G_{ji}^{*}}{|G| \cdot D_{jj}} \qquad (i = 1,2,\dots,n;\ j = 1,2,\dots,n; i \neq j) \tag{5.4}$$

2. Design of PPDCN
Designing PPDCN is used by a direct method of solving inverse matrix. Supposing PPDCN to be the inverse matrix of sensor transfer function, the decoupling and compensating matrix D_{ppdc} is given by equation (5.5).

$$D_{ppdc} = \begin{bmatrix} D_{11} & D_{12} & \dots & D_{1n} \\ D_{21} & D_{22} & \dots & D_{2n} \\ \dots\dots & & & \\ D_{n1} & D_{n2} & \dots & D_{nn} \end{bmatrix} \tag{5.5}$$

If D_{ppdc} is equal to the inverse matrix of sensor transfer function, the elements of D_{ppdc} can be solved.

$$D_{ij} = \frac{G_{ji}^{*}}{|G|}, (i = 1,2,\dots,n;\ j = 1,2,\dots n;) \tag{5.6}$$

3. Design of VSDCN
The mathematical equations of VSDCN are given as follows.

$$L_i = Y_i + \sum_{\substack{j=1 \\ j \neq i}}^{n} D_{ij} L_j \ ,(i = 1,2,\dots,n) \tag{5.7}$$

We can obtain

$$Y_i = L_i - \sum_{\substack{j=1 \\ j \neq i}}^{n} D_{ij} L_j ,(i = 1,2,\dots,n) \tag{5.8}$$

Equation (5.7) can be written into a matrix form.

$$Y = LT \tag{5.9}$$

Here, Y and L are line vectors; T is a matrix described in equation (5.10).

$$T = \begin{bmatrix} 1 & -D_{12} & \dots & -D_{1n} \\ -D_{21} & 1 & \dots & -D_{2n} \\ \dots & & & \dots \\ -D_{n1} & -D_{n2} & \dots & 1 \end{bmatrix} \tag{5.10}$$

If T is the regular matrix, the decoupling matrix D_{vsd} of VSDCN is given by

$$D_{vsd} = T^{-1} \tag{5.11}$$

To reach decoupling and compensation completely, the following equation must be satisfied.

$$G \cdot D_{vsd} \cdot D_{vsc} = I \tag{5.12}$$

Where D_{vsc} is a compensating matrix.
Equation (5.12) yields

$$D_{vsd} \cdot D_{vsc} = G^{-1} \tag{5.13}$$

Therefore

$$G = D_{vsc}^{-1} \cdot T = \begin{bmatrix} \frac{1}{D_{11}} & -\frac{D_{12}}{D_{11}} & \dots & -\frac{D_{1n}}{D_{11}} \\ -\frac{D_{21}}{D_{22}} & \frac{1}{D_{22}} & \dots & -\frac{D_{2n}}{D_{22}} \\ \dots\dots & & & \\ -\frac{D_{n1}}{D_{nn}} & -\frac{D_{n2}}{D_{nn}} & \dots & \frac{1}{D_{nn}} \end{bmatrix} \tag{5.14}$$

The model of VSDCN can be obtained from solving equation (5.14).

$$D_{ii} = \frac{1}{G_{ii}}, D_{ij} = -\frac{G_{ij}}{G_{ii}}, (i = 1,2,...,n; j = 1,2,...,n; i \neq j) \tag{5.15}$$

4. Design of VPDCN

The mathematical equations of VPDCN are described by

$$L_i = D_{ii}(Y_i + \sum_{\substack{j=1 \\ j \neq i}}^{n} D_{ij} L_j), (i = 1,2,...,n) \tag{5.16}$$

From equation (5.16), we obtain

$$Y_i = \frac{L_i}{D_{ii}} - \sum_{\substack{j=1 \\ j \neq i}}^{n} D_{ij} L_j \ ,(i = 1,2,...,n) \tag{5.17}$$

Equation (5.17) can be written in a matrix form

$$Y = LT \tag{5.18}$$

Where, Y and L are line vectors, and T is a $n \times n$ matrix.

$$T = \begin{bmatrix} \frac{1}{D_{11}} & -D_{12} & \cdots & -D_{1n} \\ -D_{21} & \frac{1}{D_{22}} & \cdots & -D_{2n} \\ \cdots\cdots & & & \\ -D_{n1} & -D_{n2} & \cdots & \frac{1}{D_{nn}} \end{bmatrix} \tag{5.19}$$

If T is the regular matrix, the D_{vpdc} of VPDCN is

$$D_{vpdc} = T^{-1} \tag{5.20}$$

In order to achieve decoupling and compensation, we have

$$G \cdot D_{vpdc} = I \tag{5.21}$$

Therefore

$$G = D_{vpdc}{}^{-1} = (T^{-1})^{-1} = T \tag{5.22}$$

The model of VPDCN can be solved from equation (5.22).

$$D_{ii} = \frac{1}{G_{ii}}, D_{ij} = -G_{ij}$$
$$(i = 1,2,...,n; j = 1,2,...,n; i \neq j) \tag{5.23}$$

5. Designs of decoupling and compensating networks for non-minimum phase system

If the wrist force sensor is a non-minimum phase system, the above-mentioned method which designs the dynamic decoupling and compensating networks will result in the result to be unsteady. Therefore, before the dynamic decoupling and compensating networks are designed, the dynamic compensating digital filters are designed for non-coupled paths. The design of dynamic compensating digital filter can adopt the pole-zero configuration method or system identification method [11, 12]. The result F of dynamic compensation for non-coupled paths is

$$F = \begin{bmatrix} f_{11}g_{11} & 0 & \cdots & 0 \\ 0 & f_{22}g_{22} & \cdots & 0 \\ \cdots\cdots & & & \\ 0 & 0 & \cdots & f_{nn}g_{nn} \end{bmatrix} \tag{5.24}$$

Where, g_{ii} (i=1,2,…,n) is the transfer function for the ith path of sensor, f_{ii} (i=1,2,⋯,n) is the transfer function of dynamic compensating digital filter for the ith path of sensor.
In the design process of four kinds of dynamic decoupling and compensating networks, supposed the product of sensor transfer function and the matrix of decoupling and compensation to be equal to F, the corresponding decoupling and compensating network are obtained. The deducing procedure is similar with the previous section. The models of dynamic decoupling and compensating networks are as follows.
(1) PSDCN

$$D_{ii} = \frac{f_{ii} \cdot G_{ii}^{*} G_{ii}}{|G|}, \quad D_{ij} = \frac{G_{ji}^{*}}{|G| \cdot D_{jj}} \quad (i = 1,2,...,n; j = 1,2,...n; i \neq j) \tag{5.25}$$

(2) PPDCN

$$D_{ii} = \frac{f_{ii} \cdot G_{ii}^{*} G_{ii}}{|G|}, D_{ij} = \frac{G_{ji}^{*} f_{ii} G_{ii}}{|G|} \quad i = 1,2,...,n; j = 1,2,...n; i \neq j \tag{5.26}$$

(3) VSDCN

$$D_{ii} = f_{ii}, D_{ij} = -\frac{G_{ij}}{G_{ii}}, (i = 1,2,...,n; j = 1,2,...,n; i \neq j) \tag{5.27}$$

(4) VPDCN

$$D_{ii} = f_{ii}, D_{ij} = -\frac{G_{ij}}{f_{ii} G_{ii}}, (i = 1,2,...,n; j = 1,2,...,n; i \neq j) \tag{5.28}$$

5.3 Determination of orders and parameters

A FLANN-based method is used to determine the orders and parameters of the dynamic decoupling and compensating network. The system identification method can also been used to do this work, but it sometimes makes the model orders too high or decoupling and compensating results divergent because of modeling error. The FLANN method overcomes these shortcomings.
The designs of decoupling parts in PSDCN, VSDCN and VPDCN have nothing to do with the mix output signal of sensor, which includes non-coupled and coupled output signals. Therefore using input and output signals of sensor under no coupled condition at first sets up the models of decoupling parts. In design process of compensating part, the decoupling model is used for decoupling coupled signal, and then the compensating parts are designed in according with decoupled signal. Thus it can bring decoupling error in the design of compensating parts, and correct decoupling error in the design of compensating parts.
Designing PSDCN and VSDCN can adopt the system identification method or the FLANN method. Designing VPDCN only utilizes the FLANN method because it is a parallel structure with internal feedback, and modeling error may result in divergent. Designing PPDCN is complex, the models of compensating parts for non-coupled paths are set up at first by using the input and output signals of sensor under no coupled condition. The models of decoupling parts are trained by adjusting the difference between the

compensating result of mix output signal and input signal. Thus we can bring the compensating error in the design of decoupling part, and the compensating error is corrected in the decoupling part.

Suppose the input signals of sensor are $X_i(k)$; the output signals are $Y_i(k)$. The $X_i(k-1),..., X_i(k-r), Y_i(k-1),..., Y_i(k-s)$ are obtained by the functional expansion technique, which are used as the inputs of FLANN. The k expresses number of data, $k = 1,..., N$. The inputs are weighted and summed, and the output $L_i(k)$ are yielded. The difference between $L_i(k)$ and $X_i(k)$ is regarded as error $e_i(k)$ to adjust the weights $W_i(k)$ of FLANN. A schematic diagram of FLANN for determining parameters is shown in Fig.5-2. An equation describing the neural algorithm can be written as

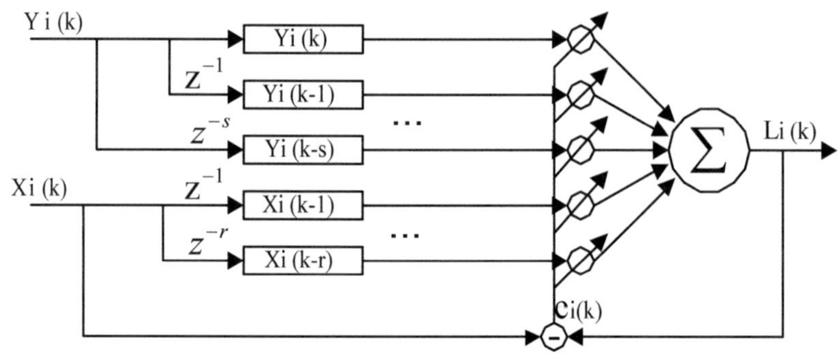

Fig. 5-2 Schematic of modeling by FLANN

$$L_i(k) = \sum_{p=1}^{r} w(j)X_i(k-p) + \sum_{q=1}^{s} w(q+r+1)Y_i(k-q) \tag{5.29}$$

The error is

$$e_i(k) = X_i(k) - L_i(k) \tag{5.30}$$

The weight update equations are given by

$$w_i(p) = w_i(p) + \alpha \cdot e_i(k)X_i(k-p)$$

$$w_i(q+r+1) = w_i(q+r+1) + \alpha \cdot e_i(k)Y_i(k-q) \tag{5.31}$$

Where $L_i(k)$, $Y_i(k)$, $e_i(k)$, $W_i(k)$ stand for the desired output of the ith path, estimating output, error and the pth or the qth linking weight in the kth step of the FLANN. The α denotes the learning constant which connects with the stability and the rate of convergence, usually is selected about 0.1. In training process, initial values of weights are chosen about 0.1. After training for many times, when the average mean square error achieved a minimum value, the weights of FLANN are the parameters of dynamic decoupling and compensating network.

5.4 Dynamic decoupling and compensating results

1. Evaluating indexes

To evaluate the decoupling and compensating results, the indexes are adopted as the follows.

(1) The variance is

$$\sigma = \sqrt{\frac{(L_i - X_i)^2}{N-1}} \tag{5.32}$$

where X_i are the input signals of sensor, L_i are corresponding decoupling and compensating output signals, and N is total number of data.
(2) The relative error is

$$e_r = \frac{(L_i - X_i)_{\max}}{|X_i|_{\max}} \times 100\% \tag{5.33}$$

2. Results of simulation
In order to examine the decoupling and compensating methods, we carry out the simulation. The results of simulation indicate that the VSDCN and VPDCN can achieve the good effectiveness, but the results of PSDCN and PPDCN have error because these methods use the low order model to substitute for the high order model.
3. Dynamic decoupling and compensating result of wrist force sensor
We decouple and compensate the dynamic output signals of wrist force sensor. The sensor is a six-axis device, i.e. n=6, the decoupling parts of PSDCN and PPDCN are very complex with high order. For examples, on the assumption that the model order of element in transfer function matrix is 3, the order of decoupling and compensating model in PPDCN will high to 33, so it will result in the bad convergence and a big error because of simplification. In generally, PSDCN and PPDCN can only be used for the dimensional number less than 3.
When the dimensional number larger than 3, VSDCN and VPDCN can only be used. The models of these networks do not vary with the number of variables, so do not have the problem of over-high order. For the wrist force sensor, we prefer VSDCN and VPDCN. In brief, we only introduce the result of VSDCN. The decoupling and compensating results between direction Z and direction X is shown in Fig.5-3 (a) and (b). The decoupling and compensating results between direction Z and direction Y is shown in Fig.5-4 (a) and (b). In the figures the orders of both decoupling and compensating models are 3, curve 1 expresses the input signal of sensor, curve 2 expresses the decoupled and compensated result.

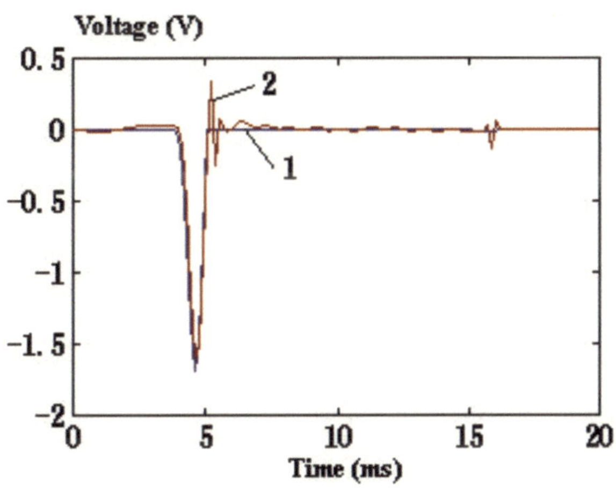

Fig. 5-3 (a) Decoupling and compensating result of direction Z

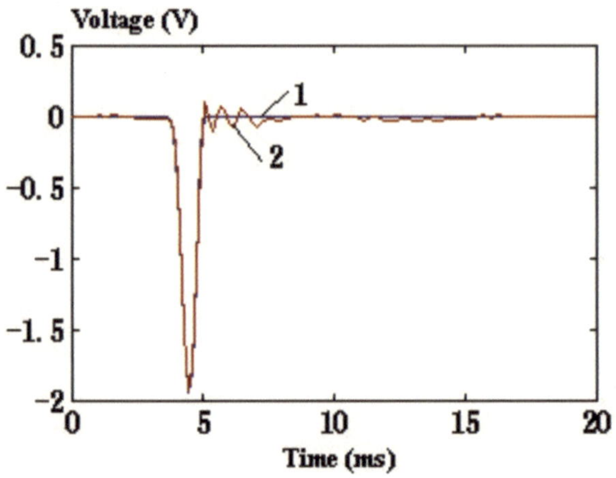

Fig. 5-3 (b) Decoupling and compensating result of direction X

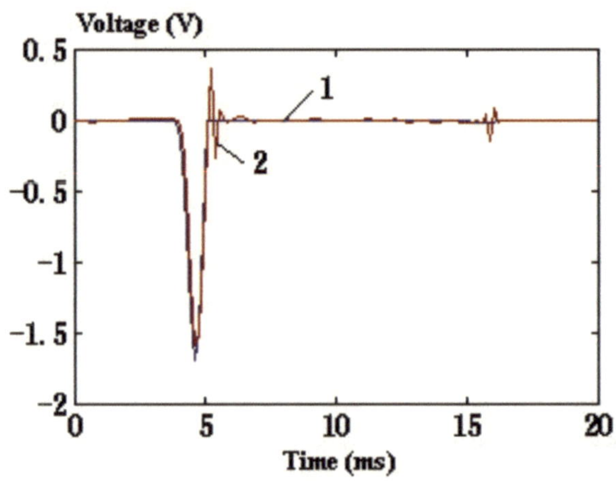

Fig. 5-4 (a) Decoupling and compensating result of direction Z

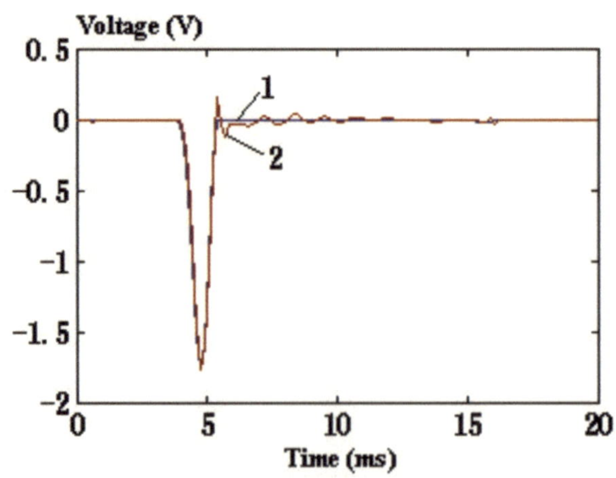

Fig. 5-4 (b) Decoupling and compensating result of direction Y

Comparisons of the decoupling and compensating errors between direction Z and direction X is seen in Table 5-1. Comparisons of the decoupling and compensating errors between direction Z and direction Y are seen in Table 5-2. Analyzing Table 1 and Table 2, the results of PSDCN are the worst and the results of VSDCN are the best.

Error indexes	PSDCN	PPDCN	VSDCN	VPDCN
σ in direction Z	0.0880	0.0567	0.0543	0.0668
e_r in direction Z	18.50%	19.35%	15.70%	14.25%
σ in direction X	0.0717	0.0276	0.0270	0.0538
e_r in direction X	19.71%	5.62%	5.53%	10.54%

Table 5-1 Comparisons of the decoupling and compensating errors between direction Z and direction X

Error indexes	PSDCN	PPDCN	VSDCN	VPDCN
σ in direction Z	0.0731	0.0585	0.0586	0.0946
e_r in direction Z	18.91%	20.59%	21.60%	25.10%
σ in direction Y	0.0960	0.0449	0.0301	0.0479
e_r in direction Y	23.70%	13.37%	9.23%	13.10%

Table 5-2 Comparisons of the decoupling and compensating errors between direction Z and direction Y

4. Conclusions

Four kinds of decoupling and compensating networks are presented in this section. Analyzing from principle, compared with PSDCN and PPDCN, VSDCN and VPDCN are more concise, they have definite physical meaning and can achieve full decoupling and compensation. Judging from the construction, parallel networks are better than serial networks because the decoupling and compensation are combined into one unit.

The decoupling and compensating error are caused mainly by the following reasons. (a) There are modeling errors in four kinds of networks. (b) The simplification of molder result in the errors in PSDCN and PPDCN.

The decoupling and compensating methods can also be applied to other multi-axis force sensors.

6. Nonlinear dynamic characteristics

There is the non-linearity in the dynamic characteristics of sensors under some conditions. In order to describe accurately the dynamic behavior of sensors some researchers studied the nonlinear dynamic characteristics of sensors. Waldemar Minkina presented nonlinear models that adequately describe the dynamic state of temperature sensor within the temperature increase range [13,14]. Ping Wang et al. discussed the analysis of nonlinear dynamic state of accelerometer transducers and its applications in the dynamic modeling [15]. S. Beling et al. approximated the dynamic behavior of nonlinear gas sensors using the feed-forward neural networks [16]. Haixia Zhang et al. studied the transient process of the sensor probe, and developed a nonlinear model based on equivalent electrical circuit techniques [17]. Ke-Jun Xu et al. studied the nonlinear dynamic characteristics of the wrist force sensor in the time and frequency domains [18,19]. These researches described the nonlinear dynamic models of sensors only using one block. On the basis of these models the nonlinear dynamic responses of sensors are compensated to improve the dynamic performances of sensors. Antonio Pardo et al. built a nonlinear inverse dynamic system to solve the non-linearity of gas sensing system [20]. Ke-Jun Xu et al. designed a dynamic compensating system for the wrist force sensor using FLANN [21]. The nonlinear dynamic compensations achieve good results under some certain conditions. Since the nonlinear dynamic system is not satisfied with the homogeneity and superposition, the dynamic compensations based on the above-mentioned nonlinear dynamic models are effective for the certain response of sensors, but are not suitable for different form and different amplitude responses of sensors.

6.1 Hammerstein model based modeling

Previous researchers present the nonlinear dynamic models of sensors only using a block, which make it difficulty to compensate the nonlinear dynamic responses of sensors. The models of sensors with the nonlinear dynamic characteristic may be divided into two blocks, that is, a nonlinear static part and a linear dynamic one. The linear dynamic part is first compensated and then the nonlinear static part is corrected. Thus the problems of previous nonlinear dynamic compensations of sensors are solved. In this section a Hammerstein model is adopted to describe the nonlinear dynamic models of sensors, and a one-stage identification algorithm is proposed to simplify the calculation. On this basis a two-step compensation method is present for the nonlinear dynamic responses of sensors.

1. Deduction of one-stage identification algorithm

The Hammerstein model is composed of a nonlinear static block $N(\cdot)$ followed by a linear dynamic one $h(t)$, which is shown in Fig. 6-1 [22,23]. The $u(t)$ and $y(t)$ are the input and output of the Hammerstein model respectively, $\xi(t)$ is white noise, and $x(t)$ stands for the output of nonlinear static block.

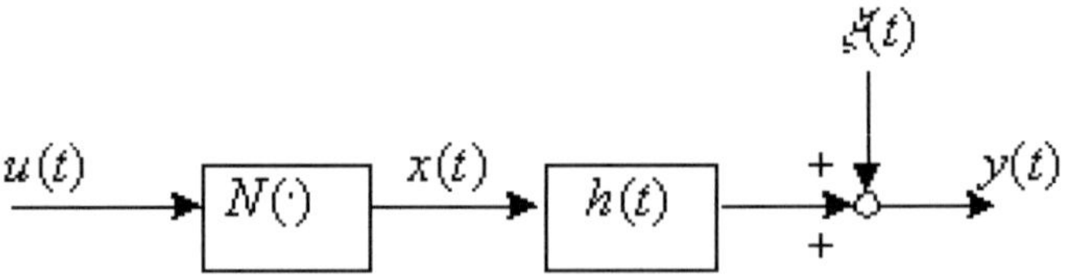

Fig. 6-1 Hammerstein model

Assuming the nonlinear static block can be approximated by a polynomial, and can be written as

$$N(\cdot) = \sum_{j=1}^{l} c_j u^j(t) \tag{6.1}$$

And

$$x(t) = N\left[u(t)\right] \tag{6.2}$$

The difference equation of the Hammerstein model is

$$A(q^{-1})y(k)=B(q^{-1})N\left[u(k)\right]+\xi(k) \tag{6.3}$$

Where

$$A(q^{-1})=1+a_1q^{-1}+\cdots+a_nq^{-n} \tag{6.4}$$

$$B(q^{-1})=b_0+b_1q^{-1}+\cdots+b_mq^{-m} \tag{6.5}$$

$$N\left[u(k)\right]=\sum_{j=1}^{l}c_ju^j(k) \tag{6.6}$$

The transfer function of the linear dynamic block can be given as

$$H(z^{-1})=\frac{b_0+b_1z^{-1}+\cdots+b_mz^{-m}}{1+a_1z^{-1}+\cdots+a_nz^{-n}}. \tag{6.7}$$

Considering Equations (6.3), (6.4), (6.5) and (6.6), we can obtain

$$y(k)=-\sum_{p=1}^{n}a_py(k-p)+\sum_{j=1}^{l}\sum_{v=0}^{m}b_vc_ju^j(k-v)+\xi(k) \tag{6.8}$$

Assuming

$$w_{vj}=b_vc_j \qquad v=0,1,\cdots m,\, j=1,\cdots l\,. \tag{6.9}$$

Then Equation (6.8) becomes

$$y(k)=-\sum_{p=1}^{n}a_py(k-p)+\sum_{j=1}^{l}\sum_{v=0}^{m}w_{vj}u^j(k-v)+\xi(k) \tag{6.10}$$

Equation (6.10) can also be expressed as

$$y(k)=\Phi^{\mathrm{T}}(k)\theta(k)+\xi(k) \tag{6.11}$$

Where

$$\theta(k)=[a_1,\cdots,a_n,w_{01},\cdots,w_{m1},\cdots,w_{0l},\cdots,w_{ml}]^{\mathrm{T}}$$

$$\Phi(k)=[-y(k-1),\cdots,-y(k-n),\cdots,u(k),\cdots,u(k-m),\cdots,u^l(k),\cdots,u^l(k-m)]^{\mathrm{T}}$$

Equation (6.11) is a parameter model describing the relation between the input and output. The parameters of the model are obtained using the least square method (LSM) or the FLANN [24].

Assuming that the gain of steady state of the linear dynamic unit is 1, that is

$$H(\infty)=\frac{b_0+\cdots+b_m}{1+a_1+\cdots+a_n}=1 \tag{6.12}$$

Thus the gain of steady state of the nonlinear dynamic system stems from the nonlinear static block.

When $\forall j = 1, 2, \cdots, l$, we can obtain the following equation from Equation (6.9).

$$\sum_{v=0}^{m} w_{vj} = (b_0 + \cdots + b_m) c_j \qquad v = 0, 1, \cdots m, j = 1, \cdots l . \tag{6.13}$$

Then

$$c_j = \frac{\sum_{v=0}^{m} w_{vj}}{b_0 + \cdots + b_m} = \frac{\sum_{v=0}^{m} w_{vj}}{1 + a_1 + \cdots + a_n} \tag{6.14}$$

Thus the coefficients of the nonlinear static block $c_1, \cdots, c_l$ are solved from $a_1, \cdots, a_n, w_{01}, \cdots, w_{ml}$ obtained by identification. On this basis we can yield

$$b_v = \frac{w_{vj}}{c_j} \qquad v = 0, 1, \cdots m, j = 1, \cdots l . \tag{6.15}$$

Now the coefficients of the linear dynamic unit are also obtained. Due to the inevitable iterative error of the LSM or the FLANN, b_v $(v = 0, 1, \cdots m)$ obtained from Equation (6.15) may not satisfy Equation (6.12). However, the dynamic characteristics of a linear system, such as what can be expressed as Equation (6.7), mainly depend on its poles instead of zeros. While b_v $(v = 0, 1, \cdots m)$ is the zero, and has little effect on the dynamic characteristics of a system. Therefore Equation (6.7) can also be expressed as

$$H(z^{-1}) = \frac{A}{1 + a_1 z^{-1} + \cdots + a_2 z^{-n}} \tag{6.16}$$

Where

$$A = 1 + \sum_{i=1}^{n} a_i \tag{6.17}$$

This viewpoint may be proved as the following. Assuming $a_1 = -1.95974$, $a_2 = 0.98681$, $b_0 = 0.00677$, $b_1 = 0.01353, b_2 = 0.00677$, and $m = n = 2$, the step responses of the system are obtained from Equations (6.7) and (6.16) respectively, and are shown in Fig. 6-2. Two response curves are identical. Even if we have not solved the parameters b_v $(v = 0, 1, \cdots m)$, the dynamic characteristics of the sensor can also be obtained. Therefore, using one-stage identification method, we can obtain the coefficients of both the nonlinear static block and the linear dynamic unit according to the inputs and outputs of the nonlinear dynamic system.

2. Simulations of modeling

In order to examine the one-stage identification algorithm, the simulations are carried out. The step signal and impulse signal are chosen as input signals of modeling because they are usually applied to the experimental calibrations of sensors. Since a second-degree polynomial is commonly used in describing the nonlinear static characteristics of sensors in practice engineering, and a second-order linear dynamic unit is often admitted, we make our simulations based on the second-degree nonlinear static model and the second-order linear dynamic model. So let $m = n = 2$ in Equations (6-4) and (6-5), and $l = 2$ in Equation

(6-6), then a_1, a_2, b_0, b_1, c_1 and c_2 are parameters that will be estimated. We should first obtain a_1, a_2, $w_{01},\cdots,w_{22}$ through Equation (6-11).

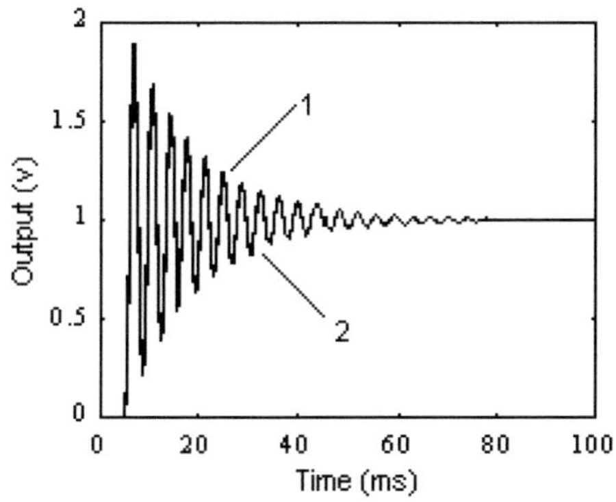

Fig. 6-2 Comparison of two response curves: (1) step response of Equ.(6-7), and (2) step response of Equ.(6-16).

Suppose a nonlinear static subsystem is

$$x(k) = u(k) + 0.5u^2(k) \tag{6.18}$$

A linear dynamic unit is given by

$$y(k) = 1.95974y(k-1) - 0.98681y(k-2)$$

$$+0.00667x(k) + 0.01353x(k-1) + 0.00667x(k-2) \tag{6.19}$$

A step input signal and a nonlinear dynamic response of the system are shown in Fig. 6-3. According to the inputs and outputs of the system, $a_1, a_2, w_{01}, \cdots, w_{22}$, parameters in equation (6.11), are obtained using the LSM. Afterwards b_0, b_1, b_2, c_1, c_2 can be easily obtained through Equations (6.14) and (6.15). Thus a nonlinear dynamic model is set up using the one-stage identification algorithm. The response of this model is compared with that of the supposed system which is shown in Fig. 6-4. The supposed and identified parameters are listed in Table 6-1.

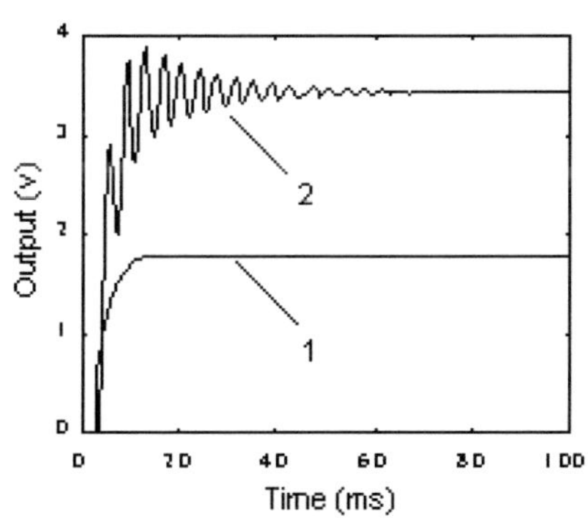

Fig. 6-3 Step input and supposed nonlinear dynamic response: (1) step input, and (2) nonlinear dynamic response .

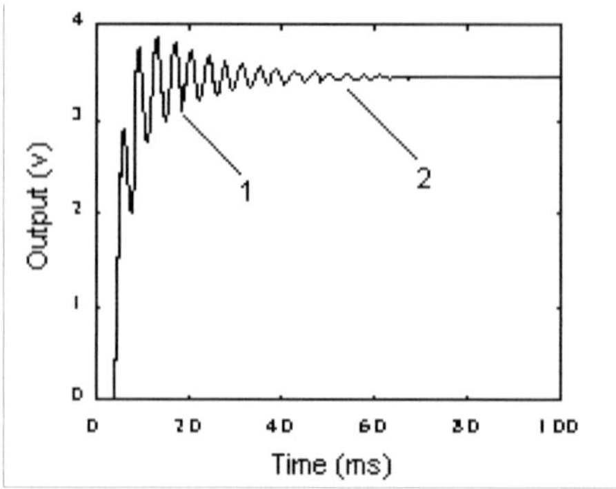

Fig. 6-4 Comparison of identification result and supposed response: (1) supposed response, and (2) identification result.

Parameter	a_1	a_2	b_0	b_1	b_2	c_1	c_2
Supposed	-1.95974	0.98681	0.00677	0.01353	0.00677	1.00	0.50
Identified	-1.95971	0.98679	0.00437	0.01323	0.00948	0.99663	0.50188

Table 6-1 Comparison between supposed and identified parameters

The simulation results of the impulse signal are shown in Fig. 6-5 and Fig. 6-6. The supposed and identified parameters are listed in Table 6-2.

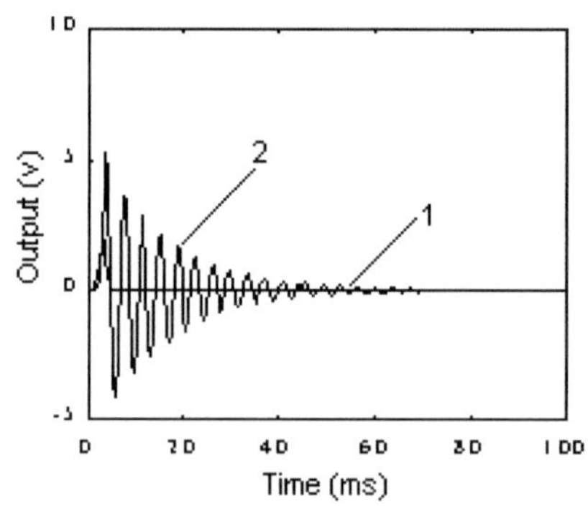

Fig. 6-5 Impulse input and nonlinear dynamic response: (1) impulse input, and (2) nonlinear dynamic response.

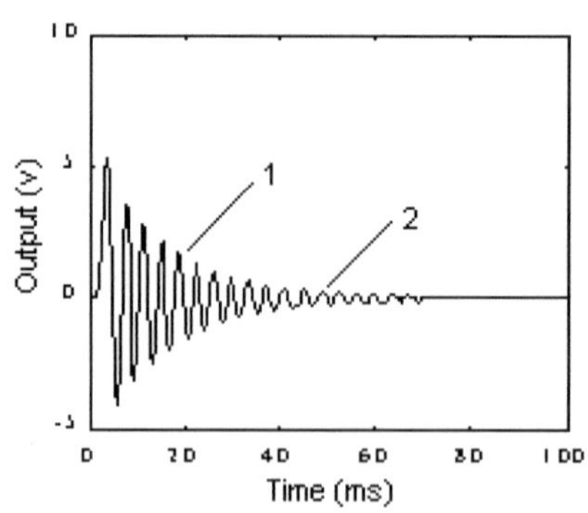

Fig. 6-6 Comparison of identification result and supposed response: (1) supposed response, and (2) identification result.

Parameter	a_1	a_2	b_0	b_1	b_2	c_1	c_2
Supposed	-1.95974	0.98681	0.00677	0.01353	0.00677	1.00	0.50
Identified	-1.95974	0.98681	0.00676	0.01350	0.00675	0.99995	0.50014

Table 6-2 Comparison between parameters supposed and identified

All above simulation results show that the performance and convergence of the algorithm presented in this section are good.

3. Modeling of wrist force sensor

The impulse response method is easily done and works well in the dynamic calibrations of sensors. So we adopt this method to make the dynamic calibration experiments of the wrist force sensor. In the calibration, a wrist force sensor is mounted on a testing platform. If no load is placed on the wrist sensor in the dynamic calibration, we call it no-load-calibration (NLC); while when there is some load laid on the wrist force sensor, we call it having-load-calibration (HLC). An impulse force is applied to the wrist force sensor with a hammer, that is, the hammer strikes vertically on the sensor directly in the NLC or on the load placed on the sensor in the HLC in a very short interval. In the head of the hammer, a piezoelectric sensor is installed to transform the impulse force into the electric charge signal. This signal is amplified by a charge amplifier and sent to a computer based data acquisition system. The wrist force sensor outputs six channel signals, of which three channels express force components of x, y, and z directions, and three channels express moment components of x, y, and z directions. These six channel signals of the wrist force sensor are also collected by the data acquisition system.

In the NLC, the zero point of work of the wrist force sensor is located in the middle part of the linear working range, so the dynamic response of the sensor is linear. But in HLC, the position of the zero point of work is moved from the linear working range to the nonlinear working range because of the applied load. Therefore the effect of the nonlinear factor becomes serious. The dynamic response of the sensor in HLC is nonlinear. The impulse response of NLC is shown in Fig. 6-7, and that of HLC in Fig. 6-8.

In order to demonstrate the superiority, in the modeling of sensors, of the algorithm presented in this section, the following work is done. First assume the models of sensors in NLC and HLC are linear, we carry out linear modeling (LM) using the LMS. Secondly we regard the models of sensors as nonlinear ones, and identify them with the algorithm presented in this section, which is nonlinear modeling (NLM). Finally we compare these identification results. Fig. 6-9 and Fig. 6-10 show their difference in terms of curves, and Table 6-3 and Table 6-4 in terms of parameters. The sum of square error $\sum e^2$ of each identified curve to the real impulse response is used to evaluate the accuracy of model, which is shown in Table 6-3 and Table 6-4. The smaller is the value of $\sum e^2$; the better is the identification result.

The two curves in Fig. 6-9 are almost overlapped to each other, and the identified parameters with the algorithm presented in this section contain a small coefficient value c_2. It shows that the nonlinear factor of the impulse response in NLC is not very serious or there lies a quite weak non-linearity. But in Fig. 6-10, two curves have a little difference, and the value of coefficient c_2 is not a very small one. The nonlinear factor should be considered under this circumstance. Judging from the values of $\sum e^2$ in Table 6-3 and 6-4, we come to the conclusion that the nonlinear modeling method presented in this section is better than that of the linear modeling method in describing the model of the wrist force sensor.

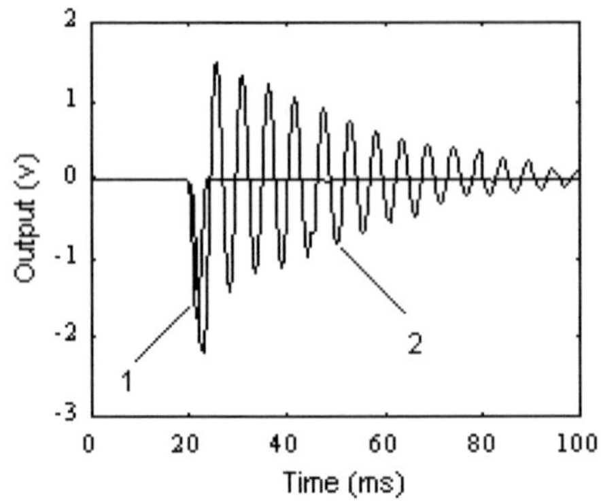

Fig. 6-7 Impulse response in NLC: (1) impulse input, and (2) dynamic response of the sensor.

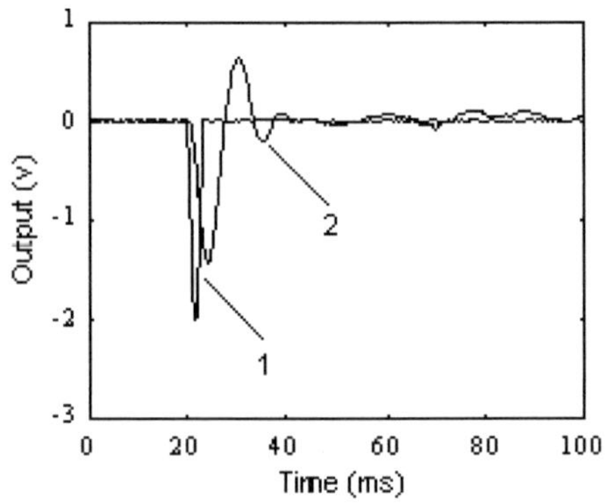

Fig. 6-8 Impulse response in HLC: (1) impulse input, and (2) dynamic response of the sensor.

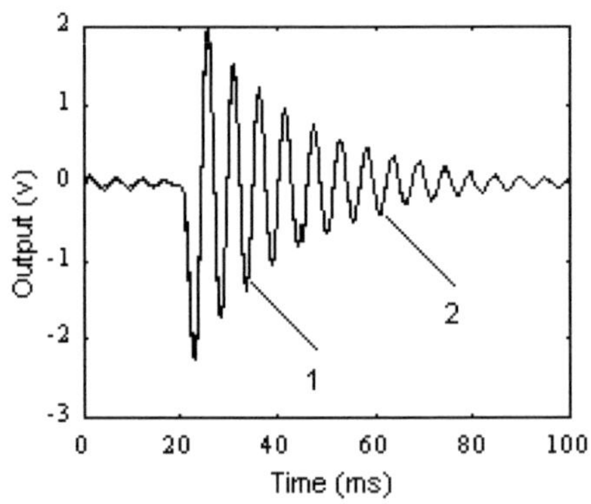

Fig. 6-9 Comparison of modeling in NLC with two kinds of methods: (1) modeling with LM, and (2) modeling with NLM.

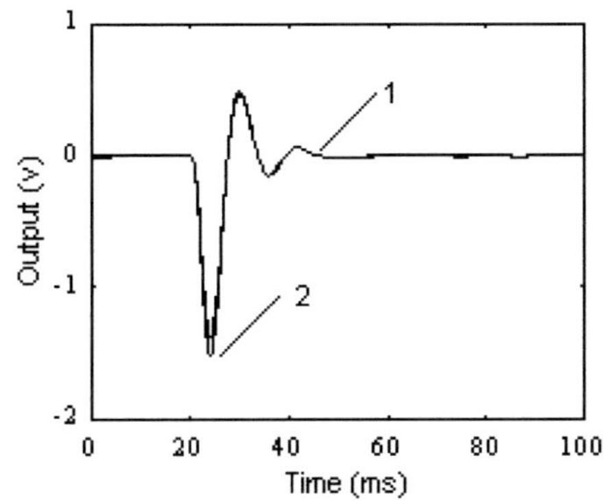

Fig. 6-10 Comparison of modeling in HLC with two kinds of methods: (1) modeling with LM, and (2) modeling with NLM.

Parameter	a_1	a_2	c_1	c_2	$\sum e^2$
LM	-1.97785	0.99095			15.97152
NLM	-1.97818	0.99138	0.89319	0.05800	12.08818

Table 6-3 Comparison between linear modeling and nonlinear modeling in NLC

Parameter	a_1	a_2	c_1	c_2	$\sum e^2$
LM	-1.96237	0.96528			3.45087
NLM	-1.96144	0.96438	1.48733	-0.19036	3.38959

Table 6-4 Comparison between linear modeling and nonlinear modeling in HLC

4. Discussions

The one-stage identification algorithm has advantages as follows: (1) One-stage identification simplifies the algorithm; (2) It depends only on the data of input and output of the system, not needing to introduce the auxiliary variables that could not be measured in practice; (3) It only needs dynamic calibration experimental data of systems, not needing to do static calibration experiments. On the basis of identification, the nonlinear dynamic compensation is easily completed.

6.2 Hammerstein model based correction

With the increasing higher requirement of the dynamic measurement, it is more and more important to improve the dynamic performances of sensors. We brought forward a nonlinear compensation method for the Hammerstein model. The Hammerstein model is composed of two parts, one linear dynamic unit and one nonlinear static subsystem, therefore the compensation includes two steps accordingly: The first step is linear dynamic compensation and the second one is nonlinear static correction. Thus we call it two-step compensation. Fig. 6-11 shows a block diagram of this method. The linear dynamic compensation unit is $h'(t)$, and the inverse unit of the nonlinear static subsystem $N(\cdot)$ is $N^{-1}(\cdot)$. The ultimate compensated output is $u'(t)$.

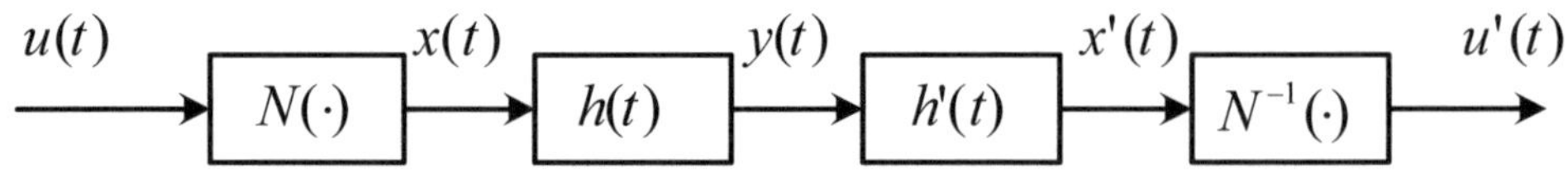

Fig. 6-11 A block diagram of two-step compensation

A linear dynamic compensation unit $h'(t)$ is designed using the pole-zero configuration method or system identification method [24]. Through the linear dynamic compensation, we get $x'(t)$. A nonlinear static correction unit $N^{-1}(\cdot)$ should be designed. The nonlinear static subsystem can be expressed by a second-degree polynomial.

$$N[u(k)] = x(k) = c_1 u(k) + c_2 u^2(k) \tag{6-20}$$

Its inverse system is assumed as

$$N^{-1}[\hat{x}(k)] = \hat{u}(k) = d_0 + d_1 \hat{x}(k) + d_2 \hat{x}^2(k) \tag{6-21}$$

Where $\hat{u}(k)$ and $\hat{x}(k)$ are the predictive data, $\hat{x}(k) = c_1 \hat{u}(k) + c_2 \hat{u}^2(k)$. Though c_1, c_2 have been obtained, $N^{-1}(\cdot)$ is still difficult to be solved from Equation (6-16). We adopt the FLANN to get the parameters d_0, d_1, d_2 of $N^{-1}(\cdot)$ as the artificial neural network has the excellent approximation property. A schematic diagram of the FLANN for training parameters is shown in Fig. 6-12.

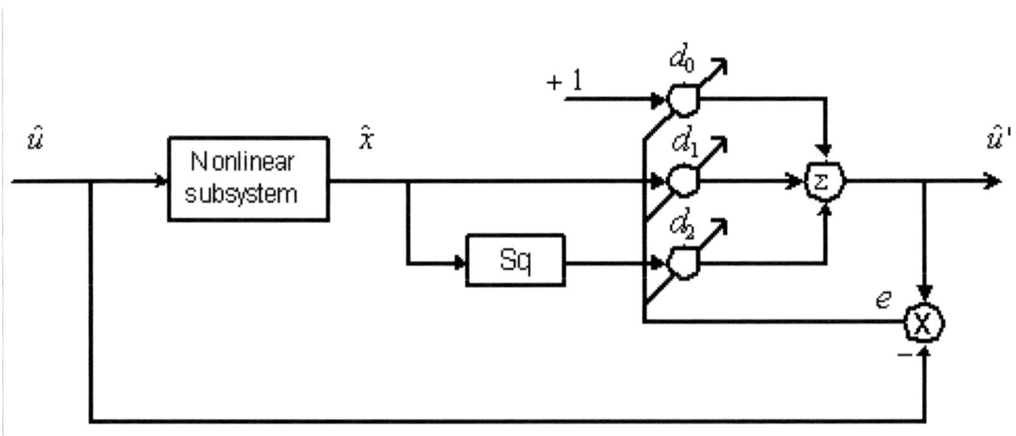

Fig. 6-12 A training schematic diagram of the FLANN

1. Simulations of nonlinear dynamic compensation

Using the compensation method stated above, the simulation results of the step response and impulse response are shown in Fig. 6-13 ~ Fig. 6-16. Fig. 6-13 and Fig. 6-15 show the results of the first step, that is linear dynamic compensation, compared with the output signal of sensors. Fig. 6-14 and Fig. 6-16 show the results of the second step, that is nonlinear static correction, compared with the input signal of sensors. It can be seen that the method of nonlinear dynamic compensation is effective.

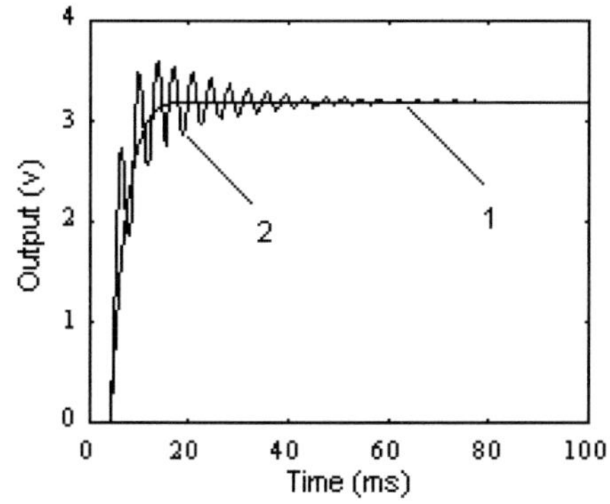

Fig. 6-13 The first step of compensation: (1) nonlinear dynamic response, and (2) dynamic compensation of the first step.

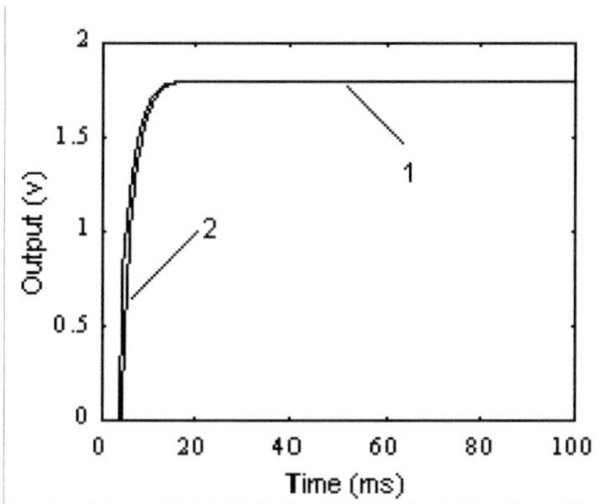

Fig. 6-14 The second step of compensation compared with the input signal (1) step input, (2) nonlinear static correction of the second step.

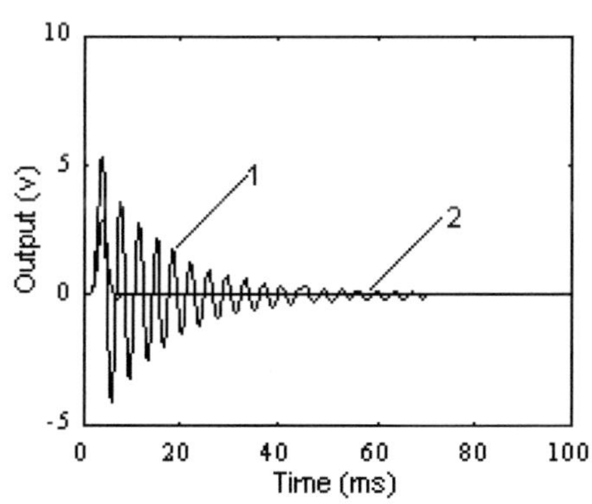

Fig. 6-15 The first step of compensation (1) nonlinear dynamic response, (2) dynamic compensation of the first step.

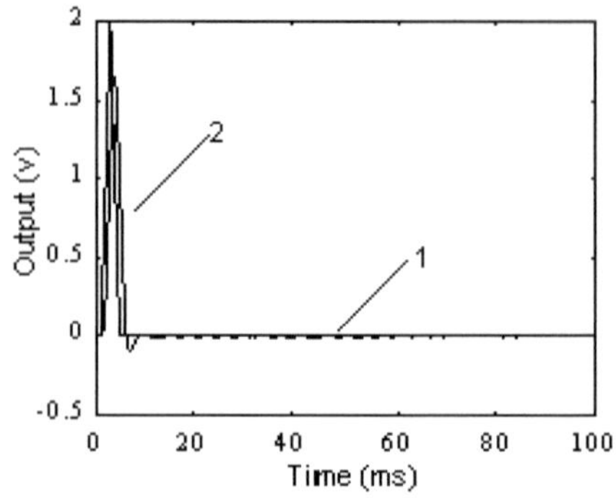

Fig. 6-16 The second step of compensation compared with the input signal (1) impulse input, (2) nonlinear static correction of the second step.

2. Compensation of the impulse response of wrist force sensor

The impulse responses of the wrist force sensors in NLC and HLC are compensated using the two-step nonlinear dynamic compensation method. Fig. 6-17 shows the result of the nonlinear dynamic compensation for NLC, and Fig. 6-18 shows the result of the linear dynamic compensation using the linear compensation method. Comparing the two results, we find that the compensation result in Fig. 6-17 is not better than that in Fig. 6-18, because the nonlinear dynamic factor is a weak one, as we have analyzed in above section.

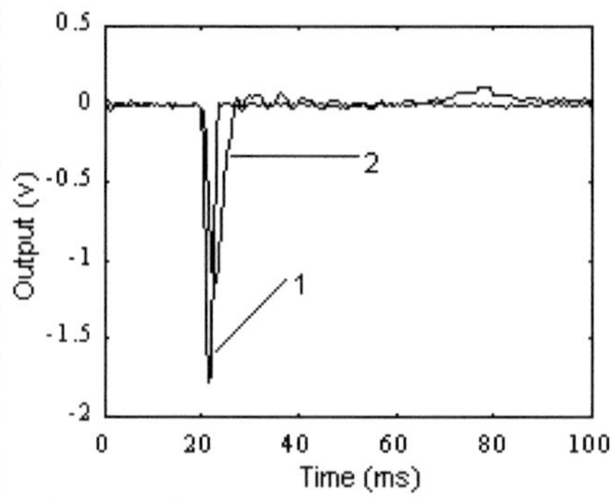

Fig. 6-17 Compensation result of NLC compared with the input signal (1) impulse input, (2) nonlinear dynamic compensation result

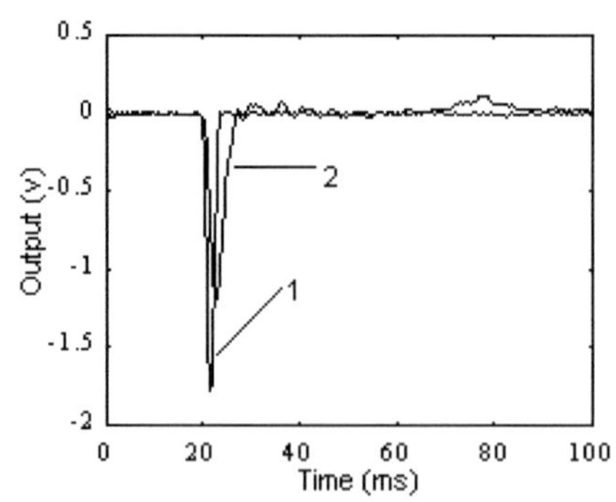

Fig. 6-18 Compensation result of NLC using linear approach compared with the input signal (1) impulse input, (2) compensation result

Fig. 6-19 shows the result of the nonlinear dynamic compensation for HLC, and Fig. 20 shows the result of the linear dynamic compensation for HLC. The nonlinear compensation method is better than the linear one in HLC.

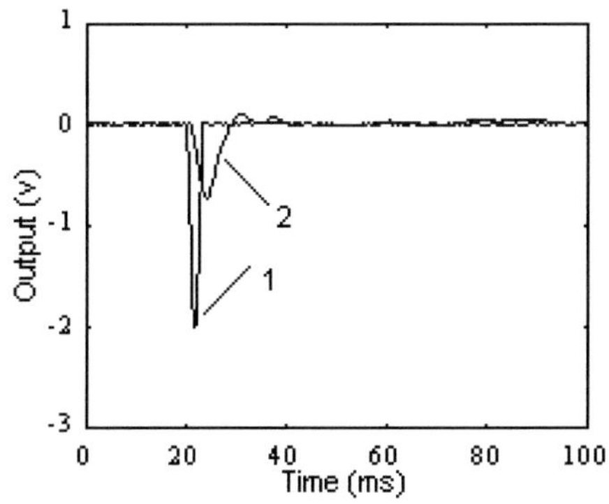

Fig. 6-19 Compensation result of HLC compared with input signal (1) impulse input, (2) nonlinear static correction of the second step

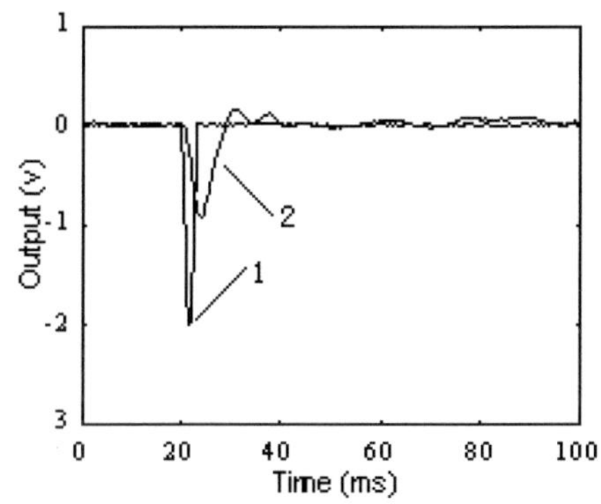

Fig. 6-20 Compensation result of HLC using linear approach compared with the input signal (1) impulse input, (2) compensation result

6.3 Wiener model based modeling and correction

A kind of nonlinear dynamic compensation method is proposed based on the Wiener model. Sensors with the nonlinear dynamic characteristics are describing as the Wiener model that is the cascade connection of a linear dynamic subsystem followed by a nonlinear static part. The nonlinear static characteristic of sensors is first corrected, and then the linear dynamic response is compensated. A DSP-based nonlinear dynamic compensating system and a sensor simulator are developed, and the experiments are carried out to demonstrate the effect of the nonlinear dynamic compensation method.

1. Principle of nonlinear dynamic compensation

Some sensors with the nonlinear dynamic characteristics can be divided into a linear dynamic subsystem and a nonlinear static part, which is shown in Fig. 6-21. They can be described by the differential equation as the following.

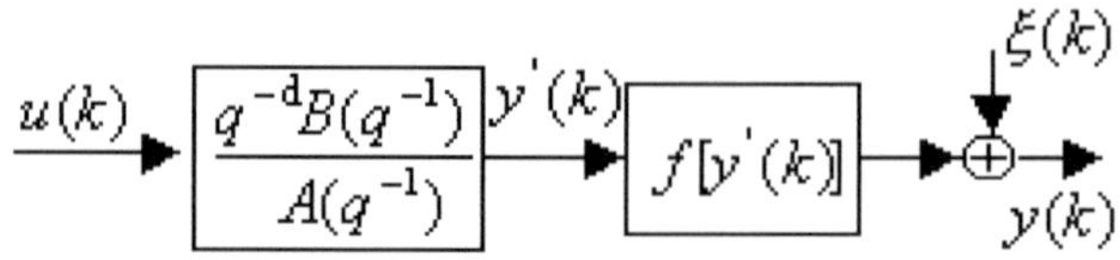

Fig. 6-21 Structure scheme of sensor after decomposition

$$\begin{cases} A(q^{-1})y^{'}(k) = q^{-d}B(q^{-1})u(k) \\ y(k) = f[y^{'}(k)] + \xi(k) \end{cases} \tag{6-22}$$

Where

$$A(q^{-1}) = 1 + a_1 q^{-1} + \cdots + a_n q^{-n}$$
$$B(q^{-1}) = b_0 + b_1 q^{-1} + \cdots + b_m q^{-m}$$

Where, $A(q^{-1})$ and $B(q^{-1})$ are polymerizations of n and m order, d is the time delay of the system, $f(\cdot)$ is the nonlinear static part, $y(k)$ and $u(k)$ are the output and input of the sensor, respectively, $\xi(k)$ is noise at the output end of the sensor, $y^{'}(k)$ is the output of linear dynamic part, k is discrete time variable.

We design a nonlinear dynamic compensating system shown in Fig. 6-22. $G(s)$ is a transfer function of the linear dynamic subsystem of sensor, and $y^{'}$ is the output of the linear dynamic subsystem. $f(\cdot)$ expresses the nonlinear static relationship, which is monotonous, for examples,

$$y_i = c_0 + c_1 y_i^{'} + c_2 (y_i^{'})^2 + c_3 (y_i^{'})^3 + \cdots, i = 0, 1, \cdots, M \tag{6-23}$$

Where i is the amplitude variable.

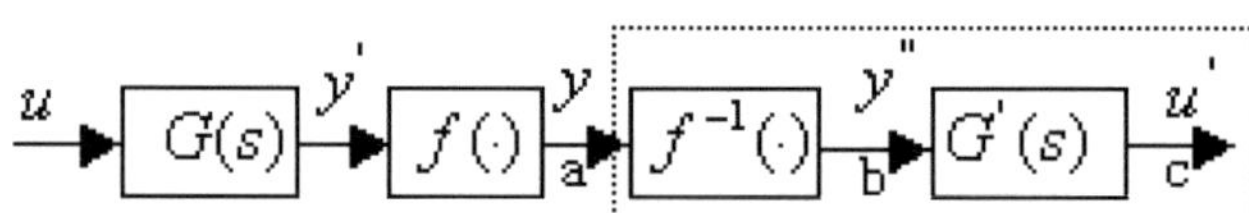

Fig. 6-22 Schematic diagram of sensor and compensating system

The frame that is dashed line shows the nonlinear dynamic compensating system. In the frame, $f^{-1}(\cdot)$ is the nonlinear static correcting part. After correction, the dynamic linear response $y^{''}(k)$ is obtained. $G^{'}(s)$ is the linear dynamic compensation part. After compensation, the output signal $u^{'}(k)$ should express the measured signal $u(k)$ accurately.

2. Design of nonlinear dynamic compensation system

The design procedure of the nonlinear dynamic compensating system is given as follows.

a. Static calibration experiment

For example, a force sensor is load different weights in its measuring rang, and the sensor output are tested and recorded. Thus the sensor data of input and output $(y_i^{'}, y_i), i = 0, 1, \cdots, M$, are collected, and the nonlinear static characteristic $f(\cdot)$ is obtained in Fig. 6-23.

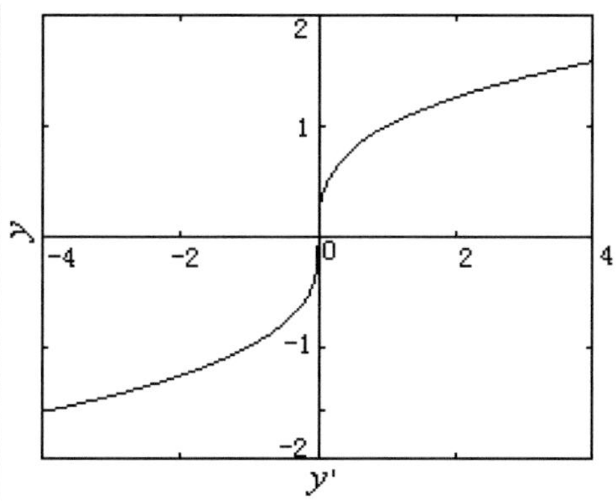

Fig. 6-23 Nonlinear curve

b. Design of nonlinear static correction subsystem

Assuming $f^{-1}(\cdot)$ is a reverse function of $f(\cdot)$. In some conditions, it is difficulty to obtain the reverse function $f^{-1}(\cdot)$. We adopt the method of looking up table. According to the nonlinear input and output of sensors, a table of correcting nonlinear is determined. In order to improve the precision of looking up table, the number of data in the table is increased using the interpolating method. Assuming the total number of data is $M+1$, the data in the table form are restored in the memory of a real time compensating system of the sensor, its distribution is shown in Fig. 6-24. The compensation system is connected with the sensor. The output signals of the sensor are firstly processed by the nonlinear static correction method. After correction, $y^{*} = y^{'}$, i.e. the nonlinear output become the linear output.

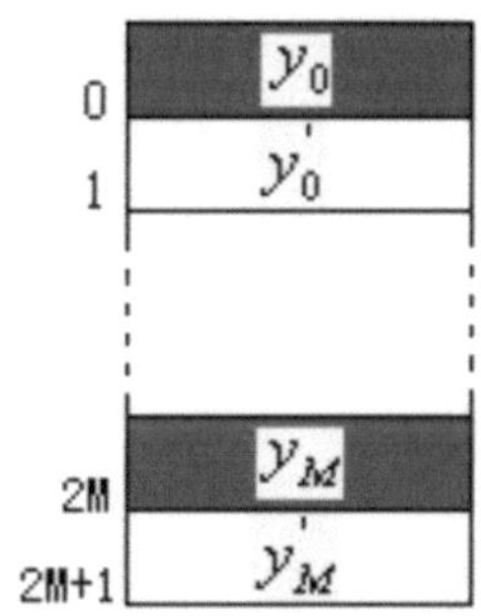

Fig. 6-24 Date distribution in the memory

c. Dynamic calibration experiment

For examples, the force sensor is applied to a negative step form force through sudden removing weights attaching to the sensor, and the sensor response is collected by a real time compensating system. The dynamic response acquired is linear because the compensation system has a function of correcting static non-linearity.

d. Design of linear dynamic compensation subsystem

According to the step input u of the sensor and the step response y' of the compensation system, a linear dynamic model in the form of differential equation is set up using the system identification or the artificial neural network. On the basis of the linear dynamic model, a linear dynamic compensation subsystem can be designed using the pole-zero configuration method.

(1) The linear model of the sensor in the form of differential equation is transformed into the transfer function in the continuous domain.

(2) The zeroes of compensation part are designed as equal to the poles of linear dynamic part of the sensor. Thus the poles of sensor are canceled out completely or partly.

(3) According to the criterion that the damp ratio is 0.707 and natural frequency is not changeable, the poles of linear dynamic compensation subsystem are determined.

3. Development of nonlinear dynamic compensation system

The nonlinear dynamic compensating system is shown in Fig. 6-25. This system mainly includes an ADSP-2181 EZ-KIT Lite, an analog input part, an output part and the logic control circuit. The analog input part consists of eight sample and hold circuits (S/H), a multiplexer (MUX), an amplifier (AMP) and an analog to digital converter (A/D). The output part contains six digital to analog converters (D/A) and six RC filters. The logic control circuit mainly consists of a decoder. The ADSP-2181 EZ-KIT Lite is a minimal implementation of an ADSP-2181 processor designed by ADI Inc., and includes an ADSP-2181, an EPROM and a serial communication port et al. The outputs of sensors are connected to the inputs of S/Hs of the dynamic compensating system. It is controlled with the sampling frequency by ADSP-2181 that the sample mode is switched to the hold mode. The signals are switched and connected sequentially by MUX, amplified by the AMP, and sent to the A/D. A busy pin of A/D is connected to a programmable input/output pin. ADSP-2181 determines the reading time according to the state of busy pin.

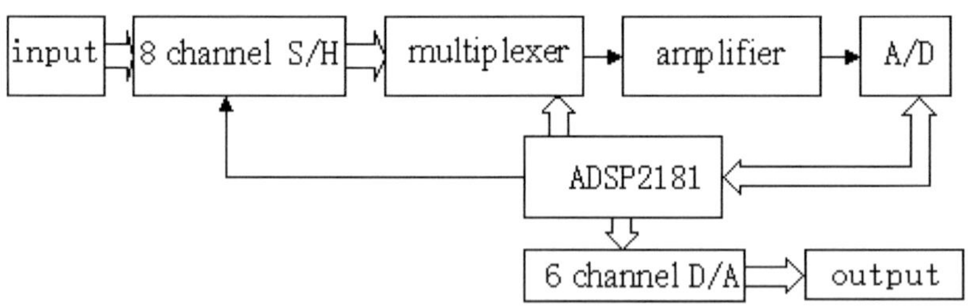

Fig. 6-25 Schematic block diagram of nonlinear dynamic compensating system

After multi-channel signals at the same time are acquired by ADSP-2181, they are processed by the nonlinear dynamic compensation method, and then are output by D/As. Under program control of ADSP-2181, the logic control circuit determines chip selects of A/D and D/A.

Software of the system includes data acquisition, data processing and result output. The sampling frequency of the system is determined by interrupt of timer, and is between $20\,K\ Hz \sim 25\,K\ Hz$. When power is applied to the system, the system start initialization, then enter the state of waiting for interruption. When the timer generates an interruption, the system begins a circle of data acquisition, processing and output.

4. Sensor simulator

In order to verify the effectiveness of nonlinear dynamic compensation method and system, a DSP-based simulating system of sensors is developed to produce the nonlinear dynamic responses in various forms. The sensor simulator can also produce noise to exam the anti-disturbance of the nonlinear dynamic compensation system.

Fig. 6-26 shows the hardware schematic diagram of the sensor simulator. This system mainly includes an ADSP-2181 EZ-KIT Lite, an output part and logic control circuit. The output part contains eight digital to analog converters (D/A) and eight RC filters. The logic control circuit mainly consists of a decoder.

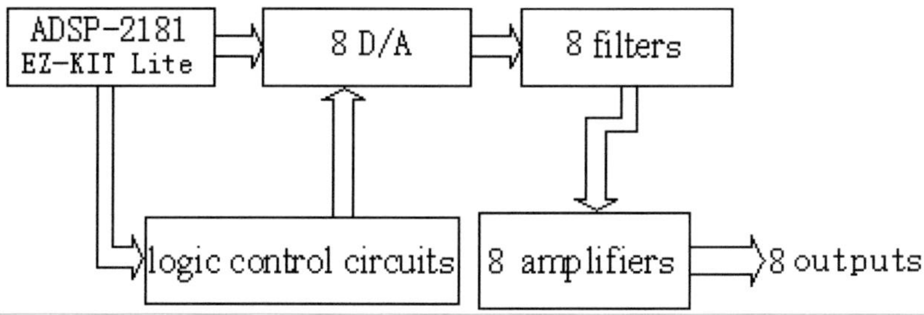

Fig. 6-26 Schematic block diagram of simulating system of sensors

The software flow chart of the sensor simulator is shown in Fig. 6-27. It includes an initialization, reading the input data, generation of the linear dynamic response by solving the differential equation, and generation of the nonlinear dynamic response through solving the nonlinear equation.

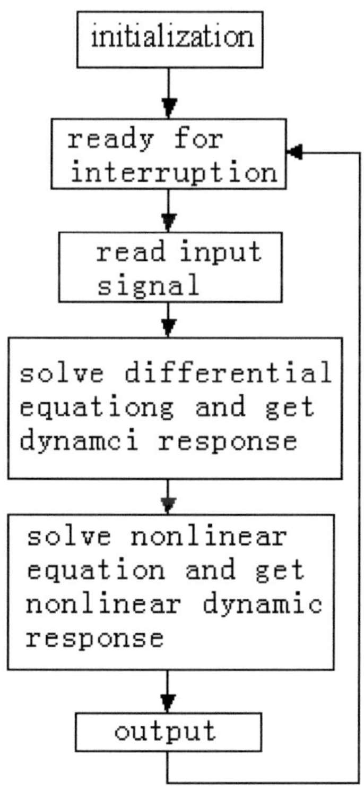

Fig. 6-27 Flow chart

5. Experiments

(1) Experimental setup

An experimental setup is shown in Fig. 6-28. PC1, PC2 and PC3 are personal computers. PC1 is in communication with the sensor simulator through a RS232. PC2 controls a scope through a GPIB to sample the outputs of the sensor simulator and compensating system. PC3 is connected with the compensating system through a RS232. The sensor simulator produces three channel signals as shown in Fig. 6-28. The channel 1 is the nonlinear output, the channel 2 is the linear output, and the channel 3 is the input signal. The compensating system has one input channel, i.e. the channel 4 that samples the nonlinear output of the sensor simulator. It has four output channels, the channel 5 is a direct output of sampled signal, the channel 6 is the nonlinear static correcting result, and the channel 7 is the dynamic compensating result.

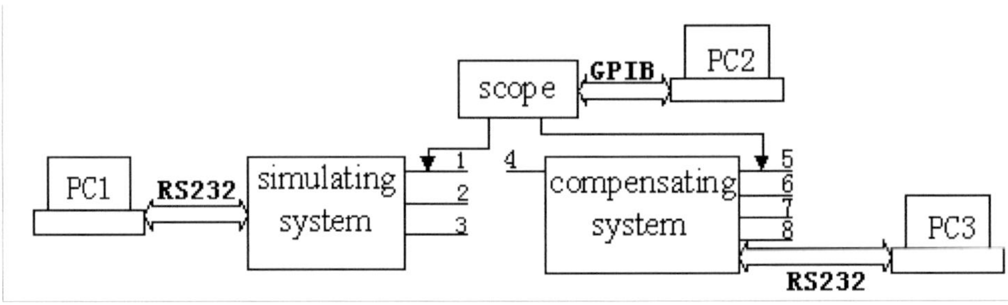

Fig. 6-28 Experimental setup

Assuming the linear part of the sensor is

$$y'(k) = 1.97097y'(k-1) - 0.99139y'(k-2) + 0.00527u(k) + 0.00233u(k-1) + 0.01282u(k-2) \quad (6\text{-}24)$$

The nonlinear part of the sensor is

$$y(k) = (y'(k))^{\frac{1}{3}} \quad (6\text{-}25)$$

(2) Nonlinear dynamic compensation process

The nonlinear dynamic compensating system is connected with the sensor simulator. It samples the output signal of the sensor. In one sampling interval, when one date is acquired, the corresponding linear output is obtained through the table of nonlinear correction, and then is handled using the linear dynamic compensation method.

(3) Experimental results

The step input and response of the sensor simulator are shown in Fig. 6-29. The result of nonlinear dynamic compensation, i.e. the output of the compensation system is shown in Fig. 6-30. The impulse input and response of the sensor simulator are shown in Fig. 6-31. The result of nonlinear dynamic compensation, i.e. the output of the compensation system is shown in Fig. 6-32. It is clear that the output of the compensation system is approximated to input of the sensor simulator, which proves that the nonlinear dynamic compensation method is effective.

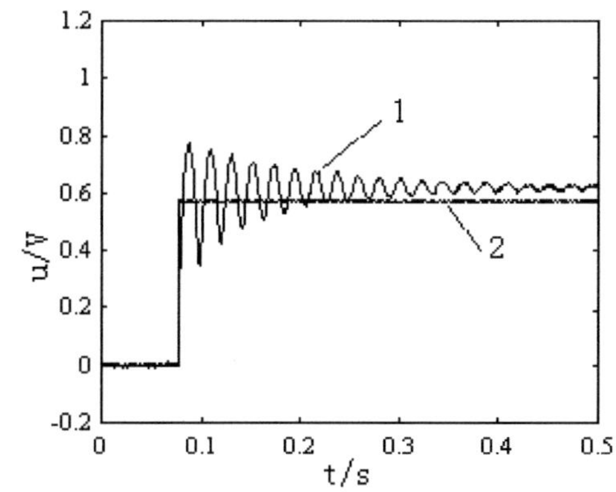

Fig. 6-29 Input and nonlinear output (1)nonlinear output, (2) input signal

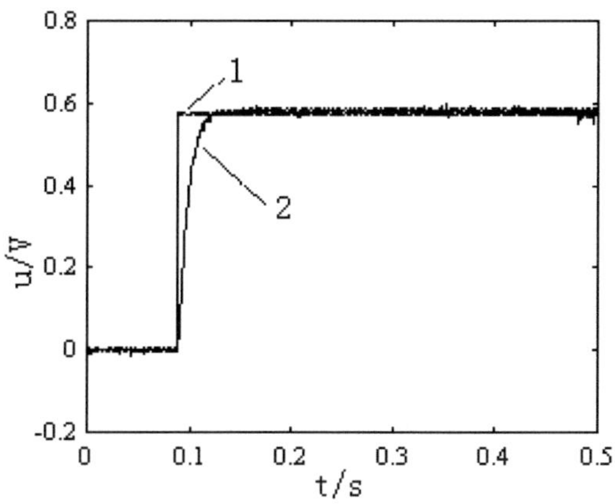

Fig. 6-30 Nonlinear output and compensating result (1) input signal, (2) compensated result

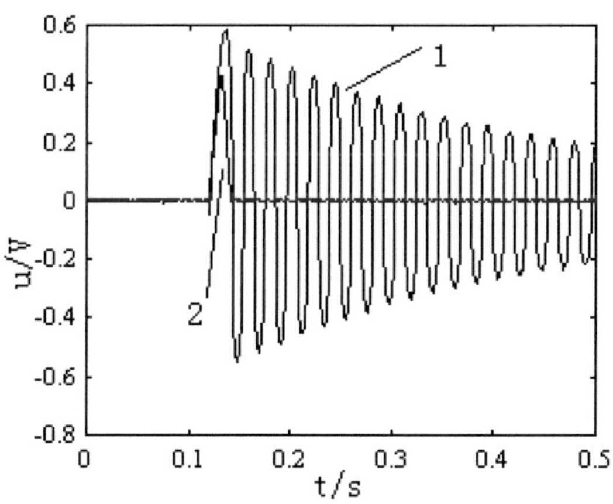

Fig. 6-32 Input and nonlinear output (1) nonlinear output, (2) input signal

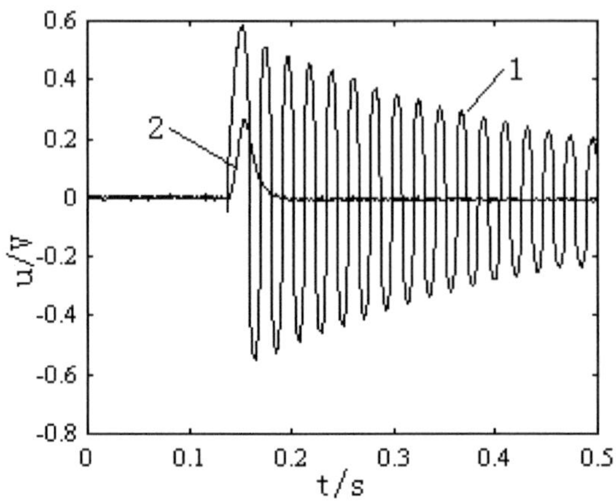

Fig. 6-33 Nonlinear output and compensation result (1) nonlinear output, (2) compensation result

6. Conclusions

(1) The dynamic response of sensors that posse the nonlinear dynamic characteristics is first handled using the nonlinear static correction method to obtain the linear dynamic response, and then is processed using the linear dynamic compensation method to short the time of reaching the steady state.

(2) This kind of method is applicable for different form and amplitude nonlinear dynamic responses of sensors.

(3) If there are noises in the nonlinear dynamic responses of sensors, a digital filter with two-order may be added into the compensation system. The place and order of the digital filter have been studied. The filter may be put the behind of the nonlinear static correction part or the linear dynamic compensation part. The cut-off frequency of the filter should be 2 times as large as the natural frequency of sensors.

7. References

Dwayne M. Perry, "Multi-axis force and torque sensing", Sensor Review, Vol.17, No.2, pp.117-120, 1997

Lu-Ping Chao,Kuen-tzong Chen , "Shape optimal design and force sensitivity evaluation of six-axis force sensors", Sensors and Actuators A, ol.63,pp. 105-112,1997

Maurice Dubious , "Six-component stain- gage balances for large wind tunnels", Experimental Mechanics, No.11, pp.401- 407,1981

Dirk Diddens, Dominiek Reynaerts , Hendrik Van Brussel , "Design of a ring- shaped three-axis micro force/torque sensor", Sensor and Actuators A , Vol.46-47,pp.225-232,1995

T. C. Hsia, "System Identification—Least Square Methods", Lexington Books, Lexington, Mass, 1977

P. Daponte et al. "Artificial neural networks in measurements", Measurement, Vol.23, No.2, pp.93-115,1998

J. C. Patra, G. Panda, R. Baliarsingh, "Artificial neural network-based non-linearity estimation of pressure sensors", IEEE Trans. Instru. Meas., Vol.63, No.6, pp.874-881, 1994

J. C. Patra, "An artificial neural network-based smart capacitive pressure sensor", Measurement, Vol.22, No.2, pp.113-121,1997

Ke-Jun Xu and Ting Tang, "A pole-zero offset method for sensor's dynamic compensation", Chinese Science Bulletin, Vol.39, No.16, pp.1407-1408, 1994

Ke-Jun Xu, Yin. Zhang and Chong-Wei Zhang , "Investigation of dynamic compensation for wrist force sensor", ACTA Metrologica Sinica, in Chinese, Vol.18, No.2,pp.116-121,1997

Waldemar Minkina, "Non-linear models of temperature sensor dynamics," Sensors & Actuators A, vol.30, pp.209-214, 1992

Waldemar Minkina, "Theoretical and experimental identification of the temperature sensor unit step response non-linearity during air temperature measurement," Sensors & Actuators A, vol.78, pp.81-87, 1999

Ping Wang, Shangshu Duan & Xinming Zhao et al, "Analysis the dynamic nonlinearity of transducers and its application in dynamic modeling," (in Chinese) Acta Metrologica Sinica, Vol.14, No.4, pp. 302-307, 1993

S. Beling, G. Blaeser, J. Bock, L. Heineit et al, "Signal conditioning for semiconductor gas sensors being used as detectors in gas-chromatographs and similar applications," Sensors & Actuators B, Vol.52, pp.15-22, 1998

Haixia Zhang & Hejun Li, "Nonlinear model of A^2PI^1 magnetic field sensor," Sensors & Actuators A, Vol.86, pp.206-210, 2000

Ke-Jun Xu and Cheng Li, "An estimation of nonlinear transfer function for wrist force sensor in frequency domain," (in Chinese) Journal of Applied Sciences, Vol.17, No.4, pp. 457-462,1999

Ke-Jun Xu and Cheng Li, "Dynamic nonlinear analysis of a wrist force sensor in the time and frequency domains," Proceedings of the 3rd World Congress on Intelligent Control and Automation, pp.1483-1487, Hefei, P. R. China, June 28 - July 2, 2000

Antonio Pardo, Santiago Marco, and Josep Samitier, "Nonlinear inverse dynamic models of gas sensing system based on chemical sensor arrays for quantitative measurements," IEEE Trans. on IM, vol.47, no.3, pp.644-651, 1998

Ke-Jun Xu, Zhi-Neng Zhu and Jia-Jun Liu, "Research on dynamic non-linearity compensation of sensor," (in Chinese) Proceedings of the 3rd World Congress on Intelligent Control and Automation, pp.1501-1504, Hefei, P. R. China, June 28 - July 2, 2000

Jozef Voros, "Iterative algorithm for parameter identification of Hammerstein systems with two-segment nonlinearities," IEEE Transactions on Automatic Control, Vol. 44, No.11, pp 2145-2149, 1999

Zhengliang Huang, Baiwu Wan and Chongzhao Han, "A two-stage identification technique for Hammerstein model," Control theory and applicatioon,Vol.12, No.1, pp.34-39, 1995

Ke-Jun Xu, "Applied research methods for dynamic characteristics of sensors," (in Chinese) Press of University of Science and Technology of China, 1999

11

Performance Analysis and Optimization of Sizable 6-axis Force Sensor Based on Stewart Platform

Y. Z. Zhao, T. S. Zhao, L. H. Liu, H. Bian and N. Li
Robotics Research Center, Yanshan University
P. R. China

1. Introduction

The Stewart platform, originally proposed for a flight simulator by Stewart (1965) has been suggested for a variety of applications by Hunt (1978), Fichter (1986) and Portman (2000). The advantage of the compact design with six degrees of freedom prompts one to consider the mechanism for force-torque sensor application. The parallel 6-axis force sensor is a kind of measure instrument which has the ability of detecting the forces and moments in x, y, and z directions simultaneously. The 6-axis force-torque sensor has been widely used in the situation of force/force-position control, such as parts teaching, contour tracking, precision assembly, etc. in addition to the applications in thrust testing of rocket engines and wind tunnel by Gaillet (1983) and Kaneko (1996).

Performance analysis and optimization design are important during the design of the sensor. There are a lot of literatures available on the design of force-torque sensor. Kerr (1989) analyzed an octahedral structure and enumerated a few design criteria for the sensor structure. Uchiyama and Hakomoic (1985) studied the isotropy of force sensor. Bicchi (1992) discussed the optimization of force sensor. Xiong (1996) defined the isotropy of force sensor on the basis of the information matrix. Jin (2003) presented the indices design method for 6-axis force sensor used on a dexterous hand. Ranganath (2004) studied the performances of the force sensor in the near-singular configuration. Tao (2004) optimized the performances of force sensor with finite element method. Theoretical and experimental investigations of the Stewart platform sensor were carried out by various authors, namely Romiti and Sorli (1992), Zhmud (1993) and Dai (1994) etc. So far, the researchers have obtained many achievements in the field of 6-axis force sensor, but the performances of the sizable parallel 6-axis force sensor prototype based on Stewart platform varies largely in different directions. The further application of the sizable parallel sensor is blocked by the existent performance anisotropy. So, the performance analysis and optimization design is significant to evaluating performances and the conceptual design of the sizable parallel sensor based on Stewart platform.

This paper presents the performance analysis and optimization design of the sizable parallel 6-axis force sensor with Stewart platform. The paper is organized as follows. Section 2 presents the static mathematics model of the 6-axis force sensor with screw theory. The static force influence coefficient matrix and the generalized force Jacobian matrix of the 6-axis force sensor are derived. Based on the screw theory and the theory of physical model of the solution space, some performances indices are defined. The force isotropy, torque isotropy, force sensitivity isotropy and torque sensitivity isotropy indices atlases of the 6-axis force sensor are plotted, and the rules how structure parameters affect the performances indices are summarized in Section 3. The optimization method of sizable parallel 6-axis force sensor's structure parameters is proposed, and an optimization numerical example is demonstrated in nonlinear single objective and multi-objective in section 4, respectively. Based on the result of the performances analysis and optimization, the section 5 presents a novel sizable 6-axis force sensor with flexible joints, which can avoid effectively the friction and the clearance in general spherical joint and has a wider application foreground. The research result reported of the chapter is concluded in section 6, future research in section 7, acknowledgement in section 8, and references in section 9.

2. Static mathematics model of 6-axis force sensor

The Stewart platform 6-axis force sensor is a kind of special parallel mechanism that is symmetrical design. Fig.1 is the sketch of the mechanism and forces acted on the platform. The platforms of the upper and lower platform are shown in Fig.2. O_u-$X_uY_uZ_u$ is the coordinate system fixed on the center point **P** of the upper platform, when the upper platform and the lower are both in the horizontal position. The spherical joints connecting links and upper platform at the upper ends are signed $a_i\ (i=1,2,\cdots,6)$ while the spherical joints in the lower and the corresponding position vectors are $\boldsymbol{A}_i\ (i=1,2,\cdots,6)$ and $\boldsymbol{B}_i\ (i=1,2,\cdots,6)$ respectively. Each link will be subjected only to the axial force, ignoring the links' gravitation and the friction between joints.

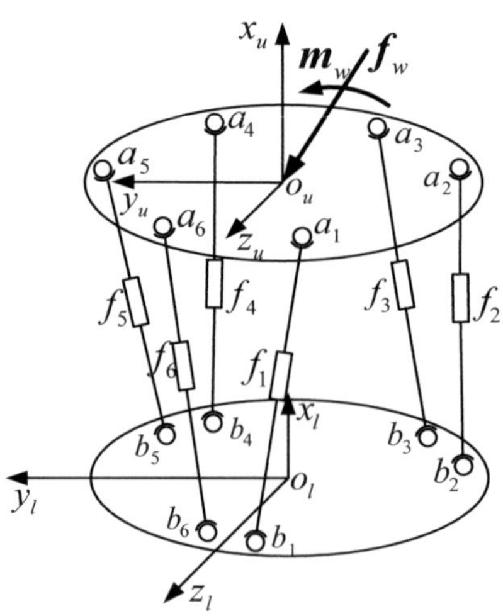

Fig. 1. The sketch of 6-axis force sensor based on Stewart platform

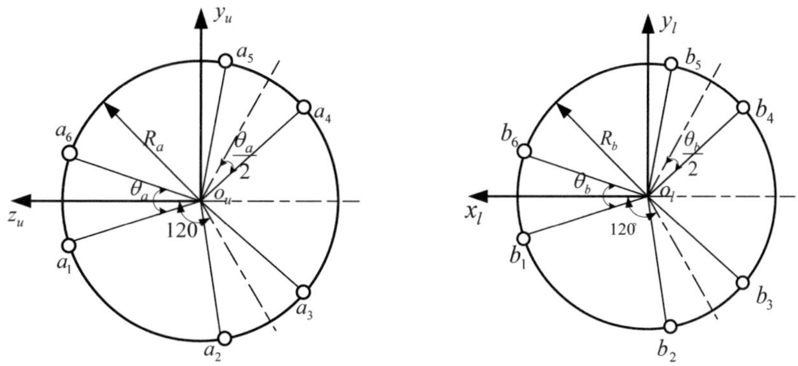

Fig. 2. The upper and lower platform of 6-axis force sensor's

Investigating the upper platform, the force equation based on the screw theory and static equilibrium can be obtained as

$$F_w = \sum_{i=1}^{6} f_i \$_i \tag{1}$$

where, f_i is magnitude of the ith link's axial force, $\$_i = (S_i ; S_{0i})^T$ expresses the unit vector of ith link's direction, and $F_w = (f_w \ m_w)^T$ is the generalized external force applied to the upper platform. $f_w = (f_{wx} \ f_{wy} \ f_{wz})^T$ and $m_w = (m_{wx} \ m_{wy} \ m_{wz})^T$ are the external force and moment. The above equation can be disintegrated as

$$\begin{aligned} f_w &= \sum_{i=1}^{6} f_i S_i \\ m_w &= \sum_{i=1}^{6} f_i S_{0i} \end{aligned} \tag{2}$$

where, $S_i = (a_i - b_i) / |a_i - b_i|$ and $S_{0i} = (b_i \times a_i) / |a_i - b_i|$. So, the equation (1) can be also expressed as $F_w = G f$, where $f = (f_1 \ f_2 \ f_3 \ f_4 \ f_5 \ f_6)^T$. The static force influence coefficient matrix of the parallel 6-axis force sensor can be expressed as

$$G = \begin{bmatrix} S_1 & S_2 & \cdots & S_6 \\ S_{01} & S_{02} & \cdots & S_{06} \end{bmatrix} \tag{3}$$

The former three rows of the matrix G is the force transmitting factor of the parallel sensor, while the latter three rows is the torque-transmitting factor. The factors having different unit, which the former is dimensionless, while the latter has length unit, the matrix G is disintegrated into the static force influence coefficient matrix G_1 and the static torque influence coefficient matrix G_2. That is $G = [G_1 \ G_2]^T$. The transformational relation between the generalized external force in 6 dimensions and the link's axial force can be given as

$$\boldsymbol{J}\,\boldsymbol{F}_w = \boldsymbol{f} \tag{4}$$

where, $\boldsymbol{J} = \boldsymbol{G}^{-1}$ is the generalized force Jacobian matrix of the parallel 6-axis sensor. Similarly, the generalized force Jacobian matrix $\boldsymbol{J}$ is disintegrated into the force Jacobian matrix $\boldsymbol{J}_1$ and the torque Jacobian matrix $\boldsymbol{J}_2$, that is $\boldsymbol{J} = \begin{bmatrix} \boldsymbol{J}_1 & \boldsymbol{J}_2 \end{bmatrix}$.

3. Performance analysis of parallel 6-axis force sensor

3.1 Physical model of the solution space theory

The physical model of the solution space theory has the ability to show all possible size combination of the mechanism. It is convenient to obtain the law of the sensor's indices following the changing of the element structure parameters. From the static mathematics model of the force sensor above, the 6-axis force sensor based on Stewart platform contains four structure parameters. That is the radius R_a of the upper platform, the radius R_b of the lower platform, the height H between platforms, and the angle difference $\theta_{ab} = \theta_a - \theta_b$ between the corresponding twin link of the upper and the lower platform. With the precondition of θ_{ab} is changeless, let $R_a + R_b + H = T$, then

$$\frac{R_a}{T} + \frac{R_b}{T} + \frac{H}{T} = 1 \tag{5}$$

Let $r_a = \dfrac{R_a}{T}$, $r_b = \dfrac{R_b}{T}$ and $r_c = \dfrac{R_c}{T}$, the equation (5) gives

$$r_a + r_b + r_H = 1 \tag{6}$$

where, $0 < r_a < 1$, $0 < r_b < 1$, $0 < r_H < 1$. Thus, the physical model of the solution space theory of the 6-axis force sensor based on Stewart platform is developed. For displaying conveniently, the physical model can be transformed into two dimension *O-XY* plane as shown in Fig. 3. The transformation between the coordinates can be expressed as

$$x = \frac{2\sqrt{3}}{3} \cdot r_b + \frac{\sqrt{3}}{3} \cdot r_H \text{ and } y = r_H \tag{7}$$

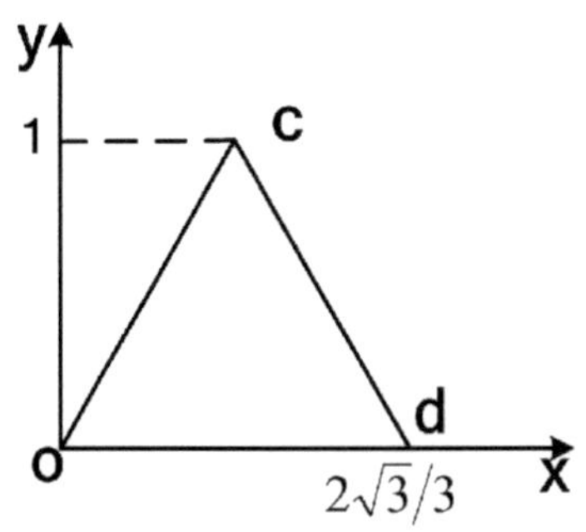

Fig. 3. The ichnography of the sensor's spacial model

Therefore, all possible parameters combination of the 6-axis force sensor based on Stewart platform are included in the triangle *ocd*. In other words, each point in the triangle ocd corresponds with a set of structure parameters. With the physical model of the solution space theory, selecting parameters and optimization structure design are convenient greatly.

3.2 Performances atlases analysis

The indices evaluating the performances of the 6-axis force sensor are the foundation of the performance evaluating and the optimization design. As for the parallel 6-axis force sensor, it should have high force isotropy, torque isotropy, and force/torque (F/T) sensitivity isotropy, in addition to the high sensitivity, precision, signal noise ratio (SNR) and speedy response. The performances atlases are plotted in the area of the physical model triangle ocd, based on the static mathematics model above and the defining of the performances indices given by Uchiyama and Hakomori (1985), Xiong (1996) and Jin (2003) with the force isotropy $u_1 = 1/\text{cond}(\boldsymbol{G}_1)$, the torque isotropy $u_2 = 1/\text{cond}(\boldsymbol{G}_2)$, the force sensitivity isotropy $u_3 = 1/\text{cond}(\boldsymbol{J}_1)$ and the torque sensitivity isotropy $u_4 = 1/\text{cond}(\boldsymbol{J}_2)$. From the sensor's physical model of the solution space theory developed above, the performances atlases varies with the angle θ_{ab}. It is unpractical to show all existent performances atlases. Considering the latter optimization design of the structure parameters, the performances spacial and planar atlases are plotted as shown in Fig. 4-11, respectively, when the coordinate system fixed on the center point of the lower platform and $\theta_{ab} = 60^\circ$. It can be easily gotten the indices distributing laws with the performances atlases of force isotropy, torque isotropy, force sensitivity isotropy and torque sensitivity isotropy, especially in the planar atlases of as shown in Fig.5, Fig.7, Fig.9 and Fig.11.

From the influence that the structure parameters act on the sensor's performances indices shown in Fig. 4-11, the laws guiding the optimization design can be concluded as following. The plot of the force isotropy distributes parabola approximately in the area of the physical model as shown in Fig. 4 and Fig. 5. The force isotropy will becomes higher in the middle and lower area of the physical model. The corresponding structure parameters can be selected, when the index of the force isotropy should be attached importance to design.

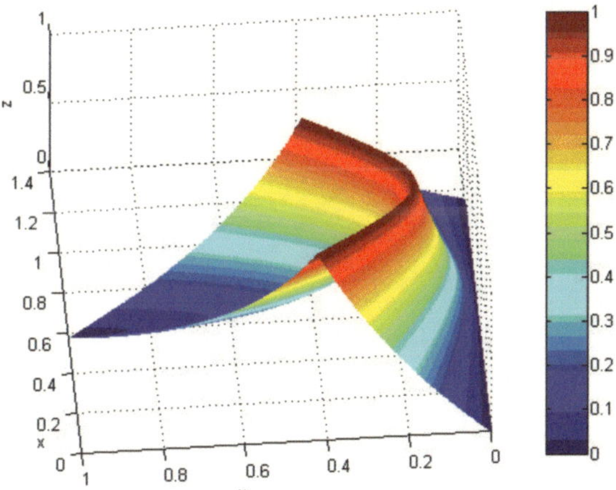

Fig. 4. Force isotropy spacial atlas with respect to $\theta_{ab} = 60^\circ$

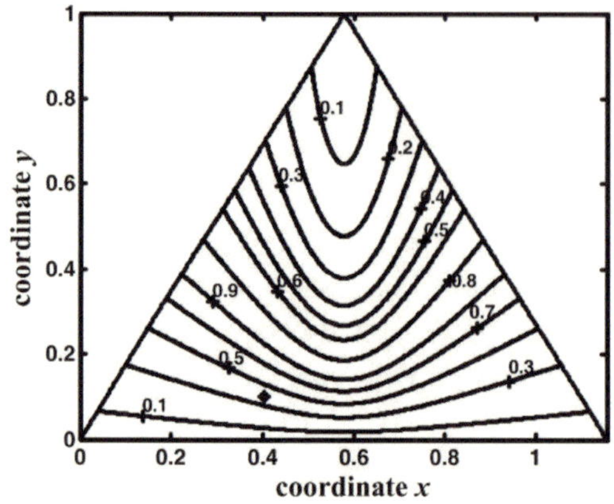

Fig. 5. Force isotropy planar atlas with respect to $\theta_{ab} = 60^{\circ}$

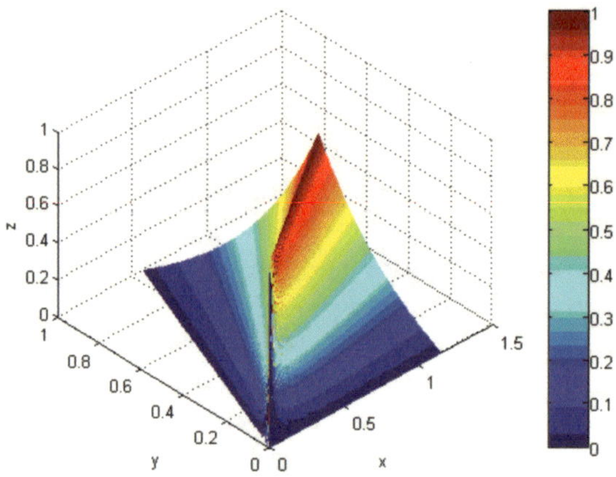

Fig. 6. Torque isotropy spacial atlas with respect to $\theta_{ab} = 60^{\circ}$

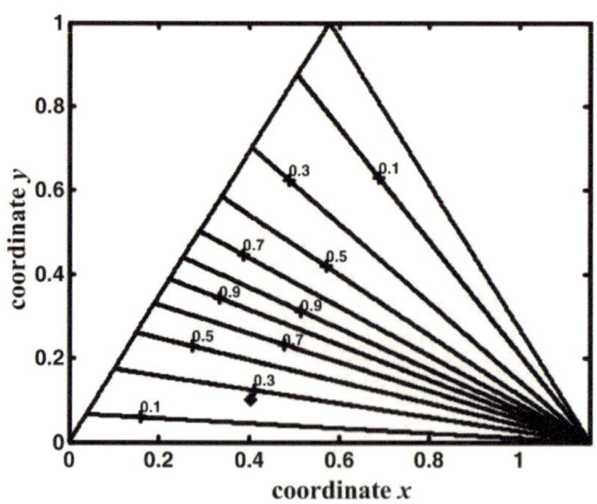

Fig. 7. Torque isotropy planar atlas with respect to $\theta_{ab} = 60^{\circ}$

The plot of the torque isotropy distributes beeline approximately in the area of the physical model as shown in Fig. 6 and Fig. 7. The torque isotropy will becomes lower in the right side and upper area of the physical model. The corresponding structure parameters should be eliminated, when the index of the torque isotropy should be attached importance to design.

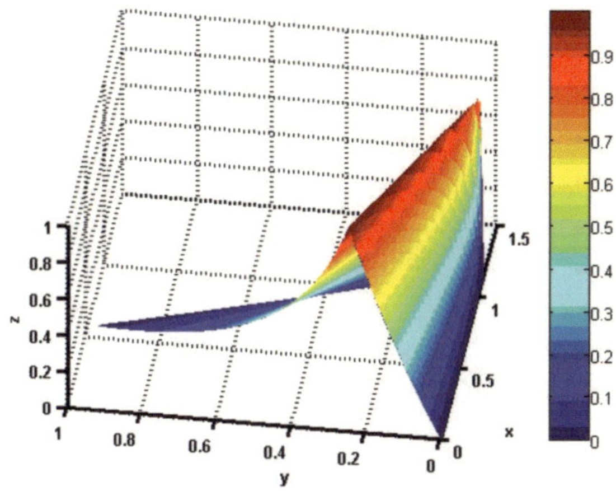

Fig. 8. Force sensitivity isotropy spatial atlas with respect to $\theta_{ab} = 60^{\circ}$

The plot of the force sensitivity isotropy distributes beeline approximately in the area of the physical model as shown in Fig. 8 and Fig.9. The force sensitivity isotropy will change rapidly by the x axis in the physical model. The corresponding structure parameters should be eliminated in design. In the upper most area, the index of the force sensitivity isotropy is smaller. The force sensitivity isotropy distributing resembles the torque isotropy distributing of the force sensor.

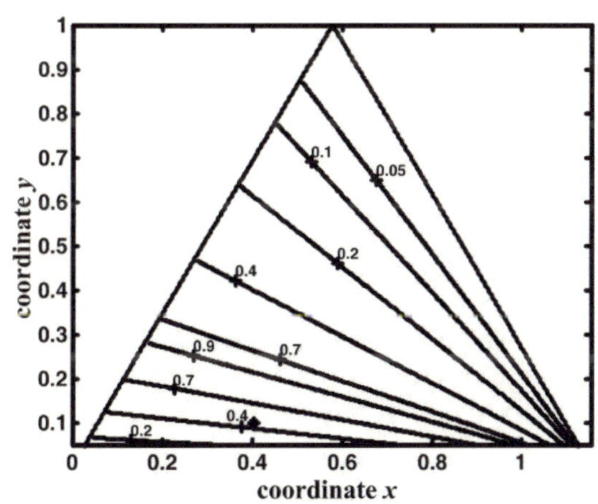

Fig. 9. Force sensitivity isotropy planar atlas with respect to $\theta_{ab} = 60^{\circ}$

The plot of the torque sensitivity isotropy distributes parabola approximately in the area of the physical model as shown in Fig. 10 and Fig. 11. The torque sensitivity isotropy will becomes higher in the middle part of the physical model. The corresponding structure parameters can be selected, when the index should be attached importance to design.

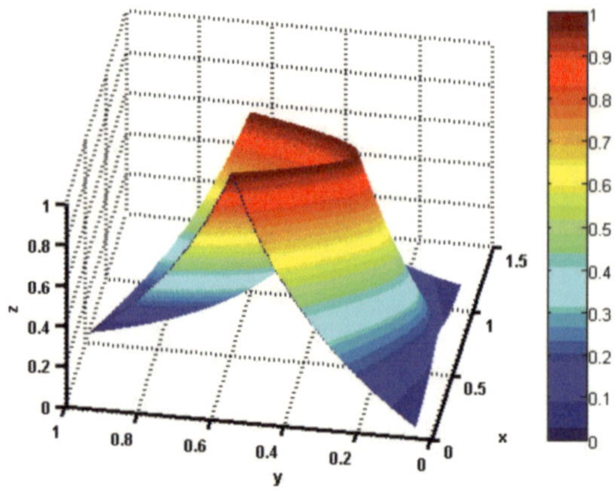

Fig. 10. Torque sensitivity spacial isotropy atlas with respect to $\theta_{ab} = 60^{\circ}$

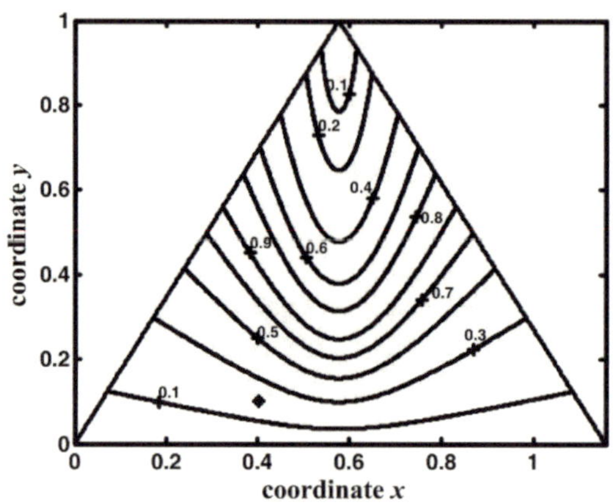

Fig. 11. Torque sensitivity planar isotropy atlas with respect to $\theta_{ab} = 60^{\circ}$

4. Optimization design of sizable parallel 6-axis force sensor

4.1 Optimization objective function

In the sensor's practical application, the request for the performances indices varies with the practical application cases. Some performance index should be considered principally in some cases, while the comprehensive performance index is pivotal in some cases. The paper optimizes the existing sensor's structure parameters in nonlinear single objective and multi-objective respectively, in order to obtain better performances than that of the initial ones. As the restriction of mechanical special model, the constraint equation $0 \le |\theta_{ab}| \le 120°$ should be applied. In the single objective optimum, the objective functions are chosen as following

$$f_{FD}(R_a \quad R_b \quad H \quad \theta_{ab}) = [\lambda_{\max}(\boldsymbol{G}_1^T \cdot \boldsymbol{G}_1)]^{1/2} / [\lambda_{\min}(\boldsymbol{G}_1^T \cdot \boldsymbol{G}_1)]^{1/2} \tag{8}$$

$$f_{MD}(R_a \quad R_b \quad H \quad \theta_{ab}) = [\lambda_{\max}(\boldsymbol{G}_2^T \cdot \boldsymbol{G}_2)]^{1/2} / [\lambda_{\min}(\boldsymbol{G}_2^T \cdot \boldsymbol{G}_2)]^{1/2} \tag{9}$$

$$f_{FS}\left(R_a\ \ R_b\ \ H\ \ \theta_{ab}\right)=[\lambda_{\max}(\boldsymbol{J}_1^T\cdot\boldsymbol{J}_1)]^{1/2}/[\lambda_{\min}(\boldsymbol{J}_1^T\cdot\boldsymbol{J}_1)]^{1/2} \tag{10}$$

$$f_{MS}\left(R_a\ \ R_b\ \ H\ \ \theta_{ab}\right)=[\lambda_{\max}(\boldsymbol{J}_2^T\cdot\boldsymbol{J}_2)]^{1/2}/[\lambda_{\min}(\boldsymbol{J}_2^T\cdot\boldsymbol{J}_2)]^{1/2} \tag{11}$$

The objective functions in the above equation (8)-(11) are the reciprocals of the force isotropy u_1, the torque isotropy u_2, the force sensitivity isotropy u_3 and the torque sensitivity isotropy u_4, respectively. When the objective function reaches the minimum, the corresponding performance index attains the maximum. When the comprehensive performance is pivotal, the multi-objective optimum would be executed to obtain the sensor with the high performances. The corresponding objective function can be expressed as

$$f_G\left(R_a\ \ R_b\ \ H\ \ \theta_{ab}\right)=\min\left(k_{FD}\cdot f_{FD}+k_{MD}\cdot f_{MD}+k_{FS}\cdot f_{FS}+k_{MS}\cdot f_{MS}\right) \tag{12}$$

where, k_{FD}, k_{MD}, k_{FS}, and k_{MS} are the weights of the corresponding indices. During the practical optimization, the weight matrix $\boldsymbol{k}=\left[k_{FD}\ k_{MD}\ k_{FS}\ k_{MS}\right]$ can be set as the weights of the corresponding performances indices.

4.2 Optimization numerical examples

Considering the practical structure parameters of the sizable parallel 6-axis force sensor, the initial parameters are set as $R_a=720$ mm, $R_b=360$ mm, $H=120$ mm, $\theta_{ab}=60^\circ$. With the Matlab optimization toolbox and the defined objective functions, the corresponding optimal design parameters are obtained. Based on the force isotropy single objective optimization, the optimal parameters can be obtained as $R_a=692$ mm, $R_b=378$ mm, $H=228$ mm and $\theta_{ab}=17^\circ$. The force isotropy of the sensor with the optimal parameters $u_1=1.0000$ with respect to the initial $u_1=0.3804$. Based on the torque isotropy single objective optimization, the optimal parameters can be obtained as $R_a=715$ mm, $R_b=360$ mm, $H=144$ mm and $\theta_{ab}=16^\circ$. The torque isotropy u_2 of the sensor with the optimal parameters improves to 1.0000 from the initial $u_1=0.2357$. When the force sensitivity isotropy is optimized, the optimal parameters can be obtained as $R_a=713$ mm, $R_b=373$ mm, $H=95$ mm and $\theta_{ab}=60^\circ$. Whereafter, the force sensitivity isotropy u_3 improves to 1.0000 from the initial 0.4714. When the torque sensitivity isotropy is optimized, the optimal parameters can be obtained as $R_a=634$ mm, $R_b=424$ mm, $H=357$ mm and $\theta_{ab}=31^\circ$. Whereafter, the torque sensitivity isotropy u_4 improves to 1.0000 from the initial 0.2116.

In the multi-objective optimization, the comprehensive performances indices should be taken into account synthetically. With the weight matrix **k**=[1111], the optimal performance indices with respect to the comprehensive parameters and the corresponding initial indices are shown in Table 1. It is obvious that the performances of the sizable parallel 6-axis force sensor are improved. The corresponding performance indices of the initial structure parameters in Table 1 are shown in the planar atlas Fig. 5, Fig. 7, Fig. 9, and Fig. 11 with the "*" symbol.

	R_a	R_b	H	θ_{ab}	u_1	u_2	u_3	u_4
Initial	720	360	120	60°	0.380	0.235	0.471	0.190
Optimal	614	450	418	85°	0.701	0.712	0.701	0.712

Table 1. Contrast multi-objective optimal result with the initial design

5. A novel sizable 6-axis force sensor with flexible joints

Based on the above analysis, optimization design and considering the machining technics synchronously, we design the a novel sizable 6-axis force sensor structure with flexible joints as shown in Fig. 12. Each branch is composed of UUR flexible joints and a standard pull and press force sensor. The axe of the flexible R joints go through the near U flexible joint, which can be considered as a sphere joint. The flexible joints here are the novel flexible joints which can carry the biggish loading. The six branches are divided the same 3 groups. The first U joints of the branches in some group are made in a whole material, similary as the last R joints of the branches in some group. Another design project with the same 6 unitary branch is shown as in Fig. 13.

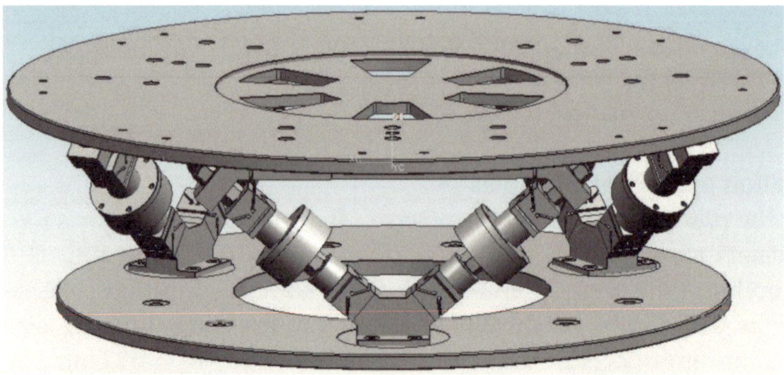

Fig. 12. A novel sizable 6-axis force sensor prototype with flexible joints

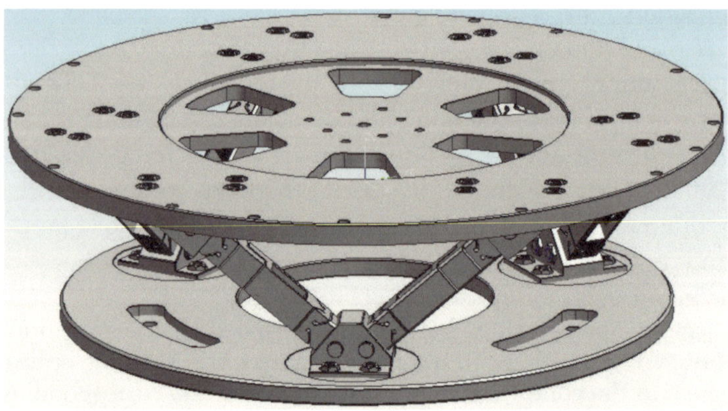

Fig. 13. Another sizable 6-axis force sensor prototype with flexible joints

6. Future research

The performance indices of the 6-axis force sensor based on Stewart platform shouled be further analyzed, especially dynamic performance index. The novel sizable 6-axis force sensor with flexible joints should be futher optimized, especially the stucture parameters of the flexible joints. The manufacture and calibration of the sizable 6-axis force sensor with flexible joints are also the future research.

7. Conclusion

The paper plots the indices atlases based on the screw theory and definition of the performances indices, and summaries the law how structure parameters affect the indices. With the constructed optimization objective functions, the sizable parallel 6-axis force sensor's structure parameters are optimized in nonlinear single objective and multi-objective respectively. The corresponding optimal structure parameters are obtained. A novel sizable 6-axis force sensor with flexible joints is developped. So, the powerful basis and method are raised for design and optimization of sizable parallel 6-axis force sensor based on Stewart platform.

8. Acknowledgment

The research work reported here is supported by National Natural Science Foundation of China (NSFC) under Grant No.50375134 and No.50675191.

9. References

Stewart D. (1965). A Platform with Six Degrees of Freedom. *Proc. Inst. Mech.*, Part 1 180 (15), 371-386.

Hunt K. H. (1978). *Kinematic Geometry of Mechanisms*, Clarendon Press, Oxford.

Fichter E. F. (1986). The Stewart Platform Manipulator: General Theory and Practical Construction. *Int. J. Robot. Res*, 5(2), 157-182.

Portman V. T.; Sandler B. & Zahavi E. (2000). Rigid 6×6 Parallel Platform for Precision 3-D Micro-manipulation: Theory and Design Application. *IEEE Trans. Robot. Automat*, 16(6), 629-643.

Gaillet A. & Reboulet C. (1983). An Isostatic Six Component Force and Torque Sensor. *Proc. 13th Int. Symposium on Industrial Robotics*, pp. 783-792.

Kaneko M. (1996). Twin-head Six-axis Force Sensor. *IEEE Trans. Robot. Automat*, 12(1), 146-154.

Kerr D. R. (1989), Analysis, Properties and Design of Stewart Platform Transducer. *Trans. ASME. J. Mech. Transm. Automn*, 111, pp. 25-28.

Uchiyama M. & Hakomori K.A. (1985). Few Considerations on Structure Design of Force Sensors. *Proceedings of the Third Annual Conf. Japan Robotics Society*, pp. 17-18.

Bicchi A. (1992). A Criterion for Optimal Design of Multi-axis Force Sensors. *Robotics and Autonomous Systems*, 10(4), 269-286.

Xiong Y. L. (1996). On Isotropy of Robot's Force Sensors. *Acta Automatica Sinica*, 22(1), 10-18. (in Chinese)

Jin Z. L.; Zhao X. C. & Gao F. (2003). The Research on Link Length Design of a Novel Dexterous Hand's 6-axis Force Transducer. *Chinese Journal of Scientific Instrument*, 24(4), 371-374.

R. Ranganath, et al (2004). A Force-torque Sensor Based on a Stewart Platform in a Near-Singular Configuration. *Mechanism and Machine Theory*, 39, pp. 971-998.

Tao L.; Yoshi I. & Kyoko S. (2004). A Six-dimension Parallel Force Sensor for Human Dynamics Analysis. *Proceedings of the IEEE Conference on Robotics, Automation and Mechatronics*, pp. 208-212.

Romiti A. & Sorli M. (1992). Force and Moment Measurement on a Robotic Assembly Hand. *Sensors Actuators A*, 32, pp. 531-538.

Sorli M. & Zhmud N. (1993). Investigation of Force and Moment Measurement System for a Rotating Assembly Hand. *Sensors Actuators A*, 37, pp. 651-657.

Dai J. S.; C. Sodhi & Kerr D. R. (1994). Design and Analysis of a New Six-component Force Transducer for Robotic Grasping. *Proceeding of the Second Biennial European Joint Conference on Engineering Systems Design and Analysis*, ASME PD, pp. 809-817.

12

Grip Force and Slip Analysis in Robotic Grasp: New Stochastic Paradigm Through Sensor Data Fusion

Debanik Roy
Bhabha Atomic Research Centre
India

1. Introduction

Algorithmic *data fusion* is instrumental in evaluating the quantitative output of a multi-sensory system and the same becomes extremely challenging, especially when the elemental sensory units do vary in type, size and characteristics. Truly, fusion of such *heterogeneous* sensory data remains an open-research paradigm till date, especially in the field of robotics, owing to its inherent characteristics in quantifying the output response of the system. The problem gets even critical when we need to contour with a limited number of elemental sensor-cells (*taxels*), in contrast to traditional theories dealing with large agglomeration of (identical) sensor units. In fact, fusion models used hitherto have been found to be largely inappropriate for the distinct object-groups, e.g. from point-mass to small-sized ones. Besides, paradigms of grasp synthesis (grip force & slippage) were largely unattended. Although traditional theories on sensory data fusion fit quite satisfactorily in searching a pre-defined object with a tentative dimension and depth perception, they fail to do justice in cases where profile of the object do vary from micro-scale to a finite spatial dimension. In answering these lacunas, the present article dwells on modeling, algorithm and experimental analysis of three novel fusion rule-bases, which are implemented in small-sized tactile array sensor to be used in robot gripper. A new proposition has been developed for assessing the *decision threshold*, signaling the *presence of object* inside the grasp-zone of the gripper. Besides, the developed model evaluates the approximate planar area of the grasped object alongwith its shape in real-time. The model also provides estimate for the gripping force required to sustain a stable grasp of the object vis-à-vis slippage characteristics, if any.
Signal detection with multiple sensors, either all similar or dissimilar or any arbitrary combination, can be performed in two manners. In the traditional method, the local sensors communicate all *observations* (raw data) directly to a *centralized detector* (e.g. system controller board) where decision processing is performed. This method, although incorporates parallel channels for data communication, often requires a large bandwidth for the communication channels in order to obtain real-time results. In contrary, the second method deals with each sensor individually, by associating a *detector* module to each of the sensor-cells, which *decides locally* whether a signal is detected or not. These *local decisions* get transmitted to the main controller unit (traditionally called "Data Fusion Center" in the literature), where those get *unified* for *global* decision. Although this method suffers from

loss of information, yet it is the most optimal choice for sensory system design because of high reliability, compact hardware, lower cost and a user-friendly operative environment. In fact, this group of signal processing via localized decision vis-à-vis the field of 'Decentralized / Distributed Decision Making' has been an active area of research, wherein the realization has come out in the manner that these very problems are qualitatively different from the corresponding decision thematic with centralized information. It is, perhaps, wise to conjecture that the prohibitive factor in decentralized problems is not so much the inadequacy of the mathematical tools presently been used, rather the inherent complexity of the problems that have usually been formulated.

The classical theory of optimal sensor signal processing is based on 'Decentralized Testing & Augmentation', using statistical estimation and hypothesis testing methods. The logically driven coherent *unified* output of the said aggregation is being used for processing allied control system signals of the robotic /gripper system. Unlike most of the decentralized control problems, the hypothesis-testing problem can be solved in a relatively straightforward way. This is due principally to the fact that since the decisions made do not get looped back into the system dynamics, those do not affect the information of other decision makers either. However, even in the case of independent observations, several types of unusual behaviour can occur. For example, the threshold computations can yield locally optimal thresholds, which are far from the globally optimal values. The paradigm of decentralized sensor fusion has hitherto been attributed largely by Bayesian Theory, which deals quite robustly the situations involving probabilistic hypothesis testing but fails to address the cases where fuzziness is involved in the main process itself. On the contrary, Dempster-Shafer Theory tackles only those problems where system caters for fuzzy concepts. Unfortunately both of the theories are inadequate so far as the data fusion in mechatronic system is concerned.

We propose a new fusion theory wherein the threshold for fusion can be suitably adapted depending upon the end-application. The proposed schemata provides insight to two aspects, namely evolution of new rule-bases towards data fusion and an optimized inference about object's presence or absence based on stochastic hypothesis testing model. This *dynamic thresholding* of the proposed hypothesis helps fusing the sensory data from the physical device (a multi-input heterogeneous tactile array sensor in the present case), based on the requirement of the user. Moreover, the *fusion rules*, do represent a unique strategy for assimilating the *raw* sensor data. The threshold estimation has been based on using the *variable limits*, exploiting the metrics of Type I error (i.e. rejecting the Alternative Hypothesis when true), as well as Type II error (i.e., accepting the Null Hypothesis when false), corresponding to three different fusion rule-bases. The aim of our work in developing a tailor-made fusion-based hypothesis is concentrated on two vital aspects, viz. it should be able to i] *cater large number of sensor-cells*, which are heterogeneous in nature and ii] *sense the presence of tiny 'point-objects'* on the gripper surface. It may be mentioned that both of these two paradigms were overlooked in the researches hitherto and thus, the existing fusion cum hypothesis testing models are unsuitable to real-life applications in robotics. In the contrary, our model of data fusion and statistical hypothesis testing with new threshold thematic will ensure reliable measure towards overall *qunatization* (e.g. overall external shape, surface area and approximate contour) of the object(s) present in the vicinity of the gripper. In our model, hypotheses are postulated corresponding to different types of sensory outputs. Here, we will differ from the traditional nomenclatures for Null Hypothesis (H_0) as "Signal is absent" against Alternative Hypothesis (H_1) as "Signal is present", in order to suit our requirement towards robotic grasp-based situations. We, therefore, define the hypotheses as *Significant Object Present* [SOP] vs. *Significant Object Absent* [SOA] respectively for 'H_1' and 'H_0'. We define "Significant Object" as those objects whose surface area is larger than that of the graspable area of the gripper or in other words, larger than the sensing area of the gripper. We prefer to adhere to the bi-modal hypothesis paradigm and represent the inherent *fuzziness* in decision-making process with *white noise*, having a relatively higher value of Signal-to-Noise Ratio (SNR). Three novel *Fusion Rule-bases*, viz. [a] *Multiplicative*, [b] *Additive* and [c] *Preferential Selection*, have been formulated in order to reveal the inter-cell relationship of the array (matrix) sensor. In other words, these rule-bases are devised to represent the exact way the elemental cells (*taxels*) are 'reacting' with one another. We will finally have a logical *unified* output from the system controller using these rule-bases and individual signal-output from the taxels. The *dynamic (global) threshold*, as proposed in our model, is to be selected optimally using the user-specified value of either probability of Type I error or Type II error. The filtered sensory data is used to estimate the optimal value of the grip force, which is required to be applied by the jaw to maintain stable vis-à-vis slip-free farm grasp. Nonetheless, the developed model will also ensure the user regarding the characteristics of the post-grasp slip, if any.

Although traditional surveillance problem with multiple sensors and optimization therein using log-likelihood function have been addressed substantially in last two decades (Chair & Varshney, 1986) & (Gustavo & Grajal, 2006), it lacks generality in situations of sensing 'point-objects'. These algorithms are based on an approach wherein maximum likelihood of remotely located unknown signal(s) can be estimated under white Gaussian noise, but the techniques can't be adopted for sensing *localized* sensor-signals.

Field-sensor outputs in a distributed tracking vis-à-vis surveillance system are categorized in two broad groups, depending on a] the *modus operandi* or activation syntax (parallel or serial) and b] the physical layout (staggered or synchronous). Unlike the case of distributed decision making in parallel, fusion problem with the configurations of sensors in serial chain (Viswanathan et al, 1988), (Hashemi & Rhodes, 1989) & (Swaszek, 1993) may have better performance over the parallel distribution case for two sensors. However, the methods perform poorly for large sensor-cells, which is the typical case in real-life applications. Likewise, a comparative study on the system performance was made using temporally staggered sensors as well as synchronous sensors, using a novel metric, namely, average estimation error variance (Niu et al, 2005). Irrespective of the *modus operandi*, serial or parallel, selection of *local node* in a distributed sensor network is crucial, as it will govern the decision-making system regarding the incorporation of the corresponding data for surveillance (Kaplan, 2006).

Tracking of remote target using multiple field-sensors is a well-researched field, irrespective of its genesis; vide *static target* [e.g. a rigid sensor rig] (Chroust & Vincze, 2004) or *maneuvering target* (Jeong & Tugnait, 2005). In case of static targets, e.g. presence of objects in the vicinity of the gripper-sensor, an estimation of the bias in outputs of the asynchronous field-sensors is important. The decoupling between the numerical estimations for the target state and the sensor bias is attained, considering, a] the cross-covariance between the state & bias estimates (Lin et al, 2005) and b] the reduced bias estimate of the joint probabilistic data association (JPDA) algorithm (Kalandros & Pao, 2005). However, for large number of sensor-cells, like the case of ours, the performance of the distributed tracker has been found degrading in comparison to centralized estimation, despite using optimal track-to-track fusion algorithm (Chen et al, 2003).

In contrast to log-likelihood method, the fusion tests to be adopted for achieving maximized probability of detection (for a fixed probability of false alarm) should ideally be Neyman – Pearson [N-P] (Srinivasan, 1986). Nonetheless, the threshold of N-P test becomes data dependent if the conditional impedance is removed (i.e. hypotheses are not truly statistically independent) and does not yield any easy solution for optimization (Tsitsiklis & Athans, 1985). Moreover, N-P fusion rule needs the sensor error probabilities (i.e. probability of false alarm & probability of miss) to be known a-priori and those must not be altered during the fusion process. These are very stringent preludes to the decision paradigms that degrade the performance of the data fusion (El-Ayadi, 2002).

The solution of data fusion problem for fixed binary local detectors with statistically independent decisions (Chair & Varshney, 1986) was amplified and extended for i] correlated local binary decisions (Moshe, 1992) and ii] team hypothesis testing and environmental simulation (Tenney & Sandell, 1981a), (Sadjadi, 1986), (Reibman, 1987), (Papastavrou & Athans, 1992) or iii] multidimensional data association (Kirubarajan et al, 2001), (Gan & Harris, 2001); iv] covariance control (Kalandros & Pao, 2002) and v] new design for low-bandwidth track fusion (Ruan & Willett, 2005). In fact, the fused decision in a distributed data fusion problem for similar as well as dissimilar sensors using N–P test is transmitted alongwith a quality information in numerical terms, viz. 'degree of confidence' (Thomopoulos et al, 1987). Various methodologies, such as *extended Kalman filter* (Nabaa & Bishop, 1999), *layered neural networks* (Karniely & Siegelmann, 2000) or *Bayesian estimation* (Okello & Challa, 2004) have been postulated towards fixing sensor alignment problems (known as sensor registration in the literature) in distributed fusion, unlike the method of maximum likelihood estimator, used hitherto.

Another school of thought in fusion optimization is using time-varying global threshold, which essentially calls for the solution of two coupled sets of dynamic programming equations for computation (Tenney & Sandell, 1981b,c), (Teneketzis & Varaiya, 1984) & (Tang, 1991). The case of decentralized detection system with feedback and memory using the Bayesian formulation is investigated (Alhakum & Varshney, 1996), wherein the optimization gets summed up in a likelihood ratio test at the local detectors for statistically independent observations. However, the process gets computationally intensive once the system has a large number of sensors (Moshe et al, 1991). Although the universality of Bayesian approach is recognized for computational transparency (Moshe et al, 1999), yet a few specific situations are better analyzed either through Dempster - Shafer theory (Murphy, 1998) or another novel theory (Thomopoulos, 1990), which gives a good trade-off between the Bayesian and D-S approaches. Interestingly, the inherent uncertainty in a two-hypothesis model is also alleviated in a recent research, by considering a discrete decision zone between Null & Alternate Hypothesis (Wang, 1998). However, irrespective of the major three techniques used in hypothesis testing, e.g. Log-likelihood Ratio [LLR], N-P or D-S, one lacuna is surfacing that the potential of Type I and Type II error shielding is not utilized to a proper extent. But these two cut-offs, i.e. Type I & Type II error can be of significant relevance in defining global thresholds and to be specific, this potential has been used in our architecture. Our methodology essentially involves a *non-parametric stochastic adaptive decision fusion*, wherein fusion center knows only the number of the sensors under each sub-types, but does not consider their error probabilities. In contrast to non-stochastic

fusion algorithms, e.g. (El-Ayadi, 2002), our technique uses a hypothesis error-based global threshold for the final decision regarding the *target* (i.e. presence of object near the graspable zone of the robot gripper).
The first two rule-bases and the developed hypothesis, described here, have been tested with an indigenous tactile array-sensor, comprising three types of taxels, namely, resistive cells (*R-cells*), capacitive cells (*C-cells*) and piezo cells (*P-cells)* placed in matrix. The sensor has got 61 elemental taxels in total, each of which provides calibrated output (in mV) when excited with external forcing. However, we will consider the readings from 57 taxels, which will be sufficient for experimental investigation. Out of these 57 taxels, 32 taxels belong to R-cells, while C-cells & P-cells constitute 21 & 4 taxels respectively. The sensor has been simulated for two-jaw grasp (i.e. considering it as a gripper-sensor) and experimented with a variety of objects impinging over it. The sensory output is processed for the determination of object's presence alongwith its size & shape and used for the evaluation of grip as well as slip force. On the other hand, the *Preferential Selection* rule-base was tested through a two-jaw sensor-instrumented robotic gripper, having 18 heterogenous taxels in total, distributed in three categories, viz. load cells (*L-cells*), thin-beam sensors (*TBS-cells*) & infrared sensors (*IR-cells*), bearing 2, 10 & 6 taxels respectively.

2. Hypothesis testing and proposed schemes of data fusion

2.1 Formulation of statistical hypothesis

We prefer to adhere to the bi-modal hypothesis paradigm and represent the inherent *fuzziness* in decision-making process with *white noise*, having a relatively higher value of Signal- to -Noise Ratio (SNR). Nonetheless, these two hypotheses have been re-modeled from real-life perspective as shown below,

$$H_0 : X_i = N_i \tag{1}$$

$$H_1 : X_i = S + N_i, \ \forall\ i = 1,2,3,\ldots..,n$$

where, X_i : Observation vector of the $i^{th.}$ sensor; N_i : Noise vector at the $i^{th.}$ sensor; S: Actual detectable signal vector; n: Total number of sensors in the system. The a-priori probabilities of 'H_0' and 'H_1' are: $P(H_0) = P_0$ and $P(H_1) = P_1$. We assume all of these 'i' detectors ($\forall$ i=1,2......,57 in case of matrix sensor & i=1,2..,18 for jaw-gripper) have observations at the individual detector level, denoted by, X_i. Now, each detector employs a "Decision Rule", in order to make a decision-vector 'u_i', which is the *localized* logistic metric. Let, logistic parameter, $\{u_i\}$, $\forall$ i=1,2......57 or 18, be defined against individual sensor-cells, such that, u_i= -1, if H_0 is true and = + 1, if H_1 is true. We also consider the activation syntax of the data fusion to follow *serial* path. Nonetheless, the set $\{u_i = +1\}$ is to be arrived at by considering a cut-off value, ζ, in the following manner,

$$\{u_i = +1\} = \{\forall u_i | \in u_{S+}\} \tag{2}$$

where, u_{S+} is mapped as,

$$u_{S+} \xrightarrow{mapping} (X_i - N_i) \geq \zeta \tag{3}$$

and,

$$[X_i] - [N_i] = [Y_i] \tag{4}$$

We further assume that the observations at the individual detectors are statistically independent and the conditional probability density function is described by, P (Y_i / H_k), $\forall i=1,2,\ldots,57$ or 18 &$\forall k= 0,1$. The stated propositions are equally valid for $\{u_i\} = [0,1]$ tuple $\forall i=1,2,\ldots,n$, wherein "0" signifies truth of H_0 and "+1" is correlated to occurrence of H_1. Nonetheless, the Global U, viz. 'U_G' will be a function of all the elemental fused data, i.e., 'u_i'. In other words, it essentially means that the global fused data, $\{U_G\}$, is an extrinsic function of $\{u_i\}$, i.e. $\{U_G\} = f(u_1, u_2, \ldots, u_N)$, where N=57 or 18, depending upon the total number of taxels. The proposed method of inference relies on a 'variable limit' or *dynamic threshold,* which is to be estimated through a mathematical model. The location of the dynamic threshold limit is dependent on the confidence level for rejecting 'H_1', chosen a-priori, i.e. on the numerical value of Type I or Type II error.

2.2 Development of fusion rules

2.2.1 "k-out-of-n" logic revisited

Before detailing out the proposed application-oriented fusion rules, we would examine the "k-out-of-n" logic, used hitherto as a quick reference to the evaluation of fusion hypothesis. The rule verdicts "presence of object" if 'k' or more detectors select 'H_1' at the elemental detection level out of total 'n' detectors. As a matter of fact, considering the set of 'u_i' as $\{u_i\}=[-1,1]$, for first two rule-bases, U_G is syntaxed as,

$$U_G = +1, \text{ if } (u_1 + u_2 + \ldots + u_{57}) \geq 2k - n$$

$$= -1, \text{ otherwise.}$$

Although the rule clearly demarcates the acceptance or rejection of 'H_1' in case of large objects (i.e. objects having planar area sufficiently more than that of the sensing area), it fails to demarcate adequately the occurrence of 'point-force', unless we define k=1 a-priori. It may be stated that barring its marginal limitations, this logic has been imbibed by most of the decentralized fusion metric, by and large. However, the bottleneck in fusion problem with "k-out-of-n" logic can be tackled more elegantly using the concept of threshold, as used in Bayesian estimation theory. Our rules will hence follow the *Global Thresholding* principle in evaluating the test hypotheses. However, in contrast to "k-out-of-n" logic, the new rules will propose acceptance or rejection of 'H_1' on the basis of numerical value of U_G and the value of *dynamic threshold* considered in that situation. By definition, it is the global threshold only, but we call it *dynamic* as its value gets changed depending upon the application environment. This dynamic threshold can be evaluated numerically from the system parameters, known a-priori. Thus, in true sense, the new rules are functionally in inverse proposition with respect to "k-out-of-n" logic.

2.2.2 Syntax of the fusion rules developed

The first two novel *Fusion Rule-bases,* viz.*Multiplicative* and *Additive,* have been formulated in order to reveal the inter-cell relationship of the matrix sensor, while the third one, i.e. *Preferential Selection,* is aimed at revealing the inter-cell relationship of the semi-matrix layout of the jaw gripper sensors. In other words, these rule-bases have been devised to represent the exact way the taxels are 'reacting' with one another, i.e. in what way these taxels are 'influencing' the neighbouring taxels and/or getting influenced by those. We will finally have a logical *unified* output from the system controller, considering the set of 'u_i' as

$\{u_i\}=[-1,1]$. All the three rule-bases culminate in a non-zero value of U_G for $\{u_i\}=[-1,1]$ and the evaluation is *unbiased* so far as the object-size is concerned.
The Multiplicative model defines 'U_G' as,

$$U_G = \prod_{i=1}^{i=n} e^{u_i} = (e)^{n-2k} \tag{5}$$

while, the Additive model postulates the definition of 'U_G' as,

$$U_G = \sum_{i=1}^{i=n} u_i = (n - 2k) \tag{6}$$

where, $\{u_i\}$: localized decision for the $i^{th.}$ taxel, $\forall i = 1,2,\ldots.n$; 'n': total number of taxels activated in the sensor and 'k': number of taxels giving output as "-1", i.e. "signal absent".
Now, the *fused* decision regarding the selection of test hypothesis will be ruled by the evaluation paradigm, decided a-priori. In our model, we use the numerical value of the *dynamic threshold* ($\lambda_{Threshold}$) as the evaluation metric. We define the evaluation metric for both of the models as: if $U_G \geq \lambda_{Threshold}$, then accept H_1, otherwise reject H_1. In a similar manner, we can obtain the values of U_G using $\{u_i\}=[0,1]$ tuple for multiplicative and additive models also. For multiplicative model, the value of U_G will be e^{n-k} and for additive model the value of U_G will be equal to (n-k).
Now, we define 'U_G' under Preferential Selection model as,

$$U_G = \sum_{i=1}^{i=n} u_i (1 + u_{i+1})^p (1 + u_{i-1})^q \tag{7}$$

where, $\{u_i\}$: localized decision for the $i^{th.}$ taxel, $\forall i = 1,2,\ldots.n$; 'n': total number of taxels activated in the gripper sensor system; p: relative weightage of the succeeding taxel, i.e. $(i+1)^{th.}$ cell and q: relative weightage of the preceding taxel, i.e. $(i-1)^{th.}$ cell, where $0 \leq p,q \leq 2$. Exact numerical values of p & q need to be ascertained from experimentation with the sensory modules of the jaw-gripper. In that respect, eqn 7 is in a generalized format, which can be adapted for other similar systems as well. We also assume that in eqn. 7, $\{u_{i-1}\}|_{i=1} = 0$ and $\{u_{i+1}\}|_{i=n} = 0$. It may be noted that this model is essentially taxel-specific, unlike the previous models (vide eqn. 5& 6), wherein cumulative effect of the taxels are reflected only. In contrast, the *relative dependency* of one taxel over the neighbouring ones is getting priority in the present model. Hence, we have christened this model as *'preferential selection'*, as the effect of adjoining taxels can be taken into consideration towards computing U_G, depending upon their relative influence/ importance. It may be stated that, this model, by definition, is best suited for taxels arranged in a row or column-wise fashion, i.e. applied for *row /column matrix*. Also, taxels may or may not be equally likely; nonetheless, we are unsure about the outcome (-1 or +1) of a specific taxel in the *grid*. It may be mentioned additionally here that the effect of relative dependency of the taxels could also be considered by another *sister-model* of U_G, viz. $U_G = \sum u_i (1- u_{i+1})^p(1- u_{i-1})^q$, but it would have been rather difficult to be interpreted graphically. Now, so far as the evaluation is concerned, we use the *dynamic threshold band* and the numerical value of the *mean threshold* ($\lambda_{Threshold\text{-}mean}$) as the evaluation metric in this rule-base. We define the evaluation metric as: if $U_G \geq \lambda_{Th\text{-}mean}$, then accept H_1, otherwise reject H_1. But, alongwith discrete acceptance / rejection, we will also encounter one *fuzzy-zone*, which will signify *in-decision* regarding the acceptance or rejection of H_1. Numerically, this in-decision zone will be directly proportional to the width of the threshold-band. It may be noted from eqns. 5 & 6 that the value of 'k' carries significance so far as the relative implications of it on P (H_1) is concerned. Figure 1 shows the schematic view of the variation of P(H_1) with 'k'. The first plot we considered is a simple straight-line plot while the other two are exponential fits.

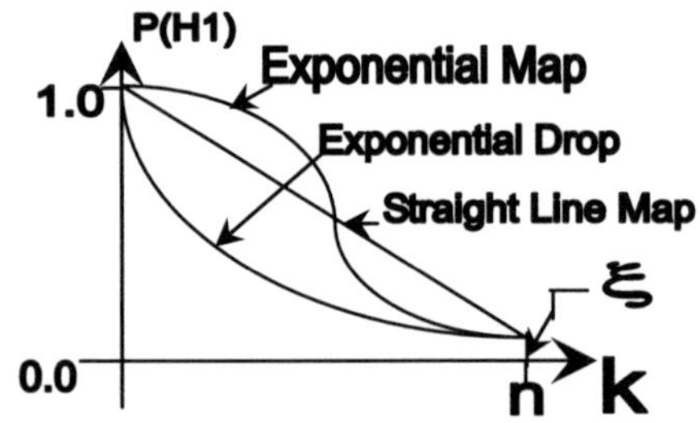

Fig. 1. Probability curves as per the fusion rules proposed

Based on the curves shown above, the mathematical formulae for evaluating P(H_1) or in short, 'p' will be as follows,

$$p = \left(\frac{\xi - 1}{n}\right)(k) + 1 \quad \text{[For straight-line map]} \tag{8}$$

and,

$$p = \exp\left(\frac{\ln \xi}{n}\right)(k) \quad \text{[For exponential curves]} \tag{9}$$

2.2.3 HEBTEM: new strategy for selecting dynamic threshold

The *dynamic threshold*, as proposed in our model, is to be selected optimally using the user-specified value of either probability of Type I error or Type II error. It is termed as "Hypothesis Error Based Threshold Evaluation Method" **(HEBTEM)**. The proposed method relies on the selection of the confidence level, as stipulated by probability of Type I or Type II error. However, we use probability of Type I error as the confidence level for additive model of fusion rule-base and probability of Type II error as the same for multiplicative model. For example, in case of multiplicative model of fusion, estimation using "2% non-confidence level" (i.e. probability of Type II error as 0.02) essentially declares the situation of 'object presence' with 98% certainty. We shall now investigate the situations in order to select the dynamic threshold using HEBTEM.

2.2.3.1 *Using Multiplicative Model*

Here, in-line with eqn. 9, an exponential curve has been fitted for the plot of P(H_1) vs. U_G. The plot uses $\{u_i\}$=[-1, +1] tuple and the generic representation of the *probability curve* is shown in fig. 2a. Similarly, we can get the plot of P(H_1) vs. U_G for $\{u_i\}$ = [0,1] tuple too (refer fig. 2b). We assume that the probability of alternative hypothesis, i.e. P(H_1) to be 0.5 when nearly half of the sensor-cells (i.e 'n/2') will show "signal present", i.e. the value of 'k' becomes 'n/2' and that of U_G is 1.0. The final decision about "acceptance" and "rejection" of alternative hypothesis will be based on the location of global threshold (λ_{Th}) and the observed value (represented as x′ in fig. 2) of U_G. The scale along the X-axis of the plot also depicts the increment of U_G from e^{-n} to e^n, corresponding to the variation of 'k', from n to 0. We also assume that at k = n, i.e. when all the sensor-cells are giving $\{u_i\}=-1$, the occurrence of an object is *minimal*, which is quantified through 'ξ'. The numerical value of 'ξ' is kept very small, in the order of 0.002 to 0.004, as found suitable for the matrix sensor. In fact, 'ξ' is the true indication of the presence of 'point-mass' or 'point-force' over the gripper surface. Thus, we can identify two representative points in the plot, viz. (e^{-n}, ξ) and $(e^n, 1.0)$, which will be decisive in using the plot analytically.

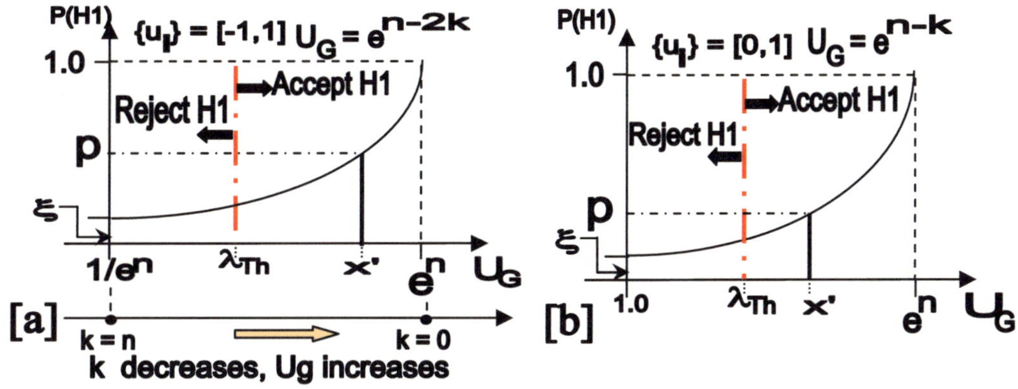

Fig. 2. HEBTEM with multiplicative model of data fusion for [a] $\{u_i\}=[-1,1]$ & [b] $u_i\}=[0,1]$

As part of two-point interpolation for the exponential curve shown in fig. 2a, we consider the following transcendental equation involving ξ, viz.

$$P(H_1) \equiv p = a\exp^{mU_G} \tag{10}$$

It may be observed from the above equation that the value of both 'm' and 'a' can be evaluated numerically if 'ξ' is known, vide,

$$m = \frac{e^n \ln\left(\frac{1}{\xi}\right)}{\left(e^{2n}-1\right)} \tag{11a}$$

and

$$\ln\left(\frac{1}{a}\right) = \frac{e^{2n} \ln\left(\frac{1}{\xi}\right)}{\left(e^{2n}-1\right)} \tag{11b}$$

2.2.3.2 *Using additive model*

Additive model of fusion rule-base considers probability of Type I error as the basis for ascertaining the presence or absence of object on the gripper. A graphical representation of the probability curve, using HEBTEM, is plotted in fig. 3 considering $\{u_i\}=[0,1]$ tuple. Considering an exponential fit for the probability distribution shown in fig. 3, the probability of alternative hypothesis becomes,

$$P(H_1) \equiv p = \xi \exp^{\frac{\ln\left(\frac{1}{\xi}\right)}{n} U_G} \quad (12)$$

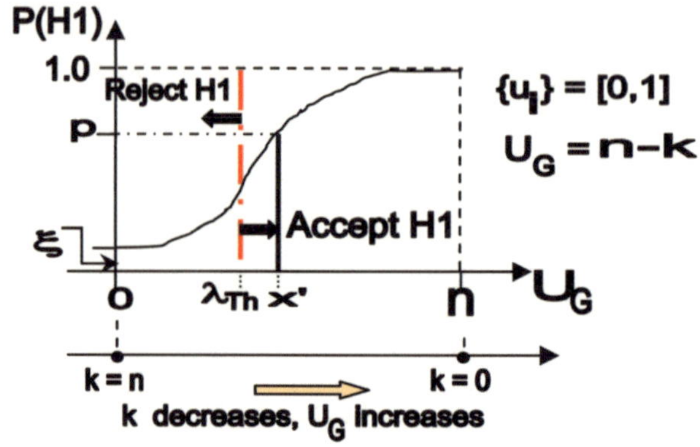

Fig. 3. Probability curve for additive model of fusion rule with $\{u_i\}$=[0,1] using HEBTEM

However, paradigms of P (H_1) plot changes significantly for $\{u_i\}$=[-1,+1] tuple (refer fig. 4).

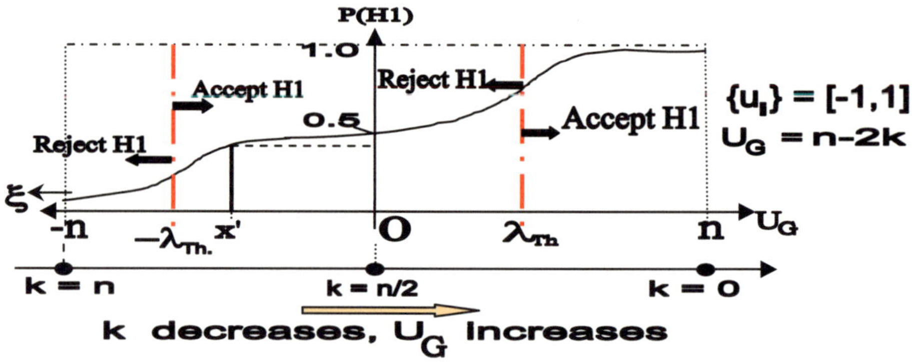

Fig. 4. Probability curve for additive model of fusion rule with $\{u_i\}$=[-1,1] using HEBTEM

As articulated before, ξ symbolizes the presence of 'point-mass' or 'point-force' over the gripper surface. Here we can identify three representative points in the plot, viz. (-n, ξ); (0, 0.5) and (n, 1.0), which will be used in formulating the transcendental equation involving ξ in order to evaluate P (H_1). We will use a generic equation for the exponential curve, viz. p =a. e^{mx} + cx, where 'p'=P(H_1); 'x' =U_G and 'm' & 'c' are constant functions of ξ. The final equation, after three-point interpolation as stated above, becomes,

$$P(H_1) \equiv p = 0.5 \exp^{\frac{\ln\left[(\xi+1)+\sqrt{\xi(\xi+2)}\right]}{n} U_G} + \left[\frac{1-0.5\{(\xi+1)+\sqrt{\xi(\xi+2)}\}}{n}\right] U_G \quad (13)$$

It is to be noted from eqn. 13 that we will use only positive values of 'm' and 'c', discarding the theoretically possible negative values of those. Also, unlike fig. 4, here we can use either of the two thresholds, namely '$\lambda_{Th.}$' or '-$\lambda_{Th.}$', depending upon the application environment. In case we use a positive value for '$\lambda_{Th.}$', i.e. right-hand side of the curve, then the decision regarding the rejection /acceptance of 'H_1' will be restricted to the right-hand-side zone of the curve only. Likewise, if we intend to use the negative value for '$\lambda_{Th.}$', our decision has to be within the left-hand-side zone of the curve. Thus, in a way, additive model with $\{u_i\}$=[-1,+1] tuple advocates *symmetric thresholding*, but not occurring simultaneously.

2.2.3.3 *Using preferential selection model*

A graphical representation of the probability curve for 'H_1' as per this model is plotted in fig. 5 considering $\{u_i\}$=[-1,1] tuple. Here we will characterize the threshold-band with three parameters, namely: 'A': $\lambda_{Th\text{-}initial}$; 'B': $\lambda_{Th\text{-}final}$ & the mid-point of the band as: $\lambda_{Th\text{-}mean}$.

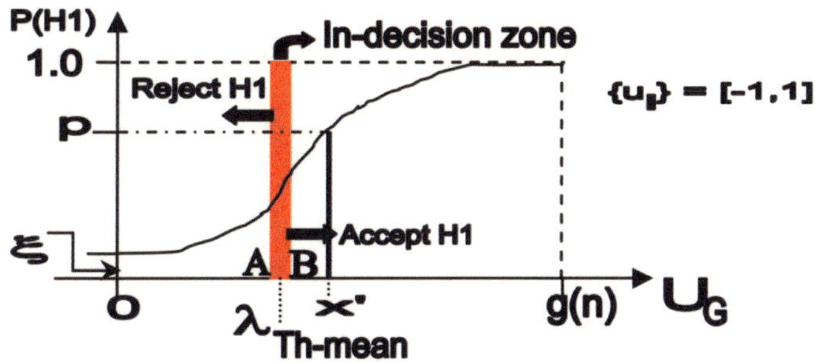

Fig. 5. Probability curve for the preferential selection model using HEBTEM

Using an exponential fit for the probability distribution shown in fig. 6, we have,

$$P(H_1) \equiv p = \left[\frac{1-\xi}{\{g(n)\}^2}\right] U_G^{\ 2} + \xi \tag{14}$$

In eqn. 14, ξ symbolizes the presence of 'point-mass' or 'point-force' over the gripper surface and g(n) signifies the gamut of the maximum possible values of U_G for various numerical combinations of (p,q). The function g(n) can be computed using the format of U_G, viz.

$$U_G = (2^p + 2^q) + (n-2)[2^{(p+q)}], \quad \forall u_i = +1 \tag{15}$$

As per eqn 15, the values of g(n), in ascending order will be, {g(n)} : {(2n-1), (4n-5), (4n-4), (8n-10), (16n-24)…..} for various combinations of (p,q), where $\{p,q\} \subset [0,2]$. Figure 6 explains the geometric interpretation of g(n), in evaluating U_G. Successive values of g(n) are interpreted serially as $[g(n)]_1$, $[g(n)]_2$,….$[g(n)]_k$, which becomes the yard-stick for the flattening of the probability curve, keeping ξ unaltered through-out.

2.2.3.4 *Analysis of the Rule-bases and Evaluation of Dynamic Threshold*

Major advantage of the proposed method, viz. HEBTEM, lies with the fact that it doesn't include the concept of *Bayesian Risk*, which inherently involves the computation related to Probability of False Alarm (P_f) and Probability of Miss (P_m). The concept of *Bayesian Risk* is suitable only to cases where a large group of field-sensors are either attempting to evaluate the presence or absence of a single object or *tracking* a single *target*. In such a situation, all of the field-sensors do participate in the decision-making process and the output of each one of those will definitely be biased by its own [$\{P_f\}$-$\{P_m\}$] tuple. But, in situations like robotic grasping or stand-alone tactile sensing, the use of sensor-cells will be governed by the actual size, shape and contour of the object to be 'sensed' and/or grasped.

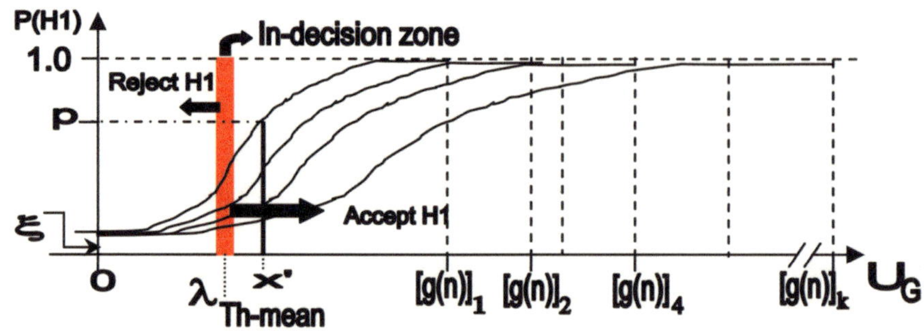

Fig. 6. Variation of probability curves under different tuples of the model

The value of the *dynamic threshold* (viz. λ_{Th} in figs. 2 to 4) can be estimated using either of the models, namely, multiplicative or additive. However, the paradigm of evaluation is different in these two cases. While multiplicative model of the fusion rule-base considers a non-confidence level (β), the additive model considers a confidence level (α), both chosen a-priori. It may be noted that the confidence level (α) for the additive model is nothing but the accepted level of probability of Type I error and 'β' is the probability of Type II error. Figure 7 illustrates the philosophy of HEBTEM, wherein we have shown the zones of certainty and uncertainty for selecting 'H_1'. Here by 'f (x)' we mean the exponential curve for $P(H_1)$, vide eqns. 10, 12 & 13, i.e. corresponding to multiplicative & additive models using [-1,1] tuple.

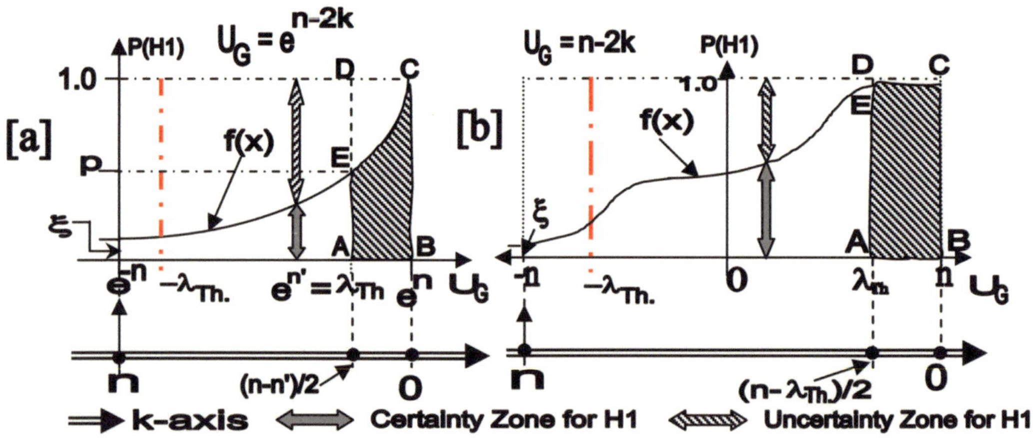

Fig. 7. Computation of dynamic threshold using [a] multiplicative & [b] additive model

It may be stated here that using an exponential fit for the multiplicative model is advantageous, because we can use a wider range of 'λ_{Th}' with almost the same numerical value of 'β'. Thus in order to interpret the presence of *point-force /point-object*, it is wiser to consider such a 'λ_{Th}', which is *closer* to the origin (i.e. the point with k=n). For these cases of point-force selection, the usual value of 'β' will be quite large, because there will be a high tendency of selecting 'H_0' (refer the plots of fig. 2). With reference to fig. 7, the dynamic threshold is evaluated mathematically using multiplicative model as,

$$\int_{\lambda_{Th}}^{e^n} f(x)dx = [1-\beta]\left(e^n - \lambda_{Th.}\right) \tag{16}$$

and the same is computed using additive model as,

$$\int_{\lambda_{Th}}^{n} f(x)dx = [1-\alpha](n-\lambda_{Th.}) \tag{17}$$

where, n: Total number of sensor-cells in the system and x: Individual decision vector of the sensor-cells. Both of these two equations are powerful in the sense that on simplification, these culminate in transcendental equations for '$\lambda_{Th.}$', which will finally produce two values for '$\lambda_{Th.}$', one positive and the other negative. So, depending upon the nature of our application, we can select either '$+\lambda_{Th.}$' or '$-\lambda_{Th.}$'. The details of the computations for the above two equations are presented in *Appendix I*.

It may be noted that the formulation of the multiplicative model inherently leads to the inference for 'H_0', especially when the object-size is very small (i.e. 'k' is large). That means, there is always a natural tendency towards selecting 'H_0' *wrongly* and thereby committing a *Type II error*. Hence we need to consider the probability of Type II error, 'β' in computing the required area (under hatch, refer fig. 7), and subsequently, '$\lambda_{Th.}$'. In contrast to this, the additive model suggests equally likely outcome, i.e. the model can select 'H_1', but may commit a *Type I error* by *rejecting* it too. That means, here we must consider the probability of Type I error, 'α' in computing the desired area and the factor, (1-α) in eqn. 17 denotes the level of confidence we have in selecting the 'H_1'.

However, choice of 'α' or 'β' will largely depend upon the value of 'n', i.e. how large the sensor-cell system is or to that extent, how big is the sensor-matrix. Nevertheless, a *stricter* level (i.e. lower value of α or β) will lead to a tougher strategy for accepting the alternate hypothesis, viz. "signal is present". So, depending upon the exact use of the sensor system, 'α' or 'β' can be selected. For example, for a gross gripping of a comparatively large object we need not have a stricter confidence level and hence, even a probability of Type I error of 10% may be allowed in such a case. However, that won't be the right choice when the same sensory system is being used as a *stand-alone system* detecting *point-force*, instead of being augmented with the robot gripper. Hence, the user can take a final decision regarding the actual presence of object by analyzing $P(H_1)$, as obtained from the models.

Unlike the distinct value of the dynamic threshold as per the evaluation pattern proposed in fig. 7, the preferential selection model dictates the *dynamic threshold-band* (viz. $\lambda_{Th\text{-}mean}$ in fig. 5), as described earlier. It can be estimated using the statistical confidence level (α), i.e. probability of Type I error, as stated below. Here 'x' & 'f(x)' represent individual decision-vector of the taxels and the exponential curve for $P(H_1)$, vide eqn. 14 respectively, while $[g(n)]_s$ signifies the maximum value of U_G ($\forall s=1,2,\ldots k$).

$$\int_{\lambda_{Th-mean}}^{g(n)_s} f(x)dx = [1-\alpha][g(n)_s - \lambda_{Th-mean}] \tag{18}$$

The planar area of the threshold-band, i.e. the *fuzzy-area* of *in-decision* can be computed as,

$$\int_{\lambda_{Th-initial}}^{\lambda_{Th-final}} f(x)dx = \int_{\lambda_{Th-initial}}^{g(n)_s} f(x)dx - \int_{\lambda_{Th-final}}^{g(n)_s} f(x)dx = [1-\alpha][\lambda_{Th-final} - \lambda_{Th-initial}] \tag{19}$$

and the width of the threshold-band ($\delta\lambda_{Th}$) is defined as the numerical difference between $\lambda_{Th\text{-}final}$ and $\lambda_{Th\text{-}initial}$, i.e. ($\lambda_{Th\text{-}f} - \lambda_{Th\text{-}i}$) and modeled as,

$$\delta\lambda_{Th} = \frac{\alpha}{4} g(n) \tag{20}$$

However, choice of 'α' will largely depend upon the value of 'g(n)', i.e. how intense is the effect of relative dependency in the sensor-matrix or to that extent, how large is (p,q) tuple. Nevertheless, a *stricter* level (i.e. lower value of α) will lead to a tougher strategy for accepting the alternate hypothesis, viz. "object is present".

2.2.3.5 *Fusion optimization: advantage HEBTEM*

HEBTEM, augmented with our fusion models, has been found advantageous over the traditional optimization techniques followed hitherto, namely, [a] Log-likelihood Ratio Test (LLR) and [b] Neyman-Pearson (N-P) Test. These techniques, though used vastly in situations concerning distributed data fusion problems, do possess inherent drawbacks. Both of these methods simply rely on the basic fact that both 'P_0' and 'P_1' are un-biased, i.e. in case of a new data fusion problem, such as the detection of the object-presence by the gripper sensor, both 'H_0' and 'H_1' are equally likely to occur. However, this pre-assumption of equally likeliness of the hypotheses is not valid for situations like robotic grasping and sensing the presence of object in the gripper-jaw. With regard to grasp analysis, there can be two domains of hypothesis metric, which has got *in-built biasing,* e.g. a] situations where either 'H_0' or 'H_1' is biased and b] situations where 'H_0' or 'H_1' are associated with penalty coefficients. In fact, neither LLR nor N-P test is fit for handling these two environments with impeding bias. We shall now investigate these situations in detail.

a. Situations where either 'H_0' or 'H_1' is biased

For an un-biased equally likely situation, the ratio between P_0 and P_1 is always 1.0, because the probabilities of both null and alternate hypotheses are 0.50. Mathematically it implies,

$$P_0 / P_1 = P(H_0)/ P(H_1) = 1.0 \tag{21a}$$

and

$$\text{we denote, } \lambda_0 = P_0 / P_1 \tag{21b}$$

This ratio between P_0 and P_1 is being used to compute *global threshold* (λ_0) for the fusion problem, in general. As per equation (21b), the working formula for λ_0 will simply be the ratio between P_0 and P_1 in all *equally likely* cases, wherein the numerical value for the global threshold will be 1.0. However, since the numerical value of λ_0 is the deciding factor for both LLR and N-P tests, we need to judge the validity of the same under *biased situations*. For example, if for a typical case, H_0 is biased with a higher probability to occur due to some application-specific reason, the numerical value of λ_0 will not be equal to 1.0. For example, consider the situation, wherein the global threshold is computed as $\lambda_0 = P_0 / P_1 \equiv P(H_0) / P(H_1) = 0.7/0.3$, i.e. $\lambda_0 > 1.0$. Similarly, the value of λ_0 will be less than 1.0, in case a situation gets biased with H_1. Thus, we surely need some other measurand for λ_0 that will be applicable for non-equally likely and/or biased hypothesis cases, as LLR and N-P are not suitable in such situations. The reason being in case of robotic gripping, we generally come across situations wherein only a 'point-object' or very small object is being grasped, leaving majority of the *taxels* free from getting exited. Hence, in such a situation, we need to bias the environment with the alternative hypothesis (H_1) in order to get fusion paradigms successfully. Likewise, biasing of hypothesis is a must for related situations, e.g. exciting the gripper sensor with a 'point-force'. Nonetheless, in situations wherein our main concern is to disassociate the noise effects from the valid sensor data so that no erroneous reading can crop in during situations where no object is present physically, we need to bias the environment vis-à-vis optimization process with the null hypothesis (H_0).

b. *Situations where 'H_0' or 'H_1' are associated with penalty coefficients*

Here the general expression for *global threshold* (λ_0) is given as,

$$\lambda_0 = \frac{P(H_0)[C_{10} - C_{00}]}{P(H_1)[C_{01} - C_{11}]} \tag{22}$$

where, 'C_{ij}' refers to the *cost* (penalty) of accepting 'H_i' when 'H_j' is true ($\forall$ i,j = 0,1). Now, for a non-specific case, it goes very fine if we assume the values of the cost-coefficients, viz. numerically C_{10}, C_{00}, C_{01} & C_{11} to be 1,0,1 & 0 respectively. As a matter of fact, these are the standard values of the cost-coefficients used in majority of the fusion problems and hence the numerical value of the global threshold becomes 1.0. However, the *impended bias* situation does arise not only because of the probabilistic values of the null and alternative hypotheses but also the cost involved in accepting the incorrect hypothesis. For example, if a specific situation demands inherent bias to be incorporated, then, λ_0 will be a function of both P (H_k , $\forall$ k =0,1) and C_{ij} ($\forall$ i,j=0,1). In other words, we can certainly have non-zero values for C_{00} & C_{11} and non-unity values for C_{10} & C_{01}. In a way, this metric of evaluating global threshold is not unique and cannot be tackled by LLR or N-P method, when all 'C_{ij}' s are numerically different. In case of robotic grasp, we often need to use one or more penalty coefficients, as the exact model for λ_0 should account for instances like detecting point-force, grasping micro-objects, rectifying the readings from faulty taxels or correcting noise-levels of the taxels. In such situations, we need to use eqn.22, wherein the value of λ_0 will not be equal to unity and thereby that value of λ_0 will be unfit for LLR or N-P to process further.

Thus, we can observe that both of the well-accepted optimization tests, namely, LLR and N-P, do possess hindrances while tackling fusion problem pertaining to robotic gripping and these tests fail to address the divergent situations, such as biased or non-equally likely hypotheses and penalty factors. And, it is apparent that we need to have suitable metric for global threshold, which will be *dynamic* and able to handle such *biased hypotheses*. As a matter of fact, fusion problem in such situations should ideally be considered with *on-line observation* values and *posteriori processing* using suitable model. In that respect, our model (HEBTEM) proves to be a viable option in processing on-line sensory data from the *taxels* using the metric of dynamic threshold (λ_{Th}). The developed method is capable of tackling the situation of *selective biasing*, as the method is solely based on experimental observations and not any a-priori assumption.

3. Spotlight on the matrix sensor and jaw-gripper used for case-studies

3.1 Matrix sensor: case-study I for additive & multiplicative models

The prototype version of the sensory system (external dimension: 175 mm. x 160 mm. x 20 mm), used in the case study, has been optimally designed for a moderately spaced layout to house three categories of sensor units, viz. resistive ('R'), capacitive ('C') and Piezo ('P') cells in a matrix layout. The "R-cells", spaced in 4x4 array, have been designed in the form of small slender 'struts', with a rectangular cross-section (5 mm. x 4 mm., with a 2 mm. diameter blind hole inside). A pair of strain gauges is pasted on the opposite walls of the struts. The capacitive cells, on the other hand, have been placed in a 5x5 matrix and these cells (15 mm. x

12 mm. x 5 mm.) have been designed taking into account the compliancy and overall sensitivity of the individual units, i.e. top & bottom plates and the dielectric layer. Figure 8 illustrates the schematic of the plan view of the of the matrix sensor used for the present study. A quasi-compliant protrusion pad atop protects these planar matrix-cells, having triangular / trapezoidal /spherical *serrations* embedded in it. Replaceable type pads are used, having triangular, trapezoidal or spherical serrations embedded. The protrusion pad has been designed in a way to make it quasi-compliant, shear stress-resistive and light-weighed. In the *assembled* version of the sensor system, a direct contact is being established between the struts and the serrations through slender pins, enabling the transmission of force(s) to the respective R-cells. Figure 9 schematically presents the internal disposition of the sensory assembly. 4 nos. PVDF sensors (P-cells, dimension: 25 mm. x 13 mm. x 205 μm) are mounted on the underneath of the protrusion (rubber) pad in a customized manner, so as to arrest micro-strains in both X & Y planes of the pad. The placement layout of the P-cells is illustrated in fig. 10.

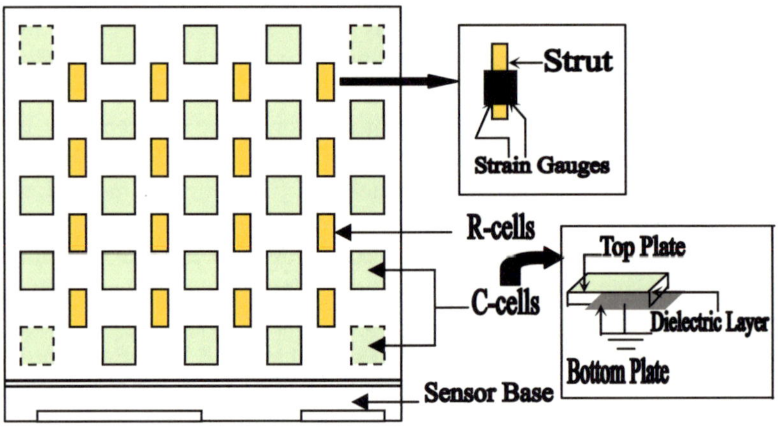

Fig. 8. Design metrics of the matrix sensor used in the case study I

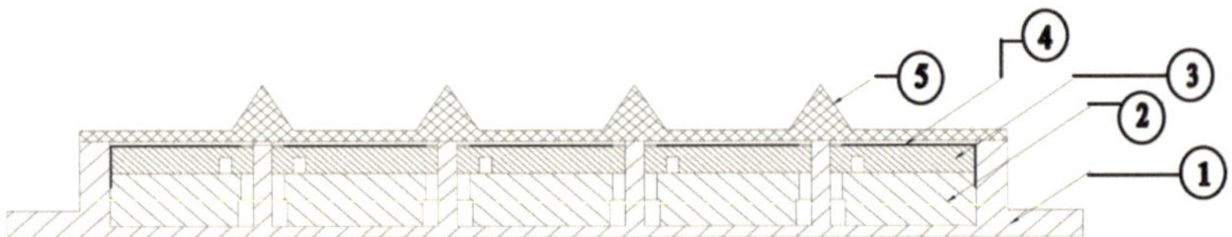

Legends:
1: Base Plate 2: Bottom Support Plate [for the struts] 3: Top Guide Plate 4: Top Support Plate 5: Protrusion Pad

Fig. 9. Sectional view of the sensor assembly in case study I

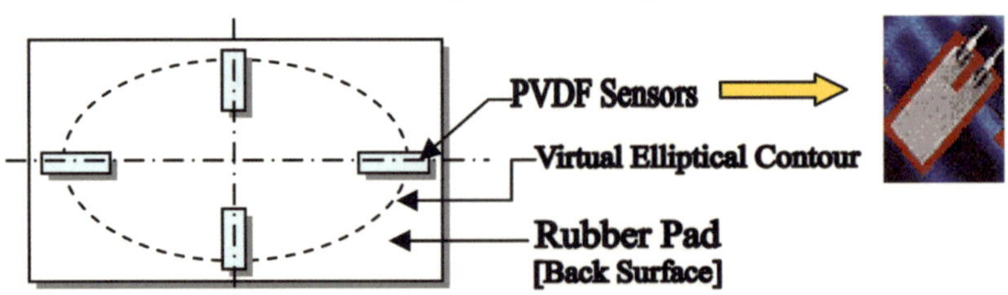

Fig. 10. Layout of the protrusion pad of the matrix sensor in case study I

The sensory system is being interfaced with a control circuitry, developed indigenously. The design is conceived as embedded system, with the provision for real-time processing of sensory signal. The control circuitry is conceptualized as optimally suited size, which necessarily compels minimum numbers of the measuring components for both R & C-cells. Sensory signals in analog form (mV) are being generated, as and when the cells are being *activated* through external force / *excitation.* The force-induced excitation gets manifested either in the form of *strains* generated in the strain gauges of the R-cells or by a change in capacitance in the C-cells, through the *instantaneous deformation* of the dielectric layer. The analog signals, so generated, are transferred to the control circuitry board (in the form of PCB) and are processed through stabilized circuits. The signals generated from the individual R & C-cells, then pass through micro-controller card and generate the final output signal. These *raw* output signals can be displayed over the VDUs (through serial communication with a PC or by using CRO) in real-time. The most appropriate fusion model is then superimposed on these raw data to get *unified* output.

3.2 Instrumented jaw gripper: case-study II for preferential selection model

The mechanical assembly of the planar parallel two-jaw robotic gripper, used in this case-study, comprises six major functional elements, namely, a] drive mechanism & associated drive train; b] motion transferring mechanism (through servomotor system); c] jaw assembly; d] drive for jaw movement; e] sensor assembly and related hardware and f] mounting structure (for assembling with the robot wrist). Nonetheless, the sensor system, augmented with the gripper body, includes the following four types, viz. a] *miniaturized load cell* for grip force evaluation [one per jaw, i.e. 2 in total]; b] *Infrared LED* for detection of object's presence /absence (three per jaw, i.e. 6 in total); c] *force sensor* for auxiliary measurement of grip-force (strain gauge, four in total, symmetrically placed over the moveable link) and d] *Thin-beam sensor* (TBS) for 'slip' measurement (five per jaw, i.e. 10 in total). Figure 11a presents a photographic view of the fabricated instrumented jaw gripper, with sensory interfaces (✥) fitted inside the jaws, while the snapshots of the load cell (LC) & TBS (used for the slip sensor grid) are illustrated in fig. 11b,c respectively.

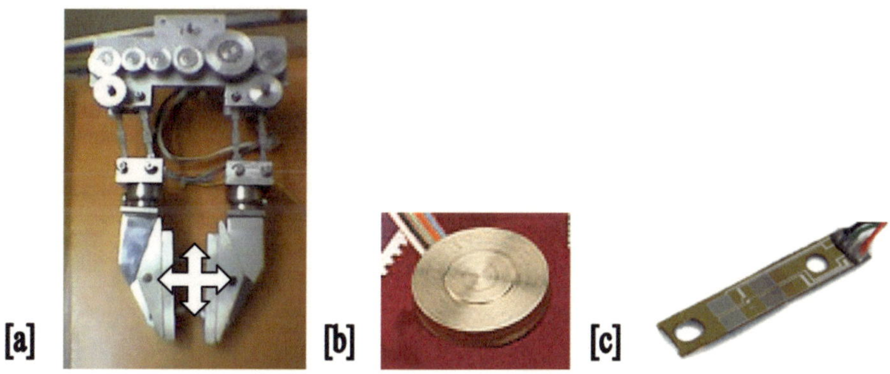

Fig. 11. Photographic view of [a] jaw gripper assembly [b] LC & [c] TBS used in case-study II

The fabricated hardware of the gripper system has suitable provision for easy mounting as a 'stand-alone' unit as well as while being interfaced with the wrist of the robotic manipulator. The drive system is through electrical d.c. servo-motor and the generated rotational motion gets transmitted finally to the pair of jaws by means of gear-train and actuating system. The jaws are being activated through mechanical linkage-train, which follow semi-straight line path. The jaw assemblies are proportionate enough to house the sensory modules. The *sub-miniature* load cell has a cylindrical 'button' like shape in exterior (overall height: 3.2 mm), employing tiny metallic foil strain gauges with teflon insulated lead wires. On the other hand, thin-beam sensors employ specially developed instrumented strain gauge *assembly* (of 4 gauges), which is laminated to the *beam* to provide excellent stability vis-à-vis reliability.

4. Fusion model for evaluation of grip and slip force

The developed rule-bases are used to sense the external excitations on the gripper sensor or the jaw gripper (depending on the fusion model), operated remotely with an unknown loading. On assimilating raw sensory data, a *fusion model* is applied over in order to evaluate grip and slip force. Associated parameters, like slip velocity and slipped distance are also evaluated subsequently. Although the syntax of the fusion model is similar, it has been duly fine-tuned for a] additive & multiplicative rule-bases (i.e. for case-study I) and b] preferential selection rule-base (i.e. for case-study II), as detailed below.

4.1 Model for matrix sensor-based study

Here the model first evaluates tangential force on each of the affected taxels and afterwards, total tangential force or the *slip* force coming upon the sensor, based on the raw sensory signals. The gripping force is computed from the slip force and the allowable range of object loadings is decided thenafter. A straight-line interpolation is used to evaluate tangential forces on affected taxel(s), wherein the 'slope' ('m') and 'intercept' ('c') of the calibration line will jointly indicate the effect of other spurious force-components, largely non-tangential. The equation of the calibration line is given below,

$$y_j = m F_t^k + c \tag{23}$$

where, y_j: Raw output (in milli-volts) of the affected sensory cells due to object loading; F_t^k: Tangential force, generated on the $k^{th.}$ resistive cell .The total tangential force is evaluated by adding the individual forces vectorially. However, by virtue of the design of the matrix sensor, tangential forces on the cells are co-planar and parallel to one another. The exact number of R-cells affected will be ascertained from the object size, shape (i.e. convex hull profile) and weight. Hence, the gross tangential force is calculated numerically as,

$$Tot\vec{F}_t = \sum_{k=1}^{k=m} \vec{F}_t^k \equiv \left[\vec{F}_t^{(1,2)} + \vec{F}_t^3 + ...upto\,'m'\,terms \right] \tag{24}$$

Also,

$$\vec{F}_t^{(1,2)} = \left[\left(F_t^1\right)^2 + \left(F_t^2\right)^2 + 2\left(F_t^1\right)\left(F_t^2\right) Cos\alpha \right]^{1/2} \equiv \left(\left|F_t^1\right| + \left|F_t^2\right| \right) \tag{25}$$

where, $TotF_t$: Total tangential force on the sensor / gripper surface; F_t^k: Tangential force on the $k^{th.}$ cell; m: Total number of cells affected; $F_t^{(1,2)}$: The vectorial sum of the tangential forces on the 1st. & 2nd. Cell; α: The inclusion angle between any two coplanar tangential force-vectors. The upper threshold of the gripping force (F_g^{cal}) vis-à-vis maximum allowable load (W) are evaluated as,

$$F_g^{cal} \le \frac{\sum_{\forall k} F_t^j}{\mu} \tag{26a}$$

and

$$W = 2\mu F_g^{cal} \tag{26b}$$

where, 'F_t^j' is the tangential force coming on the kth. resistive taxel at jth. time-instant & 'μ' is the coefficient of kinematic friction between sensor & the object surface. The inherent design of the matrix sensor generates tangential forces on the affected R-cells in case of gripping a large object, but, for objects with serrated contours, it will give rise to both tangential as well as normal reaction force on R- & C-cells respectively.

4.2 Model for jaw gripper-based study

In this case, the model relies on the characteristics of the external excitations on the jaw gripper. Based on the raw sensory signals from the thin-beam sensor grid (*TBS-cells*), the model first evaluates tangential force on each of the TBS-sensors and thenafter, total tangential force or the *slip force* coming upon the gripper. The gripping force is computed directly from the raw data of the load cells. A straight-line interpolation is used to evaluate tangential forces on affected TBS-taxel(s), wherein the 'slope' (m) and 'intercept' (c) jointly indicate the effect of other non-tangential spurious force-components. However, for load cell data, the calibration is straight-forward, with only a mapping factor (M). The equations of the calibration lines for the TBS and LC are given below,

$$y_j = m F_t^k + c \tag{27a}$$

$$Y_j = M F_g^k \tag{27b}$$

where, y_j: Raw output of the affected TBS-cells due to object loading; F_t^k: Tangential force, generated on the kth. TBS-cell; Y_j: Raw output from the LCs; F_g^k: Grip force, generated on the kth. LC. The total tangential force is evaluated by adding the individual forces vectorially. However, by virtue of the design of the gripper-sensor grid, tangential forces on the TBS-cells along one specific axis (e.g. X or Y) are co-planar and parallel to one another. The exact number of TBS-cells affected will be ascertained from the object size, shape (i.e. convex hull profile) and weight. The schematic layout of the TBS-grid in a single jaw is presented in fig. 12a (TBS-cells are numbered as #1,…,#5), while the overall sensory layout of the jaw and grasp force signature are shown in fig. 12b & 13c respectively.
Hence, the gross tangential force is calculated numerically as the vector-sum of 'X' & 'Y' components (i.e. F_{t-x} & F_{t-y}) using the following model, viz.

$$Tot\vec{F}_{t-y} = \sum_{k=1}^{k=4} \vec{F}_{t-y}^{\,k} \equiv \left[\vec{F}_{t-y}^{(1,2)} + \vec{F}_{t-y}^{(3,4)}\right] \tag{28a}$$

where,

$$\overrightarrow{F}^{(1,2)}_{t-y} = \left[\left(F^1_{t-y}\right)^2 + \left(F^2_{t-y}\right)^2 + 2\left(F^1_{t-y}\right)\left(F^2_{t-y}\right) Cos\,\alpha \right]^{1/2} \equiv \left(\left|F^1_{t-y}\right| + \left|F^2_{t-y}\right| \right) \tag{28b}$$

and,

$$Tot\,\overrightarrow{F}_t \equiv \left[\overrightarrow{F_{t-x}} + \overrightarrow{F_{t-y}} \right] = \left[\left(TotF^1_{t-y}\right)^2 + \left(F_{t-x}\right)^2 \right]^{1/2} \tag{29}$$

where, $TotF_t$: Total tangential force on the sensor-grid / gripper surface; $F_{t-y}{}^k$: Tangential force on the $k^{th.}$ TBS-cell along y-axis; $F_{t-y}{}^{(1,2)}$: The vectorial sum of the tangential forces on the $1^{st.}$ & $2^{nd.}$ Cell along y-axis; α: The inclusion angle between any two coplanar tangential force-vectors along y-axis; F_{t-x}: Tangential force on the TBS-cell along x-axis.

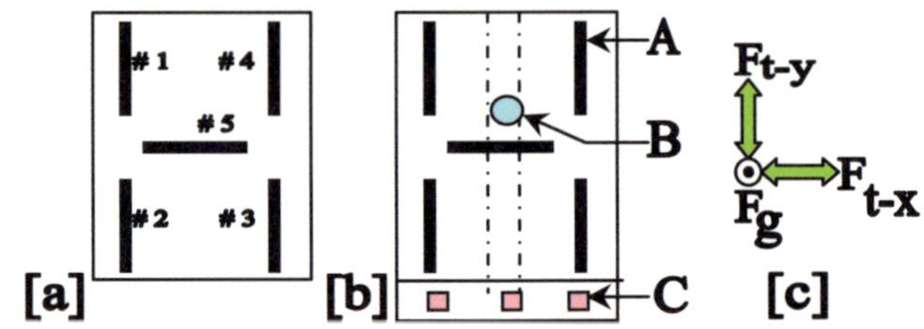

Index: A: TBS-cell; B: Load Cell; C: IR-cell; $F_{t-x,y}$: Tangential Force in x & y axes; F_g: Grip Force

Fig. 12. Layout of [a] TBS-sensor grid [b] overall sensory system & [c] grasp force signature

4.3 Overview of the software developed for case-study I

4.3.1 Schematic layout

The software, developed indigenously, has been customized for the statistical hypothesis testing of fused data evolved from the matrix sensor. It has a *core* module, like *backbone*, supported by four sub-modules, viz. [a] *Grip Force*, [b] *Slip Facets*, [c] *Planar Area* and [d] *Hypothesis Testing-cum-Optimality Check*. Each sub-module acts independently and provides quantitative output in terms of desired process parameters. The software is interlinked with 15 nos. *data files*, which either supply necessary input for the analytical computation or used for storing output data. The general layout of the software has been illustrated in fig. 13, highlighting functional dependencies between various sub-modules and data files.

The *beta* version of the software can incorporate a maximum of 200 sensor-cells, for which various categories of data (e.g. local co-ordinates, output response, noise threshold etc.) can be processed simultaneously. The entire code is framed considering [a] each of the resistive cell has got two R-taxels and [b] the number of rows and columns of R-cells are always one less than those of the C-cells.

The software also provides an interactive session as *front-end*, wherein user can feed design data relevant for the statistical analysis. In general, these user-input data are of two types, one related to the physical hardware of the sensor and the other one incorporates variables needed for fusion analysis. The former group notes the information, such as number of columns and rows in C & R-cells, pitch distance of the C- & R-taxels and evaluates data like total number of sensor-cells (taxels) and total number of data points in the matrix sensor as output. Similarly, the latter group takes the numerical values for ξ, k, $P(H_1)$, $P(H_0)$ and the delay (in seconds) before the next output, as desired by the user.

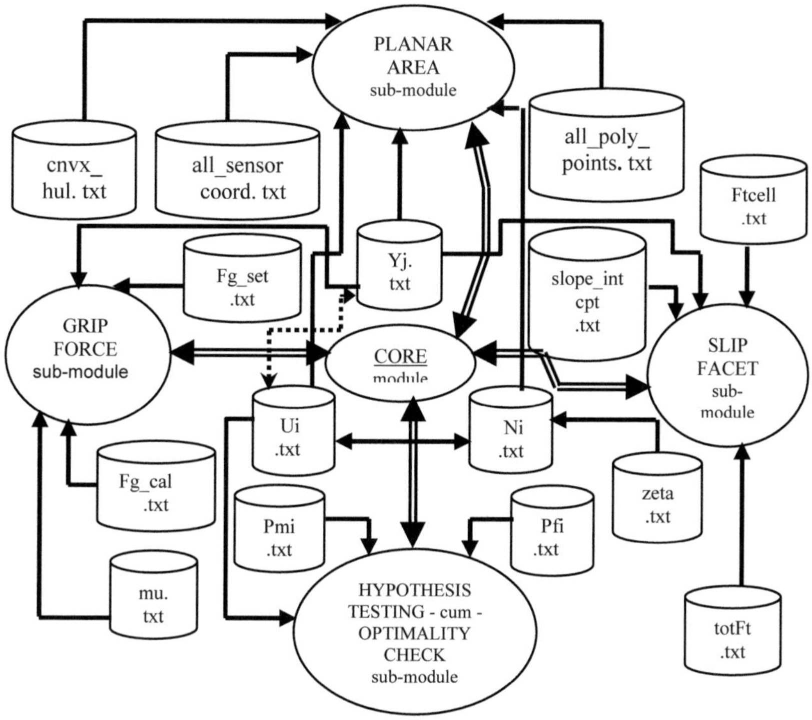

Fig. 13. Organizational layout of the developed software using fusion models

It may be noted that the sub-modules concerning evaluation of grip force, slip force and planar area of the object being grasped work coherently to certain extent and mutually share a group of data files (e.g. y_j.txt). However, the optimality check sub-module needs different inputs, such as data files for 'probability of false alarm' (P_{fi}.txt) and 'probability of miss' (P_{mi}.txt), alongwith other files. 'P_{fi}.txt' and 'P_{mi}.txt' are required only for analyzing the case-study while using LLR or N-P test. These are not needed for our method, viz. HEBTEM.

4.3.2 Sub-modules for the evaluation of grip force, slip facets and planar area

For interim processing of grasp and slip force, the program is able to tackle a maximum of 150 data-points at a time, of which, a set of 100 points go critical mainly for the evaluation of planar area of the object grasped. The program code uses dummy variable in order to indicate the number of data points stored in a file and also keep a record of respective Cartesian co-ordinates. This set of *data map*, viz. taxel number and its co-ordinates, has been used for the computation of planar area of the object being grasped.

The program takes the 'y_j' values from a text file, wherein 'y_j' values are put in the order of increasing value of the 'ordinate' (i.e. 'y' co-ordinate) and the points having same 'y' are stored in the order of increasing 'x'. It takes the noise data, viz, 'N_i', in a similar manner from another text file. The program performs a confirmatory check to ascertain the correctness of the number of data-point entries in the corresponding files. It then calls for the value(s) of the noise threshold (i.e. ξ, as per eqns. 3 & 4), both for the R- and C-cells. The program then evaluates $\{u_i\}$ matrix as per the model, described earlier. It stores 'u_i' values $[\forall u_i, \in(-1, 1)]$ in a text file, in the same way as 'y_j'. The program then decides the presence or absence of the object by calculating 'U_G' as per the fusion rules. Using the calibration data of the R-cells (analog signal in mV vs. force applied), the code calculates the total tangential force ($totF_t$) acting on the sensor-pad and controls the gripping force. Essentially, the program takes the calibration values from a text file (namely, 'slope_intcpt.txt'), having a pair of entries ('slope' & 'intercept') per R-cell alongwith 'y_j' values. It then calculates the instantaneous tangential force on each active R-taxel as,

$$F_t^j[i] = \frac{\left(y_j[i] - incpt[i]\right)}{slope[i]} \tag{30}$$

where, $F_t^j[i]$: Instantaneous tangential force for $i^{th.}$ resistive taxel at $j^{th.}$ time-instant; $y_j[i]$: Sensory reading (in mv.)for $i^{th.}$ resistive taxel at $j^{th.}$ time-instant; *incpt [i]*: Intercept value of the calibration line for $i^{th.}$ resistive taxel; *slope[i]*: Slope value of the calibration line for $i^{th.}$ resistive taxel. Once the numerical value of the tangential force at each R-taxel at a given time-instant is evaluated, the program finds out the total tangential force generated vis-à-vis the required gripping force by using the model (viz. eqns. 24 & 25). It may be noted that 'slope_intcpt.txt' file must have dummy entries corresponding to C-cells. Ideally, the 'slope' of the C-cells should be zero, signifying infinite tangential force (to be applied) on that C-cell. This is logical, as C-cells are comparatively more sensitive towards normal force in the prototype design.

The software stores the Cartesian co-ordinates of all the taxels in a text file (all_sensor_coord.txt) in the order of increasing 'y' (if two or more taxels have got same ordinate, then it is put in the order of increasing 'x'). As per the design, R-cells are assumed to be at the center of the *rectangle* formed by four C-cells and their co-ordinates are being evaluated accordingly. The next part takes the 'u_i' values from 'u_i.txt' and coordinates all the points from 'all_sensr_coord.txt' and finds those points only which are sending recognizable signal (above the threshold, as described by 'zeta.txt') and prints those points to 'all_poly_points.txt'. The module then examines the points that are on the *convex hull* from the previous file and puts those in 'cnvx_hul.txt' in either clockwise or counter-clockwise manner. The planar area of the polygon, determined by the convex hull so formed, is calculated thereafter.

4.3.3 Hypothesis testing -cum- optimality check sub-module

This module checks the presence or absence of the object in the gripper through the testing of the hypothesis using the fusion rule-bases described a-priori. After hypothesis testing, the program evaluates the value of the global threshold (λ_0) using HEBTEM as well as *Log-Likelihood Ratio* (LLR) and *Neyman-Pearson* (N-P) tests. The LLR & N-P tests are incorporated in the module so as to make a comparative judgment between the existing methods and our method. Now, in order to evaluate the global threshold using HEBTEM ($\lambda_{Th.}$), we will use eqn. 16 or 17. Then after, the final decision on the selection of hypothesis will be made by evaluating the numerical value of 'U_G', as described in section 2.2.2. On the contrary, the numerical value of the global threshold by LLR test is evaluated as per the standard formulae, viz.

$$\lambda_{threshold}^{LLR}(U) = \log\frac{P(H_1|U)}{P(H_0|U)} = \log\frac{P_1}{P_0} + \sum_{S+}\log\frac{1-P^i{}_M}{P_F{}^i} + \sum_{S-}\log\frac{P_M{}^i}{1-P^i{}_F} \quad (31)$$

where, U: The *global* event that data is generated out of the sensor-cells; P_1: The un-biased estimate of $P(H_1)$; P_0: The un-biased estimate of $P(H_0)$; $P_M{}^i$: The probability of miss for the $i^{th.}$ sensor-cell; $P_F{}^i$: The probability of false alarm for the $i^{th.}$ sensor-cell; 'S+':The set of all 'i', such that $\{u_i = +1\}$ and 'S-':The set corresponding to $\{u_i=-1\}$. The final selection of hypothesis is being governed by the numerical value obtained for $\lambda_{threshold}{}^{LLR}$, namely, if $\lambda_{threshold}{}^{LLR} > 0$ then accept 'H_1'. Likewise, the global threshold by N-P test is computed as,

$$\lambda_{threshold}^{N\text{-}P}(U) = \frac{P(U|H_1)}{P(U|H_0)} = \prod_{i=1}^{i=n}\left(\frac{1-P_M^i}{P_F^i}\right)^{u_i}\left(\frac{P_M^i}{1-P_F^i}\right)^{1-u_i} \equiv \prod_{S+}\left(\frac{1-P_M^i}{P_F^i}\right)\prod_{S-}\left(\frac{P_M^i}{1-P_F^i}\right) \quad (32)$$

where all the legends bear the same definition, as of eqn. 31. The final selection of the hypothesis is made through the following proposition, viz. if $\lambda_{threshold}{}^{N\text{-}P}(U) > \lambda_{threshold}$, then '$H_1$' is true and if $\lambda_{threshold}{}^{N\text{-}P}(U) < \lambda_{threshold}$ then 'H_0' is true. Here, '$\lambda_{threshold}$' signifies the threshold under un-biased condition of the hypotheses, i.e. the value of '$\lambda_{threshold}$' is 1.0.

5. Synthesis of the model under various grasp postures

One of the salient aspects of the developed model is its ready applicability in practical situations, pertaining to different grasp postures. Apart from using HEBTEM on a stand-alone tactile sensor, we need to explore its functional utility while interfaced with a robotic gripper. In this section, we will examine three major types of contact between the object and the gripper-jaw and draw an analogy between the contact semantics of those grasp-postures and the HEBTEM attributes. We will design analytical metric to classify the contact-types and retrofitting the rule-bases of HEBTEM thereon.

5.1 Jaw-based gripping re-visited

We will discuss the case with reference to a planar parallel two-jaw robotic gripper, which is capable to gripping objects, as restricted by its shape, size and payload threshold. A pair of matrix sensors (refer section 3, fig. 8) is mounted on the jaw surfaces at suitable location and soft rubber padding is provided over it. Figure 14 schematically shows the disposition of this sensor-instrumented jaw gripper, in course of grasping an object. The figure also illustrates the contact metrics between the jaw and object, highlighting on the *asperity contact*. It may be mentioned that it is the geometry of the contact between the asperities of both the object and the jaw surface that defines the type of contact, prior to a stable grasp.

We shall now examine the three major types of contacts, prevailing in the domain of planar grasp. We may categorize these as: a] *Point Contact (without friction)*; b] *Point Contact (with friction)* and c] *Soft Finger Contact*. However, this classification is in turn dependent on the *relative nature* of the jaw and object surface, e.g. we can have soft jaw vs. hard jaw vis-à-vis soft object vs. hard object. Thus, theoretically speaking, we can get four combinations in total, namely, I] soft jaw_ soft object; II] soft jaw_ hard object[1]; III] hard jaw_ soft object and IV] hard jaw_ hard object. Now considering various feasible combinations between the

[1] A combination of soft jaw and hard object is impractical for real-life use.

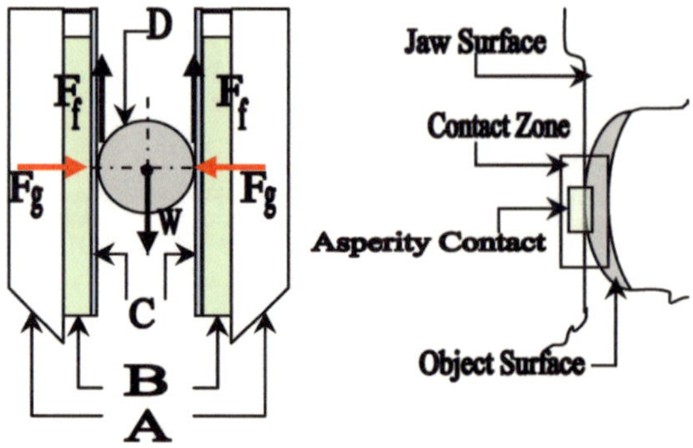

Legends: **A:** Gripper Jaws; **B:** Matrix Sensors; **C:** Rubber Pads; **D:** Object Gripped; **W:** Weight of the Object; $\mathbf{F_g}$**:** Grip Force; $\mathbf{F_f}$**:** Friction Force.

Fig. 14. Disposition of the grasp condition and contact for a planar two-jaw robotic gripper

nature of jaw and object, we can define a parameter, which will measure the *conjugate softness* of the *jaw-object pair* in a relative scale. We will term this parameter as *relative softness index* (RSI) and will be defined as,

$$RSI = \frac{Softness_{Object}}{Softness_{Jaw}} - 1 \equiv \frac{1/H_O}{1/H_J} - 1 = \frac{H_J}{H_O} - 1 \tag{33}$$

where, 'H_J' and 'H_O' represent the hardness values for the jaw and object respectively, measured through an uniform *Rockwell* Scale. It may be mentioned here that in a dynamic condition, governed by the Coulomb's laws of friction, the variation of the three groups of contact is essentially dependant on the value of RSI. For example, a very low numerical value of RSI will target towards point contact (without friction) while a high value of RSI hints for a soft finger contact. Obviously the zone for point contact (with friction) lies in-between these two extremities, with a fairly large range. Figure 15 illustrates an idealistic distribution of the contact zones with respect to the variation of relative softness index. It also highlights a case of *Locked Contact*, with RSI=0, where there is no movement even between the asperities of both object and jaw.

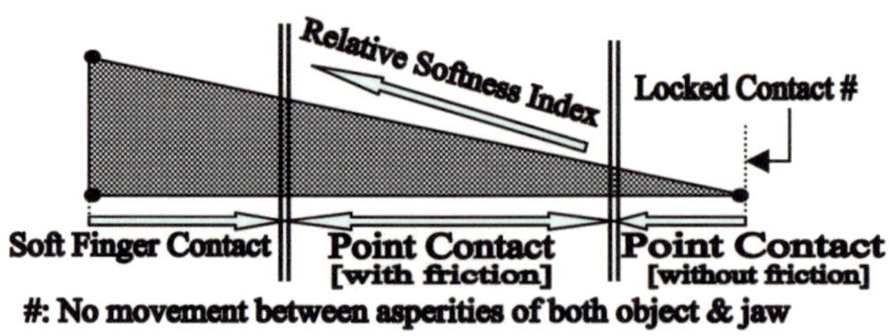

Fig. 15. Schematic view of the distribution of contact zones as contrasted with RSI

5.2 Mapping of contact criteria to HEBTEM rule-bases

The mechanics of contact between the object and the jaw starts zooming in once the object is grasped between the jaws of the gripper. The taxels of the matrix sensor get excited by the external loading as a consequence of the contact. However, we need to have a thorough mapping between different contact-types and the level of taxel-excitation in order to arrive at a decision regarding the nature of the object. Once we get the data regarding the excited taxels, alongwith their location on the sensor, we can use HEBTEM for the optimality test and can infer about the presence / absence of the object in the gripper-zone. Keeping in mind the basics of the three contact-types referred here, qualitatively we can infer the nature of taxel-excitation. For example, the situation of frictionless point contact is equivalent to the case of excitation of the taxels by 'point-force', while point contact with friction is analogous to impingement by small or very tiny objects and soft finger contact is mapped through an external loading by a comparatively large object.

With this qualitative paradigm, we will now introduce the concept of *friction surface* and *friction plane*, in relation to the taxel geometry of the matrix sensor. Enveloping all the excited taxels of the matrix sensor generates a *friction surface* and the corresponding projection in the 2D plane is denoted as the *friction plane*. As soon as a grasp takes place the contact-type between the jaw-object interface is frozen and the excitation of the taxels begins thereof. The extent of such excitation is imaged on the friction surface and its effect is ascertained by measuring the area of the friction plane. Figure 16 pictorially explains this mapping scenario, wherein all three types of contact criteria have been studied through a] physical view at the gripper; b] spatial view at the matrix sensor in order to represent 3D friction surface and c] planar view of the friction plane, as generated.

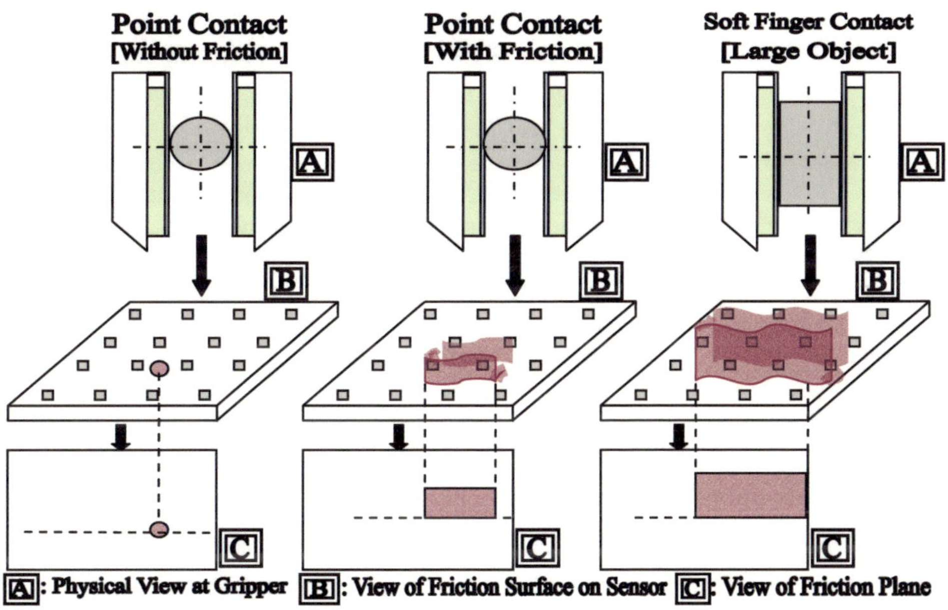

Fig. 16. Schematic view of the mapping of contact criteria during grasping

We propose two methods in order to map the contact-types with the equivalent object description, in terms of taxels. These are: 1] *Taxel_Count* (t_c) and 2] *Friction_Plane_Area_Count* (fp_c). The computations of the two methods are as follows,

$$t_c = \frac{N_{affected}}{N_{total}} \tag{34a}$$

and

$$fp_c = \frac{\Delta_{friction\_plane}}{\Delta_{total}} \tag{34b}$$

where, $N_{afftected}$: Number of taxels excited by the grasp, i.e. the contact between the object and the jaw; N_{total}: Total number of taxels present in the sensor; $\Delta_{friction\_plane}$: Area of the *friction_plane* and Δ_{total}: Total area of the taxels, i.e. the planar area of the sensor. It is apparent that both of these methods essentially tells us the comparative range of the taxels affected, which will be used finally in judging the hypothesis by knowing the threshold for HEBTEM ($\lambda_{Th.}$) a-priori. Table 1 describes the mapping index between the three groups of contact-types and the corresponding signature in using fusion logic and HEBTEM.

Contact Type	Equivalent Object Nomenclature	Taxel Signatures & Analytical Limits
Point Contact [Without Friction]	"Point-force"	Taxel Signature: "*Least*" Limits: $0.01 \le \{t_c, fp_c\} \le 0.02$
Point Contact [With Friction]	Small /Very Tiny Object	Taxel Signature: "*Minimum*" Limits: $0.01 \le \{t_c, fp_c\} \le 0.1$
Soft Finger Contact	Bigger Object	Taxel Signature: "*Moderate to High*" Limits: $0.1 \le \{t_c, fp_c\} \le 1$

Table 1. Mapping index for different contact-types during jaw-based grasp

Thus, we get a working thumb-rule for distinguishing different contact-types mathematically, by evaluating *taxel_count* and/or *friction_plane_area_count*. Once the numerical values for 't_c' and 'fp_c' are obtained, those can be mapped (as per Table 1) and the contact-type vis-à-vis equivalent object nomenclature, as prevailed in the specific grasp action, is to be determined. Also, since *txael_count* is known numerically, we get the exact number of taxels affected by the loading, which will give us $\{u_i\}$ matrix for the sensor and the gripper in action. The subsequent procedure will follow thereon, i.e. use of rule-bases and confidence level and finally, we can judge the hypothesis as per HEBTEM, as stated earlier. It may be stated that although identical fusion rules are applied in all the three contact situations, different values of λ_{Th} may be adopted based on the equivalent object nomenclature.

6. Simulation studies

6.1 Perspective

The proposed hypothesis, HEBTEM has been tested thoroughly using simulated results, prior to real-life experimentation on the matrix sensor & jaw gripper respectively. Nonetheless, the simulated sensor-data have been assumed to represent the basic functioning of the sensor & gripper, only difference being that those are not real-time data. Considering the case-study for matrix sensor, We analyze HEBTEM vis-à-vis traditional optimization techniques (namely, LLR and N-P tests) for checking the condition of presence / absence of object, aided by requisite matrices, viz. 'P_{fi}' (Probability of False Alarm), 'P_{mi}' (Probability of miss), 'Y_i' (raw sensor output) and 'N_i' (sensor noise). Each matrix is composed of data, arranged row-wise in increasing order[2], pertaining to the *active taxels* of the matrix sensor. However, four C-cells, located at the four corners of the sensor-matrix (marked with dashed-line border in fig. 8) are not included in the data fusion algorithm. This has been done deliberately as none of the objects tried under real-life testing are covering up those corner C-cells. Hence, in a way we are only considering 21 *C-cells*, 32 *R-cells* and 4 *P-cells* in action.

6.2 Simulation results

We are having the following four matrices, namely, 'P_{fi}', 'P_{mi}', 'Y_i' and 'N_i' as input, each of which has got 57 readings, pertaining to 57 taxels in action.
Probability Matrix of *False Alarm* (P_{fi}):

$$\begin{bmatrix} 0.0005 & 0.0023 & 0.0010 & 0.0002 & 0.0002 & 0.0004 & 0.0005 & 0.0013 & 0.0006 & 0.0004 & 0.0066 & 0.0021 & 0.0005 & 0.0033 & 0.0005 & 0.0034 & 0.0002 & 0.0005 & 0.0033 \\ 0.0056 & 0.0052 & 0.0056 & 0.0032 & 0.0064 & 0.0013 & 0.0024 & 0.0075 & 0.0013 & 0.0044 & 0.0013 & 0.0035 & 0.0046 & 0.0003 & 0.0076 & 0.0021 & 0.0005 & 0.0033 & 0.0005 \\ 0.0034 & 0.0002 & 0.0005 & 0.0033 & 0.0056 & 0.0052 & 0.0050 & 0.0020 & 0.0064 & 0.0013 & 0.0024 & 0.0075 & 0.0013 & 0.0044 & 0.0013 & 0.0035 & 0.0046 & 0.0003 & 0.0076 \end{bmatrix}$$

Probability Matrix of *Miss* (P_{mi}):

$$\begin{bmatrix} 0.0015 & 0.0022 & 0.0030 & 0.0052 & 0.0062 & 0.0074 & 0.0085 & 0.0093 & 0.0076 & 0.0054 & 0.0036 & 0.0023 & 0.0004 & 0.0005 & 0.0006 & 0.0008 & 0.0032 & 0.0025 & 0.0013 \\ 0.0036 & 0.0052 & 0.0006 & 0.0007 & 0.0008 & 0.0009 & 0.0054 & 0.0008 & 0.0006 & 0.0043 & 0.0023 & 0.0055 & 0.0066 & 0.0043 & 0.0006 & 0.0023 & 0.0004 & 0.0005 & 0.0006 \\ 0.0008 & 0.0032 & 0.0025 & 0.0013 & 0.0036 & 0.0052 & 0.0006 & 0.0007 & 0.0008 & 0.0009 & 0.0054 & 0.0008 & 0.0006 & 0.0043 & 0.0023 & 0.0055 & 0.0066 & 0.0043 & 0.0006 \end{bmatrix}$$

The off-line data matrix for the raw sensory output (in milli-volts) (Y_i):

$$\begin{bmatrix} 2.3 & 7.8 & 0.6 & 0.8 & 15.6 & 5.8 & 2.4 & 18.0 & 4.3 & 5.5 & 4.2 & 0.5 & 3.7 & 2.7 & 5.1 & 5.4 & 8.8 & 19.3 & 14.6 \\ 11.7 & 7.8 & 3.6 & 8.5 & 4.4 & 5.69 & 5.5 & 2.6 & 6.4 & 8.8 & 4.88 & 17.0 & 2.9 & 13.4 & 9.0 & 5.6 & 6.5 & 7.5 & 8.5 \\ 9.5 & 7.5 & 6.4 & 5.3 & 4.3 & 3.5 & 2.6 & 2.6 & 4.4 & 4.5 & 5.6 & 6.7 & 7.8 & 8.4 & 9.6 & 0.4 & 9.5 & 4.6 & 6.7 \end{bmatrix}$$

The system noise (in milli-volts.) (N_i):

$$\begin{bmatrix} (0.1) & (0) & (0) & (1) & (0.8) & (1) & (1) & (1) & (0) & (0) & (0.9) & (1) & (1) & (0) & (0) & (0.5) & (0.4) & (1) & (1) \\ (0.6) & (1) & (0) & (0) & (1.2) & (1) & (1) & (1.3) & (0) & (0.3) & (0.7) & (1) & (0) & (1) & (1) & (0) & (0) & (0) & (0) \\ (0) & (1) & (1) & (1) & (1) & (0) & (0) & (1) & (1) & (0) & (0) & (1) & (1) & (0) & (1) & (0) & (1) & (0) & (0) \end{bmatrix}$$

Now, using these simulated sensor-data, $\{u_i\}$ matrix $[u_i \in (-1,1)]$ computed as,

$$\begin{bmatrix} -1 & -1 & -1 & -1 & 1 & -1 & -1 & 1 & -1 & -1 & -1 & -1 & -1 & -1 & -1 & -1 & -1 & 1 & 1 \\ 1 & -1 & -1 & -1 & -1 & -1 & -1 & -1 & -1 & -1 & -1 & 1 & -1 & 1 & -1 & -1 & -1 & -1 & -1 \\ -1 & -1 & -1 & -1 & -1 & -1 & -1 & -1 & -1 & -1 & -1 & -1 & -1 & -1 & -1 & -1 & -1 & -1 & -1 \end{bmatrix}$$

Based on the simulated observation values of *taxels*, we have the final value for the global threshold ($\lambda_{threshold}^{LLR}$) as -135.9376, which signifies a strong case for rejecting H_1, inferring no object is present within the graspable zone of the gripper. In fact, this evaluation is strongly influenced by the number of sensor-cells having $\{u_i = 1\}$, which, in this case is only seven (vide cell nos. 5, 8, 18, 19, 20, 31 & 33). Hence, by this test we are likely to exclude those options wherein a relatively tiny object is being gripped by the gripper or for that reason, a point-object / force is impinging on the sensor / gripper body.
Nonetheless, the pre-conceived value for the global threshold under LLR method (which is 1.0, as stated in section 2.2.3.4) also plays a deciding role in this regard. Even if we assume a non-unity value for global threshold, we need to check the values of the cost-coefficients,

---

[2] Two taxels positioned at one R-cell, giving 41 different locations for a total of 61 taxels.

such that they follow at least one of the conditions, viz. [a] $C_{10} > C_{00}$ & $C_{01} > C_{11}$; [b] $C_{10} < C_{00}$ & $C_{01} < C_{11}$. In all other cases, the test will fail, which is certainly a major bottleneck of the method itself. Moreover, the tests will lead to undefined zones of hypothesis, when the value of $\lambda_{threshold}^{LLR}$ will be equal to zero or other non-zero value.

That's why, although computationally simpler, this test lacks in handling the specific situations concerning very small object, point-force, instant impulse etc. and also the undefined cases.

For N-P test, we get the final value for the $\lambda_{threshold}^{N\text{-}P}$ as 26.19898x10^{-143}, which again signifies a strong case for rejecting H_1 (assuming $\lambda_{threshold}$ to be 1.0 in an un-biased case). However, this test also fails to identify the hypothesis in case the numerical value of $\lambda_{threshold}^{N\text{-}P}$ becomes exactly equal to $\lambda_{threshold}$. As argued earlier, this test too fails to satisfy the varied situations of robotic grasping involving miniaturized objects and point-force.

Thus, we have found from the simulation results that although traditional optimization methods, such as LLR & N-P, may be used for *larger objects* but there is no guarantee those would give correct inference for *miniaturized objects* or *'point-objects'*. As a matter of fact, fusion problem in such situations should ideally be considered with on-line observation values and posteriori processing using suitable model. In that respect, our model proves to be a viable option.

6.3 Optimized solution using HEBTEM

Evaluation of global (dynamic) threshold ($\lambda_{Th.}$) is relatively easier using HEBTEM, which also clearly demarcates the acceptance/rejection regions of 'H_1'. For example, considering multiplicative model of fusion, if 83% Type II error has been set as the accepted limit (i.e. β= 0.83), then, in the present case with n=57, we get the numerical value for $\lambda_{Th.}$ as 1.663x10^{-20} (=$e^{-45.543}$, refer eqn. 16). Similarly, in additive model of fusion, we get the value for $\lambda_{Th.}$ with n = 57 as ± 49.753 (refer eqn. 17), respectively against 3% Type I error (i.e. α = 0.03) and 47.3% Type I error (i.e. α = 0.473). Once the dynamic threshold is evaluated by either of the fusion models, the next phase what HEBTEM takes care is determination of the following facets, viz. [a] inference about the presence or absence of object /point-force in the vicinity of the gripper; [b] evaluation of the grip and slip force and [c] determination of the planar area of the object being grasped and/or impinged upon, as explined in detail below.

6.3.1 Inference about presence of object using fusion logics

Considering the simulated situation, i.e. wherein seven cells are giving acceptable signal (vide cell nos. 5, 8, 18, 19, 20, 31 & 33), U_G has been computed. For example, using additive model, U_G becomes – 43 (vide eqn. 6) and likewise multiplicative model gives the value of U_G as 2.11513x10^{-19} (=e-43, vide eqn. 5). Now, the dynamic threshold obtained for additive model being – 49.753 (which is less than – 43, i.e. the value of U_G)[3], the presence of object is confirmed. Also, the value of dynamic threshold for multiplicative model being 1.663x10^{-20} (which is less than 2.11513x10^{-19}, i.e. the value of U_G), the presence of object is inferred. It is interesting to note that the value of $P(H_1)$, corresponding to U_G in additive and multiplicative model comes out to be 0.124835 and ≈0.003 respectively (using the curve in figs. 2a & 4). Hence, although the presence of object is confirmed, yet different fusion logic infers the situation with different *levels of conformity*, viz. the value of $P(H_1)$.

---

[3] We consider left hand side of the curve shown in fig. 4, since the sign of U_G is negative.

6.2.3.2 *Evaluation of Slip and Grip Force*

We consider the following data for 'slope' and 'intercept', as required for the calibration of tangential force. These are represented in conjugate inside the bracket ( ), the first term being the 'slope' and the second term is the 'intercept' (i.e. 'm' and 'c' respectively, as per eqn. 23). The 'slope' & 'intercept' matrix:

$$\begin{bmatrix} (0,6) & (0,6) & (0,6) & (0,3) & (8,0) & (7,1) & (9,2) & (5,3) & (6,1) & (3,1) & (9,8) & (9,1) & (9,4) & (0,2) & (6,3) & (2,4) & (3,7) & (4,2) & (3,4) \\ (2,4) & (2,1) & (3,6) & (3,0) & (4,7) & (3,4) & (8,2) & (9,2) & (7,3) & (4,2) & (3,3) & (4,2) & (0,5) & (6,1) & (5,1) & (5,1) & (4,4) & (3,2) & (6,3) \\ (2,4) & (3,7) & (2,1) & (4,2) & (3,4) & (0,4) & (3,6) & (2,0) & (4,7) & (5,4) & (8,2) & (9,2) & (7,3) & (4,2) & (3,3) & (0,5) & (0,2) & (0,1) & (0,1) \end{bmatrix}$$

With the 'm' & 'c' values for the calibration of tangential (slip) force are known a-priori, the tangential forces on the affected taxels are computed using eqns.23 & 25. The calculated values of 'F_t' for taxel no. 5, 8, 18, 19, 20, 31 & 33 are 1.95, 3.00, 4.325, 3.5333, 3.85, 3.75 & 2.06667 units respectively and total tangential/slip force is 22.475 units. The upper threshold for grip force can thus be evaluated knowing the value of coefficient of dynamic friction (eqn. 26a). Similarly, maximum limit of allowable loading may be ascertained from eqn. 26b.

6.2.3.3 *Determination of planar area of the graspable object*

The size of the object (which is being gripped remotely and impinging on the tactile sensor matrix) is determined using the raw sensor data. After getting filtered for the sensor noise, these data are used to ascertain the location of the taxels and subsequently, Cartesian co-ordinates of those taxels are recorded. Figure 17 shows a typical case wherein 'n' taxels are registered for giving $\{u_i = +1\}$.

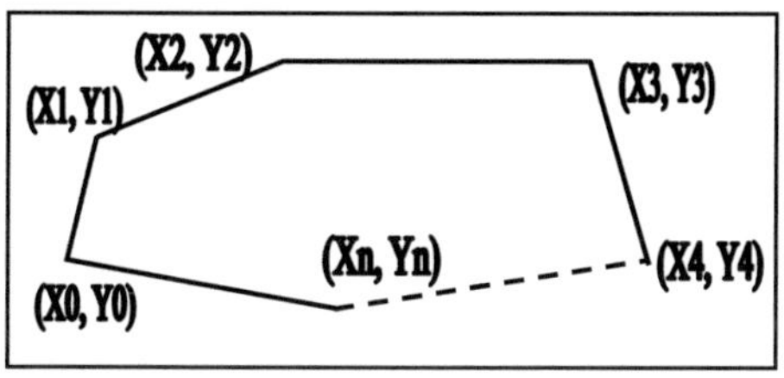

Fig. 17. Convex hull generated from the readings of the taxels of the matrix sensor

The planar area (A) of the convex hull generated thereon is computed as,

$$A = \frac{1}{2} \sum_{i=0}^{i=N-1} (X_i Y_{i+1} - X_{i+1} Y_i) \tag{35}$$

The associated program code first finds the co-ordinates of those taxels giving *recognizable* signals. Now, a *point* (i.e. taxel location) is selected having maximum Y-co-ordinate, which will always lie on the convex hull. Then it checks every double combination of the remaining points with that point, selected a-priori to find the maximum included angle. Thus the 'second' such point will be selected as a point on the convex hull.

Figure 18 illustrates the schematics of the layout of the 57 taxels of the matrix sensor and their nomenclature, in order to develop a one-to-one mapping between the taxel number and its co-ordinates. In general, 'T_x' denotes the $x^{th.}$ taxel ($\forall x = 1,2,\ldots\ldots,57$), but it is not indicative of the *location* of that taxel precisely. In contrast, '$T_{j,k}$' denotes $j^{th.}$ & $k^{th.}$ taxels placed at same location. Hence in the pictorial layout 'T_x' represents a C-cell, '$||T_y||$' represents a P-cell and '$T_{j,k}$' represents a R-cell ($x,y \neq j,k$). This nomenclature is essential. The legend, '*d*' (dummy) signifies the four corner locations of the C-cells, which are not participating in the fusion algorithm.

d		*T*54		*T*55	‖*T*57‖	*T*56		*d*
	*T*46,47		*T*48,49		*T*50,51		*T*52,53	
*T*41		*T*42		*T*43		*T*44		*T*45
‖*T*31‖	*T*32,33		*T*34,35		*T*36,37		*T*38,39	‖*T*40‖
*T*26		*T*27		*T*28		*T*29		*T*30
	*T*18,19		*T*20,21		*T*22,23		*T*24,25	
*T*13		*T*14		*T*15		*T*16		*T*17
	*T*5,6		*T*7,8		*T*9,10		*T*11,12	
d		*T*2		*T*3	‖*T*1‖	*T*4		*d*

Fig. 18. Nomenclature of taxels of the matrix sensor for simulation

The planar co-ordinates of the taxels have been arrived at after considering the locational matrix of the 41 taxels (21 C-cells, 16 R-cells & 4 P-cells), assigned proportionately with the design and dimensions of the prototype sensor, as detailed out below.

(1,104)		(21,104)		(41,104)	‖(51,107)‖	(61,104)		(81,104)
	(11,91.5)		(31,91.5)		(51,91.5)		(71,91.5)	
(1,79)		(21,79)		(41,79)		(61,79)		(81,79)
‖(1,66.5)‖	(11,66.5)		(31,66.5)		(51,66.5)		(71,66.5)	‖(82,66.5)‖
(1,54)		(21,54)		(41,54)		(61,54)		(81,54)
	(11,41.5)		(31,41.5)		(51,41.5)		(71,41.5)	
(1,29)		(21,29)		(41,29)		(61,29)		(81,29)
	(11,16.5)		(31,16.5)		(51,16.5)		(71,16.5)	
(1,4)		(21,4)		(41,4)	‖(51,1)‖	(61,4)		(81,4)

For example, the off-line data for the taxel observations (X_i), noise (N_i) and subsequently $\{u_i\}$ lead to the taxels at the locations (11,16.5), (31,16.5), (11,41.5), (11,41.5), (31,41.5), (1,66.5) & (11,66.5) to be the points on the desired polygon. These locations correspond to the taxel nos. 5, 8, 18, 19, 20, 31 & 33. The convex hull, so generated, will have its planar area of 625 sq. units, considering five major vertices.

7. Case studies and experimentation

HEBTEM has been tested using field-results of the experimentation on the matrix sensor as well as sensor-augmented jaw gripper. Through the field-trials, global threshold was evaluated, alongwith other model-specific parameters. We will present these findlings below.

7.1 Testing with matrix sensor

The sensor has been calibrated with a series of *object*, having pre-assigned contour but varying weight in real-time. The pictorial representation of the object, used for calibration as well as final testings, is shown in fig. 19.

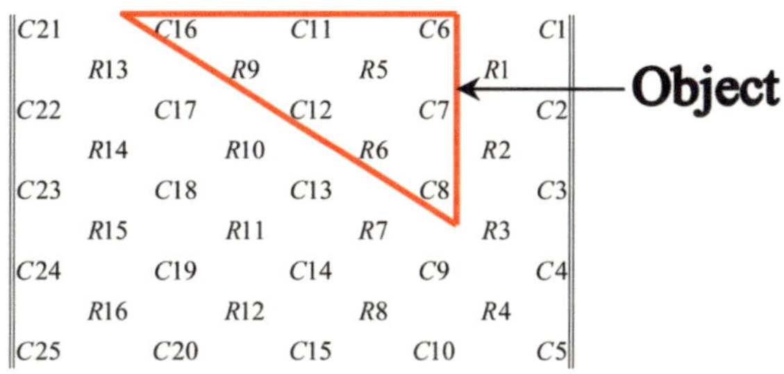

Fig. 19. Pictorial view of the test scheme with the objects for case-study I

The sensor is loaded with the *object-sets,* having weights ranging from 0.9 kg. to 4.3 kg. However, due to serrated skin, it is not possible to have simultaneous excitation of the R & C-cells. As a matter of fact, only R-cells, namely, R5, R6 & R9 are affected in this test (refer fig. 19). The calibration of the respective struts reveals the regression equations (considering straight line trend), which were used to evaluate tangential force on each R-cell numerically. Based on the testing, the calibration equations for the struts are determined as follows,

$$\begin{bmatrix} Y\_R5 \\ Y\_R6 \\ Y-R9 \end{bmatrix} = \begin{bmatrix} 0.1661 \\ 0.307 \\ 0.3185 \end{bmatrix} \{F\_T\} - \begin{bmatrix} 0.0155 \\ 0.0163 \\ 0.0133 \end{bmatrix} \quad (36)$$

where, $[Y\_Rj]$, ∀j=5,6,9 signifies the sensory output (in mV) from the respective R-cells, against the loadings (manifested in tangential force, '*F_T*').

The matrix sensor was then tested for *unknown* loadings (using objects having same contour as used in the calibration of the three R-cells) and the excitations (i.e. y_j values, refer eqn. 30) at the *affected taxels* were recorded. However, due to serrated *skin*, R-cells were found to be affected at large. The 'slope' and 'intercept' values of the R-cells were obtained from the corresponding calibration curves. A sample test-data (in mV) is analyzed, which shows *effective excitation* (i.e. x_i values) from six *taxels* (located on three struts, i.e. corresponding to R5, R6 & R9), viz. $[(0.256, 0.064), (0.514, 0.128), (0.488, 0.122)]^T$, where the data inside each bracket signifies the signals from a single strut. The 'slope' & 'intercept' data for the corresponding struts are $[(0.1661, -0.0155), (0.307, -0.0163), (0.3185, -0.0133)]^T$ (refer eqn. 36). Hence, the computed value of tangential force for each taxel becomes $[(1.6152, 0.4038), (1.7168, 0.4292), (1.5648, 0.3912)]^T$ and the total tangential /*slip force* becomes 6.121 kgf. Considering the coefficient of dynamic friction (μ) between the object and sensor surface to be 0.3, we get the upper limit of the gripping force to be 20.41 kgf., which concludes that the gripper may be loaded with a maximum weight of 12kgf. (approx.).

7.2 Testing with sensor-instrumented jaw gripper

Like earlier case, we consider HEBTEM for checking the condition of presence / absence of object inside the gripper, aided by requisite matrices, viz. 'Y_i' (raw sensor output) and 'N_i' (sensor noise). Each matrix is composed of data, arranged row-wise in increasing order, pertaining to the 18 *active taxels* of the sensory-grid. However, four strain gauge-based force sensor data are not used in the test and data fusion algorithm as well. This has been done deliberately as load cells and strain gauges represent same category of sensor elements; hence those can be treated *homogeneous* from the view-point of data fusion. Hence, in a way we are only considering 2 *LC-taxels*, 10 *TBS-taxels* and 6 *IR-taxels* in action, as per the layout-schematic given in fig. 20. Although IR-cells are sufficient in their own metric for the purpose of object detection, we adopt the extended sensor model in order to induce redundancy, in case of any malfunction in the IR-cells.

$$\begin{matrix} u_1 & u_2 & u_3 & u_4 & u_5 & u_6 \\ u_7 & u_8 & u_9 & u_{10} & u_{11} & u_{12} \\ u_{13} & u_{14} & u_{15} & u_{16} & u_{17} & u_{18} \end{matrix}$$

Index: $\{u_{1,2}\}$: LC-taxels; $\{u_{3,\ldots,12}\}$: TBS-taxels;$\{u_{13,\ldots,18}\}$: IR-taxels.

Fig. 20. Schematic layout of the taxels in the sensory grid of the gripper

The off-line data for the raw sensory output $[Y_i]$ & system noise $[N_i]$, both expressed in milli-volts are shown below alongwith the computed $\{u_i\}$ matrix, $\forall\ [u_i \in (-1,1)]$,

$$[Y_i] \equiv \begin{bmatrix} 3.93 & 3.89 & 0.0226 & 0.0218 & 0.0245 & 0.0214 \\ 0.0259 & 0.0228 & 0.0234 & 0.0219 & 0.0245 & 0.0227 \\ 429 & 424 & 410 & 407 & 388 & 387 \end{bmatrix} [N_i] \equiv \begin{bmatrix} 0.1 & 0 & 0 & 0.01 & 0 & 0.01 \\ 0.2 & 0 & 0.01 & 0 & 0 & 0.01 \\ 0 & 0.02 & 0.1 & 0 & 0.01 & 0.1 \end{bmatrix} \{u_i\} \equiv \begin{bmatrix} 1 & 1 & 1 & -1 & 1 & -1 \\ 1 & 1 & 1 & -1 & 1 & 1 \\ 1 & 1 & 1 & 1 & -1 & -1 \end{bmatrix}$$

Based on the observation values of the taxels, the computation for the mean global threshold ($\lambda_{Th\text{-}mean}$) was done using eqns. 14 & 18 with $g(n) = 4(n-1)$ [i.e. for $(p,q) = (1,1)$], and finally we got one transcendental equation in 'λ', viz.

$$7.187x10^{-5}\lambda^3 - 0.517\lambda + 12.558 = 0 \tag{37}$$

considering, ξ=0.003; n=18 and α=0.48, in order to solve for '$\lambda_{Th\text{-}mean}$'. We need to consider a comparatively larger value of 'α', as the level of uncertainty is more in the preferential selection model, compared to that in additive & multiplicative models. Finally, from eqn. 37, we get the optimized value for '$\lambda_{Th\text{-}mean}$' as 27.02 and thenafter, using eqn. 20, the value of the threshold-band becomes 8.16. We observed that in total 13 taxels are giving acceptable signal (i.e. u_i=1) in this testing, barring 5 taxels, vide taxel nos. 4, 6,10,17 & 18. Hence, the computed value of U_G (using eqn. 7) becomes 37, which is more than the mean threshold. Thus, the presence of the object in the gripper-jaw in very well inferred by HEBTEM. Now, based on the 'Y_j' values, pertaining to 18 taxels in order, we can calculate the total tangential force on each jaw (refer eqns. 28 & 29) and finally we can evaluate the final value of the tangential force, averaged over both the jaws. As per fig. 20, the readings $\{u_3,.., u_6\}$ and $\{u_7\}$ are mapped to '$F_{t\text{-}y}$' & '$F_{t\text{-}x}$' of the left-hand-side jaw respectively and likewise, $\{u_8,.., u_{11}\}$ and $\{u_{12}\}$ correspond to '$F_{t\text{-}y}$' & '$F_{t\text{-}x}$' of the right-hand-side jaw. Using the data, we get the final average tangential force on the object as η (0.09464) kgf., where 'η' is the calibration coefficient in kgf/mv. Also, $\{u_1, u_2\}$ is mapped to the grip force on the jaws, which finally amounts to (3.91/M) kgf, where 'M' is the calibration factor, defined in eqn. 27b.

8. Conclusions

A new approach for sensor data fusion of heterogeneous sensor-cells has been described in his article. It proposes a fusion theory wherein the threshold for fusion can be suitably adapted within a finite zone. The threshold estimation for the present work has been based on using *variable threshold*, exploiting the metrics of *Type I* & *Type II* error. The filtered

sensory data is used to estimate the optimal value of the slip force, which is generated in the jaw-object contact zone that ought to be combated for a farm grasp. Nonetheless, HEBTEM also ensures the characteristics of the post-grasp slip, if any.
The proposed schemata provides insight to two aspects, namely evolution of new rule-base towards data fusion and an optimized inference about object's presence or absence based on stochastic hypothesis testing model, using relative dependency of the sensor-cells. HEBTEM has been tested for various objects with different shape, size and contour using the sensor-instrumented gripper and the output is used for processing feedback control signals for its remotized operation, devoid of camera systems.

9. References

Alhakum, S. & Varshney, P.K. (1996). Decentralized Bayesian Detection With Feedback. *IEEE Transactions on Systems, Man & Cybernatics Part A: Systems & Humans*, Vol. SMC-26, No. 4, July 1996, pp 503 - 513, ISSN: 1083-4427.

Chair, Z. & Varshney, P.K. (1986). Optimal Data Fusion in Multiple Sensor Detection Systems. *IEEE Transactions on Aerospace & Electronic Systems*, Vol. AES-22, No. 1, January 1986, pp 98 - 101, ISSN: 0018-9251.

Chen, H.; Kirubarajan, T. & Bar-Shalom, Y. (2003). Performance Limits of Track-to-Track Fusion versus Centralized Estimation: Theory and Application. *IEEE Transactions on Aerospace and Electronic Systems*, Vol. 39, No. 2, April 2003, pp 386 - 400, ISSN: 0018-9251.

Chroust, S.G. & Vincze, M. (2004). Fusion of Vision and Inertial Data for Motion and Structure Estimation. *Journal of Robotic Systems*, Vol. 21, Issue 2, February 2004, pp 73 - 83, ISSN: 0741-2223.

El-Ayadi, M.H. (2002). Nonstochastic Adaptive Decision Fusion in Distributed-Detection Systems. *IEEE Transactions on Aerospace and Electronic Systems*, Vol. 38, No. 4, October 2002, pp 1158 -1171, ISSN: 0018-9251.

Gan, Q. & Harris, C.J. (2001). Comparison of Two Measurement Fusion Methods For Kalman-Filter-Based Multisensor Data Fusion. *IEEE Transactions on Aerospace and Electronic Systems*, Vol. 37, No. 1, January 2001, pp 273 -280, ISSN: 0018-9251.

Gustavo, L.R. & Grajal, J. (2006). Multiple Signal Detection and Estimation using Atomic Decomposition and EM. *IEEE Transactions on Aerospace and Electronic Systems*, Vol. 42, No. 1, January 2006, pp 84 -102, ISSN: 0018-9251.

Hashemi, H.R. & Rhodes, Ian B. (1989). Decentralized Sequential Detection. *IEEE Transactions on Information Theory*, Vol. 35, No. 3, May 1989, pp 509–520. ISSN: 0018-9448.

Jeong, S. & Tugnait, J.K. (2005). Multisensor Tracking of a Maneuvering Target in Clutter Using IMMPDA Filtering with Simultaneous Measurement Update. *IEEE Transactions on Aerospace and Electronic Systems*, Vol. 41, No. 3, July 2005, pp 1122-1131, ISSN: 0018-9251.

Kalandros, M. & Pao, L.Y. (2002). Covariance Control for Multisensor Systems. *IEEE Transactions on Aerospace and Electronic Systems*, Vol. 38, No. 4, October 2002, pp 1138 -1157, ISSN: 0018-9251.

Kalandros, M. & Pao, L.Y. (2005). Multisensor Covariance Control Strategies for Reducing Bias Effects in Interacting Target Scenarios. *IEEE Transactions on Aerospace and Electronic Systems*, Vol. 41, No. 1, January 2005, pp 153 - 173, ISSN: 0018-9251.

Kaplan, L.M. (2006). Local Node Selection for Localization in a Distribued Sensor Network. *IEEE Transactions on Aerospace and Electronic Systems*, Vol. 42, No. 1, January 2006, pp 136 -146, ISSN: 0018-9251.

Karniely, H. & Siegelmann, H.T. (2000). Sensor Registration Using Neural Networks. *IEEE Transactions on Aerospace and Electronic Systems*, Vol. 36, No. 1, January 2000, pp 85 - 101, ISSN: 0018-9251.

Kirubarajan, T.; Wang, H.; Bar-Shalom, Y. & Pattipati, K.R. (2001). Efficient Multisensor Fusion Using Multidimensional Data Association. *IEEE Transactions on Aerospace and Electronic Systems*, Vol. 37, No. 2, April 2001, pp 386 -398, ISSN: 0018-9251.

Lin, X.; Bar-Shalom, Y. & Kirubarajan, T. (2005). Multisensor-Multitarget Bias Estimation for General Asynchronous Sensors. *IEEE Transactions on Aerospace and Electronic Systems*, Vol. 41, No. 3, July 2005, pp 899 - 921, ISSN: 0018-9251.

Moshe Kam; Wei Chang & Qiang Zhu (1991). Hardware Complexity of Binary Distributed Detection Systems With Isolated Local Bayesian Detection. *IEEE Transactions on Systems, Man & Cybernatics*, Vol. SMC-21, No. 3, May-June 1991, pp 565-571, ISSN: 1083-4427.

Moshe Kam; Qiang Zhu & Steven Gray, W. (1992). Optimal Data Fusion of Connected Local Decisions in Multiple Sensor Detection Systems. *IEEE Transactions on Aerospace & Electronic Systems*, Vol. AES-28, No. 3, July 1992, pp 916–920, ISSN: 0018-9251.

Moshe Kam; Chris Rorres; Wei Chang & Xiaoxun Zhu (1999). Performance and Geometric Interpretation For Decision Fusion With Memory. *IEEE Transactions on Systems, Man & Cybernatics, Part A: Systems & Humans*, Vol. 29, No. 1, Jan. 1999, pp 52 - 62, ISSN: 1083-4427.

Murphy, Robin R. (1998). Dempster-Shafer Theory For Sensor Fusion in Autonomous Mobile Robots. *IEEE Transactions on Robotics & Automation*, Vol. 14, No. 2, April 1998, pp 197 - 206, ISSN:1042-296X.

Nabaa, N. & Bishop, R.H (1999). Solution to a Multisensor Tracking Problem with Sensor Registration Errors. *IEEE Transactions on Aerospace and Electronic Systems*, Vol. 35, No. 1, January 1999, pp 354 - 363, ISSN: 0018-9251.

Niu, R.; Varshney, P.K.; Mehrotra, K. & Mohan, C. (2005). Temporally Staggered Sensors in Multi-Sensor Target Tracking Systems. *IEEE Transactions on Aerospace and Electronic Systems*, Vol. 41, No. 3, July 2005, pp 794 - 808, ISSN: 0018-9251.

Okello, N.N. & Challa, S. (2004). Joint Sensor Registration and Track-to-Track Fusion for Distributed Trackers. *IEEE Transactions on Aerospace and Electronic Systems*, Vol. 40, No. 3, July 2004, pp 808 - 823, ISSN: 0018-9251.

Papastavrou, Jason D. & Athans, Michael (1992). Distributed Detection By a Large Team of Sensors in Tandem. *IEEE Transactions on Aerospace & Electronic Systems*, Vol. AES-28, No. 3, July 1992, pp 639 - 652, ISSN: 0018-9251.

Reibman, Amy R. & Nolte, L.W. (1987). Optimal Detection and Performance of Distributed Sensor Systems. *IEEE Transactions on Aerospace & Electronic Systems*, Vol. AES-23, No. 1, Jan. 1987, pp 24 - 30, ISSN: 0018-9251.

Ruan, Y. & Willett, P. (2005). A Quantization Architecture for Track Fusion. *IEEE Transactions on Aerospace and Electronic Systems*, Vol. 41, No. 2, April 2005, pp 671-681, ISSN: 0018-9251.

Sadjadi, Firoz A. (1986). Hypothesis Testing in a Distributed Environment. *IEEE Transactions on Aerospace & Electronic Systems*, Vol. AES-22, No. 2, March 1986, pp 134 - 139, ISSN: 0018-9251.

Srinivasan, R. (1986). Distributed Radar Detection Theory. *IEE Proceedings*, Vol. 133, Part F, No. 1, February 1986, pp 55 - 60, ISSN: 0956-375X.

Swaszek, Peter F. (1993). On the Performance of Serial Networks in Distributed Detections. *IEEE Transactions on Aerospace & Electronic Systems*, Vol. AES-29, No. 1, January 1993, pp 254 - 259, ISSN: 0018-9251.

Tang, Z.B.; Pattipati, K.R. & Kleinman, D.L. (1991). Optimization of Detection Networks: Part I—Tandem Structures. *IEEE Transactions on Systems, Man & Cybernatics*, Vol. SMC-21, No. 5, September / October 1991, pp 1044-1059, ISSN: 1083-4427.

Teneketzis, D. & Varaiya, P. (1984). The Decentralized Quickest Detection Problem. *IEEE Transactions on Automatic Control*, Vol. AC-29, No. 7, July 1984, pp 641-644. ISSN: 0018-9286.

Tenney, Robert R. & Sandell Jr., Nils R. (1981a). Detection With Distributed Sensors. *IEEE Transactions on Aerospace & Electronic Systems*, Vol. AES-17, No. 4, July 1981, pp 501 - 509, ISSN: 0018-9251.

Tenney, Robert R. & Sandell, Nils R. (1981b). Structures For Distributed Decision Making. *IEEE Transactions on Systems, Man & Cybernatics*, Vol. SMC-11, No. 8, August 1981, pp 517 - 527, ISSN: 0018-9251.

Tenney, Robert R. & Sandell, Nils R. (1981c). Strategies For Distributed Decision Making. *IEEE Transactions on Systems, Man & Cybernatics*, Vol. SMC-11, No. 8, August 1981, pp 527 - 538, ISSN: 0018-9251.

Thomopoulos, S.C.A.; Viswanathan, R. & Bougoulias, D.C. (1987). Optimal Decision Fusion in Multiple Sensor Systems. *IEEE Transactions on Aerospace & Electronic Systems*, Vol. AES-23, No. 5, Sept. 1987, pp 644 - 653, ISSN: 0018-9251.

Thomopoulos, S.C.A. (1990). Sensor Integration and Data Fusion. *Journal of Robotic Systems*, Vol. 7, No.3, 1990, pp 337 -372, ISSN: 0741-2223.

Tsitsiklis, John N. & Athans, Michael (1985). On the Complexity of Decentralized Decision Making and Detection Problems. *IEEE Transactions on Automatic Control*, Vol. AC-30, No. 5, May 1985, pp 440 - 446, ISSN: 0018-9286.

Viswanathan, R.; Thomopoulos, S.C.A. & Tumuluri, R.J. (1988). Optimal Serial Distributed Decision Fusion. *IEEE Transactions on Aerospace & Electronic Systems*, Vol. AES-24, No. 4, July 1988, pp 366 - 375, ISSN: 0018-9251.

Wang, X.; Shen, H.C. & Qian, W.H. (1988). A Hypothesis Testing Method For Multi-sensory Data Fusion. *Proceedings of the IEEE International Conference on Robotics & Automation*, pp 3407 - 3412, ISBN: 0-7803-4300-X, Leuven, Belgium, May 1998, IEEE publications, U.S.A., Catalogue No.98CH36146.

Appendix I

Refer to fig. 7. For both multiplicative and additive models, vide eqns. 16 & 17, the required integral is analytically equivalent to the area ABCDE. And, geometrically this area is a *factor* (which is less than 1.0) of the total area, namely ABCD. This rectangular area ABCD is computed from the geometric measure of the sides AB (=CD) and BC (=DA). Using U_G -axis as the common scale for both the plots, we have,

$$\left|\overline{AB}\right| = \left|\overline{CD}\right| = dist\left\{\left(e^{n},0\right)\ \ and\ \ (\lambda_{Th.},0)\right\} \equiv (e^{n'} - \lambda_{Th.}) = (e^{n} - e^{n'})$$ [for multiplicative model]

and,

$$\left|\overline{AB}\right| = \left|\overline{CD}\right| = dist\left\{\left(n,0\right)\ \ and\ \ (\lambda_{Th.},0)\right\} = (n - \lambda_{Th.})$$ [for additive model]

And, in both the cases using $P(H_1)$ axis, we have, $\left|\overline{BC}\right| = \left|\overline{DA}\right| = dist\left\{\left(0,0\right)\ \ and\ \ (1,0)\right\} = 1.0$

Now, for multiplicative model, the required area is '(1-β)' times the area ABCD and for additive model it is '(1-α)' times the area ABCD. Hence we get the deduced expressions on the right-hand-sides of the eqns. 16 & 17.

13

Development of Anthropomorphic Robot Hand with Tactile Sensor: SKKU Hand II

Byung June Choi, Jooyoung Chun and Hyouk Ryeol Choi
School of Mechanical Engineering, Sungkyunkwan University
Korea

1. Introduction

Recently, robots have begun to perform various tasks on replacing the human in the daily life such as cleaning, entertainments etc. In order to accomplish the effective performance of intricate and precise tasks, robot hand must have special capabilities, such as decision making in given condition, autonomy in unknown situation and stable manipulation of object. It must also possess tactile information to be able to carry out complicated manipulative tasks in a natural environment. Consequently, the tactile sensor is required to support natural interaction between the robot and the environment.

Many researches on the tactile sensing and the anthropomorphic multi-fingered robot hand have been reported up to now. Dario *et al.* developed "Artificial tactile sensing system" for a robot finger (Dario & Buttazzo, 1987). The system is able to detect the contact force, the vibration and the variation of temperature like mechanoreceptor of the human by arranging PVDF films that possess piezoelectricity and pyroelectricity. Howe *et al.* developed a dynamic tactile sensor that can detect slippage by means of the change of stresses due to deformation of the contact with the object (Howe & Cutkosky, 1993). Maeno *et al.* presented a tactile sensor, called "artificial finger skin" based on PVDF (Fusjimoto et al., 1999; Yamano et al., 2003). The sensor capable of detecting the incipient slip was designed to possess the characteristics similar to that of the human finger. Hosoda *et al.* reported a soft fingertip with two layers made of different kinds of silicon rubbers (Hosoda et al., 2003). The Utah/MIT hand developed by Jacobsen *et al.* is driven by actuators that are located in a place remote from the robot hand frame and connected by tendon cables (Jacobsen et al., 1984; Jacobsen et al. 1988). Hirzinger *et al.* developed DLR-Hand II, which build the actuators into the hand. Each finger of robot hand is equipped with motors, 6-DOF fingertip force torque sensor and integrated electronics (Butterfass et al., 2001; Gao et al., 2003). Kawasaki *et al.* presented anthropomorphic robot hand called the Gifu hand III, which has a thumb and four fingers (Kawasaki et al., 2002; Kawasaki et al., 2002). The thumb has 4 joints with 4-DOF and each of the fingers has 4 joints with 3-DOF. Moreover, the distributed tactile sensor which is made of conductive ink is arranged about 859 sensing points on the palm and the fingers. Shimojo *et al.* utilized the pressure conductive rubber as a pressure sensitive material (Shimojo et al., 2004). They attached the sensor onto a four finger robot hand and a demonstrated its grasping operations with a column, sphere, etc. Although a number of researches have been done up to now, however, their motion of robot hands is unlike that of the human because the mechanism of robot hands is different from that of human in many aspects. A study on the grasping motion of the human hand noted that the metacarpal link of the thumb plays the key role in power grasping (Calais-Germain, 1993). Despite these differences, however, many researches have been investigated about the robot hand of gripper type, which is difficult to perform dexterous grasping and manipulation of object like the human hand. Furthermore, most developed robot hands are larger than human hands. In addition, more researches are still necessary to put the tactile sensor into the practical use, because there remain many problems such as the limitations in the hardware as well as the algorithms for signal processing, the lack of the reliability, accuracy, response speed, dynamic•static characteristic, economical efficiency (Nicholls & Lee, 1989; Lee & Nicholls, 1995).

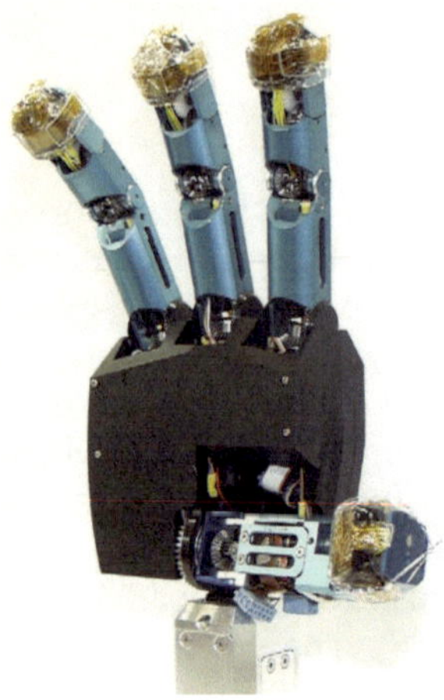

Fig. 1. SKKU Hand II with tactile sensor

In this chapter, we propose an anthropomorphic robot hand called SKKU Hand II, which has a miniaturized tactile sensor applicable to the robot hand. Thumb is at an angle opposite to its other fingers, and the thumb and fingers are orthogonal, such that it can performs dexterous grasping and manipulation like the human hand. The hand is similar to the human hand in geometry and size because inessential degree-of-freedom is abbreviated during grasping. All parts of the SKKU Hand II were composed of independent modules from each finger to the electric board for control. Moreover, SKKU Hand II's fingertip tactile sensor is composed of two functional units: a PVDF-based slip sensor designed to detect slippage such as incipient slips between sensing elements and contact surfaces, and a thin flexible force sensor that can read the contact force of and geometrical information on the object using a pressure-variable resistor ink. Both of them are integrated into a tactile sensor. All the sensing system can be embedded into the fingertip by miniaturizing the sensor and signal processing units. The proposed system is able to communicate with PC or external devices to provide information for controlling the robot hand. Also, the actuators, driving circuits of SKKU Hand II and its entire sensing system are embedded in the hand, and each driving circuit communicates with others using CAN protocol.

This chapter is organized as follows. In sections II and III, kinematic design and mechanical design of SKKU Hand II are presented. The issues on the development of the PVDF texture sensor and thin flexible force sensor are discussed in sections IV and V. Also, system schematic of robot hand is described in section VI. In section VII, experimental procedures for evaluation of the performance of the robot hand which has fingertip tactile sensors are mentioned. And finally the paper is concluded with summary in section VIII.

2. Kinematic design of SKKU hand II

To develop the SKKU Hand II, the design process started with the simulation to get optimal ratios of link lengths of finger (Kyriakopoulos et al., 1997; Wilkinson et al., 2003). We estimated an index of power grasping and fingertip grasping using kinetics model. Through the kinematic analysis and simulation, we decided that the ratio of length of link is 2-3-5 by Fibonacci sequence. Human hand is able to grasp objects by finger and thumb crossing each other to length way. The position of thumb and other finger is opposite to each other and thumb parallels other finger when gripper-type robot hand is grasping. However, in case of anthropomorphic robot hand, it grasps an object using the angle to direction of length of each finger for the power grasp and fingertip grasp. Especially, it has more powerful grasp by using the angle to direction of length of each finger in pinch grasp.

	Joint	Gear	Torque	Size [L× M][mm]	Weight [kg]
finger	J1	275:1	0.115	115×22.5	0.116
	J2	258:1	0.106		
Thumb	J3	275:1	0.285	139×28	0.242
	J4	275:1	0.285		
	J5	258:1	0.106		
	J6	64:1	0.0297		
Total	-		-	-	0.9

Table 1. Specification of SKKU Hand II

3. The mechanical design of SKKU HAND II

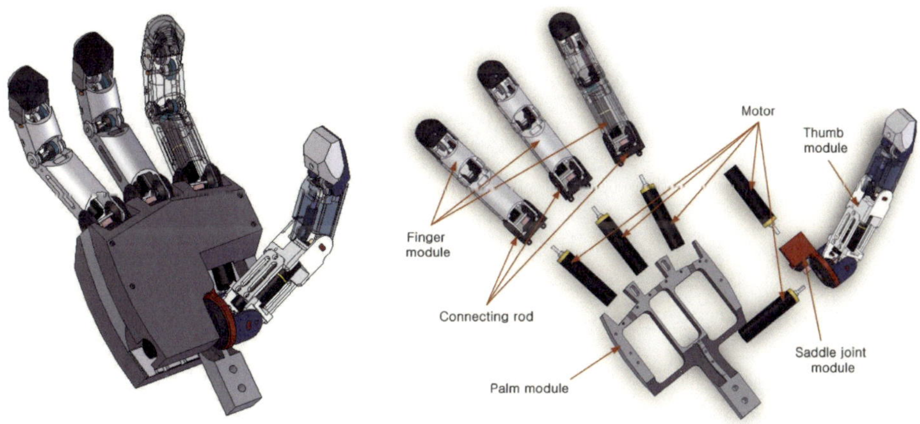

Fig. 2. The mechanical design of SKKU Hand II

As shown in Fig. 2, the SKKU Hand II is designed to be anthropomorphic in terms of geometry, size, and kinesis so that it performs power grasping and fingertip grasping as well as manipulations like the human hand. Especially, all of the parts consist of modules for easy development, maintenance and repair.

3.1 Finger module

As shown in Fig. 3, the SKKU Hand II has three fingers, and it is about 1.1 times bigger than a human hand. Each finger module has total 3-DOF, including coupled joint of the last two joints, and degree of freedom of finger of robot hand is smaller than that of human finger and the difference is due to the reduction of unnecessary degree of freedom for the ability of grasp and maximization of efficiency with size of robot hand very close to that of human hand. The actuator of finger module has two electric motors. And every motor is installed possibly close to palm module in order to consider weight balance and kinesis. And the last two joints, Distal phalange and Medial phalange joint, are mechanically coupled like a human finger by the pulley and timing belt. Also, it has some special space for being easy to install a variety sensors and the sensor processing circuit, in which for movement is more similar to human hand.

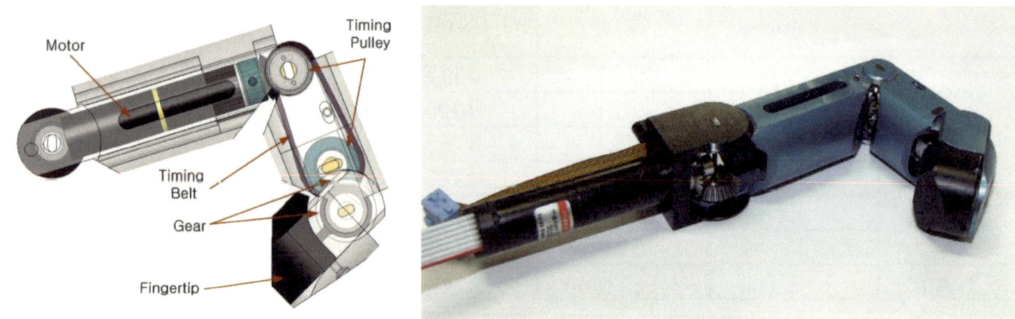

Fig. 3. Finger module

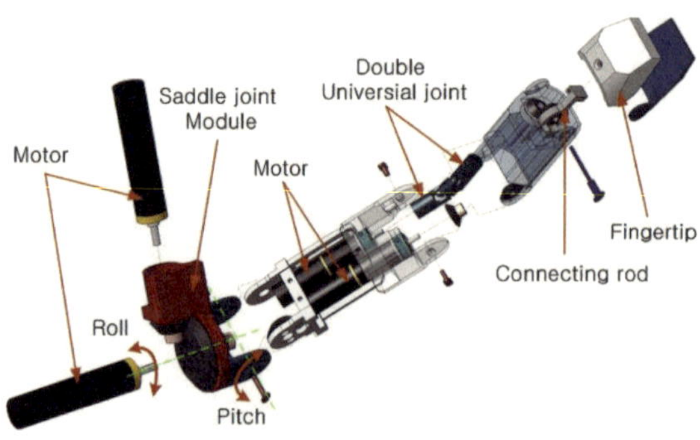

Fig. 4. Composition of thumb module

3.2 Thumb module

The thumb module has four DOF, and it is about 1.1 times bigger than a human thumb. The thumb has played a very important part in the anthropomorphic robot hand as well as human hand. The thumb can fulfil a complex work by means of saddle joint that is closest to the wrist. In general, the saddle joint of human has 3-DOF, and it is possible to manipulate any motion in the three-dimensional space because the motion of pitch, roll and yaw is performed simultaneously. The motion of pitch and roll is usually used to grasp the object, and the motion of yaw is used to circumvolve the object (precision grasp) motion like opening the cap of bottle. As depicted in Fig. 4 and 5, SKKU Hand II is realized with mechanism which imitates the role of saddle joint of human, but motion of yaw is neglected. The transmission of power is used with the bevel gear with 1:1 ratio, but distal phalange joint is composed of the double universal joint so that the position of motor stays close to palm module and for the independence of actuators.

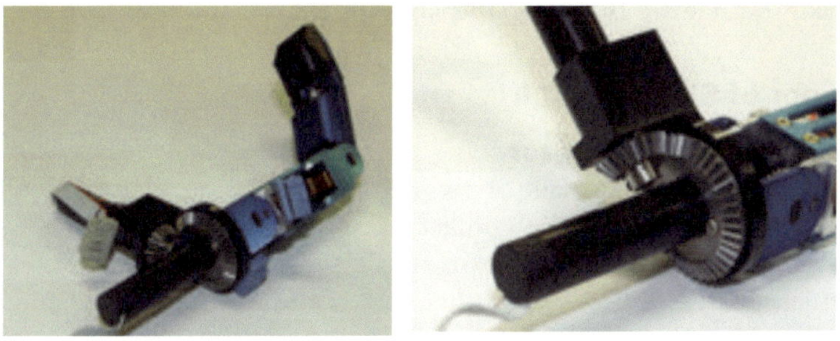

Fig. 5. Thumb module

3.3 Fingertip module

The shape of human fingertip is not just round but polyhedral. The surface of the fingertip can be discriminated into five parts depending on the grasping modalities such as pinch grasp, fingertip grasp and power grasp. As shown in Fig. 6, the fingertip grasp uses a bottom of fingernail, the pitch grasp that hold a small and long object strongly and safely uses a side of fingertip. In the power grasp which is for wrapping an arbitrary object, the object is restricted using bottom of finger, thumb, and palm, and then object is fixed by bottom of fingertip at last. Consequently, the fingertip of SKKU Hand II is designed as a unique shape which can realize composite task like fingertip of human.

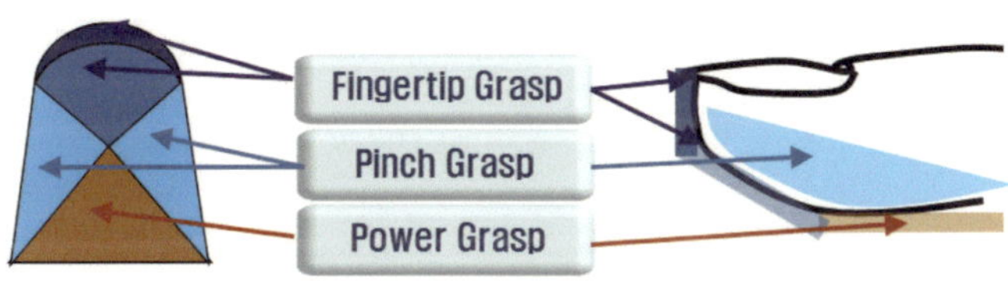

Fig. 6. The surface of fingertip depending on the grasping modalities

Fig. 7. Motor control board

3.4 Motor control board

Our anthropomorphic robot hand has the ten motor control boards. As shown in Fig. 7, each board size is 39.5 x 23.5(mm) and every board is able to control just one corresponding motor. All of this board is composed independently from each other, but they are connected by CAN protocol. Also main microprocessor of motor control board used PIC16F458 and the Motor control board includes the current sensor and counter chip to check the state of motor in real time. Each current sensor which can be utilized information of force feedback control with tactile sensor of fingertip is used to measure the torque of finger joint.

4. Tactile sensor of SKKU hand II

4.1 Integrated fingertip tactile sensor

The finger tip tactile sensor consists of two different sensing elements, which is a thin flexible force sensor for detecting the contact force and location and the PVDF sensor for incipient slip. The structure of the fingertip tactile sensor is shown in Fig. 8. Thin flexible force sensor which possesses 24 sensing elements is attached under the fingertip to detect static contact force. Also, the PVDF sensor which has two PVDF strips is located on the thin flexible force sensor to detect dynamic response such as slippage using the mechanical deformation of the silicone (Choi et al., 2005).

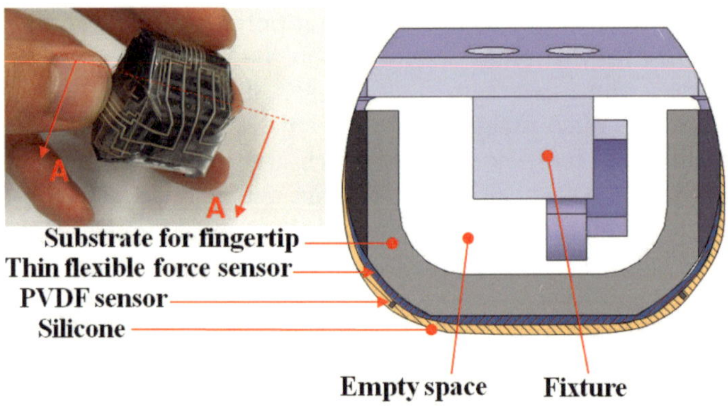

Fig. 8. Structure of fingertip tactile sensor

4.2 Principle of PVDF

Since Kawai discovered strong piezoelectricity in PVDF in 1969 (Kawai, 1969), PVDF has been used in a lot of commercial products. PVDF is a semicrystalline polymer with the 50% degree of crystallinity approximately. Its chemical structure contains the repetitive unit of doubly fluorinated ethene CF_2-CH_2. The voltage output of PVDF is 10 times higher than piezo-ceramics for the same force input. The reason for the strong piezoelectric activity is related to the large electronegativity of fluoride atoms in comparison to the carbon atoms, thus accommodating a large dipole moment (Kolesar et al., 1996; Fraden, 1997; Yu et al., 2002).

To derive fundamental equations for PVDF, as shown in Fig. 9, a frame is defined in terms of the length direction (direction 1), normal to the length direction in the plane of the film (direction 2) and normal to the plane of the PVDF strip (direction 3). The surface charge Q from the piezoelectric phenomena is proportional to the applied force F such as

$$D = \frac{Q}{A} = d_{3n}\sigma_n \tag{1}$$

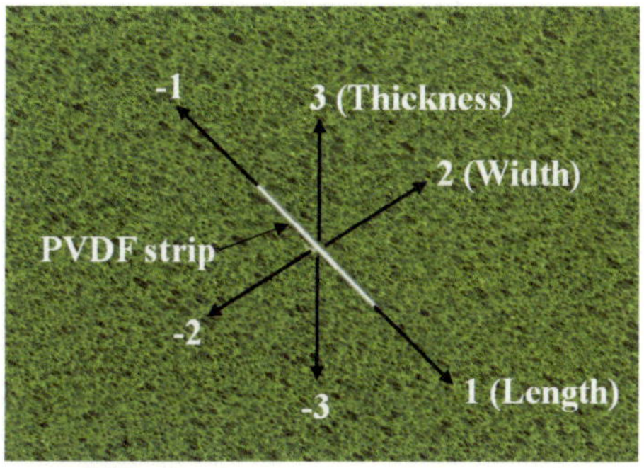

Fig. 9. Reference axes in PVDF strip

where D is the charge density, A denotes the electrode area of the PVDF strip formed in the 3-3 plane, d_{3n} is piezoelectric strain coefficient for the axis of the applied tensile force or compressive force, σ_{3n} is stress applied in the relevant direction. The piezoelectric strain coefficient d_{3n} is commonly expressed in $\left[\frac{C/m^2}{N/m^2}\right]$. When a PVDF strip is compressed by a probe on a rigid flat surface, assuming that both the flat surface and the probe are friction-free, the film is free to expand along the 1-1 and 2-2 directions, the output charge can be expressed as

$$\frac{Q}{A} = d_{33}\sigma_3 = d_{33}\left(\frac{F}{A_3}\right) \tag{2}$$

$$Q = d_{33}F \tag{3}$$

where d_{33} represents the piezoelectric strain coefficient along the 3-3 direction. Normally, the output charge is due to the combination of d_{31}, d_{32}, and d_{33}. It is important to remember that, for a given applied force, the output charge from the PVDF in the length or width direction much higher than that of thickness direction (Dargahi, 2000). It is because of the extreme thinness of the PVDF film, which is applied to develop the miniaturized high sensitivity PVDF sensor.

4.3 Principle of pressure variable resistor ink

In case that a constant load is applied to PVDF for an extended period of time, the response decreases and eventually becomes zero. Consequently, PVDF is adequate for sensing dynamic force, not the static one (Russell, 1990). In this research, pressure variable resistor ink (Creative Materials Inc.) is used to develop a force sensor. Pressure variable resistor ink is an electrically conductive ink, where its resistance decreases as the pressure goes up. Also, its minimum and maximum resistance can be adjusted by blending it with other conductive ink (Ashruf, 2002). As shown in Fig. 10 the force sensor is fabricated by sandwiching the ink between two polyester films with the pattern of electrodes. The fabricated force sensor can be modeled as a variable resistance such that its resistance decreases as the pressure increases as illustrated in Fig. 10.

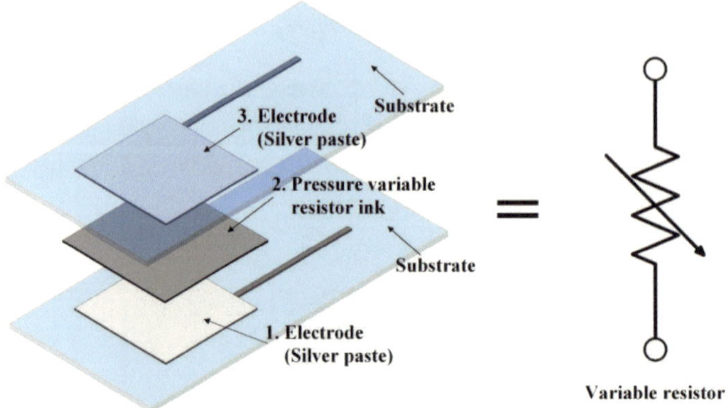

Fig. 10. Configuration of the force sensor

Fig. 11 explains the simple circuits utilized in the force sensor. When the fixed input voltage is applied to the sensor, it can read the change of the output value amplified by the voltage gain. Thus, the output is written by

$$V_{out} = -V_{cc}\frac{R_F}{R_S} = -V_{cc}R_F S_S \tag{4}$$

where R_S is resistance of force sensor, S_S denotes the conductance, that is, the reciprocal of resistance. Also, R_F is the op-amp's feedback resistance.

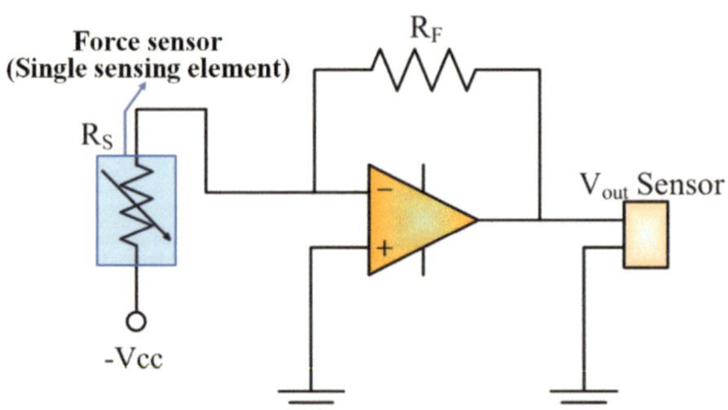

Fig. 11. Basic circuit of the force sensor

4.4 Slip sensor using PVDF

In this section, we explain the manufacturing and design of a miniaturized PVDF sensor with high sensitivity. After Polyvinylidene fluoride pellets (Aldrich Chemical Co.) is put between polyimide films, these pellets are pressed at 230°C ~ 240°C for 25 minutes with the pressure of 10MPa using a hot press machine. Then, PVDF film is allowed to cool down in the ambient temperature with the cooling speed about 1°C/sec. The thickness of fabricated films has a value between 50 and 70 μm. Then, the surface electrodes on both sides are fabricated using silk-screening technique with silver paste. By this method, a cost-effective and simple fabrication process is secured. The thickness of the electrode is about 5 μm, which can stand the temperature of 200°C. In order to exhibit high piezoelectricity in the fabricated films, the fabricated PVDF film is polarized by applying the strong electric field using the high voltage supply. Polarization is carried out at 170°C with the applied voltage of 3kV and load of 5N. The film is placed on a glass substrate while mechanically constrained and held at 170°C for 1 hour, then allowed to cool down before the applied voltage is removed. As shown in Fig. 12, the PVDF sensor consists of single PVDF strip with the thickness of 100 μm, 0.8mm width and 100mm length, where the sensing element has the size of $0.4mm \times 0.4mm$.

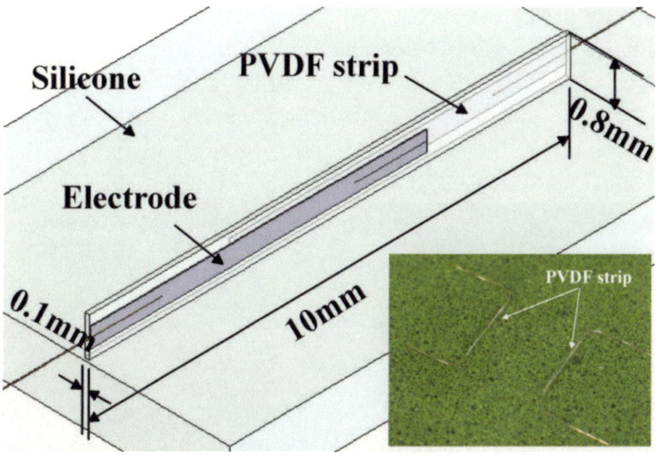

Fig. 12. Schematic of PVDF sensor and photograph

4.5 Thin flexible force sensor

In general, complexity of wirings increases largely depending on the density of the tactile sensor. To develop a force sensor with high resolution, the electrode pattern of the grid type is adopted in this research. The two polyester films are aligned as a grid while the pressure variable resistor ink layers face each other. Thus, each cross section of the grid forms single sensing element of the force sensor. As shown in Figs. 13 and 14, the size of each sensing

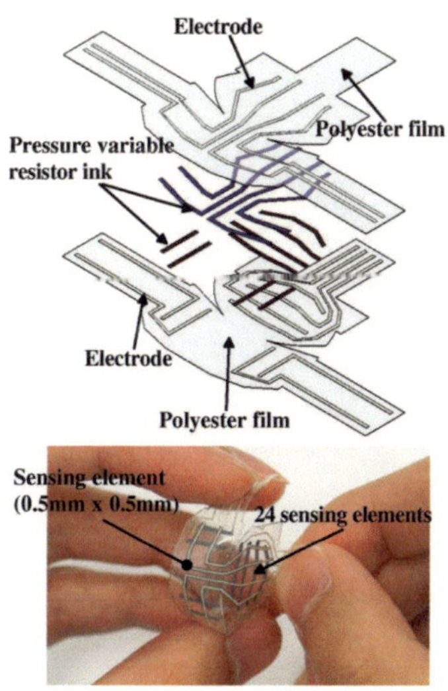

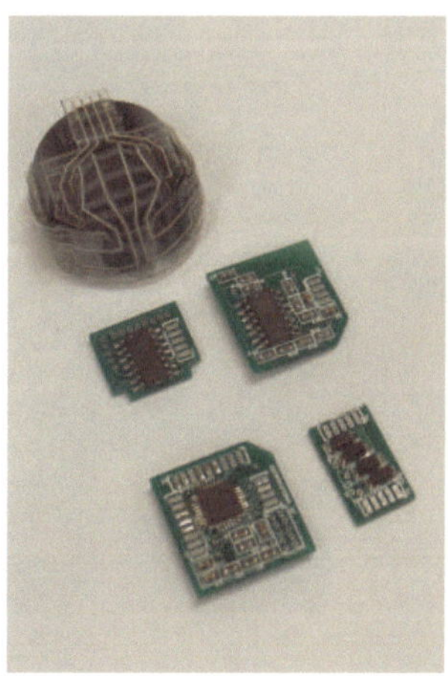

Fig. 13. Fingertip force sensor

element is $0.5mm \times 0.5mm$ and the total number of sensing elements goes up to 24. If the tactile sensor reads the output signal from the each sensing element respectively, total 48 signal lines are required. However, in the current approach, it is possible to read 24 sensing elements only with 8 input voltage lines and 4 output signal lines although the timing circuit is required.

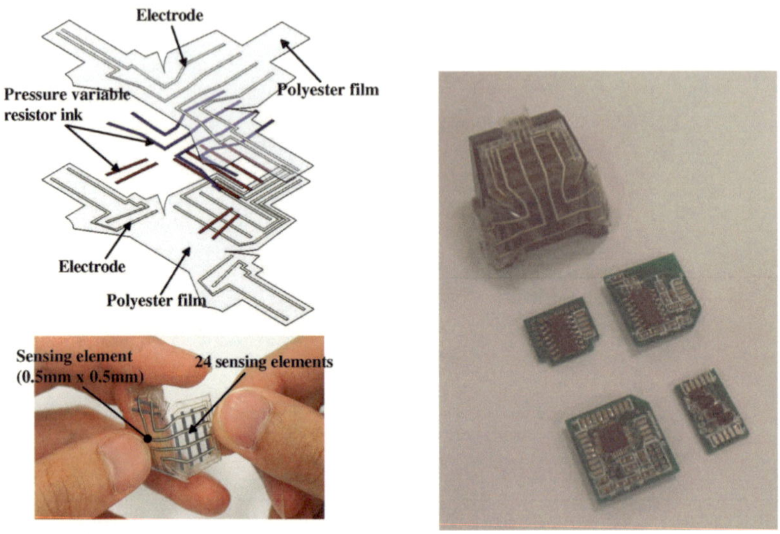

Fig. 14. Thumb tip force sensor

5. Hardware for signal processing

In this section, we introduce the miniaturized electronic hardware to be utilized for signal processing. As shown in Fig. 15, after being read from the sensing elements, the signals are

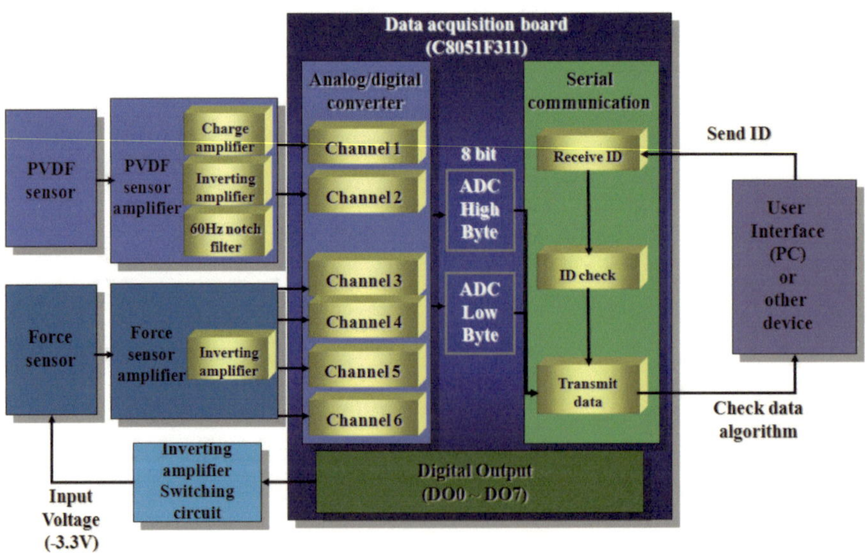

Fig. 15. Schematic of tactile sensing system

amplified and transformed into digital data with the A/D converter which is included in the microcontroller. Then the collected data is transmitted to the PC via an RS232 serial communication or input port of the external device.

5.1 Amplifiers

Two signal amplifiers have been developed in this research. One of them is for the PVDF sensor and the other for the force sensor. Because the output of PVDF depends on an instant variation of the load, it is electrically charged for a relatively short time. Therefore PVDF sensor gives the output voltage only in response to changing load rather than static one. The amplifier for PVDF sensor is basically used to convert the minute charge output from the PVDF strip into the voltage signal. The circuit for the PVDF sensor amplifier with the size of 18mm × 14mm, consists of charge amplifier, non-inverting amplifier and 60 Hz notch filter. Different from the PVDF sensor amplifier, the force sensor amplifier just amplifies the output voltage of the sensor using an op-amp.

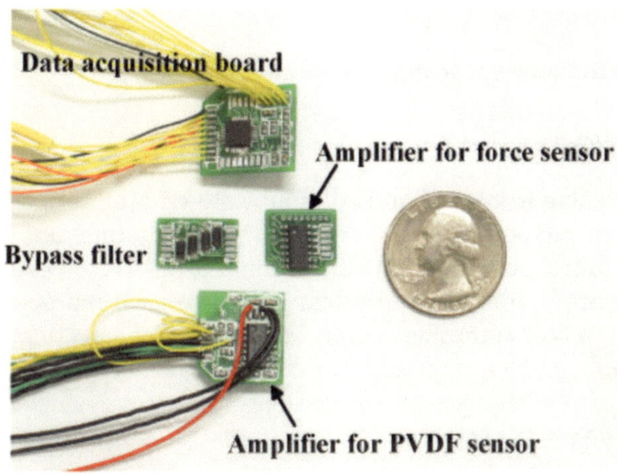

Fig. 16. Photograph of tactile sensing system

5.2 Signal processing unit

The circuit for the signal processing is designed on single board for data acquisition, control, communication. Data acquisition and communication are performed by using micro-controller (C8051F311). The signal processing board is able to transmit data to the PC via RS-232 or SMBus. The microprocessor in the signal processing unit is able to receive output signals from each sensing element by periodical scanning. Therefore, 8 digital output ports of a signal processing board become the input voltages of the force sensor, 4 A/D convert ports receive the output signal of the force sensor. Also, 2 A/D converter ports of the signal processing board receive the amplified output signal of the PVDF sensor. The scanning circuit of the force sensor is shown in Fig. 17.

In the circuit, the digital output lines of the microcontroller are changed to the active state and the microcontroller receives the outputs of the sensing elements via the AD converter. Therefore, the proposed configuration of the force sensor can effectively control all the sensing elements by using the minimum number of input and output signal lines.

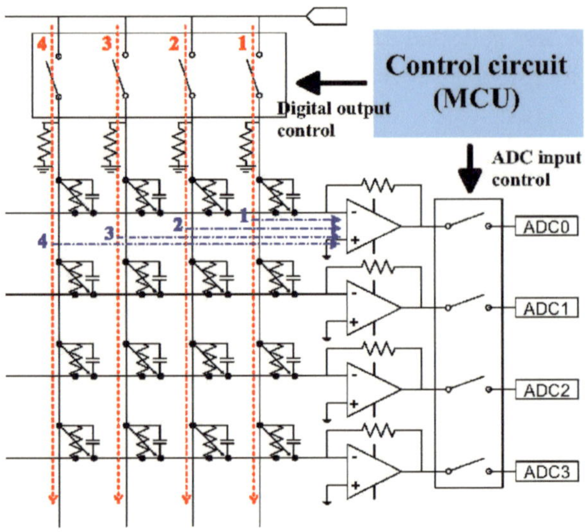

Fig. 17. Thin flexible force sensor scanning circuit

6. Control of the SKKU hand II

The SKKU Hand II is able to control and communicate with motor control boards through CAN communication method. As shown in Fig. 18, motors are controlled by each independent motor control board respectively. If main control receives a message for control of finger from other application, this message is sent to each motor control board by the main controller. Then motor controllers control the motors of each finger using PID control. Force feedback control can be interpreted in the main controller using output signal of thin

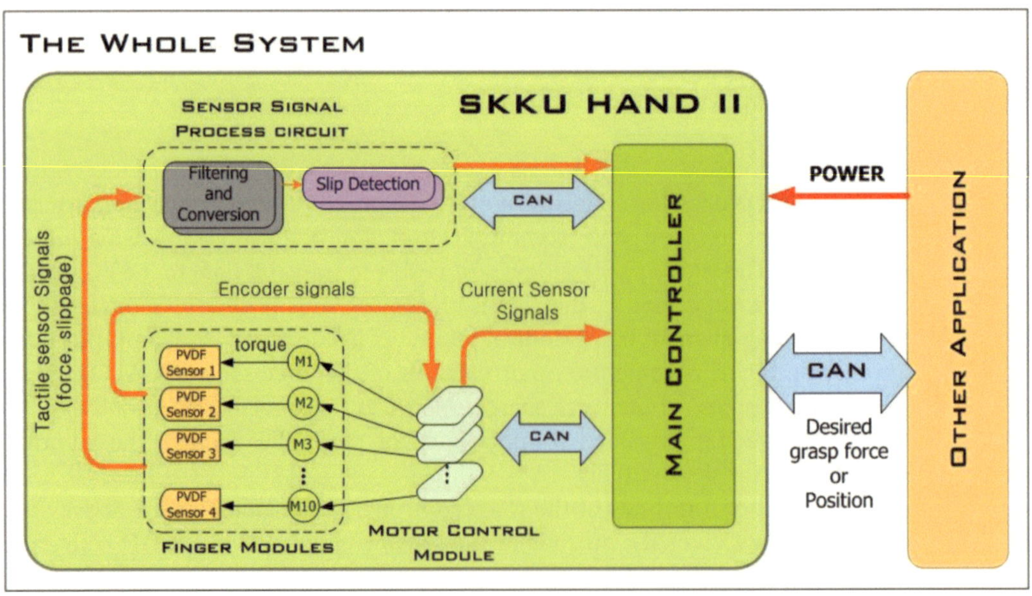

Fig. 18. System schematic of SKKU Hand II

flexible force sensor, PVDF slip sensor which is embedded each fingertip and current sensor which is integrated motor controller, and then, feedback parameters is sent to motor control module.

7. Experiments

In the first experiments, the sensor was touched and rubbed after installing on the fingertip. As shown in Fig. 19, it is noted that there exists the sharp change of signals, which implies that stick-slip occurred between the sensor and the contact surface. Also, the weight of 100g was rolled on the PVDF sensor. As shown in Fig. 20, the effect of stick-slip was not found and the smoother patterns of the signal compared to Fig. 19, was observed. When the weight of 100g rolled on the sensing element the output indicated about 1.2V constantly. Consequently, it is concluded that the characteristic of response can be discriminated depending on the surface characteristics of the object and the contact method, although calibrations are still needed.

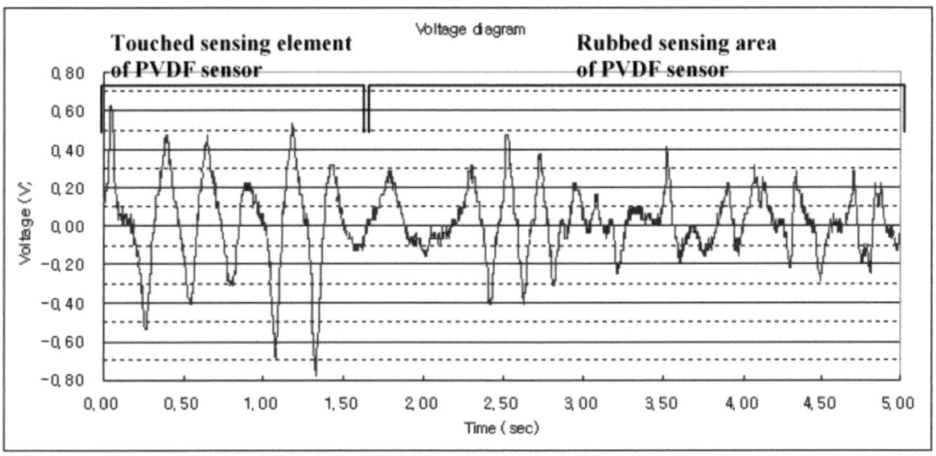

Fig. 19. Signal output from touching and rubbing.

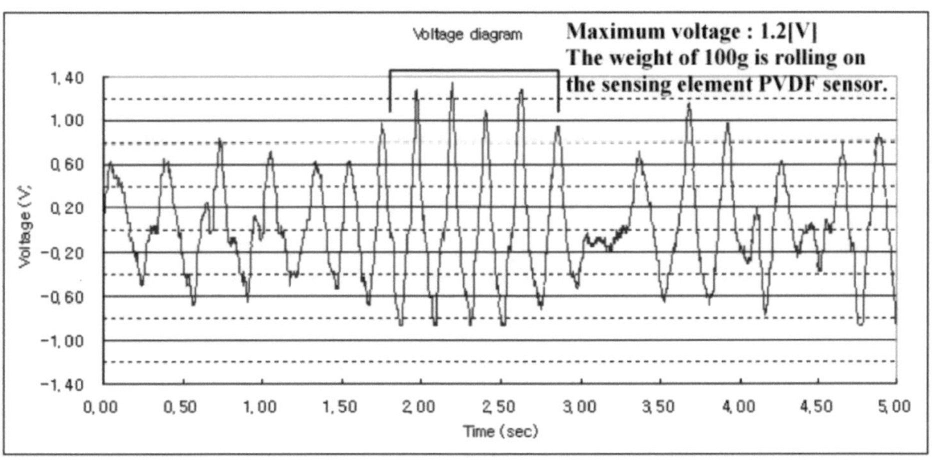

Fig. 20. Signal output from rolling of 100g weight

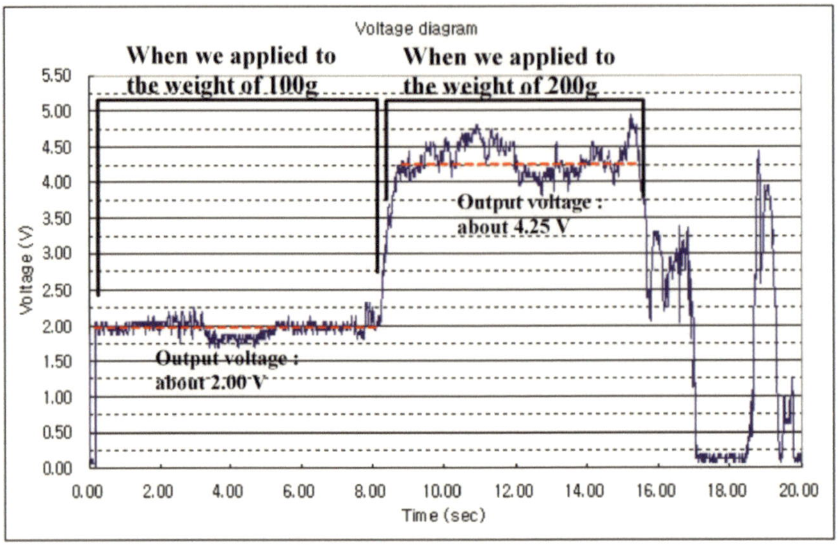

Fig. 21. The response of PVDF sensor, when we applied to the weight of 100g and 200g

In the second, the force sensor was tested. Static loads with the weight of 100g and 200g were applied, and the responses were obtained. As shown in Fig. 21, the output voltages of 2V and 4.25V were obtained for each weight. It is noted that the output has linear relation with the weight. Also, as shown in Fig. 22, user interface on the PC shows the data when the weight of 100g is rolled on the sensing elements of PVDF sensor and force sensor. Fig. 23 shows the contact information when we pressed the sensor in the fingertip. Before the

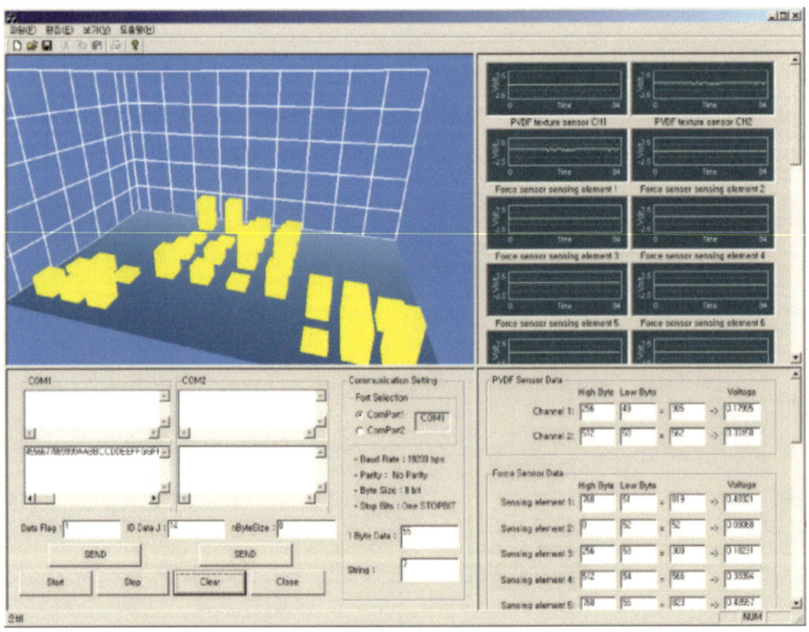

Fig. 22. The responses of PVDF sensor and force sensor, when the weight of 100g is rolled on the sensing elements

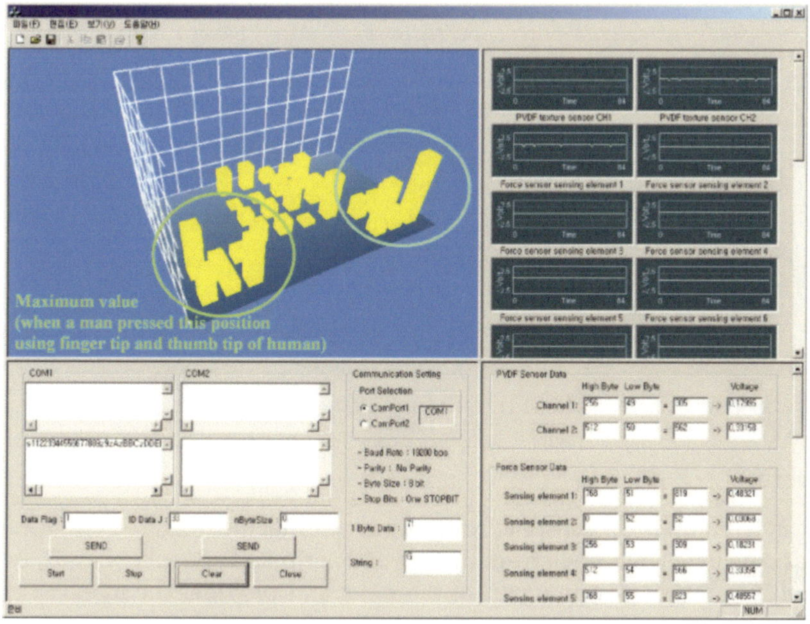

Fig. 23. The response of PVDF sensor and force sensor, when we pressed sensors using fingertip and thumb tip of human

integrated tactile sensor was attached to the robot hand, the SKKU Hand II was tested by grasping a bottle. As shown in Fig. 24, it is possible to confirm that the developed robot hand can grasp the bottle stably. Finally, the overall sensing system was tested by attaching it to the robot hand. We conformed the contact information through the user interface on the PC as shown in Fig. 25. It shows the contact information when we pressed the sensor in the fingertip. According to display of the user interface, the output distribution changes and the PVDF sensor responded to the stimuli sensitivity. In addition, Fig. 25 shows the contact information when robot hand grasped the bottle using the power grasping. According to the contact condition between sensing elements and contact surfaces of bottle, each fingertip sensor shows corresponding response. Therefore, it is possible to confirm that the each fingertip tactile sensor can detect the static force, location of contact and slippage.

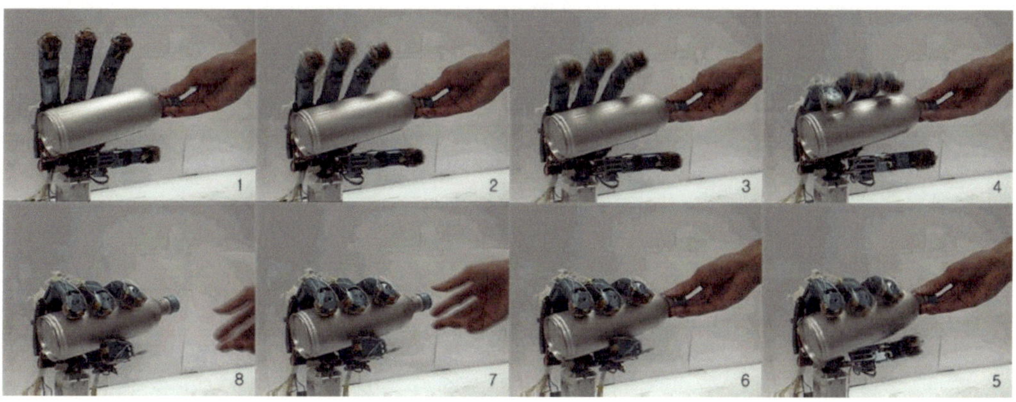

Fig. 24. Movement of SKKU Hand II

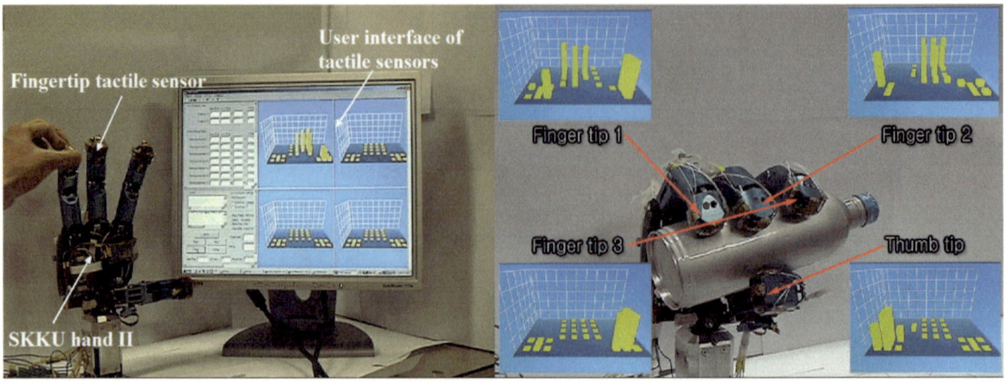

Fig. 25. Experiments of SKKU Hand II with fingertip tactile sensors

8. Conclusion

In this research, an anthropomorphic robot hand with tactile sensing system called SKKU Hand II was developed. Different from the previous gripper-type robot hands, the thumb of SKKU Hand II is designed as one part of the palm and provides the mobility of the palm. The robot hand is actuated by built-in DC motors, and fingertip tactile sensors are attached to its fingertips. A tactile sensor which can detects contact normal forces as well as slip is made of two organic materials, such as polyvinylidene fluoride (PVDF) that is known as piezoelectric polymer, and pressure variable resistor ink. Also, the tactile sensor is physically flexible and it can be deformed three-dimensionally to any shape so that it can be placed on anywhere on the curved surface. In order to detect incipient slip, a PVDF strip is arranged along the direction normal to the surface of the finger of the robot hand. Also, a thin flexible sensor to sense the static force as well as the contact location is fabricated into an arrayed type using pressure variable resistor ink. The driving circuits and sensing systems for the SKKU Hand II were miniaturized as small as to be integrated into the robot hand. The SKKU Hand II which integrated fingertip tactile sensors is validated through preliminary experiments. According to experiments on this research, it is possible to confirm that the each fingertip tactile sensor can detect the static force, location of contact and slippage. In the next research, we will control the robot hand for dexterous grasping and manipulation using the force feedback from the fingertip tactile sensors and evaluation will be performed.

9. Acknowledgment

Authors are gratefully acknowledging the financial support by Agency for Defence Development and by UTRC (Unmanned Technology Research Center), Korea Advanced Institute of Science and Technology.
We also appreciate project and equipment support from Gyeonggi Province through the GRRC program.

10. References

Ashruf, C. M. A. (2002). Thin Flexible Pressure Sensor, *Sensors Review*, Vol.22, No.4, 2002.02, pp. 322-327, ISSN: 0260-2288

Butterfass, J.; Grebensteion, M.; Hirzinger, G. & Liu, H. (2001). DLR-Hand II : Next Generation of a Dexterous Robot hand, *Proceedings of IEEE International Conference on Robotics and Automation*, pp. 109-114 , ISBN: 0-7803-6576-3, Seoul, Korea, 2001.05, IEEE, Seoul

Calais-Germain, B. (1993). *Anatomy of movement*, Eastland Press, ISBN: 0-9396-1622-X, Seattle, WA, USA

Choi, B. J.; Kang, S. C. & Choi, H. R. (2005). Development of Tactile Sensor for Detecting Contact Force and Slip, *Proceedings of IEEE/RSJ International conference on Intelligent Robots and Systems*, pp. 1977-1982, ISBN: 0-7803-8912-3, Edmonton, Alberta, Canada, 2005.08, IEEE/RSJ, Edmonton, Alberta

Dargahi, J. (2000). A piezoelectric tactile sensor with three sensing elements for robotic, endoscopic and prosthetic applications., *Sensors and Actuators A: Physical*, Vol.80, 2000.11, pp. 23-30, ISSN: 0924-4247

Dario, P. & Buttazzo, G. (1987). An Anthropomorphic Robot Finger for Investing Artificial Tactile Perception, *International Journal of Robotics Research*, Vol.6, No.3, 1987, pp. 25-48, ISSN: 0278-3649

Fraden, J. (1997). *Handbook of Modern Sensors: Physics, Designs, and Applications*, American Institute of Physics Press, ISBN: 0-3870-0750-4, USA

Fusjimoto, I.; Yamada, Y. & Maeno, T. (1999). Development of Artificial Finger Skin to Detect Incipient Slip for Realization of Static Friction Sensation, *Proceedings of IEEE International Conference on Multisensor Fusion and Integration for Intelligent Systems (MFI2003)*, pp. 15-20, ISBN: 0-7803-7987-X, Tokyo, Japan, 2003.08, IEEE, Tokyo

Gao, X. H.; Jin, M. H.; Jiang, L.; Xie, Z. W.; He, P.; Yang, L.; Grebenstein, M. & Hirxinger, G. (2003). The HIT/DLR Dexterous Hand: Work in Progress, *Proceedings of IEEE International Conference on Robotics and Automation*, pp. 3164-3168, ISBN: 0-7803-7736-2, Taipei, Taiwan, 2003.09, IEEE, Taipei

Hosoda, K.; Tada, Y.; Yamasaki, Y. & Asada, M. (2003). Sensing the Texture of Surfaces by Anthropomorphic Soft Fingertips with Multi-modal Sensors, *Proceedings of IEEE/RSJ International Conference on Intelligent Robots and Systems*, pp. 31-35, ISBN: 0-7803-7860-1, Las Vegas, Nevada, USA, 2003.10, IEEE/RSJ, Las Vegas, Nevada

Howe, R. D. & Cutkosky, M. R. (1993). Dynamic Tactile Sensing: Oerception of Fine Surface Features with Stress Rate Sensing, *IEEE Transaction on Robotics and Automation*, Vol.9, No.2, 1993, pp. 140-151, ISSN: 1042-296X

Jacobsen, S. C.; Wood, J. E.; Knutti, D. F. & Biggers, K. B. (1984). The Utah/MIT dexterous hand: Work in progress, *International Journal of Robotics Research*, Vol.3, No.4, 1984, pp. 21-50, ISSN: 0278-3649

Jacobsen, S. C.; McCammon, I. D.; Blggers, K. B.; & Phillips, R. P. (1988). Design of Tactile Sensing Systems for Dexterous Manipulator, *IEEE Control Systems Magazine*, vol. 8, issue. 1 ,1988, pp. 3-13, ISSN: 0888-0611

Kawai, H. (1969). The Piezoelectricity of Poly(vinylidene fluoride), *Japanese Journal of Applied Physics*, Vol. 8, 1969.05, pp. 975-976, ISSN: 1882-0778

Kawasaki, H.; Komatsu, T. & Uchiyama, K. (2002). Dexterous Anthropomorphic Robot Hand with Distributed Tactile Sensor: Gifu hand II, *IEEE/ASME Transactions on Mechatronics*, Vol.7, No.3, 2002, pp. 296-303, ISSN: 1083-4435

Kawasaki, H.; Komatsu, T. & Uchiyama, K. (2002). Anthropomorphic Robot Hand: Gifu Hand III, *Proceedings of International Conference on Control, Automation, and Systems*, pp. 1288-1293, Muju, Korea, 2002.06, ICASE, Muju

Kolesar, E. S.; Dyson, C. S.; Reston, R. R.; Fitch, R. C.; Ford, D. G. & Nelms, S. D. (1996), Tactile Integrated Circuit Sensor Realized with a Piezoelecric Polymer, *Proceedings of Innovative Systems in Silicon, Eighth Annual IEEE International Conference*, pp. 372-381, ISBN: 0-7803-3639-9, Austin, TX, USA, 1996.04, IEEE, Austin, TX

Kyriakopoulos, K. J.; Riper, J. V.; Zink, A. & Stephanou, H. E. (1997). Kinemaic Analysis and Posiotion/Force Control of the Anthrobot Hand, *IEEE/ASME Transactions on Systems, Man and Cybernetics*, Vol.27, 1997, pp. 95-104, ISSN: 1083-4427

Lee, M. H. & Nicholls, H. R. (1995). Tactile sensing for mechatronics-A state of the art survey, *Mechatronics*, Vol.9, 1999, pp. 1-31, ISSN: 0957-4158

Nicholls, H. R. & Lee, M. H. (1989). A survey of robot tactile sensing technology, *International Journal of Robotics Research*, Vol.8, No.3, 1989.07, pp. 3-30, ISSN: 0278-3649

Russell, R. A. (1990). *Robotic Tactile Sensing*, Prentice Hall Inc., ISBN: 0-13-781592-1 , USA

Shimojo, M.; Namiki, A.; Ishikawa, M.; Makino, R.; & Mabuchi, K. (2004). A Tactile Sensor Sheet using Pressure Conductive Rubber with Electrical-Wires Stitched Method, *IEEE Sensors Journal*, Vol.4, No.5, 2004, pp. 589-596, ISSN: 1424-8220

Wilkinson, D.; Weghe, M. V. & Matsuoka, Y. (2003). An Extensor Mechanism for an Anatomical Robotics Hand, *Proceedings of IEEE International Conference on Robotics and Automation*, pp. 238-243, ISBN: 0-7803-7736-2, Taipei, Taiwan, 2003.09, IEEE, Taipei

Yamano, I.; Takemura, K. & Maeno, T. (2003). Development of a Robot Finger for Five-fingered Hand using Ultrasonic Motors, *Proceedings of IEEE/RSJ International Conference on Robots and Systems*, pp. 2648-2653, ISBN: 0-7803-7860-1, Las Vegas, Nevada, USA, 2003.09, IEEE/RSJ, Las Vegas, Nevada

Yu, K. H.; Kwon, T. G.; Yoon, M. J. & Lee, S. C. (2002). Development of a Tactile Sensor Array with Flexible Structure using Piezoelectric Film, *KSME International Journal*, Vol.16, No.10, 2002, pp. 1222-1228, ISSN: 1738-494X

14

Design of a Tactile Sensor for Robot Hands

Giorgio Cannata[1] and Marco Maggiali[2]
[1]*University of Genova*
[2]*Italian Institute of Technology*
Italy

1. Introduction

Grasping and manipulation of objects, as well as the capability interacting safely with the environment (possibly including also humans), is a fundamental task for a humanoid robot. In order to tackle these issues a great effort has been put for over two decades to develop robot hands or mechanisms emulating the grasping capabilities of a human hand (Salisbury & Mason 1985; Jacobsen et al. 1986; Melchiorri & Vassura 1992; Lotti et al. 2004; Butterfass et al. 2001; Carrozza et al. 2003; Kargov et al. 2005; Caffaz & Cannata 1998; Lane et al. 1997; Hashimoto 1995). However, grasping and manipulation control also rely the availability of suitable contact and force feedback.

In this chapter we first present a survey of recent advances in the area of skin-like tactile sensors for both robot hand and for robot bodies. In the second part we will discuss the design of an embedded and modular tactile/force sensor to be installed on the phalanges of a robot hand. Each sensor consists of a distributed array of analog tactile elements and an integrated three-component force transducer with embedded analog and digital electronics. The tactile sensor consists of a matrix of electrodes etched on a flexible printed circuit board covered by pressure conductive rubber. The force sensor is instead an off-the-shelf integrated three components micro-joystick.

2. Tactile and force sensors for robots and robot hands

Tactile sensing in robotics has been widely investigated in the past 30 years (Webster 1988), and has received particular attention in recent years for the problems of grasping control with robot hands (Kawasaki 2002, Martin et al. 2004, Cannata 2005, Carrozza et al. 2006, Liu 2006), but also for sensing contacts over robot links (Asfour 2006, Hishiguro 2006) in order to enable safe interaction between humanoid robots and humans.

Research in this field has focused largely on transduction principles and transduction technology: as a matter of fact a wide range of transduction solutions have been proposed (Lee 1999). Most of the devices were either of the scalar single-point contact variety or were linear or rectangular arrays of sensing elements. The main transduction methods identified mostly belong to the following classes: resistive, capacitive, piezoelectric and pyroelectric, magnetic, optical, ultrasonic.

Despite the efforts made to prove the effectiveness of different type of transduction principles, various issues have strongly limited the development of embedded solutions for

both robot hands and large area tactile sensors (*robot skin* in the following) and their actual application by the robotic scientific community. On one hand developing a tactile sensor is a *very hardware* problem involving several engineering issues and often also special process equipment. Furthermore, there are at the bottom various technological difficulties which represent the bottleneck limiting the transition from a single tactile element (or a small matrix prototype) to a large scale integrated solution (e.g. the *curse of wiring* issue).
This section is a short survey of various recent and interesting papers addressing the problem of designing sensorized robot hands as well as robot skin mostly at system level. The objective is to provide a state of the art of tactile sensors from a system level point of view, emphasizing the issues of integration of the transducers with the electronic and mechanical hardware. Also, we will focus on *standard* transduction techniques (piezoresistive, piezo-electric, capacitive and optical being the most common), since we assume that these principles are currently the best candidates for in-house custom development of embedded tactile sensing devices.

2.1 Tactile sensors for robot hands

For manipulation and grasping robots must be capable controlling the forces arising at the contact points. If the geometry of the gripper or robot hand is well known and the contacts area can be modeled as point wise it is possible to compute (or estimate) the contact location as well as the applied force using a six-axes force/torque intrinsic tactile sensor as discussed in (Bicchi et al 1993). However, this minimalistic approach may fail as long as more complex interactions could arise (e.g. in the case of multiple contacts, or contacts with non rigid objects etc.). In this case the geometry of the contacts as well as pressure distributions must be measured directly using skin-like sensors.
The design of distributed tactile sensors for robotic applications has been widely discussed in recent years (Shimoyo 2004; Krishna & Rajanna 2002; Yamada et al. 2002; Engel et al. 2003).). However, there has been an only limited number of papers discussing the integration and embedding of these devices on dextrous robotic hands. As a matter of fact, miniaturization and cabling harness still represent one of the most important limitations to the design of small sized embedded sensors.
The four-fingered hand DLR-II (Butterfass et al. 2001), Figure 1, is a significant example of robot gripper integrating on board a complete force/torque sensor system. This hand has 22 degrees of freedom (DOFs) and is a complete self-contained system including motors, electronics and sensors. A sophisticated six-components force/torque transducer is installed on each fingertip and connected with the electronic modules, hosted on the palm of the hand, by ten wires (8 for data and 2 for power supply), Figure 2. The DLR-II does not make use of tactile sensors and can detect contacts located only on its fingertips.
The GIFU hands II and III (Kawasaki et al., 2002; Mouri et al 2002), Figure 3, have a commercial six axis force/torque sensor (produced by BL. AUTOTEC) integrated on each fingertip. However, these hands also feature a distributed tactile sensor based on pressure sensitive piezo-resistive ink transducers, see Figure 3. The tactile sensor has 859 taxels. formed by various grids of electrodes; the palm, the thumb, and each of the other fingers have 313, 126, and 105 taxels respectively covering about 50% of the transducer's area. The sensor can withstand a maximum pressure load of about $2.2\text{x}10^{-3}$ N/m^2, with a resolution of 8 bits. The sampling cycle is 10 ms/Frame. The sensor is about 0.2 mm thick and can cover both planar and cylindrical surfaces. The hand does not embed the sensor electronics and sensor cables from the transducers and sensors are all routed along the fingers and palm.

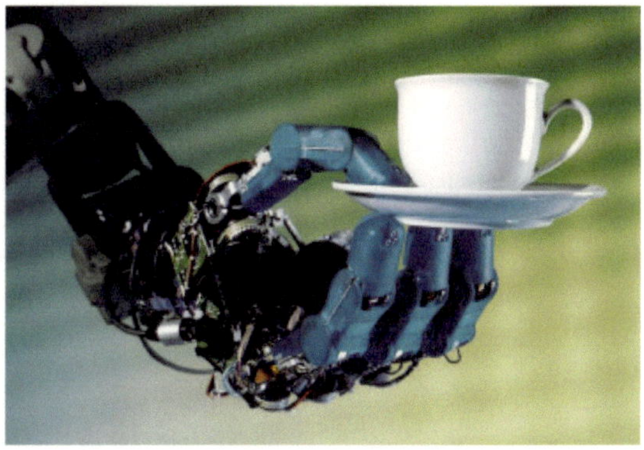

Fig. 1. DLR-II Hand (Butterfass et al. 2001).

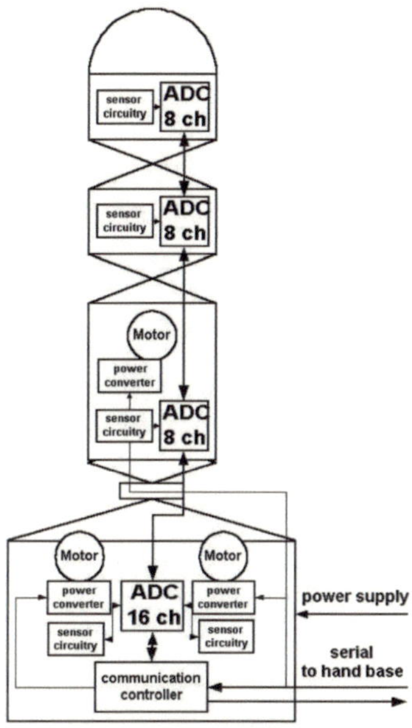

Fig. 2. Embedded electronics of the DLR-II finger (Butterfass et al. 2001).

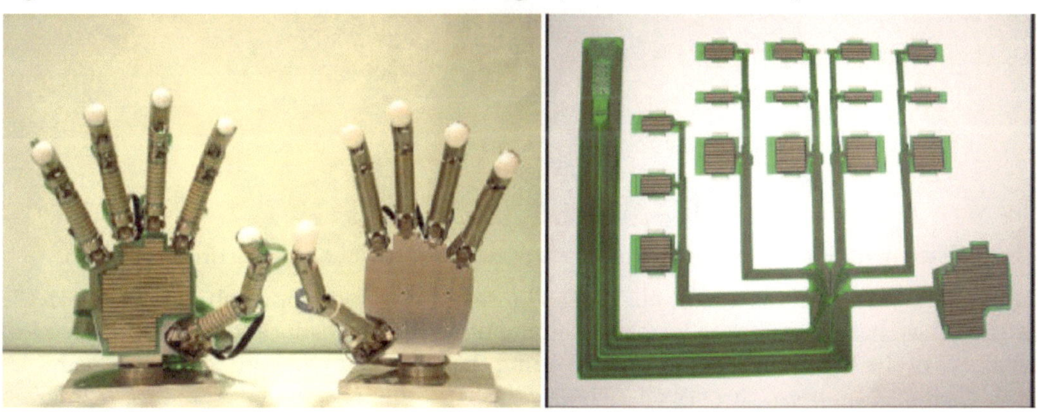

Fig. 3. Gifu hand III and skin-like tactile sensors (Kawasaki et al., 2002; Mouri et al 2002).

Another interesting example of integrated mechanical and sensing design is the robot hand CyberHand, a five fingered tendon driven under-actuated gripper, (Carrozza et al. 2006), Figure 4. Each finger has on its tip a custom three axis force transducer, while the phalanges are sensorized with custom flexible polyimide sensor formed by on-off taxels, with an activation force of about 1N. Also in this case, the design of the sensor does not feature embedded electronics on the finger. This kind of skin sensor can provide only geometrical contact information, therefore cable-tension sensors based on strain gauges have been integrated to estimate the force applied at the contact.

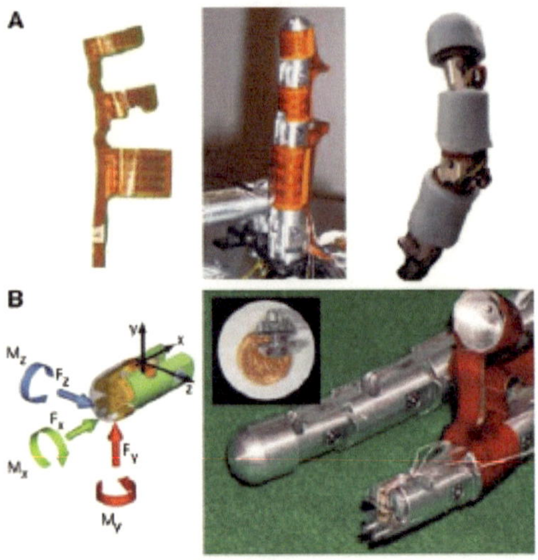

Fig. 4. Cyberhand tactile sensor.

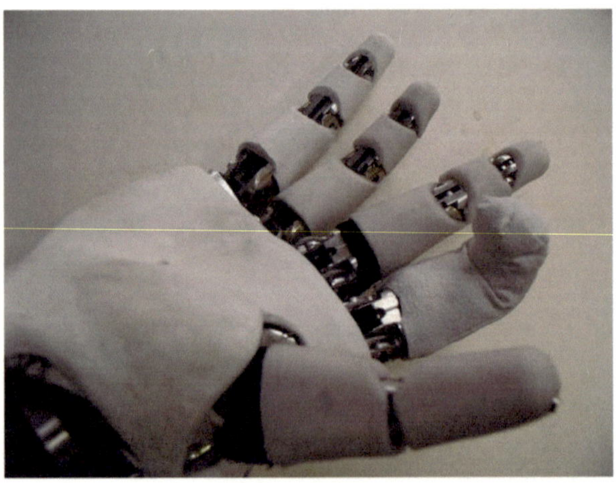

Fig. 5. Placement of the pressure transducers in the Robonaut hand (Martin et al., 2004).

In the Robonaut hand (Martin et al., 2004), Figure 6, a different approach to skin-like tactile sensing has been proposed. Instead of integrating the transducers on the mechanical structure of the hand, a *sensorized glove* covering the hand and embedding force transducers and cabling has been developed. The proposed design is based on a coarse grain distribution of pressure sensitive resistive rubber transducers (produced by QTC Ltd. and

Interlink Inc.), Figure 7. In total the glove has 33 sensitive sites where contact forces are *concentrated* by means of *plastic beads.*

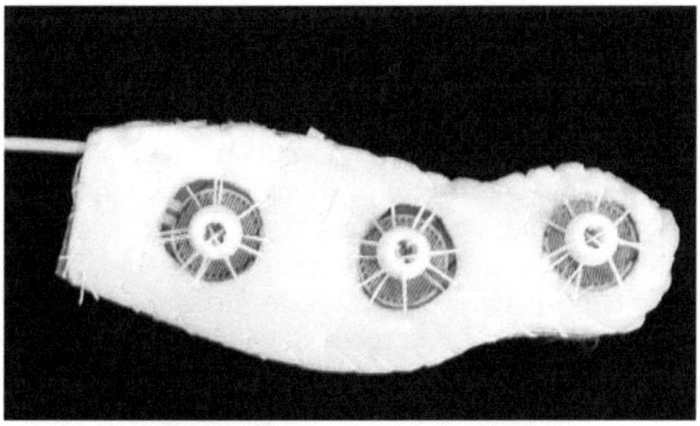

Fig. 6. Robonaut hand (Martin et al., 2004)

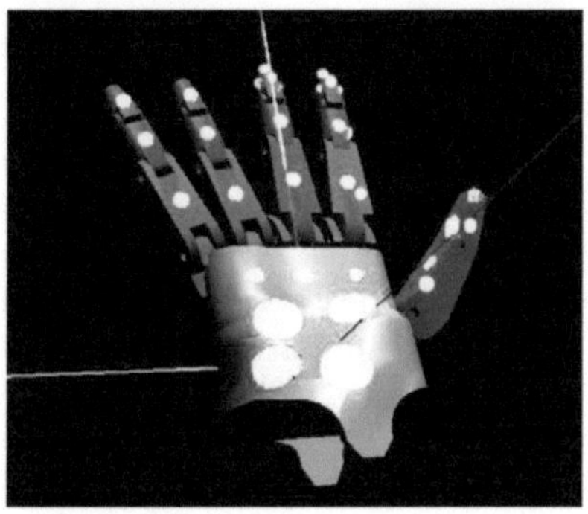

Fig. 7. Detail oft he Robonaut skin (Martin et al., 2004).

The concept is simple and the adoption of *discrete* piezo-resistive resistive transducers suggests that this solution could be used for simple custom tactile sensors, also because interface electronics consists only of a voltage divider. There are however some drawbacks quoted by the authors. In fact, as the hand opens and closes the glove itself applies forces on the transducers; furthermore the glove interfere with the hand movements. both these problems have been addressed in (Martin et al., 2004), but the authors considered them still open issues.

2.2 Large scale tactile sensors

From the previous section it emerged that a major issue for designing tactile sensors for grasping and manipulation is related to the need of tailor-made solutions. In fact, most of the solutions proposed are strongly dependent on the mechanical design and specialized for a particular platform. Other features like modularity and scalability could in fact improve the characteristics of these devices. the target is that of developing modular sensors, with embedded transducers and electronic possibly based on a common design. Networking of these devices would make possible the scalable (and incremental) development of large scale tactile sensors.

These ideas have been used for the design of a various *large scale* skin-like sensors proposed in the literature. By large scale we intend tactile sensors which can *cover* large areas of a robot body conforming to its outer shape.

The investigation on large scale robot skin has originally received attention for particular applications, as space robotics (Lumelsky et al., 2001), But it significantly rose recently with the growth of the interest in humanoid robots. A humanoid robot is expected to interact in complex and largely unpredictable way with its surrounding environment, and it is expected to be capable of safe and purposive interaction with humans (A. Billard and R. Siegwart 2004). The class of humanoid robot based tasks requiring direct interaction capabilities is certainly much larger and critical than those addressed in the past, and complex operations involving concurrent walking, interaction with humans and body manipulation (Ohmura, 2007) are strongly based on active control based also on tactile feedback. in fact, these tasks need the monitoring of the contacts ot the robot with the environment which may happen at unpredictable positions and in unpredictable ways.

One of the first examples of truly scalable robot skin systems for humanoid robots has been proposed by Ohmura and Kuniyoshi (Ohmura et al., 2006). They approached the problem at system level, and focused the issue of the wiring as th he key problem for developing a technology which could be practically exploited. Their main contribution has been that of introducing a networked architecture featuring peripheral nodes (chips) scanning (locally) a limited number of taxels. All the electronics and the transducers are embedded on a tree-shaped flex/semi-flex PCB support, Figure 8. This solution allows a simple mechanical integration of the sensor over curved surfaces and has been experimentally tested with success (Ohmura 2007), Figure 9. This system design has however some limitations. The spatial resolution is quite low since the minimum distance between the taxels is about 2.7 cm. Furthermore, the sensor is based on IR optical transducers which have (at the present status of the technology) a quite large power consumption; this means that for a complete robot skin system (which might have thousands of taxels) the power supply requirements could become critical (e.g. for autonomous robots). An attempt to limit the current requirements has been that of sequencing the switching on of the LED sand this clearly reduces the sampling rate capability of the sensor. Finally, other limitations of this design are related to the network solution adopted which does not seem to support fault tolerance mechanisms. This is important since for its own nature robot-skin could be subject to continuous or cyclic stress and impacts over long periods of time and its performance should possible have graceful degradation. The fault tolerance problem was previously addressed by Um and Lumelsky (Um et al. 1999), who tackled the problem via component redundancy for a system featuring over 1000 sensing elements.

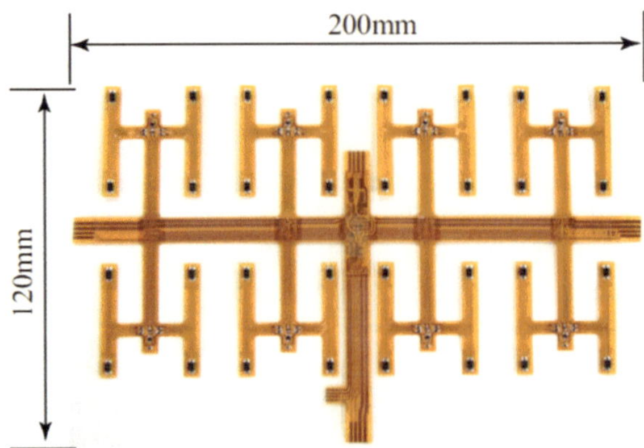

Fig. 8. Tactile sensor sheet (Ohmura 2007)

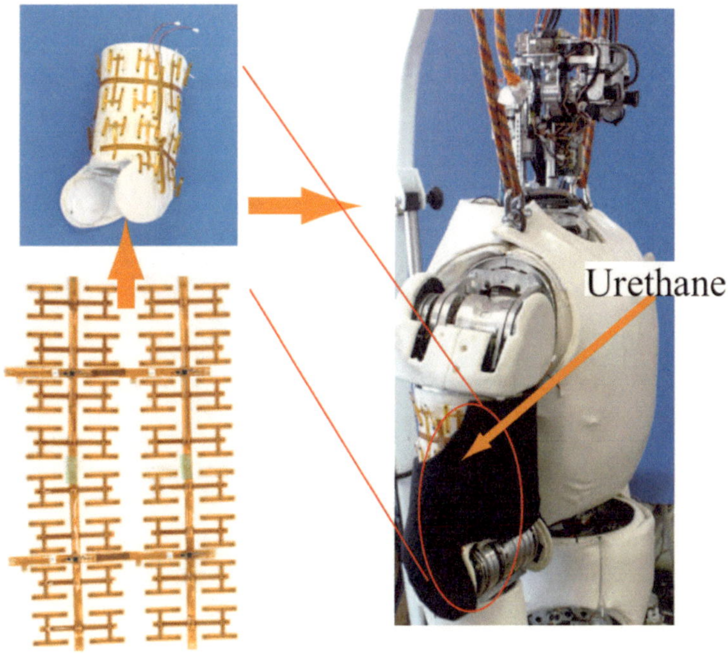

Fig. 9. Integration of the robot skin on the robot (Ohmura 2007).

A similar layout, but using piezo-resistive commercial pressure transducers has been recently proposed in (Mukai, 2008), Figure 10. The sensor is modular, and consists of 8·8 taxels with a spatial resolution of 1.8 mm which can withstand a pressure of about 12 N/cm^2. It does not embed electronics, however custom hardware modules for data acquisition and networking allow a scalable configuration of the system. The RI-MAN humanoid robot (Mukai, 2008), is equipped with five tactile modules for a total of 340 taxels and implements a tactile based feedback control operating at 50 Hz.

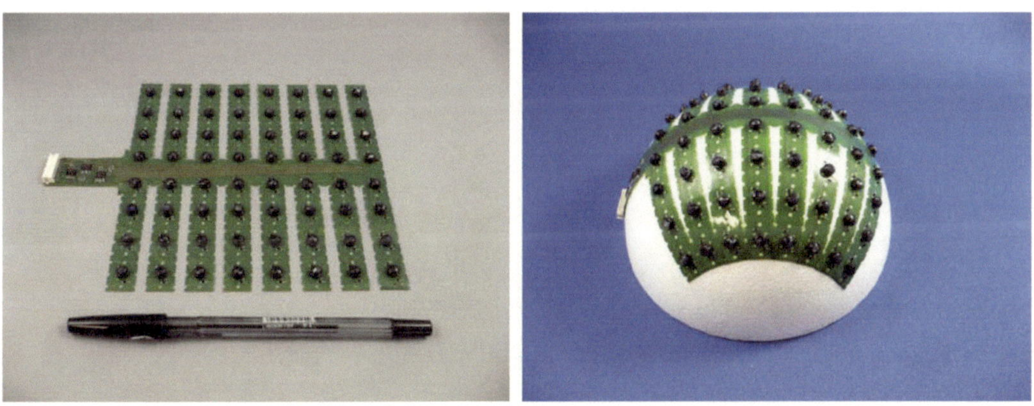

Fig. 10. Tactile sensors fort he RI-MAN humanoid robot (Mukai, 2008).

In the robot ARMAR-III (Asfour et al., 2006), Figure 11, developed by Prof. Dillmann and his team, the idea of skin patches, based on piezo-resisteive matrices of sensors, with embedded data processing electronics has been successfully implemented. Embedded electronics provides local tactile data processing in order to limit the bandwidth requirements. The patches are custom designed and have flat or cylindrical shape in 3D for covering relevant parts of the robot arms, while smaller patches are used to sensorize the fingers.

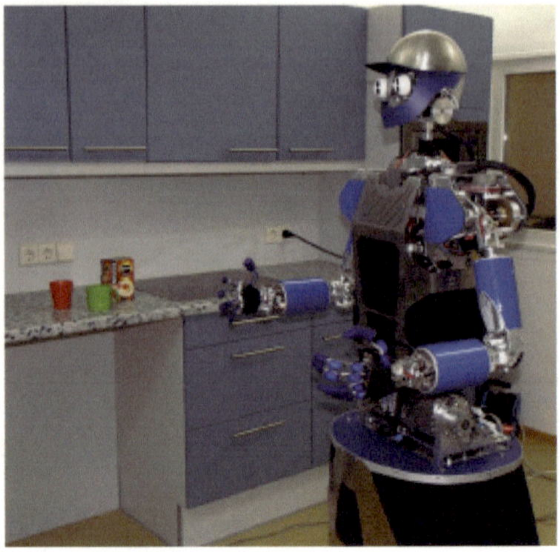

Fig. 11. The Humanoid robot ARMAR-III (Asfour et al., 2006).

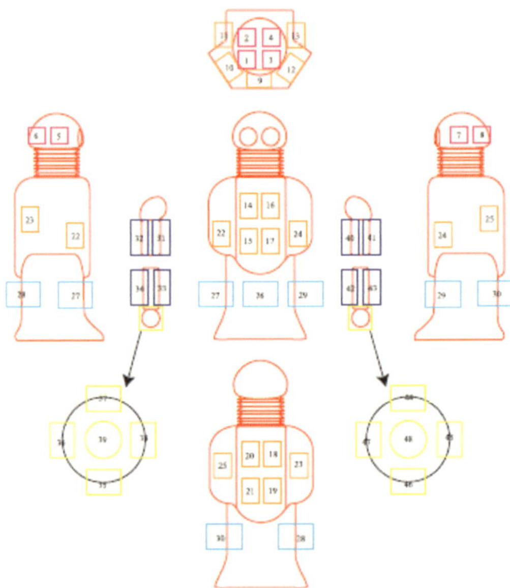

Fig. 12. Arrangement of tactile sensor in Robovie-IIS (Tajika et al. 2006).

Another example of artificial skin system for a humanoid robot has been proposed by Tajika et al. (Tajika et al. 2006) and implemented in the Robovie-IIS, Figure 13.
The tactile sensors have been designed with the aim of detecting stimuli coming from people trying to interact with the robot and it is based on PVDF based transducers. The skin has low spatial resolution (transducer area is of about 25 cm^2), and it has the possibility of measuring only stimuli at frequencies higher than 5-10 Hz, Figure 12.

Fig. 13. The humanoid robot Robovie-IIS (Tajika et al. 2006). Robovie-IIS has been developed by ATR.

A similar solution has been adopted in the CB2 humanoid robot (Minato et al., 2007). The robot it has humanlike appearance similar to a child-size boy. It is about 130cm high and weights about 33kg. Tactile sensing is obtained by embedding small thin PVDF films beneath the skin. The system is not modular, i.e. the transducers are placed ad-hoc over the robot body. The output is proportional to the rate of change of bending (deformation rate). The information equivalent to a contact force is obtained by temporal integration of sensor output. The transducers are put between a layer of urethane foam covering the mechanical parts and the outer silicone skin. On the robot there are 197 sensors providing output at a frequency of 100 Hz, Figure 15.

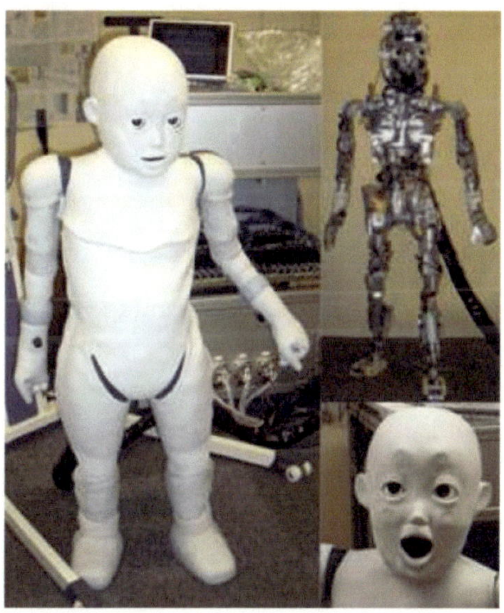

Fig. 14. Tactile sensor layout of robot CB2 (Minato et al., 2007). CB2 has been developed by JST ERATO Asada Synergistic Intelligence Project

Finally, Kotaro (Mizuuchi et al, 2006), is a very complex musculo-skeletal humanoid robot where tactile sensing is obtained using flexible with *bandages* formed by two flexible printed circuit boards with a intermediate layer of pressure-sensitive conductive rubber forming 64 taxels. The sensor can match complex shape surfaces.

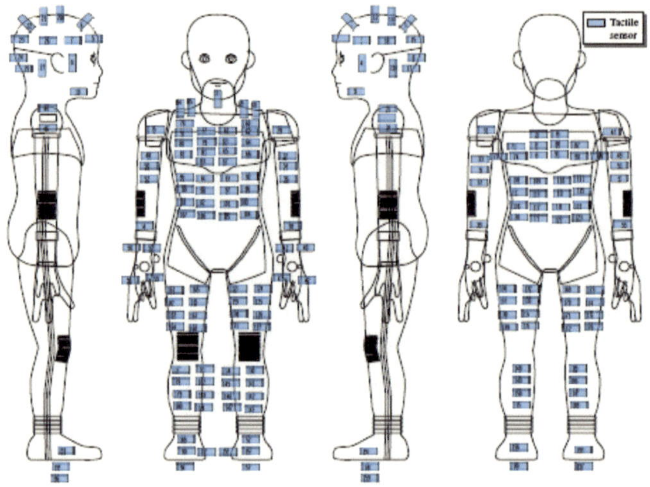

Fig. 15. Humanoid robot CB2 (Minato et al., 2007).

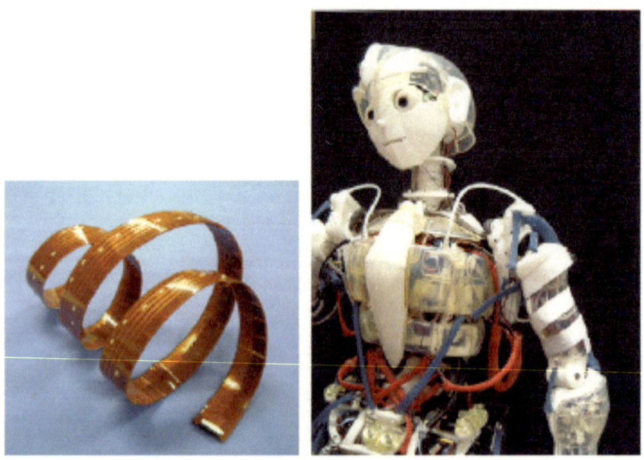

Fig. 16. Muscoloskeletal humanoid Kotaro and the bandage tactile sensors (Mizuuchi et al, 2006).

3. Tactile/force sensor design[1]

In this section we describe the prototype of a modular tactile and force sensor for a robot hand. The system is modular and the same design has been used to develop all the sensor used in different parts of the hand. Each module is composed by : a three-axis commercial force sensor(CTS 109 from CTS Corporation), a flexible tactile array based on piezo-resistive

---

[1] This section is based on the work published in (Cannata & Maggiali, 2006).

transducer and a data acquisition and processing unit consisting of analog signal conditioning electronics, a microcontroller (MCU), and communication interface, (Cannata & Maggiali 2005), Figure 17.

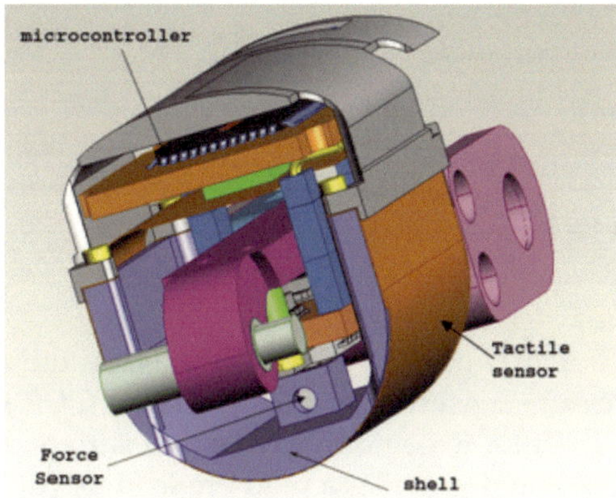

Fig. 17. CAD model of the whole phalange, with the force sensor, the tactile transducer and MCU board.

These modules are implemented on three different printed circuit boards (PCBs), as sketched in Figure 17. The first board supports the force sensor, rigidly connected with the phalange's *shell*, and its differential amplifiers. The tactile sensor module is a matrix of 64 taxels, supported by a double layer flexible PCB (flex-PCB), conformed to the finger cover. Finally, a third PCB hosts the MCU for local data processing and a CAN transceiver for communications. The whole tactile sensing system is scalable as several modules can be interconnected using a CAN bus link.
The current prototype has been designed to be integrated on each phalange of the robot MAC-HAND, Figure 18 .

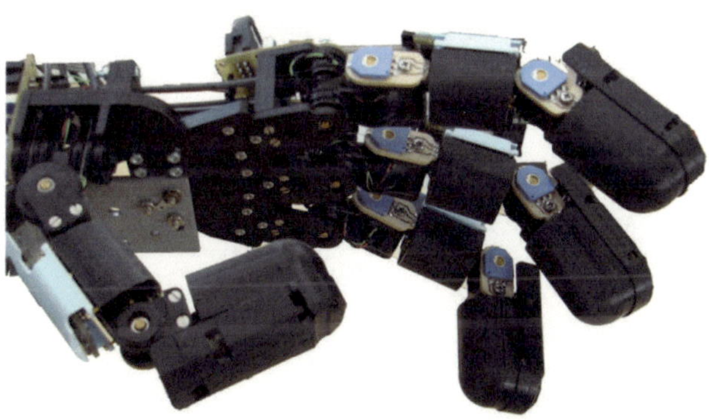

Fig. 18. The MAC-HAND robot.

3.1 Force sensor

The force sensor used is a CTS Series 109 integrated micro joystick (CTS Corporation, see for detailed specifications). This component has good sensitivity and linearity, furthermore, its SMD packaging makes its embedding and integration with other electronics much simpler than using a strain-gauge based design.

Output Linearity	1.0%
X, *Y* Axis Output Characteristics	$0.85\mu V/V/g$
Z Axis Output Characteristics	$0.125\mu V/V/g$
Maximum Overload Force	40*N*
Dimensions	10 × 7.5 × 5.5*mm*

Table 1. Specifications oft he force sensor (CTS Series 109).

3.2 Tactile sensor

The tactile transducer is a matrix of electrodes (8×8 taxels in the current implementation), covered by a layer of pressure sensitive conductive rubber (PCR Co. Ltd.). The electrodes are etched on a flex-PCB, in order to conform to a curved surface, Figure 19b. A thin elastic sheet covers the whole sensor and provides a mild pre-load useful to reduce noise. Pressure due to contacts changes the resistance among the electrodes, and can be measured with a resistive voltage divider. The scanning of the whole tactile matrix is performed using the scheme shown in Figure 20, where the columns are directly selected and driven by digital output ports of the embedded MCU, while rows are selected using a low noise and low onresistance analog multiplexer, followed by a low pass active filter. Ideally, the scanning of a resistive tactile matrix should follow the *zero potential method (Hillis 1982)*; in order to obtain fully decoupled measurements. This method requires that during scanning the taxels belonging to column *i*, all the other columns are set to ground (Shimoyo et al 2004). This approach has a major drawback since the driving electronics is too complex to be integrated

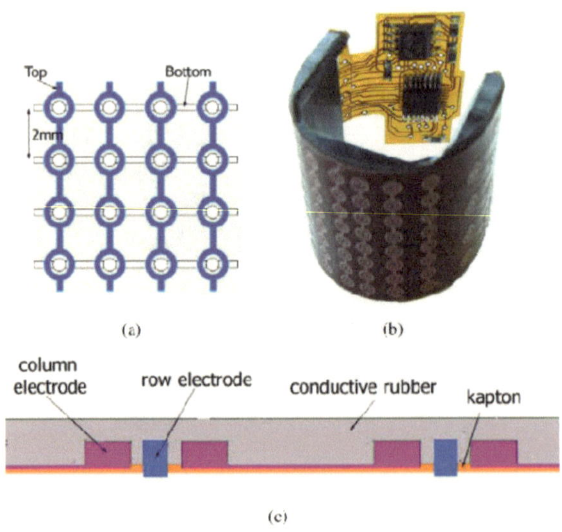

Fig. 19. The electrodes of the tactile sensor are etched on a double layer flex-PCB, and are connected in a matrix configuration (fig. a).The flex circuit conforms to the cylindrical phalange rigid cover (fig. b - resistive rubber not shown here). Fig. c: section view of the tactile sensor. Row electrodes are *encircled* by column electrodes.

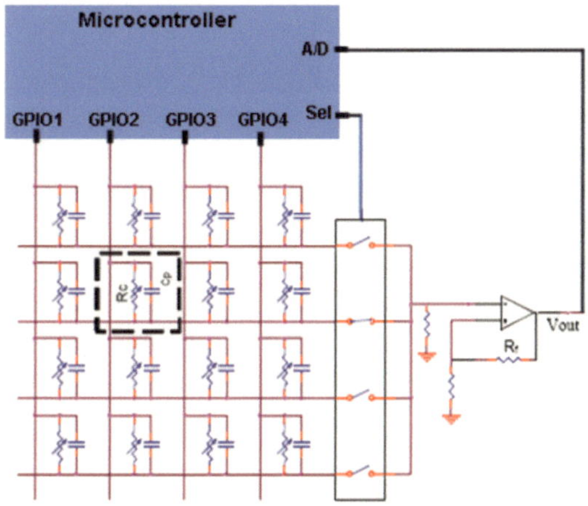

Fig. 20. Scanning Circuit Scheme

in a small volume. Therefore, in order to reduce the effects of cross-coupling among the taxels, two combined techniques have been adopted in our design. First of all, a careful definition of the geometry of the electrodes has been adopted, Figure 19a, with the goal of limiting the *parasitic* currents across the various taxels along one direction. Secondly, by exploiting the linearity of the resistive circuit characterizing the transducer, the crosscoupling is compensated performing a scanning of all the rows, keeping all the MCU driving outputs at low level. In this way, the effects of the parasitic injected currents in each column (due mainly to the characteristics of the MCU output ports) can be measured and subtracted by software from measurements obtained during the ordinary scan procedure. This method is in practice equivalent to measure data from a dummy column not providing any pressure measurements. This technique has a mild cost in term of scanning time, while the quality of the measurements is largely improved, as shown in Figure 21.

Signal conditioning is performed by using a first order active low pass filter placed on the sensor flex PCB and connected to the MCU.

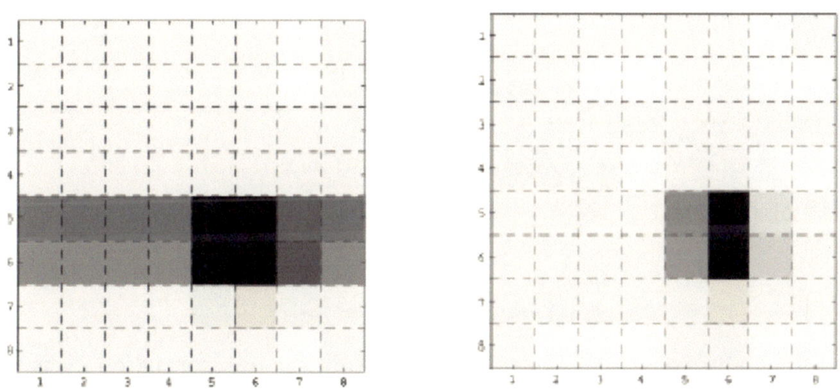

(a) (b)

Fig. 21. The tactile image refers to a small cylindrical presser. Fig. a shows the measured pressure distribution using conventional matrix scan. Fig. b shows the tactile image obtained using the proposed compensation method, based on *dummy column method*, where the false pressure distribution has been removed.

4. Local data processing

Tactile data are sampled by the on-board MCU, with a resolution of 8 bits. Each tactile image in the current prototype consists of 64 taxels requiring the transmission of 8 CAN messages (with full payload) for each module and then 64 messages for the whole hand MAC-HAND. The CAN 2.0A frame duration, operating at 1*Mbit/sec,* is nominally of 132 *μsec,* therefore, the total minimum latency is of about 8.5 *msec*. This figure does not take into account possible bus access conflicts and therefore must be considered as an optimistic lower bound.

We proposed to reduce the amount of tactile data fed back to the controller using an approach inspired by the idea of *contact centroid* introduced in (Bicchi et al 1993).Let us assume that the contact between the sensor and an object, corresponds to a pressure distribution over a connected region. The distribution in general can be assumed to be apriori unknown and therefore could be modeled as a random one. Therefore, it is reasonable to associate the contact distribution (i.e. the tactile image), to statistic moments of first and second order. In particular, it is possible to define the *pressure centroid* x_c as follows:

$$\boldsymbol{x}_c = \frac{\sum_{i+1}^{N} \boldsymbol{x}_i p(\boldsymbol{x}_i)}{\sum_{i+1}^{N} p(\boldsymbol{x}_i)} \tag{1}$$

where x_i is the position of each taxel in local surface coordinates, $p(x_i)$ is the corresponding measured pressure and N is the number of taxels of the sensor. The *shape* of the pressure distribution could be approximated as an ellipsoid, defined as follows:

$$E = \frac{\sum_{i+1}^{N} (\boldsymbol{x}_i - \boldsymbol{x}_c)(\boldsymbol{x}_i - \boldsymbol{x}_c)^T p(\boldsymbol{x}_i)}{\sum_{i+1}^{N} p(\boldsymbol{x}_i)} \tag{2}$$

where E is a positive semi-definite symmetric matrix.

Remark 1: The above quantities are valid for any sensor conforms to a regular surface.

Experimental tests, discussed in detail in the next section, have shown significant sensitivity of both x_c and E to measurements noise, in particular for low values of pressure. In order to reduce the effects of noise with respect to the actual pressure values, formulas (1) and (2) have been modified as follows:

$$\boldsymbol{x}_c = \frac{\sum_{i+1}^{N} \boldsymbol{x}_i \bar{p}(\boldsymbol{x}_i)}{\sum_{i+1}^{N} \bar{p}(\boldsymbol{x}_i)} \tag{3}$$

$$E = \frac{\sum_{i+1}^{N} (\boldsymbol{x}_i - \boldsymbol{x}_c)(\boldsymbol{x}_i - \boldsymbol{x}_c)^T \bar{p}(\boldsymbol{x}_i)}{\sum_{i+1}^{N} \bar{p}(\boldsymbol{x}_i)} \tag{4}$$

where

$$\bar{p}(\boldsymbol{x}_i) = a p^n(\boldsymbol{x}_i) \qquad i = 1 \cdots N \tag{5}$$

with a and n suitable constants. The role of the non-linear transformation (5) is to increase the signal to noise ratio.

The above formulas are implemented locally on the embedded MCU. Assuming a data resolution of 8 bits using this method a single CAN message is sufficient to transmit the force measurement and contact centroid data. In fact 3 bytes are requested for force measurements, 2 for $\mathbf{x}_c$ and 3 are sufficient for E. Of course, this condition would hold also for tactile sensors with a large number of taxels.

Remark 2: In the curent implementation the tactile sensor covers an area of 34 ×20 *mm*², then the *pressure centroid* can be computed with a spatial resolution approximately of less than 0.15 *mm*.

5. Experients

The experiments presented in this section refer to a planar tactile matrix formed by 8 × 8 taxels. The electronics used for data acquisition and processing is the same designed to be installed on the gripper MAC-HAND. The experiments have been performed using the test rig shown in fig. Figure 22. The characteristic response of each taxel has been obtained from experimental data using a least square cubic approximation, Figure 23. The experiment proposed consists of a pen pressed on the matrix. The tactile image as well as the computed pressure centroids and ellipsoids are shown in Figure 24. The time behavior of the sensor shows a mild creep of the stressed taxel, as well as the astual sensor noise, Figure 25. The computation of the centroids and the ellipsoids has an evident averaging effect, Figure 26 and Figure 27, with a better response using the non-linear algorithm (5).

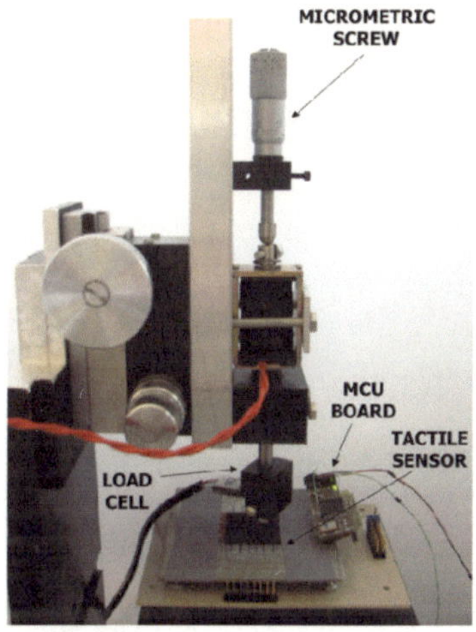

Fig. 22. Test rig for sensor calibration and tests.

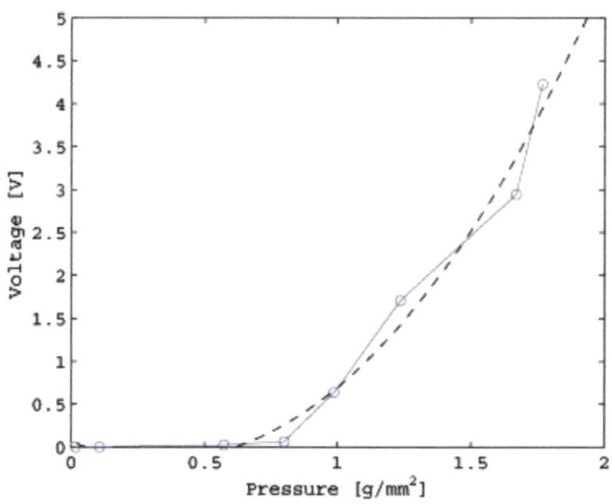

Fig. 23. Pressure to voltage ratio of a generic taxel. Dashed curve is the cubic that approximate the characteristics of the taxel

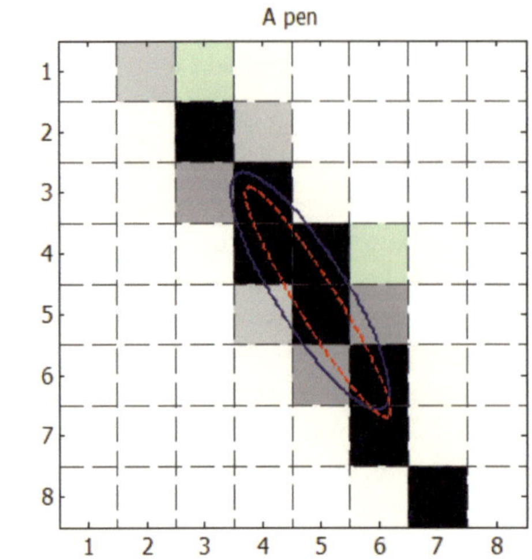

Fig. 24. Tactile image of a pressed pen.

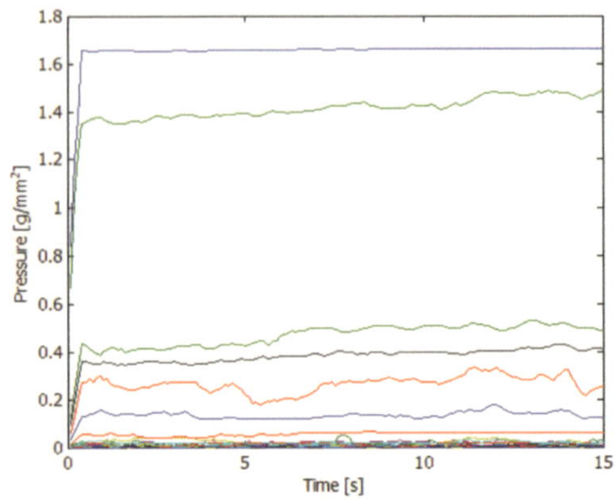

Fig. 25. Time behaviour of the taxels.

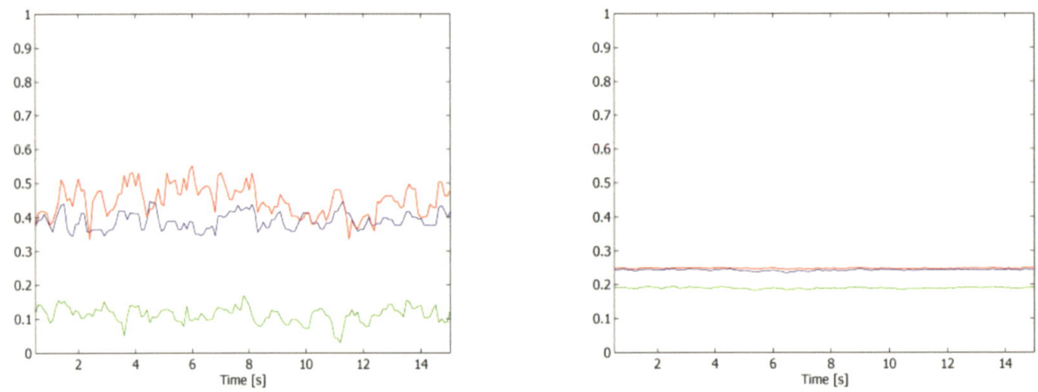

(a) (b)

Fig. 26. Components of the matrix E as functions of time. Fig. a: without using transformation (5). Fig. b using transformation (5).

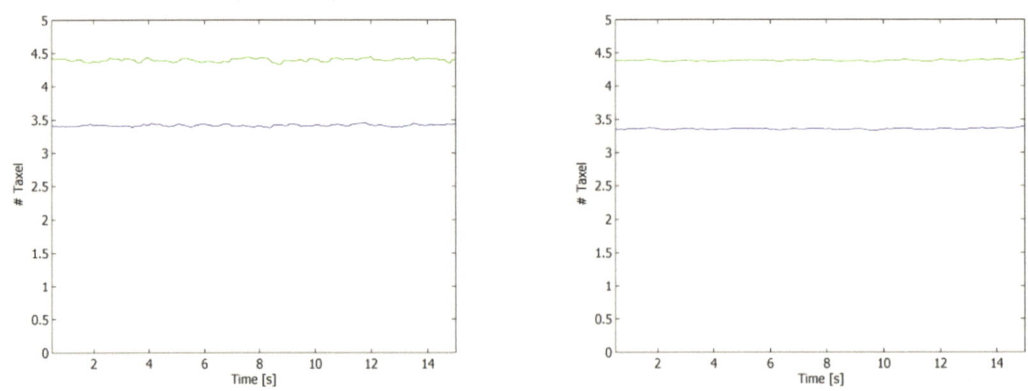

Fig. 27. Coordinates of the centroid as functions of time. Fig. a: without using transformation (5). Fig. b using transformation (5).

6. Conclusions

The use of skin like tactile sensors will play an important role for the development of humanoid robots performing complex interaction tasks possibly involving humans. In the first part of this chapter we have presented a survey of relevant recent achievements in the area of robot-skin technologies and tactile sensors for robot hand. In the second part an integrated tactile/force sensor with embedded electronics has been presented. The sensor consists of a three components commercial force sensor and of a custom matrix tactile transducer based on pressure sensitive conductive rubber. The sensor is driven by a local embedded microcontroller which implements local algorithms for the processing of raw tactile data extracting relevant parameters related to the contact force distributions.

7. References

Salisbury K., Mason M. T. (1985). Robot Hands and the Mechanics of Manipulation, M.I.T. Press. Cambridge, MA, ISBN-0-262-13205-2.

Jacobsen. S. et al. (1986). Design of the Utah/M.I.T. Dexterous Hand, Proceedings of the IEEE International Conference on Robotics and Automation, San Francisco CA. 1520-1532.

Melchiorri, C., Vassura, G. (1992). Mechanical and Control Features of the University of Bologna Hand version 2, IEEE Int. Conference of Intelligent Robots and Systems, Raleigh (North Carolina), pp. 187-193.

F.Lotti, P. Tiezzi, G. Vassura et al. (2004). UBH 3: can anthropomorphic hand with simplified endo-skeletal structure and soft continuous fingerpads, IEEE International Conference on Robotics and Automation, 26 April-1 May 2004 pp. 4736 - 4741.

Butterfass, J., Grebenstein, M., Liu, H., Hirzinger, G. (2001). DLR-Hand II:Next Generation of a Dextrous Robot Hand, Proc. IEEE Int. Conf. on Robotics and Automation, Seoul, Korea.

Carrozza, M. C., Vecchi, F., Roccella, S., Zecca, M., Sebastiani, F., Dario, P. (2003). The CyberHand: on the design of a cybernetic prosthetic hand intended to be interfaced to the peripheral nervous system, IROS 2003 vol. 3, Oct. 27-31, pp. 2642 - 2647.

Kargov A., Pylatiuk C, Oberle R., Klosek H., Schulz S., Bretthauer G. (2005). Development of an Anthropomorphic Hand for a Mobile Assistive Robot, ICORR 2005, Chicago.

Caffaz A., Cannata G. (1998). The Design and Development of the DIST-Hand Dextrou Gripper, 1998 IEEE ICRA, Leuven, Belgium, pp. 2075-2080.

Lane D., Cannata G. et al. (1997). AMADEUS: Advanced Manipulation for Deep Underwater Sampling, Robotics and Automation Magazine.

Hashimoto H., Ogawa H., Umeda T., Obama M., Kyoichi Tatsuno (1995). An Unilateral Master-Slave Hand System with a Force-controlled Slave Hand, IEEE International Conference of Robotics and Automation.

Shimojo, M. (2004). A Tactile Sensor sheet using pressure conductive rubber with electrical wires stitched method, IEEE Sensor Journal, vol. 4, no. 5.

Krishna, G. M., Rajanna, K. (2002). Tactile sensor based on piezoelectric resonance, Sensors, Volume: 2, 1643- 1647 vol.2.

Yamada, K., Goto, K., Nakajima, Y. , Koshida, N., and Shinoda, H. (2002). A Sensor Skin Using Wire-Free Tactile Sensing Elements Based on Optical Connection, Sice, Volume: 1, 131- 134, vol. 1.

Engel J., Chen J., Chang L., Flachsbart B. R., Selby, J. C., Shannon, M. A. (2003). Development of Polyimide-based Flexible Tactile Sensing Skin, Mat. Res. Soc. Symp. Proc. vol. 736.

Cannata, G., Maggiali, M. (2005). An Embedded Tactile and Force Sensor for Robotic Manipulation and Grasping, Proceeding of IEEE-RAS International Conference on Humanoid Robots, Humanoids 2005, Tsukuba (Japan), Dec. 2005.

Hillis, W. D. (1982). A high-resolution imaging touch sensor, *International Journal Robot. Res.*, vol. 1, no. 2, pp. 33-44.

Verssimo, P., Rufino J., Ming. L. (1997). How hard is hard real-time communication on fieldbuses?, Digest of Papers of the 27th IEEE International Symposium on Fault-Tolerant Computing Systems June 1997, Seattle, Washington, USA

Liu, H., Meusel, P., Hirzinger G.A. (1995). A Tactile sensing System for the Three-Finger Robot Hand, International Symposium on DLR Measurement and Control in Robotics, pp. 91-96, 1995.

Kawasaki H., Komatsu T., and Uchiyama K. (2002). Dexterous Anthropomorphic Robot Hand With Distributed Tactile Sensor: Gifu Hand II, *IEEE/ASME Transactions on Mechatronics*, VOL. 7, NO. 3, pp. 296-303, September 2002.

Martin T. B., Ambrose R. O., M. A. Diftler et al. (2004). Tactile Gloves for Autonomous Grasping with the NASA/DARPA Robonaut, Proc. 2004 IEEE Int. Conf. on Robotics and Automation, New Orleans (LA), April 2004, pp. 1713-1718.

Mukai T., Onishi M., Odashima T., Hirano S., and Luo Z. (2008). Development of the Tactile Sensor System of a Human-Interactive Robot "RI-MAN", *IEEE Transactions on Robotics*, VOL. 24, NO. 2, APRIL 2008, pp. 505-512.

Tajika, T., Miyashita, T., Ishiguro, H. et al. (2006). Automatic Categorization of Haptic Interactions-What are the Typical Haptic Interactions Between a Human and a Robot?, Proceeding of IEEE-RAS International Conference on Humanoid Robots, Humanoids 2006, Genova, 4-6 December 2006.

Bicchi. A, J.K. Salisbury and D.L. Brock, (1993). Contact Sensing from Force Measurements, *International Journal of Robotics Research*, v. 12, n. 3, pp. 249-262, June.

T. Mouri, H. Kawasaki, K. Yoshikawa, J. Takai and S. Ito (2002). Anthropomorphic robot hand: Gifu Hand III, ICCAS, 16-19, Muju Resort, Jeonbuk, Korea, October 2002.

D. Urn, V. Lumelsky, (1999). Fault Tolerance via Analytic Redundancy for a Modularized Sensitive Skin, Proceedings of the 1999 IEEVRSJ International Conference on Intelligent Robots and Systems, pp.1191-1197 vol.2, Kyongju, South Korea.

Lumelsky, V.J., Shur, M.S.; Wagner, S., (2001). Sensitive Skin, *Sensors Journal*, IEEE,Volume 1, Issue 1, June 2001 Page(s):41 - 51.

A.Billard, R. Siegwart (2004). Robot learning from demonstration, *Journal of Robotics and Autonomous Systems*, vol. 47 (2004) 65–67.

T. Minato, Y. Yoshikawa, T. Noda, S. Ikemoto, H. Ishiguro and M. Asada (2007). CB2: A Child Robot with Biomimetic Body for Cognitive Developmental Robotics, IEEERAS 7th International Conference on Humanoid Robots, November 29-December 1, Pittsburg, Pennsylvania USA 2007.

I. Mizuuchi, T. Yoshikai, Y. Sodeyama, Y. Nakanishi, A. Miyadera, T. Yamamoto, T. Niemel, M. Hayashi, J. Urata, Y. Namiki, T. Nishino and M. Inaba, (2006). Development of Musculoskeletal Humanoid Kotaro, Proceedings of the 2006 IEEE International Conference on Robotics and Automation, Orlando, Florida - May 2006

Cannata G., Maggiali M. (2006). Processing of Tactile/Force Measurements for a Fully Embedded Sensor, International Conference on Multisensor Fusion and Integration for Intelligent Systems September 3-6, 2006, Heidelberg, Germany

15

Tactile Sensing for Robotic Applications

Ravinder S. Dahiya[1, 2] and Maurizio Valle[1]
[1]*University of Genova,*
[2]*Italian Institute of Technology, Genova,*
Italy

1. Introduction

Robotic devices - limited to the structured environment of manufacturing plants until few years ago - are slowly making way into human life, which has led to the emergence of interaction and learning issues in robots. Such issues are important for future robot to be able to learn autonomously and to interact safely with the environment i.e. without causing any harm to it or to the objects with which it is interacting. In this context, it becomes important to study the ways and means of robot's interaction with the environment. Humans interact and explore the environment through five main sense modalities viz. touch, vision, hearing, olfaction and taste. Loss of any of these modalities will result in the incomplete information about the environment. Irrespective of their importance with respect to each other, the study of their individual contribution in collecting information from environment is important for improving the sensing capabilities of any future robotic system.

This chapter provides an overview of tactile sensing in robotics. This chapter is an attempt to answer three basic questions:

- What is meant by Tactile Sensing?
- Why Tactile Sensing is important?
- How Tactile Sensing is achieved?

The chapter is organized to sequentially provide the answers to above basic questions. Tactile sensing has often been considered as force sensing, which is not wholly true. In order to clarify such misconceptions about tactile sensing, it is defined in section 2. Why tactile section is important for robotics and what parameters are needed to be measured by tactile sensors to successfully perform various tasks, are discussed in section 3. An overview of `How tactile sensing has been achieved' is given in section 4, where a number of technologies and transduction methods, that have been used to improve the tactile sensing capability of robotic devices, are discussed. Lack of any tactile analog to Complementary Metal Oxide Semiconductor (CMOS) or Charge Coupled Devices (CCD) optical arrays has often been cited as one of the reasons for the slow development of tactile sensing vis-à-vis other sense modalities like vision sensing. Our own contribution - development of tactile sensing arrays using piezoelectric polymers and involving silicon micromachining - is an attempt in the direction of achieving tactile analog of CMOS optical arrays. The first phase implementation of these tactile sensing arrays is discussed in section 5. Section 6 concludes the chapter with a brief discussion on the present status of tactile sensing and the challenges that remain to be solved.

2. Tactile sensing: definition

The 'sense of touch' in humans comprises of two main submodalities - cutaneous and kinesthetic - characterized on the basis of their neural inputs. *Cutaneous sense* receives sensory inputs from the receptors embedded in the skin and *kinesthetic sense* receives sensory inputs from the receptors located within muscles, tendons and joints. In context with these submodalities, most researchers have distinguished among three sensory systems - cutaneous, kinesthestic and haptic. In the terminology of Loomis and Lederman (Loomis & Lederman 1986), *cutaneous system* involves physical contact with the stimuli and provides the awareness of the stimulation of the outer surface of body by means of receptors in the skin and associated somatosensory area of central nervous system (CNS). The *kinesthetic system* provides humans the information about the static and dynamic body posture (relative positioning of the head, torso, limbs and end effectors) on the basis of afferent information originating from the muscles, joints and skin, and the efference copy. The *haptic system* uses significant information about distal objects and events both from cutaneous and kinesthetic systems (Loomis & Lederman 1986; Klatzky & Lederman 2003). It should be noted that the sensory inputs are not only mechanical stimulation but also include heat, cooling and various stimuli that produce pain.

Human tactile sensing has generally served as a reference point for tactile sensing in robotics. A number of design parameters - discussed later - for robotic tactile sensors are derived from human sense of touch. Even though human tactile sensing has been a reference point for robotic tactile sensing, the way tactile sensing is defined in robotics falls short of what it means in humans. Most of the times, the robotic tactile sensing has been associated with detection and measurement of forces in a predetermined area only. The tactile or cutaneous sensing is associated with the detection and measurement of contact parameters which can be mechanical stimulation (force, stress, roughness etc.), temperature, moistness etc. In this context, definition of a tactile sensor by Lee et al. (Lee & Nicholls 1999) is more complete as the tactile sensor is defined as a device or system that can measure a given property of an object through contact in the world. Various studies on cutaneous sensing in humans show that some coding or pre-processing of the stimulus information takes place at the receptor or sensor level (Johannson & Birznieks 2008).

Going by these observations we define tactile sensing as the process of detecting and measuring a given property of a contact event in a predetermined area and subsequent pre-processing of the signals at the sensor level itself - before sending them to higher levels for perceptual interpretation (Dahiya, Metta et al. 2008) Similarly, the touch sensing can be termed as tactile sensing at single contact point. This distinction between touch and tactile sensing is followed throughout this chapter.

For the cutaneous and kinesthetic sensing in humans, the analogous terms in robotics are extrinsic/external and intrinsic/internal touch sensing respectively. In robotic applications, extrinsic touch sensing is achieved through tactile sensing arrays or a coordinated group of touch sensors. The extrinsic touch sensors or sensing arrays are mounted at or near the contact interface and deal with the data from localized regions. Intrinsic sensors are placed within the mechanical structure of the system and derive the contact data like magnitude of force using force sensors. Hereafter, unless otherwise stated, the term 'tactile sensing' is used for 'extrinsic touch sensing' in robotic applications.

3. Tactile sensing: why?

What happens if the humans have all sense modalities other than the sense of touch? The importance of touch/tactile sensing is implicit in this question which can be answered by performing a simple experiment of exploring the objects after putting hands on an ice block for a while or by making the hand numb through local anesthesia. In one such experiment, presented in (Westling & Johannson 1984), the skin on volunteers' hand was anesthetized so that mechanoreceptors - specialized nerve endings that respond to mechanical stimulation (force) - activity was no longer available to the brain. It was observed that even though volunteers' could see what they were doing, they could no longer maintain a stable grasp of objects. The movements become inaccurate and unstable when 'sense of touch' is lost. Real-world objects exhibit rich physical interaction behaviors on touch. These behaviors depend on how heavy and hard the object is when hold, how its surface feels when touched, how it deforms on contact and how it moves when pushed etc. 'Sense of touch' allow us not only to assess the size, shape, and the texture of objects, but, also helps in developing awareness of the body. It is also a powerful conduit for emotional connectedness. The ability to discriminate among surface textures, stiffness, and temperature, to sense incipient slip, and roll an object between fingers without dropping it are some of the reasons why touch/tactile sensing is needed.

In robotics, the touch information is useful is a number of ways. In manipulative tasks, touch information is used as a control parameter (Berger & Khosla 1991; Howe & Cutkosky 1990; Li, Hsu et al. 1989) and the required information typically includes contact point estimation, surface normal and curvature measurement and slip detection (Fearing 1990) through measurement of normal static forces. A measure of the contact forces allows the grasp force control, which is essential for maintaining stable grasps (Bicchi, Salisbury et al. May, 1990). The Grasp force along with manipulator displacement is also helpful in compliant manipulators (Cutkosky & Kao April 1989). In addition to magnitude, the direction of force is also critical in dexterous manipulation to regulate the balance between normal and tangential forces to ensure grasp stability - the so-called friction cone (Murray, Li et al. 1994). For full grasp force and torque determination, shear information is also required (Domenici, Rossi et al. 1989; Rossi, Canepa et al. 1993). The need for shear stress information is also supported by finite element analysis (FEA) (Ricker & Ellis May 1993; Ellis & Qin May 1994). Shear information is useful in determining coefficient of friction and a unique surface stress profile when the sensor is covered with elastomeric layer (Novak May 1989).

During interaction with the environment a significant portion of the information about contact objects e.g. shape (Charlebois, Gupta et al. 2000; Fearing & Binford Dec, 1991; Russell & Parkinson May 1993), surface texture (Maheshwari & Saraf 2006), slip (Howe & Cutkosky May 1989; Tremblay & Cutkosky May 1993), etc. comes through the detection of normal and shear forces. A real world contact parameters measurement also involves material properties such as hardness (Shikida, Shimizu et al. 2003), temperature (Yuji & Sonoda 2006) etc.

A range of sensors - based on various types discussed in the following section - that can detect object shape, size, presence, position, forces and temperature have been reported in (Dario & de Rossi 1985; Howe 1994; Lee & Nicholls 1999). Few examples of sensors that could detect surface texture (Maheshwari & Saraf 2006), hardness or consistency (Shikida, Shimizu et al. 2003; Omata, Murayama et al. 2004) are also reported. Very few examples of sensors that can detect force as well its direction have also been reported (Chu, Sarro et al. 1996; Torres-Jara, Vasilescu et al. 2006). Recently, the importance of dynamic events has been recognized and sensors are being developed for detecting stress changes (Howe & Cutkosky 1993; Schmidt, Mael et al. 2006), slip and other temporal contact events.

It should be noted that some of the above mentioned tasks involve both intrinsic and extrinsic touch sensing. As an example, pouring of water in a bottle requires intrinsic touch sensors for grasping the bottle and extrinsic tactile sensors to detect any slippage. Whereas, detection of parameters like smoothness may require signal from extrinsic tactile sensors only. As a matter of fact, all daily works that we perform with hands involve both extrinsic and intrinsic tactile sensors.

4. Tactile sensing: how?

The development of robots capable of operating in unstructured environments or intended to substitute for man in hazardous or inaccessible environments, demands the implementation of sophisticated sensory capabilities, far beyond those available today. Cues from the human tactile sensing system can be helpful in bringing the level of tactile sensitivity and acuity that humans possess, to the manipulators and to other human/machine interfaces. A general purpose robotic tactile system, in addition to being cost effective, should possess the following characteristics (Dario & de Rossi 1985; Howe 1994; Dahiya, Metta et al. 2008; Dahiya, Valle et al. July, 2007):

- A large number of taxels; typical estimated range should be between 25- 256 elements.
- Human like spatial sensitivity viz. 1mm.
- Sensitivity to forces spanning from 1gmf (0.01N) to 1000gmf (10N) with incremental force resolution of 1 g.
- Discrete taxel response bandwidth of 1000 Hz.

Besides these, a reasonable response linearity, negligible hysteresis and capability to measure contact parameters like hardness, temperature etc. are also desired. These design parameters are obtained with considering human sense of touch as reference. A large number of tactile sensors and sensing arrays have been reported in literature exploring nearly all possible methods of transduction - with or without above mentioned design parameters. The main transduction methods that have been reported are: Resistive/Piezoresistive, based on Tunnel Effect, Capacitive, Optical, Ultrasonic, Magnetic, and Piezoelectric. Relative merits and demerits of these methods are given in Table 1. Selected examples of robotic tactile sensors reported in the literature based on thsese transduction methods are discussed below.

4.1 Resistive sensors

Resistive sensors typically involve two conductive sheets separated by air, microspheres, insulating fabric, etc. One of the sheets carries a voltage gradient generated by applying a reference voltage and ground on its two opposite ends. The second sheet when brought in contact with the first by the applied force, serves like the slider in a linear potentiometer. A voltage divider is made at the contact point and the voltage of the sheet, which is acting as slider, can be used to find location of the contact point. Such an approach is limited to measurement of only one contact location. Touch sensors based on resistive principle are generally sensitive and inexpensive in terms of investment, but, they are expensive in terms of power consumption. An improved design of tactile sensor using resistive sensing technology is reported in (Zhang & So 2002). The design involves arranging the sensors in an array and hence enables the measurement of many contact points. But, the lack of contact force measurement still remains a critical problem.

4.2 Piezoresistive sensors

Piezoresistive touch sensors are made of materials whose resistance changes with force/pressure. Touch sensing system using this mode of transduction have been reported for use in anthropomorphic hands (Weiss & Worn 2004). Piezoresistive tactile sensing is particularly popular among the MEMS based and silicon based tactile sensors (Woffenbuttel & Regtien 1991; Beebe, Hsieh et al. 1995). FSRs (Force Sensing Resistors) based on piezoresistive sensing technology are widely used in pointing and position sensing devices such as joysticks and are commercially manufactured by Interlink (Interlink Electronics Inc. 2008). The FSR sensors have a large appeal, because of low cost, good sensitivity, low noise and simple electronics and are found in many experimental tactile systems. One of their drawbacks is the relatively stiff backing. Although examples of advanced robotic hands equipped with FSRs exist (Diftler, Platt Jr et al. 2003), these sensors generally require serial or manual assembly, provide highly non-linear response and suffer from hysteresis.

4.3 Tunnel effect tactile sensors

Tactile sensors based on Quantum Tunnel Composites (QTC) have come up recently and are commercially available from Peratech (Peratech-Ltd). QTC's have the unique capability of transformation from a virtually perfect insulator to a metal like conductor when deformed by compressing, twisting or stretching of the material. The transition from insulator to conductor follows a smooth and repeatable curve, with the resistance dropping exponentially. In QTCs the metal particles never come into contact. Rather they get so close that quantum tunneling (of electrons) takes place between the metal particles. Robot hands with QTC based tactile sensors have also been reported in literature (Walker July 2004). A sensor based on electron tunneling principle is reported in (Maheshwari & Saraf 2006). The device directly converts stress into electroluminescent light and modulation in local current density, both of which are linearly proportional to local stress. With thin film used with metal and semiconducting nanoparticles the spatial resolution better than that of the human fingertip (~40μm) is reported.

4.4 Capacitive sensor

Capacitive sensors consist of a plate capacitor, in which, the distance between plates or the effective area is changed by the applied force by shifting their relative position. Capacitive sensors can be made very small, which allows the construction of dense sensor arrays, and also allow dynamic measurements. Few examples of capacitive touch sensors are reported in (Schmidt, Mael et al. 2006). There are also commercially available capacitive-based touch sensors such as RoboTouch and DigiTacts from Pressure Profile Systems (Pressure Profile Systems 2007) and commercial products like 'iPodtouch' (Apple Inc. 2008) also use capacitive touch sensing. Touch sensors based on this mode of transduction are very sensitive but stray capacity and severe hysteresis is major drawback. An 8x8 capacitive tactile sensing array with 1 mm^2 area and spatial resolution at least 10 times better than the human limit of 1 mm is reported in (Gray & Fearing 1996). Capacitive sensing technology is also popular among the tactile sensors based on MEMS and silicon micromachining (Gray and Fearing 1996; Schmidt, Mael et al. 2006).

4.5 Optical sensors

Tactile sensors with optical mode of transduction use the properties of optical reflection between media of different refractive index. The transducer structure is composed of a clear plate, a light source and a compliant membrane stretched above, but not in close contact with, the plate. The lower surface of the plate acts as the imaging area. Light is directed along an edge of the plate and it goes through total internal reflection (when no force is applied) or diffuse reflection (when force is applied). The light coming out of plate due to diffuse reflection can be recorded by CCD or CMOS cameras placed in the imaging area. The intensity of the light (bright or dark patches on image) is proportional to the magnitude of the pressure between object and plate. Optical fiber based tactile sensors capable of measuring normal forces are reported in (Heo, Chung et al. 2006). The sensor can measure forces as low as 0.001N with the spatial resolution of 5 mm. An optical three axial tactile sensor capable of measuring normal and shear forces is reported in (Ohka, Kobayashi et al. 2006). Some cases of large area skin based on LEDs (light-emitting diodes) has also been reported (Cheung & Lumelsky 1992; Ohmura, Kuniyoshi et al. 2006). Optical based tactile sensors are immune to electromagnetic interference, are flexible, sensitive and fast, but at times they are bulky. Other problems associated with optical sensors are: the loss of light by micro bending and chirping, which causes distortion in the signal.

4.6 Ultrasonic sensors

Acoustic ultrasonic sensing is yet another technology that has been used for the development of tactile sensors. Microphones are known to be useful for detecting surface noise that occurs at the onset of motion and during slip. A device that senses contact events from their ultrasonic emission at the contact point is described in (Milighetti, Emter et al. 2006). A Polyvinylidene Fluoride (PVDF) polymer is used in a 2 x 2 array of receivers to localize the contact point on a silicone rubber sensing dome. The sensor is reported to be very effective in detecting slip and surface roughness during movement. The change in resonance frequency of PZT (Lead Zirconate Titanate), in accordance with object's acoustic impedance has been reported (Omata, Murayama et al. 2004) for detecting hardness and/or softness of objects. Tactile sensors based on ultrasonic approach have fast dynamic response and good force resolution, but materials like PZT are difficult to handle in miniaturized circuits.

4.7 Magnetism based sensors

Tactile sensors based on magnetic transduction measure the change in flux density caused by applied force on a small magnet. The flux measurement can be made by either a Hall Effect (Jamone, Metta et al. 2006) or a magneto resistive device. A few tactile sensors that use the magnetic mode of transduction have been reported in literature (Nowlin 1991). The tactile sensors based on magnetic principle have a number of advantages that include high sensitivity and dynamic range, no measurable mechanical hysteresis, a linear response, and physical robustness. Major drawback of magnetic based tactile sensor is that they cannot be used in magnetic medium and involve complex computations.

4.8 Piezoelectric sensors

The piezoelectric materials have the property of generating charge/voltage proportional to the applied force/pressure. Alternatively, they are capable of generating force due to electrical input. Thus, they can be used both as sensors and actuators and due to this property they fall under the category of 'Smart Materials'. Piezoelectric materials are suitable for use as tactile sensors. While quartz and some ceramics (PZT) have good piezoelectric properties, the polymers such as PVDF normally have been used in touch sensors because of some excellent features, such as, flexibility, workability and chemical stability (Flanagan & Wing 1993). The use of PVDF for tactile sensing was reported for first time in (Dario & de Rossi 1985) and thereafter a number of works based on PVDF or its copolymers have been reported in literature (Kolesar, Reston et al. 1992; Dargahi, Parameswaran et al. 2000; Yuji & Sonoda 2006). Temperature sensitivity of piezoelectric materials is a major cause of concern in their use as tactile sensors.

4.9 Sensors based on different physical/mechanical nature – recent trends

In the past, most devices have relied on fairly rigid, solid materials for their construction. Following studies of human tactile performance and the physical nature of the tissues and skin, it now seems that softer materials may have much to offer. Elastic overlays and compliant contact surfaces are often advocated for their frictional and other properties, although their low pass filtering behavior can be a disadvantage. But now even softer materials, such as rubber, fluids and powders, are being examined. Already, there are some commercially available touch sensors such as those from Tekscan (Pressure Sensitive Ink 2008) that use pressure sensitive ink or rubber. A number of touch sensors using conductive rubber as transducer have also been reported (Someya, Sekitani et al. 2004). They take advantage of change in impedance due to the applied force/pressure. Presence of hysteresis and non linearity are some of their drawbacks. Conductive gels having remarkable softness show a 20% change in impedance for pressure 0-400 kgf/cm^2 (Kageyama, Kagami et al. 1999). A range of materials with different consistencies have been examined in (Shimoga & Goldenberg 1992) for impact and strain energy dissipation conformability to surfaces and hysteresis effects. It is found that soft surfaces have more desirable characteristics for contact surfaces than hard materials. Among soft materials, gels are better than plastic, rubber, sponge, or paste, with powders being the second best.

5. Piezoelectric polymer – microelectrode arrays (MEA) based tactile sensing arrays

Robot's guidance and force based control has typically depended on the tri-axial or 6D force sensors placed on the robot's wrist or in other words, it has depended on the intrinsic touch sensing. However, intrinsic touch sensing method is sensitive to the accuracy of force/torque sensor calibration and can provide erroneous information as it is difficult to model dynamic forces. Further, inertia and compliance of manipulator can also generate some errors due to which intrinsic touch sensing is insufficient for the tasks that require precise manipulation. Such errors can be reduced by bringing the sensors closer to the contact points or in other words by using distributed touch sensors or tactile sensing arrays on the fingertips. As in humans, the combined signals from intrinsic and extrinsic tactile sensors can be used for various tasks. In addition to the tasks performed with hands, various safety and interaction issues call for tactile sensors distributed all over the body of robot. A number of tactile sensing arrays and artificial skin prototypes - using transduction methods, discussed earlier - have also been reported in literature. With large number of tactile sensors the number of interconnects needed to read and transfer the signals also

Type	Merits	Demerits
Resistive	• Sensitive • Low Cost	• High Power Consumption • Generally detect single contact point • Lack of Contact force measurement
Piezoresistive	• Low cost • Good sensitivity • Low noise • Simple electronics	• Stiff and frail • Non linear response • Hysteresis • Temperature sensitive • Signal drift
Tunnel Effect	• Sensitive • Physically flexibile	• Non Linear response
Capacitive	• Sensitive • Low cost • Availabilityof commercial A/D chips.	• Cross-talk • Hysteresis • Complex Electronics
Optical	• Immunity to electromagnetic Interference • Physically flexible • Sensitive • Fast • No interconnections.	• Bulky • Loss of light by micro bending • Chirping • Power Consumption • Complex computations.
Ultrasonic	• Fast dynamic response • Good force resolution	• Limited utility at low frequency • Complex electronics • Temperature Sensitive
Magnetic	• High sensitivity • good dynamic range, • no mechanical hysteresis • physical robustness	• Suffer from magnetic interference • Complex computations • Somewhat bulky • Power Consumption
Piezoelectric	• Dynamic Response • High Bandwidth	• Temperature Sensitive • Not so robust electrical connection.
Conductive Rubber	• Physically flexible	• Mechanical hysteresis • Non linear response

Table 1. Relative merits and demerits of various tactile sensor types.

increase – which is a big hurdle in the usage of distributed touch sensing or tactile sensing arrays. As for human tactile sensing, it is desirable to have tactile arrays with density and spatial distribution of taxels (tactile elements) according to the location where the sensors are installed. In this sense, the sensors can be divided in two classes: tactile sensing for body locations like fingertips and for body locations like belly, palm etc. The work presented here, focuses on the development of tactile sensing arrays for fingertips.

Considering the limited available space on the robot finger (~1 cm x 1 cm), miniaturization of sensing devices is a possible solution to accommodate large number of sensors in a small space. Miniaturization of tactile sensors has been achieved by two main approaches: MEMS based approach (Kane, Cutkosky et al. 2000) and polymer based sensors realised on organic substrate (Someya, Sekitani et al. 2004). With the MEMS based approach, it is possible to get higher spatial resolution, but, MEMS based tactile sensing devices cannot withstand large forces/pressure due to their inherent fragile nature. Also, it is difficult to realize physically flexible tactile sensing arrays by the MEMS approach. The tactile sensors realized on organic substrates have limited real time capability as they suffer from the slow time response.
As an alternative to these approaches, we proposed a novel approach for the development of tactile sensing chips for the fingertips, as shown in Fig. 1 (Dahiya, Valle et al. 2008). With this `sense and process at same place' approach, the tactile sensing arrays are developed for fingertips of robotic hand by directly coupling the "smart materials" like piezoelectric polymers with the Integrated Circuits (ICs). The working principle of the tactile sensors developed with this approach, is described in the following paragraph.

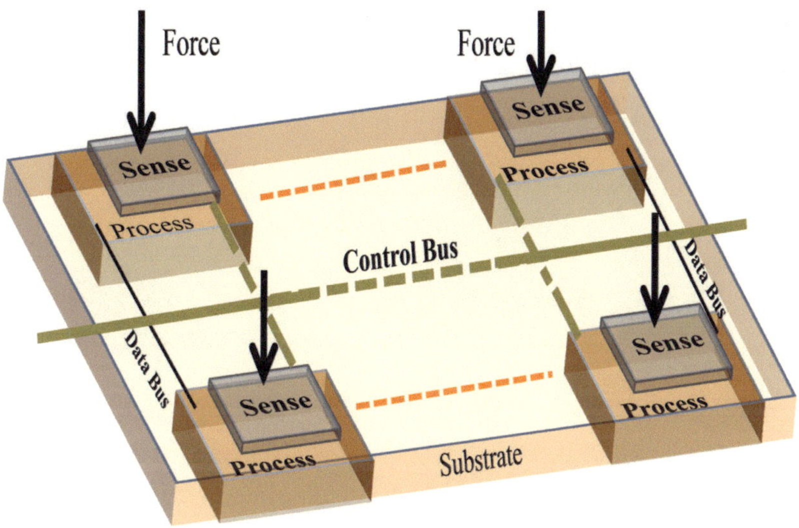

Fig. 1. 'Sense and Process at same place' approach for development of tactile sensign arrays (Dahiya, Valle et al. 2008).

A piezoelectric film working in the generating mode, produce a charge/voltage which is proportional to the applied stress. Hence by replacing the polysilicon gate of a MOSFET device with a piezoelectric polymer film, the charge in the induced channel of MOSFET can be controlled by the charge generated on the piezoelectric polymer film due to the applied stress. In other words, the charge in the channel is modulated by the applied mechanical stress. The signal is amplified by the MOSFET and is then further processed by electronic circuitry. While the piezoelectric polymer film as a sensing element would improve the speed of response; the marriage of sensing material (PVDF-TrFE) and electronics using MOS technology will improve force resolution, spatial resolution, signal to noise ratio and may help in reducing the wiring complexity - a key robotic problem. As earlier said, the lack of any tactile analog to Complementary Metal Oxide Semiconductor (CMOS) or Charge Coupled Devices (CCD) optical arrays has often been cited as one of the reasons for the slow development of tactile sensing vis-à-vis other sense modalities like vision sensing. With the proposed approach, a tactile analog of CMOS optical arrays can be obtained and hence the approach will advance the research in tactile sensing. The only disadvantage of using IC technology is lack of physical flexibility of tactile sensing chips. A possible trade off is to cover the chip with a thick and protective layer of silicone. Due to low thermal conductivity, such a layer would be helpful in reducing the effect of ambient temperature variations also - which otherwise introduces noise in the output during measurement of forces. However, a careful study is needed as such materials suffer from creep, hysteresis and in practice work as low pass filters (Shimojo 1997). Nonetheless, the advantages offered by proposed approach far outweigh the disadvantages.

As a first step towards realization of tactile sensing arrays based on the above said approach, arrays of tactile sensors were developed by directly coupling thin piezoelectric polymers films to 32 taxel MEAs realized on silicon die, as shown in Fig 2. The MEAs in this case act as the extended gates of FETs devices, which are external to the chip. Somewhat similar approach is used by Swartz et al. (Swartz and Plummer 1979) and Fiorillo et al (Fiorillo, Spiegel et al. 1990) to develop ultrasonic sensors and by Kolasar et. al. (Kolesar, Reston et al. 1992) to develop tactile sensors. While the former used the epoxy-adhered PVDF film, later one used a thin film of PVDF-TrFE directly deposited from solution on to the extended gates.

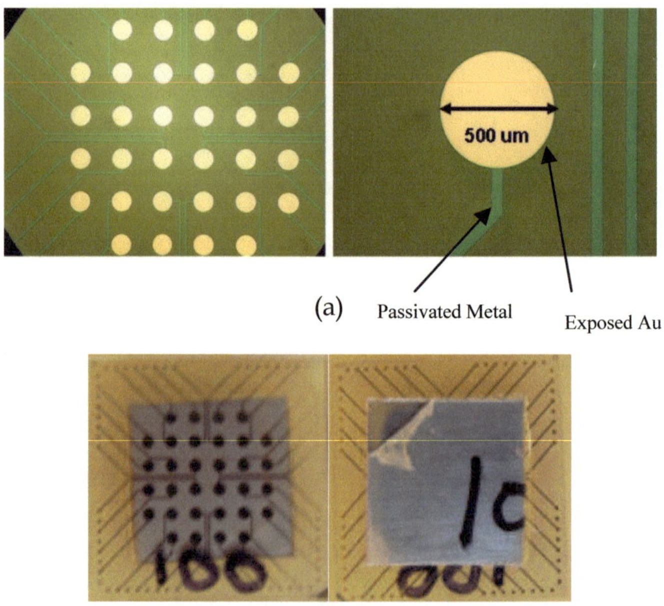

(a)

(b)

Fig 2. a) MEA for the extended gate-FET approach. Chip dimension is 1cm x 1cm. and diameter of the taxels is 500µm (b) Back and front sides of MEA with 100 µm polymer covering all the electrodes. A general purpose protecting tape can also be seen. (Dahiya, Valle et al. July, 2007)

One of the test structures devoted to characterize the polymeric material and to perform the electrical/mechanical tests to refine the read out electronics is shown in Fig. 2(a). To study the electrical response of different taxels PVDF-TrFE polymer with 25, 50 and 100 µm have been deposited. The fabrication steps for these test structures are given in Table 2. The front and backside of the MEA after depositing polymer is shown in Fig. 2(b). The fabrication of MEA is implemented on a fuse silica quartz substrate a Al:Si 1% /Ti/TiN, respectively of 410/30/140 nm thick, low resistance multilayer for both microelectrodes and electrical connections. The TiN top-layer has been introduced to guarantee a low contact resistance to the final Au/Cr (5/150 nm) seed-layer. The metal wires passivation has been guaranteed by a SiO_2/Si_3N_4 (20/210 nm) layer deposited by PECVD. These thicknesses have been chosen to keep the substrate capacitance low and hence to get the maximum of the voltage produced by polymer at the gate terminal. Fig. 3 shows the average response of three touch sensing elements - to forces up to 0.4 Kgf, when a sinusoidal force was applied at 15 Hz. The response is linear over a large range of forces. It can be noticed that unlike MEMS based sensors, the range of detectable force is higher in this case. Similarly the sensors based on silicon micromachining are known to have a mobility of three orders of magnitude higher that of organic based devices and hence they have faster response time. A further characterization is required to study the behaviour of sensing array over broad range of frequency.

Fabrication Process	
a.	Substrate: 500 µm thick quartz wafer.
b.	A Ti/TiN/Al/TiN multilayer is deposited by sputtering.
c.	This multilayer is patterned by photolithography and plasma dry etching to form the electrodes, lines and the contact pin zone.
d.	A layer of Si_3N_4 (200 nm) is then deposited by PECVD, in order to insulate the metal lines.
e.	Contacts are opened through the Si_3N_4 layer by plasma dry etching
f.	Evaporation of Cr and Au is carried out (5 nm and 150nm, respectively)
g.	The Chromium and Gold layers are patterned by wet etching.
h.	Deposition of piezoelectric polymer. The film was deposited using epoxy adhesive on the MEA, covering all 32 taxels. The film was covered with glass slide and the arrangement was again kept under vacuum to remove air between polymer and MEA and to ensure uniform thickness of the adhesive. For better adhesion the arrangement was kept at 65 degrees for thirty minutes.

Table 2. Fabrication steps of piezoelectric polymer-MEA based tactile sensing arrays.

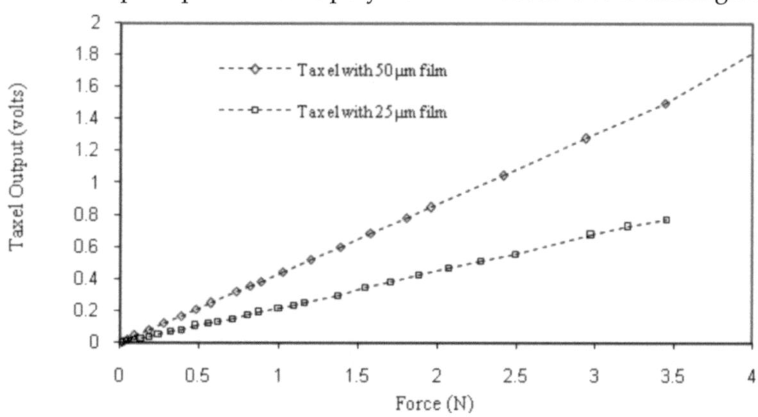

Fig. 3 Average taxels response when variable dynamic force is applied at 15 Hz (Dahiya, Valle et al. 2008).

6. Conclusion and future development

Despite an important role and being a component of robotics roughly as long as vision, the use of touch sensing in robots is lesser than other sensory modalities e.g. vision and auditory sensing, thereby restricting the cognitive capabilities of the robots and strongly limiting their real world interaction capabilities. The lesser usage of touch sensing could partly be attributed to the complex and distributed nature of tactile sensing and partly also to the non availability of satisfactory tactile sensors. The physical problems like placement, robustness of sensors, wiring complexity etc. also pose a hurdle in effective utilization of tactile sensors. The interaction of robots with environment through tactile sensing has largely been limited to the measurement of static interaction forces whereas real world interaction involves both static and dynamic forces. Similarly, most of the sensors are designed to measure static pressure or forces from which it is difficult to obtain information like friction, stickiness, texture, hardness and elasticity. In real world, one needs to measure all these contact parameters which may require use of more than one transduction method simultaneously. As an example measurement of stress and stress rate can be done by having capacitive and piezoelectric transducers. Thus, more and more multifunctional tactile sensors – very few of which have been reported (Engel, Chen et al. 2005) - are required for real world interaction.

The pursuit of tactile sensing for robotic applications, in the last two decades, has resulted in the development of many touch sensors - exploring nearly all modes of transduction -but; none could produce a tactile analog of CMOS optical arrays. Clearly, the emphasis, 'only' on the sensor development has resulted in a large number of 'bench top' sensors – suitable in a laboratory environment, and having limited practical usage in the robotic systems. This is surprising, considering the long history of gripper design for manipulative tasks. It is believed that the lack of the system approach has rendered many of them unusable, despite having a good design and performance (Dahiya, Valle et al. March, 2008). It is evident from the fact that very few works on tactile sensing have taken into account system constraints, like those posed by other sensors or by the robot controller, processing power etc. As an example, large numbers of tactile sensors put a pressure on the computing power required to process large number of data, whereas same can be solved (or at least reduced) by having distributed computing starting right from the transducer level (Dahiya, Valle et al. March, 2008). A system approach for the tactile sensing can be helpful in filling the gaps between tactile sensing and other sense modalities.

7. References

Apple Inc. (2008). "iPodtouch."

Beebe, D. J., A. S. Hsieh, et al. (1995). "A Silicon Force Sensor for Robotics and Medicine." Sensors and Actuators A 50: 55-65.

Berger, A. D. and P. K. Khosla (1991). "Using tactile data for real-time feedback." The International Journal of Robotics Research 10(2): 88-102.

Bicchi, A., J. K. Salisbury, et al. (1990). Augmentation of grasp robustness using intrinsic tactile sensing. IEEE International Conference on Robotics and Automation. Cincinnati, OH. 2: 968-973.

Charlebois, M., K. Gupta, et al. (2000). "On Estimating Local Shape Using Contact Sensing." Journal of Robotic Systems 17(12): 643-658.

Cheung, E. and V. L. Lumelsky (1992). "A Sensitive Skin System for Motion Control of Robot Arm Manipulators." Journal of Robotics and Autonomous Systems 10: 9-32.

Chu, Z., P. M. Sarro, et al. (1996). "Silicon Three-Axial Tactile Sensor." Sensors and Actuators A 54: 505-510.

Cutkosky, M. R. and I. Kao (1989). "Computing and controlling the compliance of a robotic hand." IEEE Transactions on Robotics and Automation 5(2): 151-165.

Dahiya, R. S., G. Metta, et al. (2008). "Tactile Sensing: From Humans to Humanoids." IEEE Transactions on Robotics (unpublished).

Dahiya, R. S., M. Valle, et al. (2008). Tactile Sensing Arrays for Humanoid Robots using Piezo-Polymer-FET devices. 13th National Conference on Sensors and Microsystems, Rome, Italy.

Dahiya, R. S., M. Valle, et al. (2008). Deposition Processing and Characterization of PVDF-TrFE Thin Films for Sensing Applications. IEEE Sensors 2008, Lecce, Italy. (in press).

Dahiya, R. S., M. Valle, et al. (2008). System Approach-A paradigm for Robotic Tactile Sensing. The 10th IEEE International Workshop on Advanced Motion Control. Trento, Italy.

Dahiya, R. S., M. Valle, et al. (2007). Tactile Sensor Arrays for Humanoid Robot. IEEE PRIME'07, The 3rd International Conference on PhD Research in Microelectronics and Electronics, Bordeaux, France, IEEE.

Dargahi, J., M. Parameswaran, et al. (2000). "A Micromachined Piezoelectric Tactile Sensor for an Endoscopic Grasper - Theory, Fabrication and Experiments." Journal of Microelectromechanical Systems 9(3): 329-335.

Dario, P. and D. de Rossi (1985). "Tactile Sensors and Gripping Challenge." IEEE Spectrum 22(8): 46-52.

Diftler, M. A., R. Platt Jr, et al. (2003). Evolution of the NASA/DARPA Robonaut control system. IEEE International Conference on Robotics and Automation. Taipei, Taiwan: 2543-2548.

Domenici, C., D. D. Rossi, et al. (1989). "Shear stress detection in an elastic layer by a piezoelectric polymer tactile sensor." IEEE Transactions on Electrical Insulation 24(6): 1077-1081.

Ellis, R. E. and M. Qin (1994). Singular-value and finite-element analysis of tactile shape recognition IEEE International Conference on Robotics and Automation. San Diego, CA: 2529-2535.

Engel, J., J. Chen, et al. (2005). "Polymer Micromachined Multimodal Tactile Sensors." Sensors and Actuators A 117: 50-61.

Fearing, R. S. (1990). "Tactile Sensing Mechanisms." The International Journal of Robotics Research 9(3): 3-23.

Fearing, R. S. and T. O. Binford (1991). "Using a cylindrical tactile sensor for determining curvature " IEEE Transaction on Robotics and Automation 7(6): 806-817.

Fiorillo, A. S., J. V. D. Spiegel, et al. (1990). "A P(VDF-TrFE) based Integrated Ultrasonic Transducer." Sensors and Actuators A A21-A23: 719-725.

Flanagan, J. R. and A. M. Wing (1993). "Modulation of Grip Force with Load Force during point-to-point arm movements." Experimental Brain Research 95: 131-143.

Gray, B. L. and R. S. Fearing (1996). A Surface Micromachined Microtactile Sensor Array. International Conference on Robotics And Automation, Minneapolis, Minnesota, USA, IEEE.

Heo, J.-S., J.-H. Chung, et al. (2006). "Tactile Sensor Arrays using Fiber Bragg Grating Sensors." Sensors and Actuators A 126: 312-327.

Howe, R. D. (1994). "Tactile Sensing and Control of Robotics Manipulation." Journal of Advanced Robotics 8(3): 245-261.

Howe, R. D. and M. R. Cutkosky (1993). "Dynamic Tactile Sensing: Perception of Fine Surface Features with Stress Rate Sensing." IEEE Transactions on Robotics And Automation 9(2): 140-151.

Howe, R. D. and M. R. Cutkosky (1990). Integrating tactile sensing with control for dextrous manipulation. IEEE International Workshop on Intelligent Motion Control. Istanbul, Turkey. 1: 369-374.

Howe, R. D. and M. R. Cutkosky (1989). Sensing skin acceleration for slip and texture perception. IEEE International Conference on Robotics and Automation. Scottsdale, AZ: 145-150.

Interlink Electronics Inc. (2008). "FSR Sensors."

Jamone, L., G. Metta, et al. (2006). James: A humanoid Robot Acting over an Unstructured World. 6th IEEE-RAS Int. Conf. on Humanoid Robots. Genoa, Italy.

Johannson, R. S. and I. Birznieks (2008). "First Spikes in Ensembles of Human Tactile Afferents Code Complex Spatial Fingertip Events." Nature Neuroscience 7(2): 170-177.

Kageyama, R., S. Kagami, et al. (1999). Development of Soft and Distributed Tactile Sensors and the Application to a Humanoid Robot. IEEE International Conference on Systems, Man, and Cybernetics.

Kane, B. J., M. R. Cutkosky, et al. (2000). "A Traction Stress Sensor Array for Use in High-Resolution Robotic Tactile Imaging." Journal of Microelectromechanical Systems 9(4): 425-434.

Klatzky, R. L. and S. J. Lederman (2003). Touch. Experimental Psychology. A. F. Healy and R. W. Proctor. New York, John Wiley & Sons. 4: 147-176.

Kolesar, E. S., R. R. Reston, et al. (1992). "Multiplexed Piezoelectric Polymer Tactile Sensor." Journal of Robotic Systems 9(1): 37-63.

Lee, M. H. and H. R. Nicholls (1999). "Tactile Sensing for Mechatronics - A State of the Art Survey." Mechatronics 9: 1-31.

Li, Z., P. Hsu, et al. (1989). "Grasping and coordinated manipulation by a multifingered robot hand " International Journal of Robotics Research 8(4): 33-50.

Loomis, J. M. and S. J. Lederman (1986). Tactual Perception. New York, John Wiley and Sons.

Maheshwari, V. and R. F. Saraf (2006). "High-Resolution Thin-Film Device to Sense Texture by Touch." Science 312: 1501-1504.

Milighetti, G., T. Emter, et al. (2006). Combined Visual-Acoustic Grasping for Humanoid Robots. IEEE International Conference on Multisensor Fusion and Integration for Intelligent Systems, Heidelberg, Germany.

Murray, R. M., Z. Li, et al. (1994). A Mathematical Introduction to Robotic Manipulation, CRC.

Novak, J. L. (May 1989). Initial design and analysis of a capacitive sensor for shear and normal force measurement. IEEE International Conference on Robotics and Automation. Scottsdale, AZ: 137-145.

Nowlin, W. C. (1991). Experimental Results on Bayes-ian Algorithms for Interpreting Compliant Tactile Sensing Data IEEE International Conference on Robotics and Automation. Sacramento, California.

Ohka, M., H. Kobayashi, et al. (2006). Sensing Precision of an Optical Three-Axis Tactile Sensor for a Robotic Finger. 15th International Symposium on Robot and Human Interactive Communication (RO-MAN), Hatfield, UK.

Ohmura, Y., Y. Kuniyoshi, et al. (2006). Conformable and Scalable Tactile Sensor Skin for Curved Surfaces. 2006 IEEE International Conference on Robotics and Automation, Orlando, Florida, USA.

Omata, S., Y. Murayama, et al. (2004). "Real time robotic tactile sensor system for the development of the physical properties of biomaterials." Sensors and Actuators A: Physical 112(2-3): 278-285.

Peratech-Ltd, http://www.peratech.co.uk.

Pressure Profile Systems. (2007). "RoboTouch."

Pressure Sensitive Ink. (2008). "http://www.tekscan.com/."

Ricker, S. L. and R. E. Ellis (1993). 2-D finite element models of tactile sensors. IEEE International Conference on Robotics and Automation. Atlanta, GA: 941-947.

Rossi, D. D., G. Canepa, et al. (1993). "Skin-like tactile sensor arrays for contact stress field extraction." Materials Science and Engineering C1: 23-36.

Russell, R. A. and S. Parkinson (1993). Sensing surface shape by touch. IEEE International Conference on Robotics and Automation. Atlanta,GA: 423-428.

Schmidt, P. A., E. Mael, et al. (2006). "A Sensor for Dynamic Tactile Information with Applications in Human-Robot Interaction and Object Exploration." Robotics and Autonomous Systems 54: 1005-1014.

Shikida, M., T. Shimizu, et al. (2003). "Active Tactile Sensor for Detecting Contact Force and Hardness of an Object." Sensors and Actuators A 103: 231-218.

Shimoga, K. B. and A. A. Goldenberg (1992). Soft materials for robot fingers. IEEE Int. Conf. on Robotics and Automation, Nice, France, IEEE, Robot & Automat Soc.

Shimojo, M. (1997). "Mechanical Filtering Effect of Elastic Cover for Tactile Sensor." IEEE Transactions on Robotics And Automation 13(1): 128-132.

Someya, T., T. Sekitani, et al. (2004). " A Large-Area, Flexible Pressure Sensor Matrix with Organic Field-Effect Transistors for Artificial Skin Applications." PNAS 101(27): 9966-9970.

Swartz, R. G. and J. D. Plummer (1979). "Integrated silicon-PVF2 acoustic transducer arrays." IEEE transactions on Electron Devices 26(12): 1920-32.

Torres-Jara, E., I. Vasilescu, et al. (2006). A soft touch: Compliant Tactile Sensors for Sensitive Manipulation. Cambridge, CSAIL, Massachusetts Institute of Technology: 1-8.

Tremblay, M. and M. R. Cutkosky (1993). Estimating friction using incipient slip sensing during a manipulation task. IEEE International Conference on Robotics and Automation. Atlanta, GA: 429-434.

Walker, R. (2004). Developments in Dextrous Hands for Advanced Robotic Applications. 10th International Symposium on Robotics and Applications ISORA 2004. Seville, Spain.

Weiss, K. and H. Worn (2004). Tactile Sensor System for an Anthropomorphic Robotic Hand. IEEE International Conference on Manipulation and Grasping Genoa, Italy.

Westling, G. and R. S. Johannson (1984). "Factors influencing the force control during precision grip." Experimental Brain Research 53: 277-284.

Woffenbuttel, M. R. and P. P. L. Regtien (1991). "Polysilicon bridges for the realization of tactile sensors." Sensors and Actuators A-Phys. 26: 257-264.

Yuji, J.-i. and C. Sonoda (2006). A PVDF Tactile Sensor for Static Contact Force and Contact Temperature. IEEE Sensors 2006, Daegu, Korea.

Zhang, H. and E. So (2002). "Hybrid Resitive Tactile Sensing " IEEE Transactions on Systems, Man, And Cybernetics-Part B: Cybernetics 32(1): 57-65.

16

Fast and Accurate Tactile Sensor System for a Human-Interactive Robot

Toshiharu Mukai, Shinya Hirano and Yo Kato
Bio-Mimetic Control Research Center, RIKEN
Japan

1. Introduction

With the advent of the aging society, the demand for nursing care for the elderly is becoming much larger. The application of robotics to helping on-site caregivers is consequently one of the most important new areas of robotics research. Such human-interactive robots, which share humans' environments and interact with them, should be covered with soft areal tactile sensors for safety, communication, and dextrous manipulation.

Tactile sensors have interested many researchers and various types of tactile sensors have been proposed so far. Many tactile sensors have been developed on the basis of micro-electro-mechanical system (MEMS) technology (for example, (Suzuki, 1993; Souza & Wise, 1997)). They have a high-density and narrow covering area realized by applying MEMS technology, and as a result, are not suitable for covering a large area of a robot's surface. Some tactile sensors suitable for use on robot fingers or grippers have also been developed (Nakamura & Shinoda, 2001; Yamada et al., 2002; Shimojo et al., 2004). Many of them have the ability to detect tangential stress and can be used in grasping force control. Their main target is robot fingers, and consequently they were not designed to cover a large area. There are also commercially available tactile sensors such as those offered by Tekscan (Tekscan, 2008) based on pressure-sensitive ink or rubber, and KINOTEX™ tactile sensors (Reimer & Danisch, 1999) utilizing the change in the intensity of light scattered by the covering urethane foam when deformed. However, they are not sufficiently accurate because of strong hysteresis and creep characteristics.

The idea of covering a large area of a robot's surface with soft tactile skinlike sensors is attracting researchers (Lumelsky et al., 2001). Some human-interactive robots for which a large area of their surface is covered with soft tactile sensors have actually been developed (Inaba et al. 1996; Tajima et al. 2002; Kanda et al. 2002; Mitsunaga et al. 2006; Ohmura et al., 2006; Ohmura & Kuniyoshi, 2007). However, the tactile sensors are not suitable for human-interactive robots, particularly when physical labor using tactile sensation is required. For example, one tactile sensor in (Tajima et al. 2002) has only 3 values as its output, and another tactile sensor in (Tajima et al. 2002) is gel-type and cannot be used over a long period because of the evaporation of the contained water. The tactile sensor in (Mitsunaga et al. 2006) has only 56 elements in total. Flexible fabric-based tactile sensors using an electrically conductive fabric have also been proposed for covering a robot (Inaba et al. 1996), but the sensors are binary switches, and are difficult to fabricate. To our knowledge, the tactile sensor used by (Ohmura et al., 2006) has been the most successful for covering a large area of a robot's surface. This tactile sensor is based on similar technology to KINOTEX™, but uses tiny photoreflectors under urethane foam instead of the combination of light-emitting diodes, photodetectors, and fiber-optic cables. The tactile sensor is fabricated on a flexible substrate using photoreflectors and circuitry for driving and communication. Their sensor is suitable for detecting tactile contact, but the principle of their sensor depends on the deformation of urethane foam, which inevitably causes strong hysteresis and creep characteristics. Hence, when accuracy is needed, for example, when tactile feedback control is required, we need more accurate tactile sensors.

For the realization of a robot with tactile sensation, a real-time acquisition and integration system for gathering tactile sensor data distributed all over the robot's body is also needed. When a robot is covered with tactile sensors, the problems of huge computational load and a large number of cables arise. Localized processing is an effective solution, and tactile sensors with distributed processors have been developed. For example, the robot in (Tajima et al. 2002) had multiple MPUs connected by a serial bus, and that in (Ohmura et al., 2006) used a network composed of local serial buses and a custom-designed ring-type network.

We are developing a robot named RI-MAN (Odashima et al., 2006; Odashima et al., 2007; Mukai et al., 2008), shown in Fig. 1, with soft areal tactile sensors, as a platform for physical human-robot interaction research. Our ultimate goal for RI-MAN is to help nurse elderly people in their daily lives, and this inevitably involves performing hard physical tasks in complex environments such as hospitals and homes. RI-MAN can physically interact with humans via soft and whole-body interaction. Having a similar size to a human (158 cm in height and 100 kg in weight) and a smooth surface without bumps, RI-MAN can perform the task of lifting up a dummy human in its arms. To skillfully perform such tasks that involve physical contact with humans, RI-MAN is equipped with soft areal tactile sensors in five places, the chest and the right and left of the upper arms and forearms.

Fig. 1. RI-MAN with tactile sensors for physical human-robot interaction research

We have developed soft areal tactile sensors for RI-MAN by embedding semiconductor pressure sensors as pressure-sensing elements in an elastic body (Mukai, 2004; Mukai, 2005), as well as their controllers that can realize localized processing and integration through the network in a robot. Our primary target is the achievement of physical tasks using tactile sensation, such as lifting up a human. Because semiconductor pressure sensors have little hysteresis or creep, our sensors also have an accurate response, although the elastic body covering causes a little hysteresis and creep. The sensor should be able to detect wide range of pressure from light finger touch to the load of a human held by the robot arms. The sampling speed and accuracy should be enough for usage in sensor feedback control. Our tactile sensors satisfy the requirements for human-interactive robots.

In this paper, we report our tactile sensor, as well as its implementation in a robot. After this introduction, we first describe specifications necessary for tactile sensors in human-interactive robots. Next we describe the structure and fabrication method of our tactile sensor. Next we describe the implementation of the tactile sensor system in our robot RI-MAN. Then we report experimental results of the sensor itself and sensor usage in the robot. Finally we conclude our paper.

2. Specifications

In this research, we are aiming to develop soft areal tactile sensors that cover the surface of a human-interactive robot, and provide a pressure strength and distribution that can be used to control the manipulation of objects. To make this possible, we need sensors that are soft like human skin and can cover a large area of a robot's curved surfaces, for safety and affinity. We are developing such sensors by embedding small semiconductor pressure sensors as pressure-sensing elements in an elastic body. The elastic body also contributes to the removal of undesirable insensitive regions by giving interpolation ability to the sensor. The preciseness of semiconductor pressure sensors can yield high-accuracy tactile sensors. Considering the application of tactile sensors to human-interactive robots, we determined the necessary specifications of our tactile sensor system as follows. First, we require a spatial resolution of about 10 to 20 mm, referring to the approximately 10 mm resolution of the human palm and 40 mm of the human forearm. Except for the fingertips, this resolution is sufficient to realize a resolution similar to that of a human. Next, we consider the measurable pressure range. When a robot holds a person of 60 kg in its arms with a contact area of 20×20 cm^2, the average pressure is 14.7 kPa. Using a safety factor of 6, we require a pressure range of 0 to 88 kPa. Next, we discuss the measurement resolution. Considering feedback control using tactile sensation, we require a measurement resolution in which the contact force caused by typical human-robot interaction is expressed by more than 5 bits. We believe that feedback control is smooth if the value is expressed with this resolution.

Tactile sensors for robots require tactile sensor controllers suitable to be used in robots. It is desirable that each controller be in the vicinity of the tactile sensor, with short connections between sensors and controllers, because the connection requires a large number of cables. This also reduces electric noise because tactile sensor output is analog, while we can use digital signals for connections among controllers. This requires the following. Controllers must be able to form a network for sensor data integration in the robot. Controllers must be sufficiently small to fit in a robot. Controllers must calculate abstract compressed features from sensor data. Such compressed features are carried through a narrower bandwidth, to simplify connection in robot and reduce the possibility of cable breakdown. We also require that the sampling speed be sufficiently fast for our robotic purposes. The sampling speed depends on the performance of both the sensor and its controller, so that the total system should be designed to satisfy this requirement. The sampling speed of sensor data and the communication speed among controllers should agree with the robot control period. It is empirically known that, for stable and smooth feedback control, the sampling frequency should be more than 10 times the resonance frequency of the controlled object (Paul, 1981). In general, robots have a resonance frequency of 1 to 50 Hz, so that the required sampling frequency is 10 to 500 Hz, which corresponds to a sampling period of 2 to 100 ms (RSJ, 2005).

3. Structure and fabrication method of the tactile sensor

3.1 Basic structure and components

As pressure-sensing elements, we adopted FUJIKURA FPBS-04A pressure sensors (Fig. 2). These are very small (φ5.8 mm) piezoresistive semiconductor pressure sensors that can detect absolute pressure between 42.6 and 434.7 kPa. Each sensor has a resistive bridge on its diaphragm and the output changes almost linearly with the applied pressure.

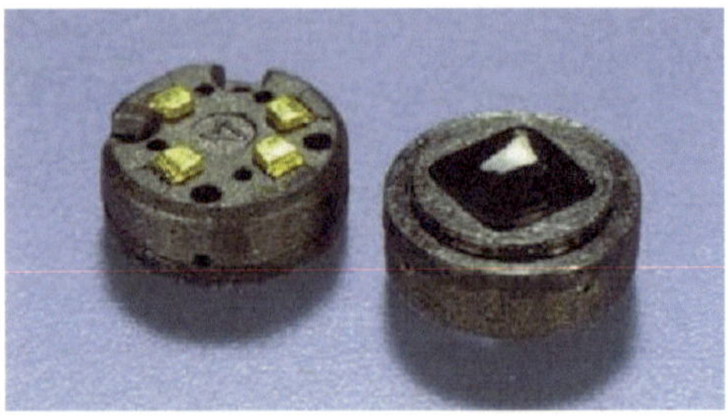

Fig. 2. Small FUJIKURA semiconductor pressure sensors

To reduce the number of cables leading out from the tactile sensors, we designed a scanning circuit, an example of which including 4x4 elements is shown in Fig. 3. By changing switches, we can select a row and a column to obtain the output from the selected element. In this example, the top-left element is selected by switching. All outputs of the pressure-sensing elements are obtained by scanning this array. The actual circuit we made has an array of 8x8 pressure-sensing elements with an 18 mm pitch. This pitch was determined to satisfy the above requirement to realize a spatial resolution similar to a human. We fabricated the circuit as a flexible printed circuit (FPC), so that it can be fit onto a 2D-curved surface (which can be expanded into a 2D plane) because of its flexibility. We call the FPC sheet with the pressure-sensing elements and other electronic parts as 'sensor sheet'. To mount it onto a free-curved surface (which cannot be expanded into a 2D plane), we designed the circuit so that all the wiring is concentrated into a comb-shaped region. Thus, the sensor sheet can be cut into a shape consisting of thin regions. After cutting, the sheet can be wound onto a free-curved surface, as shown in Fig. 4, if the curvature is not too sharp. In addition, we designed the circuit all of which wiring concentrates into a comb-shaped region, so that it works if the central part remains even when the marginal parts are cut off. Therefore it can be applied to a smaller surface than the original sensor sheet. To reduce cables, we also put analog switches and an instrumentation amplifier on the FPC. As a result, the number of cables has become 10. As the instrumentation amplifier, we adopted a digital amplifier that can select its gain from 1, 2, 4, and 8. If this amplifier gain needs to be controlled, additional two cables are needed.

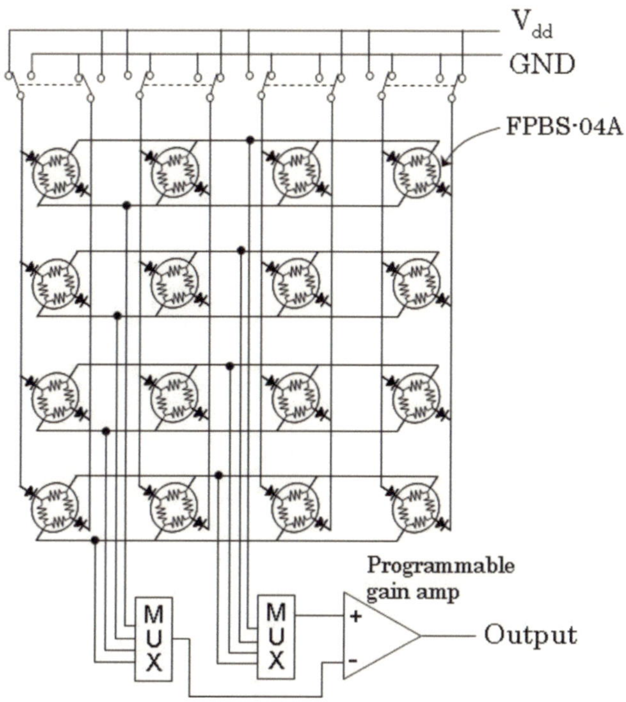

Fig. 3. Circuit of our tactile sensor

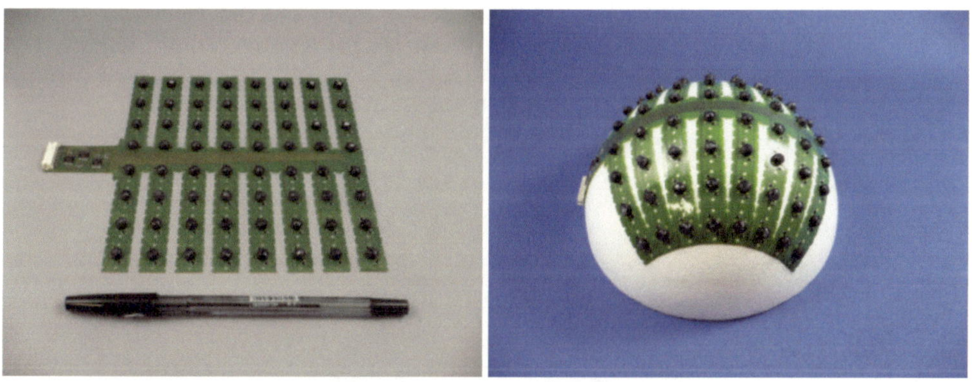

Fig. 4. Sensor sheet that can be cut into a comblike shape and wound on a curved surface

3.2 Method of fabricating a curved tactile sensor

To make the tactile sensors soft, we embedded the sensor sheet in an elastic body.

First, the sensor sheet is fixed after it is wound around the curved surface. Next, to remove gaps between the shells of pressure-sensing elements, the gaps are filled with a plastic material that hardens after the desired form is reached. Then, a small projection-shaped piece of elastic material is placed on the diaphragm of each pressure-sensing element. After that, an elastic sheet covers the above assembly. Fig. 5 shows a schematic of the structure of our curved tactile sensor. We used paper clay to fill the gaps, liquid rubber to fabricate the projection-shaped pieces, and a 5-mm-thick sponge sheet as the elastic sheet.

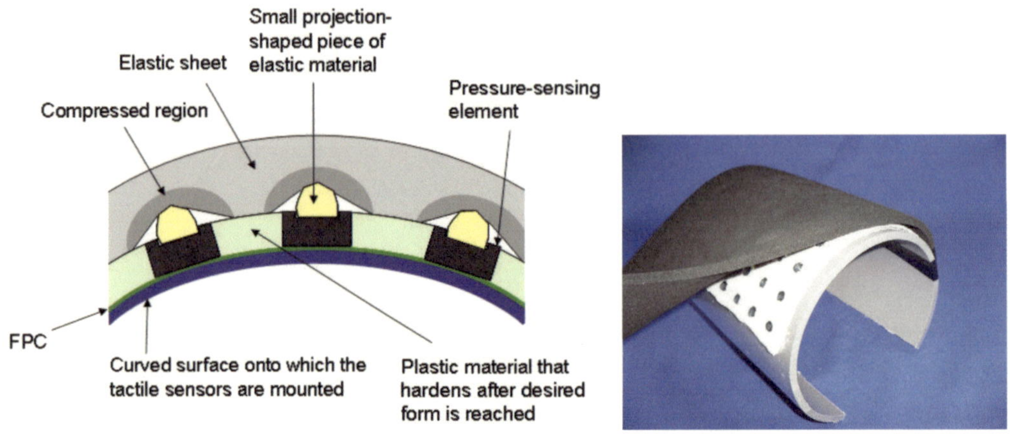

Fig. 5. Schematic structure and a photo of our curved tactile sensor

We selected the material of the elastic projection-shaped pieces to be harder than the elastic sheet. As a result, the small pieces push up against the elastic sheet, and the regions around the pieces in the sheet are compressed and become denser than the other regions. Thus, hard and soft areas appear in the elastic sheet (Fig. 5), and this structure has an amplifying effect (Mukai, 2005). A photograph of our curved tactile sensor is also shown in Fig. 5.

4. Implementation of a tactile sensor system

We installed the tactile sensors in our robot RI-MAN in five places, the chest and the right and left of the upper arms and forearms. Each sensor sheet has 8x8 elements, so that RI-MAN has 320 pressure-sensing elements in total. To process enormous sensor data in realtime in the vicinity of sensor, we developed a tactile sensor controller based on a dsPIC board (Landmark LM517), a picture of which is shown in Fig. 6.

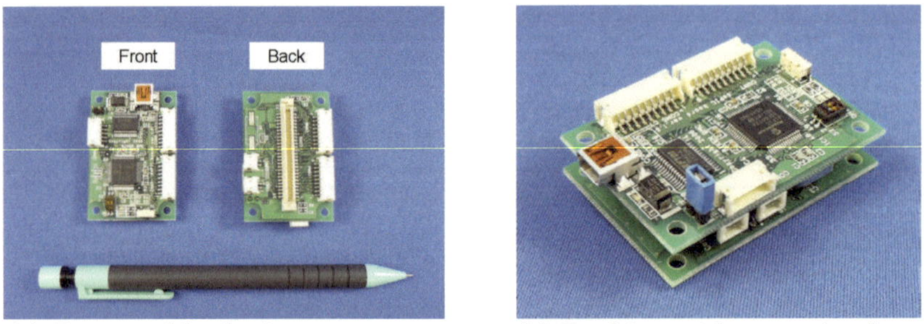

(a) Front and back view (b) Stacking with extension board

Fig. 6. dsPIC board for general sensing/controlling purposes

The dsPIC board is a general purpose sensing/controlling board developed by us, with a size of 35x50 mm^2. It is equipped with a dsPIC (Microchip technology 30F6012A) as the CPU and a USB interface IC (FTDI FT232RL) that realizes 1 Mbps communication. It also has many connectors through which many pins can be accessed. It also has a stacking connector for an extension board. The tactile sensor controller we developed uses a 12 bit A/D converter and digital I/Os in the 30F6012A. Its program is written in the C language and downloaded in the flash memory of the 30F6012A, as the firmware. The C program determines how to compress data from a tactile sensor sheet into abstract features suitable for network communication.

Using this board, up to 1 ms sampling of 1 tactile sensor sheet can be realized. We also developed an extension board for network communication, to connect the tactile sensors to the network in our robot RI-MAN. It has a network IC (Step Technica MKY40) that realizes a virtual shared memory, the maximum size of which is 512 byte. The guaranteed synchronization time of the virtual shared memory depends on the memory size, and is 1 ms for 256 byte and 2.4 ms for 512 byte. The extension board can be stacked with the dsPIC board as shown in Fig. 6(b). The small size and the network communication ability of our tactile sensor controller enable us to install them distributedly in the robot body. The communication can be realized using only 2 cables, which largely simplifies wiring and reduces maintenance cost in the robot.

In the virtual shared memory space, each node occupies an area consisting of stations one of which corresponds to 8 byte. In RI-MAN, various sensor controllers and motor controllers are connected to the network and, as a result, the area that tactile sensors can use is limited. In the current implementation, the host PC (with RT-Linux) occupies 1 station to send commands (for setting modes and parameters) through network to tactile sensors, and each of 5 tactile sensor controllers occupies 3 stations.

Each tactile sensor has only 3 stations, corresponding to 24 byte, for communication, thus all of 8x8 sensor element values (which is expressed by 12 bit) cannot be transmitted simultaneously. It is possible to send data line by line and combine them in the host PC to obtain whole output of one sensor sheet, but this takes unacceptable time (about 90ms) for feedback control. So we determined to calculate geometric moments in local controllers and transmit only the moments through network. The geometric moments are expressed, using element value I_{xy} at (x,y), as follows.

Zeroth: $M_0 = \sum_{x,y} I_{xy}$

First: $M_x = \sum_{x,y} xI_{xy}$, $M_y = \sum_{x,y} yI_{xy}$

Second: $M_{xy} = \sum_{x,y} xyI_{xy}$, $M_{x^2-y^2} = \sum_{x,y} (x^2 - y^2)I_{xy}$

The zeroth moment is the total sum itself, and the centroids (center of pressure) can be obtained by dividing the first by the zeroth. Using up to the second moments, we can obtain the axis of least inertia (Horn, 1986), which is the orientation of the object, as shown in Fig. 7. In the controller using dsPIC, all of data acquisition and calculation finishes in 2 ms.

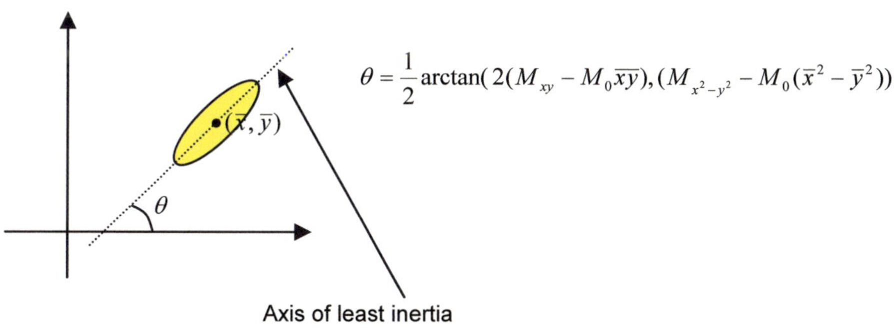

Fig. 7. The axis of least inertia

5 Experimental results

5.1 Basic performance of the sensor

We conducted experiments to investigate the basic performance of our tactile sensor system. Flat tactile sensors with the 5-mm-thick elastic sheet were used to investigate the basic performance. In experiments where a dynamic force was needed, the experimental setup shown in Fig. 8 was used. The setup consisted of a digital force gauge, a load cell (Minebea LSM-1K-B), a vibrator (Asahi SeisakusyoWaveMaker01), and an XY stage. The force gauge was used to approximately measure the applied force to ensure that an excessive force was not exerted. Actual measurement of the applied force was carried out using the load cell, which has higher accuracy. The vibrator was used to generate controlled dynamic motion, and the force was exerted using a rod with a circular tip with a diameter of 8 mm. In experiments where a force over a wider area was needed, weights were used. However, in this case, the force was limited to a static force. Outputs from pressure-sensing elements were digitized by an A/D converter in the tactile sensor controller and sent to a PC through USB connection. These values are referred to as 'sensor output'. The unit of sensor output corresponds to 0.00122 V, and the sampling period of the measurement was 1 ms. The amplifier gain of sensor sheets were fixed to 8 in all experiments.

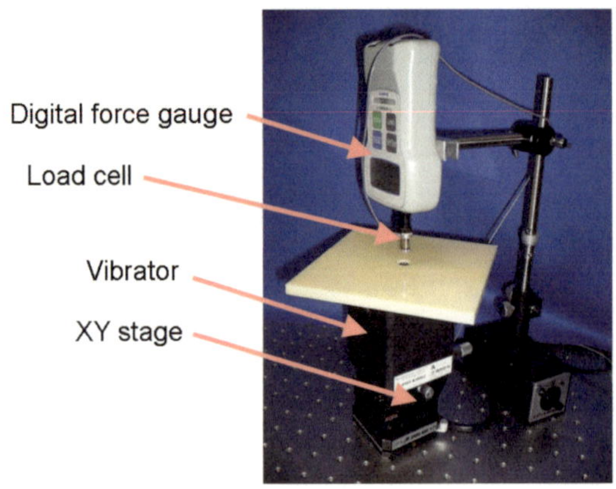

Fig. 8. Experimental setup

a) Response to a sinusoidal wave

First, we show, in Fig. 9, the measured time response of a pressure-sensing element (with the covering sheet) along with the applied force, when vibrated with a 1, 10, or 100 Hz sinusoidal wave. The force was applied from immediately above the element using the rod with the 8 mm circular tip. We can see that the time response has a smooth sinusoidal shape at 1 and 10 Hz, though the shape is slightly deformed at 100 Hz. A typical relationship between the applied force and the sensor output at 1 Hz is shown in Fig. 10. The sensor output is almost linear with the applied force. The solid line in the graph is the linear approximation $y = ax$ based on the results (from 3000 samples), where $a = 13.35$ in this case. This slope is referred to as gain in this paper. From this experiment, we can also see that our sensor has satisfactory measurable range and measurement resolution discussed in Section 2. When a force of 8 N was applied using the rod with the 8 mm circular tip, the pressure was about 159 kPa and the sensor output was more than 90. These results satisfy the required maximum range of 88 kPa and 5 bit resolution.

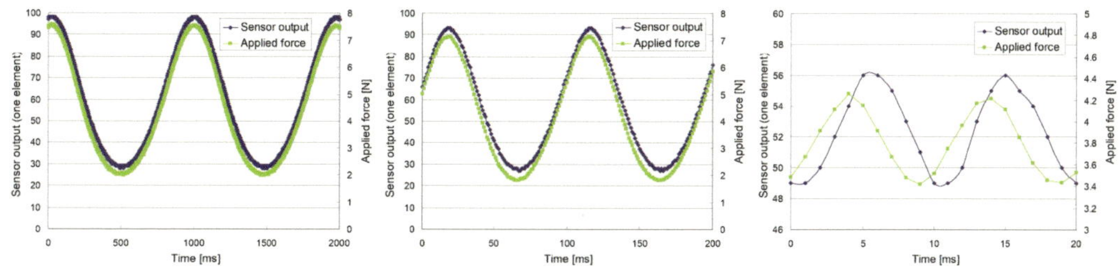

(a) 1 Hz (b) 10 Hz (c) 100 Hz

Fig. 9. Response of a pressure-sensing element to sinusoidal force

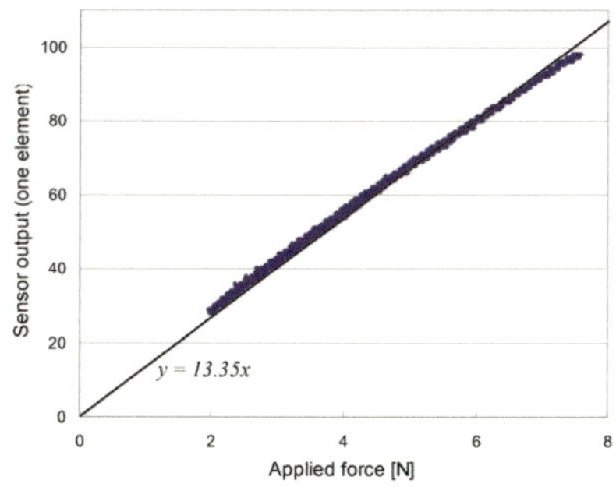

Fig. 10. Force vs sensor output

b) Variations of pressure-sensing elements

Next, we checked the variations of pressure-sensing elements on a sensor sheet. A sinusoidal force of 1 Hz was applied from immediately above the element using the rod with the 8 mm circular tip. The results of 20 elements are shown in Fig. 11. The standard deviation is about 6.5% of the averaged gain. The gains varied partly because we manually fabricated the projection-shaped elastic pieces on the diaphragms. We expect that if we make them by a more controlled method, the variation will decrease.

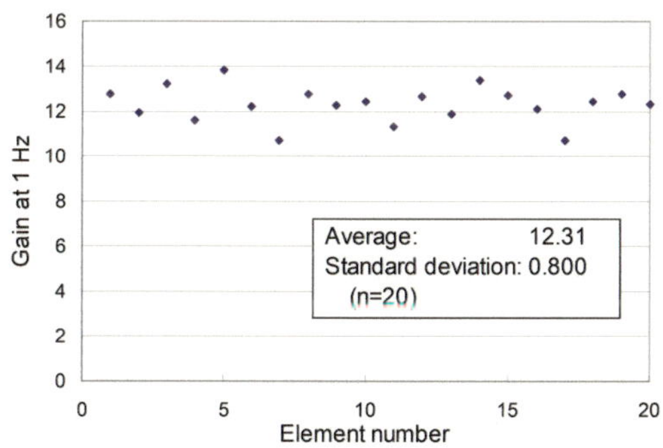

Fig. 11. Variations of elements on a sensor sheet

c) Frequency response

The results of experiments to investigate frequency response are shown in Fig. 12. The output was measured from a certain pressure-sensing element in a tactile sensor, and gain was normalized to 1 at 1 Hz. The decrease of gain at 100 Hz is 3 dB and the phase delay is 20 degree, approximately. These results show that our sensor can detect high-frequency signals.

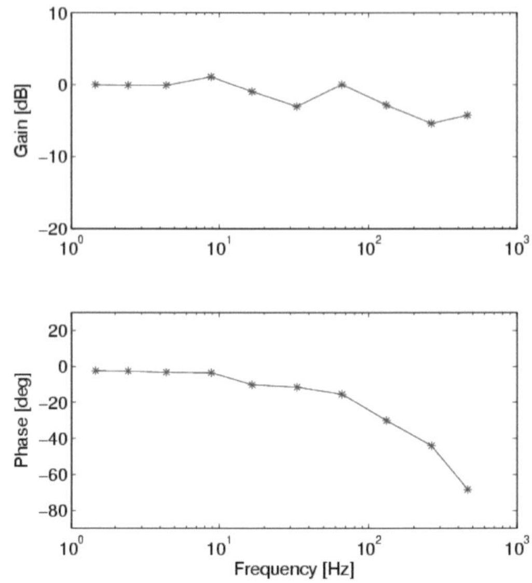

Fig. 12. Frequency response of a sensing element

d) Interpolation ability

Interpolation ability is obtained because of the covering elastic sheet even though discrete pressure-sensing elements are located separately. This removes undesirable insensitive region on the tactile sensor. Experimental results for verifying this effect are shown in Fig. 13. By applying a force using the rod with a tip diameter of 8 mm at a certain distance away from the element, gains in the graph were measured. The gain was normalized to 1 at the distance 0. The response shape agrees with the theoretically calculated values in (Shimojo, 1997).

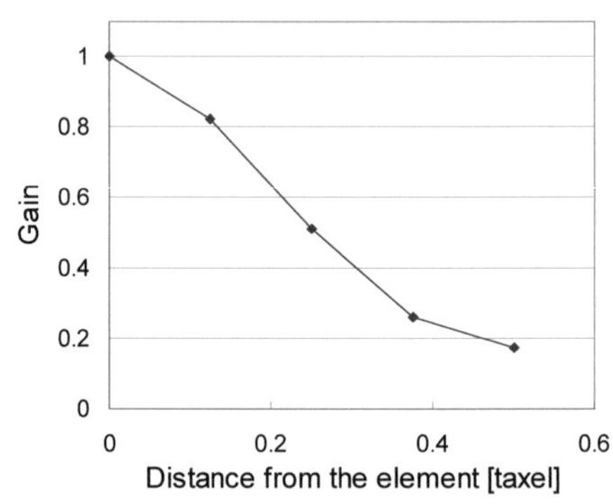

Fig. 13. Results of interpolation test

e) Hysteresis and drift

The results of experiments for investigating hysteresis are shown in Fig. 14. A load of 6.75 N approximately was applied and removed in about 5 s. Hysteresis exists but it is small. Next, to see longer-term effect (drift), a 668 g weight with an area of 52x44 mm^2 was placed on the sensor for 10 minutes. The sum of outputs from all elements is shown in Fig. 15. The output reduced slightly but it is about 4% of the initial output.

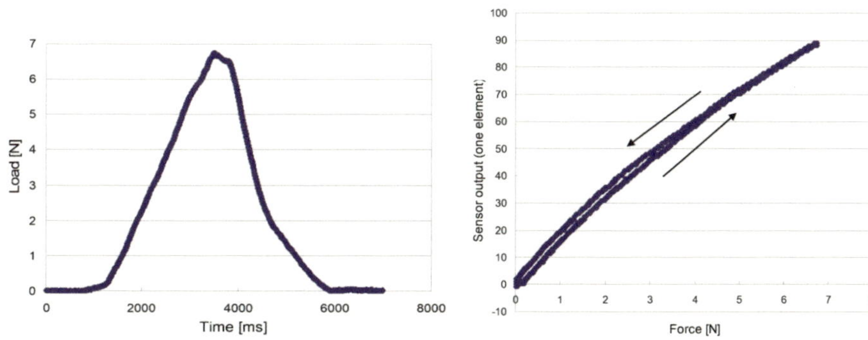

(a) Time profile of applied force (b) Force vs sensor output

Fig. 14. Results of hysteresis test

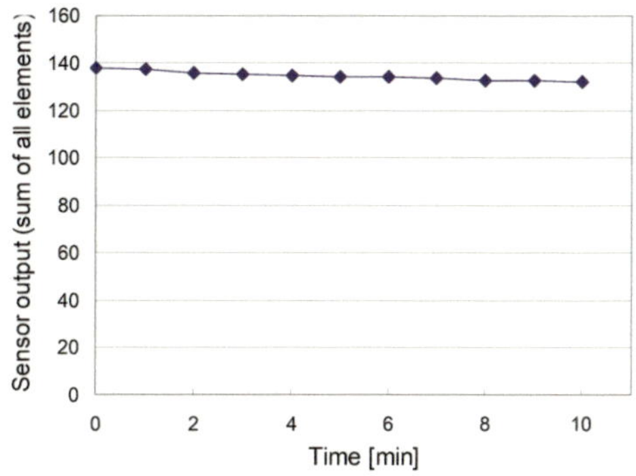

Fig. 15. Drift over 10 minutes

5.2 Tactile sensors in a robot

The output of all tactile sensors when holding a dummy human weighing 18 kg in its arms is shown in Fig. 16. The size of each square represents the amplitude of the corresponding tactile sensor element in log scale. This data were obtained and recorded in the host PC by

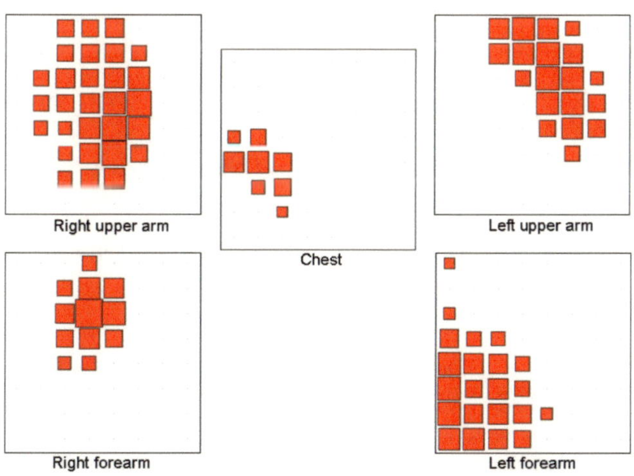

Fig. 16. Tactile sensor output when holding a dummy human weighing 18 kg

line-by-line transmission. The transmission took about 90 ms, but data acquisition itself in local tactile sensor controllers finished in 1 ms.

Next, we investigated the relationship between the roll angle and the sensor output when RI-MAN rotated its waist joint right and left 15 degrees while holding the 18 kg dummy human (Fig. 17). The results are in Fig. 18, where the realized roll angle and the pressure centroid on the left forearm are shown in graph (a), and the loads (the sum of all element outputs) detected by all tactile sensors in graph (b). In the sensor coordinates, the x-axis is along the left forearm from the base to the end, and the y-axis is from outside to inside. We can see that tactile sensors succeeded in detecting the changes of the object load and position caused by the roll angle change. The data show that the position of the dummy human shifted little by little during the cycles, even though the waist angle repeated the same motion.

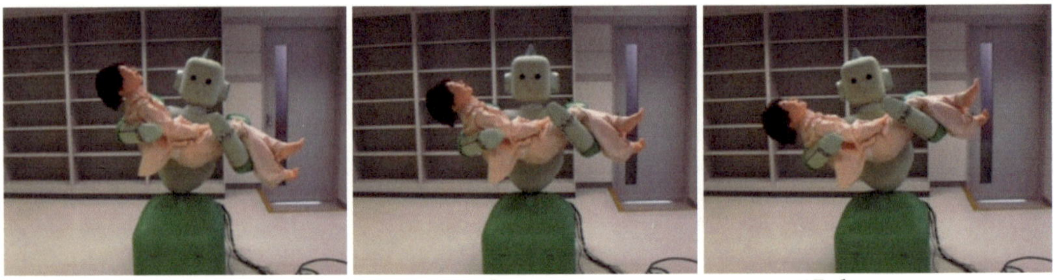

+15 degrees 0 degrees -15 degrees

Fig. 17. Experiment scenes to obtain sensor output from a dummy human

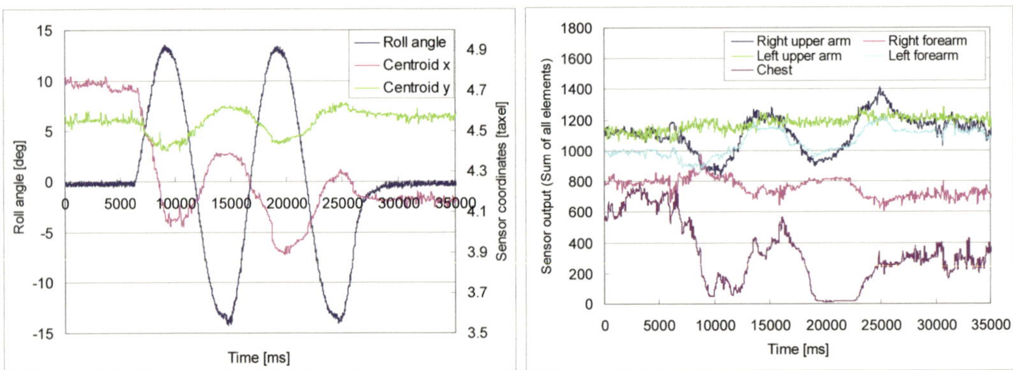

(a) Roll angle and the centroids on the left forearm

(b) Load on each tactile sensor

Fig. 18. Sensor output caused by a dummy human when waist angle is changed

7. Conclusion

In this paper, we described the successful development of the tactile sensor system of our human-interactive robot RI-MAN. The necessary specifications for RI-MAN tactile sensors are satisfied by our tactile sensor system. RI-MAN, equipped with the tactile sensors, succeeded in detecting the magnitude and position of the load when it was holding a dummy human.

For future work, we will develop control methods for manipulating objects with arms using tactile sensor feedback. We will also develop sensor signal processing methods for detecting not only physical contact but also human tactile sensations such as hitting and stroking. This enables the tactile sensor to be used in human-robot communication, as well as in physical labors.

7. References

Horn B. K. P. (1986). *Robot Vision,* MIT Press, Cambridge, MA

Inaba, M.; Hoshino, Y.; Nagasaka, K.; Ninomiya, T.; Kagami, S. & Inoue, H. (1996). A full-body tactile sensor suit using electrically conductive fabric and strings, in *Proc. IEEE/RSJ International Conference on Intelligent Robots and Systems,* pp. 450–457

Kanda, T.; Ishiguro, H.; Ono, T.; Imai, M. & Nakatsu, R. (2002). Development and evaluation of an interactive humanoid robot 'Robovie', in *Proc. IEEE International Conference on Robotics and Automation,* pp. 4166–4173

Lumelsky, V. J.; Shur, M. S. & Wagner, S. (2001). Sensitive skin, *IEEE Sensors Journal,* Vol. 1, No. 1, pp. 41–51

Mitsunaga, N.; Miyashita, T.; Ishiguro, H.; Kogure, K. & Hagita, N. (2006). Robovie-IV: A communication robot interacting with people daily in an office, in *Proc. IEEE/RSJ International Conference on Intelligent Robots and Systems,* pp. 5066–5072

Mukai, T. (2004). Soft areal tactile sensors with embedded semiconductor pressure sensors in a structured elastic body, in *Proc. IEEE International Conference on Sensors,* pp. 1518–1521

Mukai, T. (2005). Development of soft areal tactile sensor on curved surfaces, in *Proc. IEEE International Conference on Robotics and Biomimetics,* pp. 14-18

Mukai, T.; Onishi, M.; Odashima, T.; Hirano, S. & Luo, Z. (2008). Development of the Tactile Sensor System of a Human-Interactive Robot "RIMAN", *IEEE Trans. on Robotics,* Vol. 24, No. 2, pp.505-512

Nakamura, K. & Shinoda, H. (2001). A tactile sensor instantaneously evaluating friction coefficients, in *Proc. International Conference on Solid-State Sensors and Actuators,* Vol. 2, pp. 1430–1433

Odashima, T.; Onishi, M.; Tahara, K.; Takagi, K.; Asano, F.; Kato, Y.; Nakashima, H.; Kobayashi, Y.; Luo, Z. W.; Mukai, T. & Hosoe, S. (2006). A soft human-interactive robot – RI-MAN –, in *Video Proc. IEEE/RSJ International Conference on Intelligent Robots and Systems*

Odashima, T.; Onishi, M.; Tahara, K.; Mukai, T.; Hirano, S.; Luo, Z. & Hosoe, S. (2007). Development and evaluation of a human interactive robot platform 'RI-MAN', *Journal of the Robotics Society of Japan,* Vol. 25, No. 4, pp. 554–565 (in Japanese)

Ohmura, Y.; Kuniyoshi, Y. & Nagakubo, A. (2006). Conformable and scalable tactile sensor skin for curved surfaces, in *Proc. IEEE International Conference on Robotics and Automation,* pp. 1348–1353

Ohmura, Y. & Kuniyoshi, Y (2007). Humanoid robot which can lift a 30kg box by whole body contact and tactile feedback, in *Proc. IEEE/RSJ International Conference on Intelligent Robots and Systems,* pp. 1136-1141

Paul, R. P. (1981). *Robot manipulators: Mathematics, programming, and control,* MIT Press, Cambridge, MA

Reimer, E. M. & Danisch, L. (1999). Pressure sensor based on illumination of a deformable integrating cavity, U.S. Patent 5917180

RSJ (the Robotics Society of Japan) ed. (2005). *Robotics handbook - 2nd ed.,* CORONA Publishing, Tokyo, pp.652–654 (in Japanese)

Shimojo, M. (1997). Mechanical filtering effect of elastic cover for tactile sensor, *IEEE Trans. Robotics and Automation,* Vol. 13, No. 1, pp. 128–132

Shimojo, M.; Namiki, A.; Ishikawa, M.; Makino, R. & Mabuchi, K. (2004). A tactile sensor sheet using pressure conductive rubber with electrical-wires stitched method, *IEEE Sensor Journal,* Vol. 4, No. 5, pp. 589-596

Souza, R. J. D. & Wise, K. D. (1997). A very high density bulk micromachined capacitive tactile imager, in *Proc. of International Conference on Solid-State Sensors and Actuators,* pp. 1473–1476

Suzuki, K. (1993). High-density tactile sensor arrays, *Advanced Robotics,* Vol. 7, No. 3, pp. 283–287

Tajima, R.; Kagami, S.; Inaba, M. & Inoue, H. (2002). Development of soft and distributed tactile sensors and the application to a humanoid robot, *Advanced Robotics,* Vol. 16, No. 4, pp. 381–397

Tekscan (2008). http://www.tekscan.com/medical/systems.html

Yamada, D.; Maeno, T. & Yamada, Y. (2002). Artificial finger skin having ridges and distributed tactile sensors used for grasp force control, *Journal of Robotics and Mechatronics,* Vol. 14, No. 2, pp. 140–146

17

Development of a Humanoid with Distributed Multi-axis Deformation Sense with Full-Body Soft Plastic Foam Cover as Flesh of a Robot

Marika Hayashi, Tomoaki Yoshikai and Masayuki Inaba
The University of Tokyo,
Japan

1. Introduction

In order for humanoid robots to work around human, it is important to enable them to be touched by humans, and to sense its contact states throughout the body. Almost all of full-body tactile sensors of robots ever developed were sensors which detect distribution of vertical pushing force on its thin "skin". Existing whole-body tactile sensors were called "skin" (V. J. Lumelsky, 2001), "cover"(H. Iwata, 2002), "suit"(M. Inaba et.al., 1996).
About sixty percent of human body is made of soft organs such as skin and muscles. Human can sense multi-dimensional deformation on its thick deformable "flesh". As a machine modeled after human and working automatically around us, it can be considered effective to have human-like thick soft deformable exterior with multi-axis tactile sensors.
In this research, we aim to obtain a method to construct both a robot skeleton and a whole-body soft sensor exterior as "flesh" . We are developing a robot named "macra" (Hayashi et.al., 2007;) with soft thick exterior, shown in Fig.1.

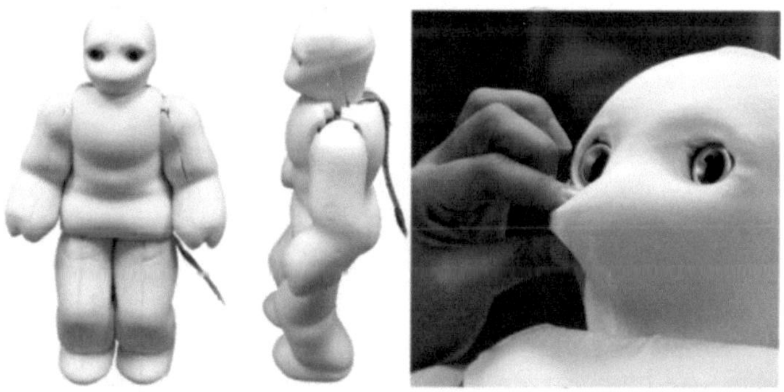

Fig. 1. A humanoid robot which has soft exterior parts with distributed force/torque sensors

2. Deformation sensing "Flesh"

Considering implementation of multi dimensional sensors, spatial limitation of soft cover is a problem. Thin tactile sensors often come up important theme. Through the "flesh" approach, it is possible to locate relatively thick sensors which sense distribution of multi-

axis force vectors, because the spatial constraint can be eased by thick exterior. There are some sensors which sense multi-axis force vector. For example, three-axis force/torque sensor which is used as sole sensor, acoustic resonant tensor cell sensor which sense six-axis stress sensor (H. Shinoda et.al.), vision-based tactile sensor(K. Kamiyama et.al. 2004).
Adopting this kind of sensor composition, it can sense not only distribution of vertical pushing force (Fig.2-a), but also shearing force (Fig.2-b) and pulling force (Fig.2-c). When the robot is grabbed its arm like Fig.2-a/b, the robot can detect whether it is pulled forward or pushed backward. The soft cover enables us to emulate local deformation such as pinching.

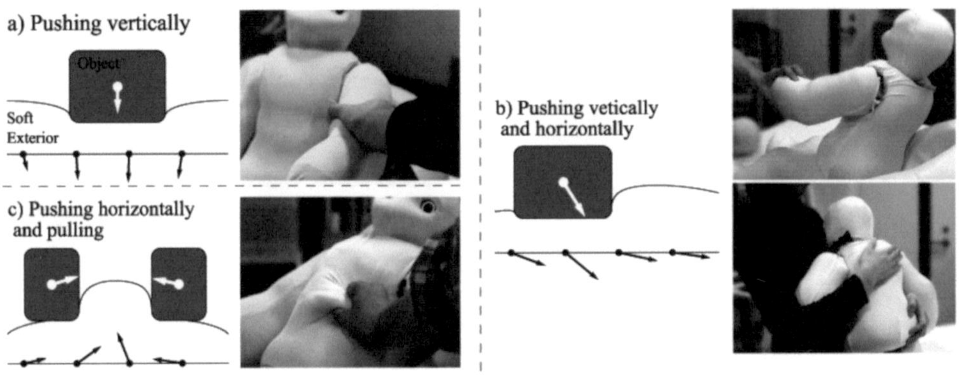

Fig. 2. If a robot has thick "flesh", it can predict its next contact state without actually changing the contact position.

3. Constitution method of full-body cover made of thick foam

3.1 Soft material which is light and possible to form its shape

Molded soft polyurethane foam is adopted for material of soft cover, because it is light and easy to fabricate. Its elastic characteristics can be customized by changing ratio of main component and curing agent. Molding method which enables precise production of exterior shape is adopted, because precise and reproducible production is important for sensing. Other representative examples of soft foam plastics are made of polyethylene, rubber. Polyethylene foam and rubber foam are produced by kneading base compound and bloating agent and then heating them, which are relatively difficult to produce experimentally. It is necessary to form films on the surface of it to avoid significant deterioration. In-mold coating and film laminating has fine texture and durability capacity. But these methods require precise temperature and pressure management, and expensive mold. Thus, we employed stretchable fabric cover to put on the molded foam.

3.2 Softness adjustment of soft plastic foam cover

Softness of thick whole-body cover including sensors affects the sensor characteristics. Movable ranges of joints are also affected. It is effective to design softness of soft cover locally around its joints. In order to adjust softness locally, there are several methods. One of these is to divide molds and mold plastic foam covers from different kinds of solution, and join them with adhesive. Other method is to control temperature or curative agent mixture locally and mold at one time. These methods are not effective for now. Adhesion plane becomes hard. Molds for local temperature control are too complex. For prototyping, adjustment in shape and slits is adopted. Bases of limbs were divided, thickness around joints were designed to be thin.

4. Implementation of distributed multi-axis deformation sense

Inside the soft exterior parts (on the surface of internal mechanical frame), forty nine three-axis capacitance-type force/torque sensors (PD3-32-05-15/40/80, NITTA Corp.) are embedded (Fig.3). Disk-shaped parts are screwed to force-sensitive sticks. When the soft exterior parts are pushed, stress is applied to the disk-shaped parts, then the sensors output voltages associated with z-axis force and x/y-axis torque(Fig.4). This sensor is not so small (diameter 18[mm], thickness 15.6[mm]), but it is possible to embed this sensor inside the thick cover. If the external force includes shearing components, stress on the disk slants, then x-axis torque and y-axis torque applied to the disk are detected. Soft exterior parts as tactile sensors have spatially interpolating characteristics, therefore there is availability that force detecting disks need not necessarily large enough to cover the entire area of robot's surface.

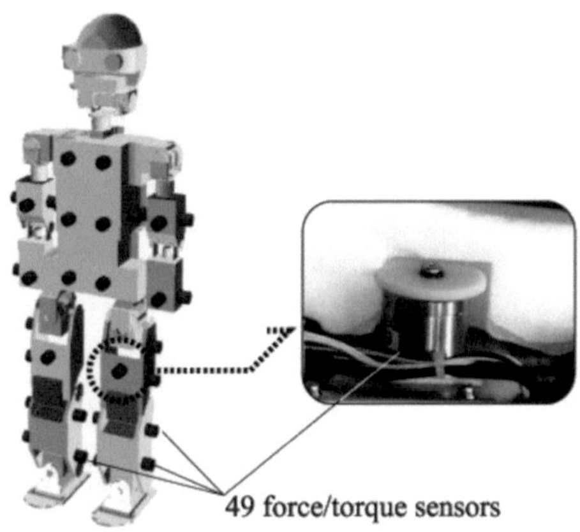

Fig. 3. Forty-nine three-axis force/torque sensors embedded inside the flesh.

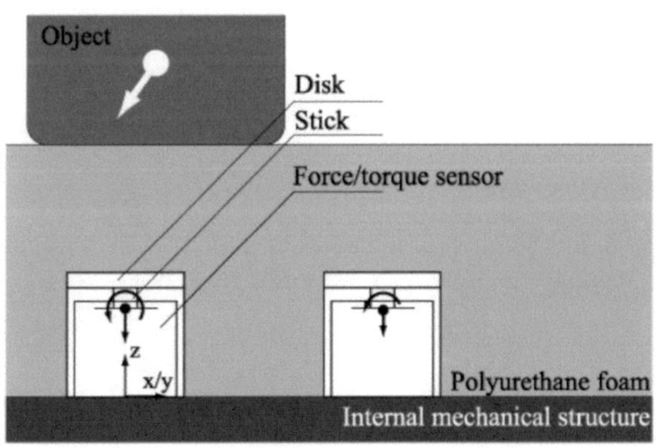

Fig. 4. Sensors are embedded in the polyurethane foam. Thicknesses of polyurethane foam varies from about 2[mm] to 60[mm]. The bottom face of the sensor is screwed shut to shell of robot skeleton.

5. Implementation to a humanoid robot

We have designed internal mechanical structure of the robot so that it has enough space to wear thick cover and joint movability. Its actuators and the sensors except tactile sensors are reused from a commercially available humanoid robot (HOAP-1, Fujitsu automation Corp.). Its height is about 700[mm], weight is about 7.5[kg]. Average thickness of soft cover is approximately 25[mm]. Specifications of the robot are shown in Table.1.

Num of Total DOF	22
Height [mm]	about 700
Weight [kg]	about 7.5
Motor Units	Smart Actuator Module, Type-I,Type-III (Fujitsu Automation)
Sensors	Incremental encoder, Accelerometer, Angular rate sensor, Force Sensing Resistor (Interlink Corp.), USB camera, Microphone,3-axis force/torque sensors (Nitta Corp.)

Table 1. Specifications of humanoid body.

5.1 Mobility of joints

Movability of Joint: Characteristics of compression and extension of polyurethane foam is not linear. Customized polyurethane foam in softness is used for the cover material. It is difficult to predict load torque from exterior to actuator at joint. Therefore load torque and maximum exterior thickness which is calculated using 25% compression hardness and breaking extension and breaking force of soft polyurethane foam were checked at first. This calculation method for design was investigated by Aono et.al.(2005). Then, for modifying its mobility of joints, slits and hollows were added. Stretchable cover envelops slits, not to expose the internal skeleton.

Currently, the robot can do several motions such as crawl, roll-over, and stepping (Fig.6). The movable range of knee joint is about 80 degrees, the maximum torque of the actuator is 45[kgf-cm].

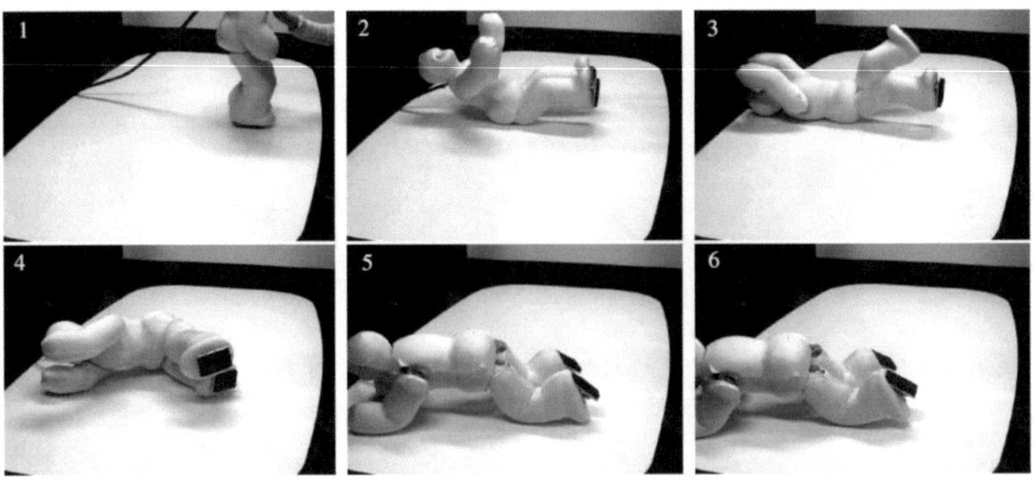

Fig. 6. Soft exterior enables a robot to fall down and roll-over more easily.

5.2 Heat exhaustion

Degree of heat conductivity of foam plastic is very low (about 0.02-0.07[W/mK]). One of solutions for heat exhaustion problem is forced cooling.

To test heat exhaustion effect of fans, two exhaust fans (1204KL-01W-B50 Minebea Corp.) and ventilation lines (width ~50[mm], thickness ~15[mm]) were embedded inside the exterior at toe and ankle. Exhaust outlet was made at crotch joints of limbs, considering occlusion caused by change in posture and contact with flat plane such as floor. Then the robot was hung and was made to bend the knee joint 40 degrees and stretch it back with a period of six seconds. Without fans, temperature on the surface of mechanical frame increased 12.5[oC] in 26 minutes. Heat elevation was reduced to 2[oC] with the two fans under the same conditions. Therefore, thick foam plastic cover was found not to be unrealistic.

5.3 Ability of distributed multi-axis deformation sense

Fig.7 shows an output of a sensor on the chest. When the robot is stroked or pushed in several ways. The output voltages were calibrated to zero. Sensitivity of x-/y-axis is 3.3-6.0[V/Nm], z-axis is 26.7-46.7[mV/N]. When the robot was pushed, z-axis output became high. When it was stroked downward, x-/y-axis output became high.

Black lines on the models of the robot in Fig.8 indicate output voltages of each actual situation. On the left of Fig.8, a person is holding and then pushing and pulling the arm of the robot. Even if the contact position does not change, the direction (pushing or pulling) can be detected.

Thick soft exterior enables to emulate various contact states including local deformation such as pinching. On the right of Fig.8, a person is pinching the robot. It can be seen that two sensors on the side of the robot react to the pinching in each direction.

7. Conclusion

In this research, a humanoid robot with thick polyurethane foam exterior and embedded distributed three-axis force/torque sensors was developed. This developed full-body tactile system can sense direction and distribution of force, so that this sense different contact states such as holding or pulling or pushing. Tactile sensor information processing using this multi dimensional distributed data is the topic of our further study.

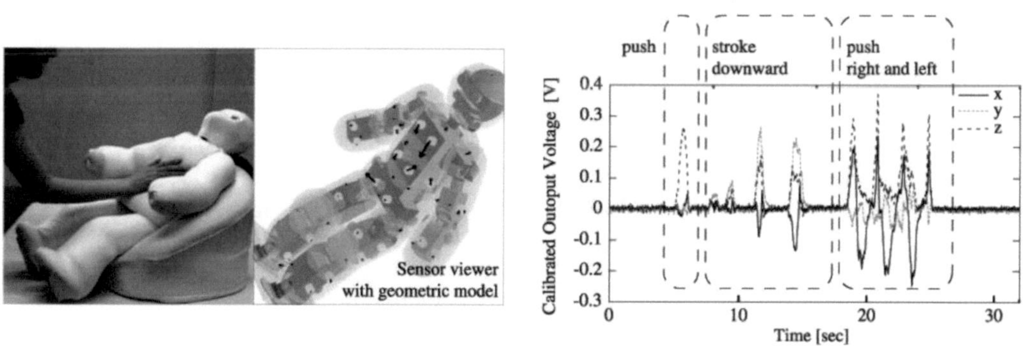

Fig. 7. The output of a three-axis force/torque type sensor which is located on front body of the robot. The sensor system is able to discriminate different states of touch from a person.

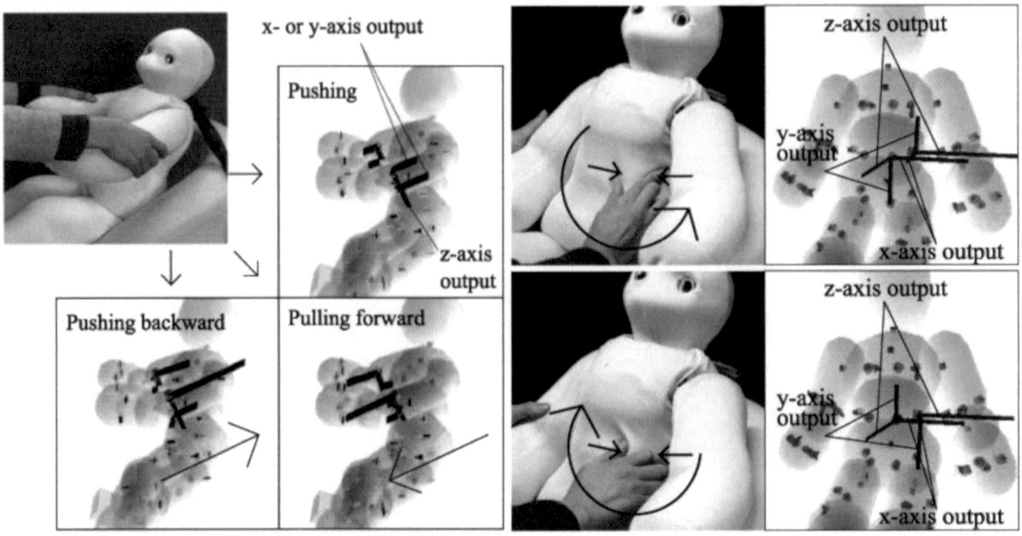

Fig. 8. Sensor output differs from its shearing force in the same contact state. Thick soft exterior enables local deformation state such as pinching. The distributed three dimensional sensors react to the direction of pinching.

8. References

Aono, T. & Nakamura, Y. (2005) Design of Humanoid with Insert-Molded Cover Towards the Variety of Exterior Design of Robots. *Proceedings of IEEE/RSJ International Conference on Intelligent Robots and Systems*, pp. 1476–1481

Inaba, M. et.al. (1996) A Full-Body Tactile Sensor Suit Using Electrically Conductive Fabric and Strings.*Proceedings of the IEEE/RSJ International Conference on Intelligent Robots and Systems*, pp. 450-457

Iwata, H. & Sugano, S. (2002) Whole-body Covering Tactile Interface for Human Robot Coordination. *Proceedings of the 2002 IEEE International Conference on Robots and Automation*, pp. 3818–3824

Kamiyama, K. et.al. (2004) Evaluation of a Vision-based Tactile Sensor, *Proceedings of 2004 International Conference on Robotics and Automation*, WP-6

Lumelsky, V.; Shur, M. & Wagner, S. (2001) Sensitive Skin, *IEEE Sensors Journal*, Vol.1, No.1, pp.41-51

Shinoda, H. et.al. (1997) Acoustic Resonant Tensor Cell for Tactile Sensing, *Proceedings of 1997 IEEE Int. Conf. on Robotics & Automation*, pp. 3087-3092

Marika Hayashi, Takashi Sagisaka, Yui Ishizaka, Tomoaki Yoshikai & Masayuki Inaba,. (2007). Development of Functional Whole-Body Flesh with Distributed Three-axis Force Sensors to Enable Close Interaction by Humanoids, *The 2007 IEEE/RSJ International Conference on Intelligent Robots and Systems*, pp. 3610—3615

18

Research and Preparation Method of Flexible Tactile Sensor Material

Ying Huang, Min Wang, Huaili Qiu, Bei Xiang and Yugang Zhang
Hefei University of Technology
Hefei, Anhui,
China

1. Introduction

A kind of new material suitable for flexible tactile sensor and its preparation method will be introduced. The tactile sensor applied this new material and method can be used in many kinds of occasions which contacting pressure measurement such as humanoid robot sensitive skin, medicine and so on.

This material is composite of carbon black and silicone rubber filled nano-SiO_2 as the main raw material, with a view to obtain sensitive material of flexible tactile sensor with both flexibility and tactile function.

In this chapter, based on the new flexible tactile sensor material, the effect that the material filled nano-SiO_2 has made to electrical and mechanical performance of pressure sensitive conductive rubber is investigated so as to theoretically analyze the property of nano-SiO_2 and the mechanism of modified by nano-SiO_2. Meanwhile, through the experiments of doping nano-particles into the flexible tactile sensor sensitive materials, the effect of nano-SiO_2 on the conductivity, resistance creep behavior and temperature stability of the sensitive materials were researched, as well as analyses are made on improving linearity, hysteresis and repeatability of tactile sensor. Moreover, the performance comparison produced by filled different content nano-SiO_2 and unfilled nano-SiO_2 into the new flexible tactile sensor material. Thereby, the optimum filled weight percent of nano-SiO_2 was obtained. The results showed that nano-SiO_2 has good modified role for flexible tactile sensor.

The preparation method will be used liquid silicone rubber solution blending molding. At room temperature and under normal pressure, the pressure sensitive liquid silicone rubber solution is filled into the arbitrary tactile sensor structure model and cured into molding. This method is suitable for various structures of the tactile sensor array.

2. Major raw materials and preparation methods

2.1 The main raw materials:

- GD-401 single component room temperature vulcanization silicon rubber;
- ECP-CB-1 high-conductive carbon black, with resistivity of 0.2 ~ 0.6 Ω·cm, average particle diameter is 30-40nm, specific surface area is 1000-1100 m^2/g;
- Nano-SiO_2, with particle diameter of 15 ~ 20 nm;
- Si-69 silane coupling agent, with density of 650 ~ 750 kg/m^3.

2.2 Sample preparation:

The preparation of laboratory samples adopts solution blending method.

2.2.1 Preparation of samples of experiment of conductive mechanism

At normal temperature and pressure, adding high-conductive carbon black ECP-CB-1 to liquid silicone rubber, a mixed solution is prepared by ultrasonic dispersion for 30 to 40 minutes with FS-150 ultrasonic disposal equipment. Injecting the mixed solution to the structural model of tactile sensor, it is molded at room temperature with time of 64 ~ 72 hours.

2.2.2 Preparation of samples of experiment of nano-modification

At normal temperature and pressure, mixing high-conductive carbon black ECP-CB-1 and Si-69 silane coupling agent, the mixed solution A is prepared by ultrasonic dispersion for 30 to 40 minutes with FS-150 ultrasonic disposal equipment. Adding nano-SiO_2 powder to mixed solution A, mixed solution B is prepared by ultrasonic dispersion for 30 to 40 minutes. Adding the mixed solution B to the liquid silicone rubber, the mixed solution C is prepared by ultrasonic dispersion for 20 to 30 minutes. Injecting the mixed solution C to the structural model of tactile sensor, it is molded at room temperature with time of 64 ~ 72 hours.

2.3 Sample components

Samples	S1	S2	S3	S4	S5	S6	S7	S8
ECP/RTV	4%	6%	8%	10%	12%	14%	16%	18%
lgρ	8.66	7.1	4.42	3.05	3.01	2.83	2.45	2.44
Region	A			B			C	

Table 1. Components of samples of experiment for conductive mechanism

Samples	A1	A2	A3	A4
ECP/RTV	8%	8%	8%	8%
SiO_2/RTV	0%	1%	3%	5%

Table 2. Components of samples of experiment for nano-modification

3. Fundamental theories

3.1 Conductive mechanism of the pressure-sensitive conductive composites

The new pressure-sensitive conductive composite is mainly composed by adding the carbon black of the compound silicon rubber and its conductive characteristics is closely relate to carbon black content added into the matrix material. Along with the increase of the carbon black content of the composite, the conductivity of the material slowly rises. When volume fraction of the carbon black particle reaches certain critical value, the volume resistivity of the composite sharply declines shown as Fig.1. Such sharp transition area from the high resistivity to the low resistivity is known as percolation zone, and this critical value is called as percolation threshold value. McLachlan and other people combined the effective medium conductive model and the percolation model, put forward the universal General Effective Media (GEM) which could explain the conductive characteristics of the carbon black filled composite as follows[1]:

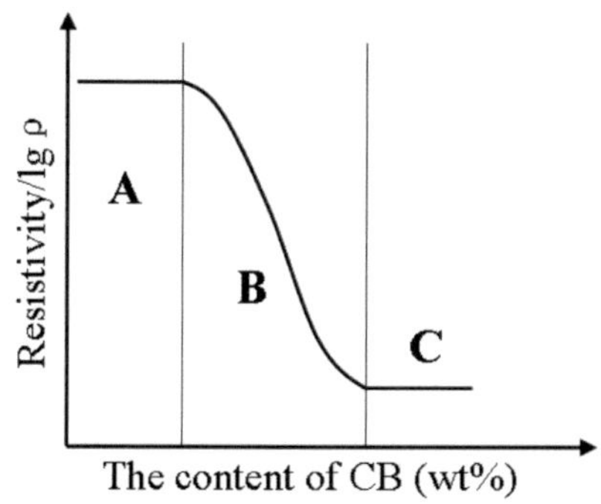

Fig. 1. Change of the resistivity along with carbon black contents

$$\frac{(1-\phi)(\sigma_1^{1/t}-\sigma_m^{1/t})}{\sigma_1^{1/t}+\{(1-\phi_c)/\phi_c\}\sigma_m^{1/t}}+\frac{\phi\left(\sigma_h^{1/t}-\sigma_m^{1/t}\right)}{\sigma_h^{1/t}+\{(1-\phi_c)/\phi_c\}\sigma_m^{1/t}}=0 \quad (1)$$

In (1): ϕ is the volume fraction of the conductive particle in composite; ϕ_c is the critical volume fraction of the conductive filler when the percolation passage is formed, which is also known as critical percolation threshold value; σ_1 is the matrix's conductivity; σ_h is the conductivity of the conductive particle; σ_m is the composite 's conductivity; t is the percolation coefficient of the composite . As for carbon black filled conductive silicon rubber, carbon black conductivity is far larger than conductivity of the silicon rubber, therefore the conductivity of the polymeric matrix could be approximately considered as zero, namely $\sigma_1=0$, $\rho_m=1/\sigma_m$, $\rho_1=1/\sigma_h$,Then (1) could be simplified as:

$$\rho_m=\rho_1\left(\frac{1-\phi_c}{\phi-\phi_c}\right)^t \quad (2)$$

In the (2): ρ_m is the volume resistivity of carbon black filled conductive composite ; ρ_1 is the carbon black's volume resistivity. Things requires attention is that (2) only applicable to condition when the volume fraction of the carbon black particle is close to critical threshold value ϕ_c.

Effective medium model describes the phenomenon of composite conductivity change along with the carbon black content, but does not explain the conductive reason from the microscopic structure of the conductive particle, fails to reveal the essential conductive mechanism of the pressure-sensitive composite. Currently researches on the explanation of the conductive mechanism are too much, but none of them has a unified theory [2]. Summary of the main conductive mechanism has two versions: Conductive path theory and quantum tunnel effect theory. Conductive path theory is the model theory only on the foundation of macro test feature instead of discussing the micro electronic movement within the material, is the modeling of the experiment phenomenon. It believes that conductive theory of the composite is due to the mutual contact of the filled carbon black particle, form conductive link inside of the silicon rubber, but this theory could not explain the problem that when filled less carbon black particle, the composite still possesses conductivity. Quantum tunnel effect theory is put forward by Sheng.P and other peoples on the foundation of the quantum tunnel effect theory, which tries to explain from the micro electronic movement in the material [3, 4]. Suppose the statistic average value of the conductive particle space is ω, relationship between the macro current density showed $J\left(\varepsilon\right)$ and ω is:

$$J(\varepsilon) = J_0 \exp\left[-(\pi\chi\omega/2)(\varepsilon/\varepsilon_0 - 1)^2\right] \tag{3}$$

In the equation, ε_0 is the field between spaces without external forces, ε is the field between spaces when giving external forces, $\chi = \sqrt{2mV_0/\hbar^2}$, m is the electronic mass, $\hbar$ is the reduced Planck constant, and V_0 is the barrier height.

This theory could explain the electronic transition pattern of the carbon black particle before direct contact within the material, which as a result explains in the sense the real reason for conduction of the composite when with less carbon black particle content. But this theory also could not unify the conductive mechanism of the composite with the different carbon black filled content, which needs to be continuously researched in the future. Currently the better explanation for the conductive mechanism for different carbon black filled content is to combine the conductive path theory and quantum tunnel effect theory. Simultaneous function of the two theories is as shown in Fig. 2. From the microscopic structure sketch of the carbon black particle filled silicon rubber, it could be found that carbon black particle structure in the silicon rubber is not evenly dispersed, instead of forming several congeries, which has spaces between each other. Carbon black particles in the congeries are mutually contacted, whereas conduction between congeries is caused by the electronic transition of the nearest carbon black particles.

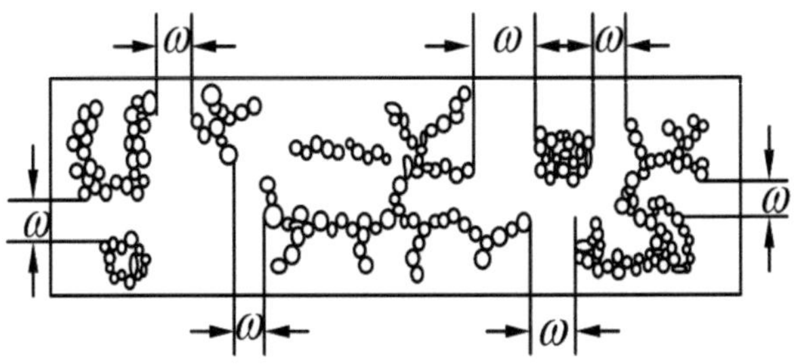

Fig. 2. Structural of the carbon black particle in the pressure-sensitive composite

Conductive characteristics curve of the composite is divided into three areas [5] (shown in Fig.1). A zone with higher resistivity is called as insulation zone; B zone with sharply decreased resistivity is called as percolation zone; stabilized C zone with lower resistivity is called as conductive zone. A zone: insulation zone, when carbon black content in silicon rubber is lower than the percolation threshold value, space between carbon black particles is large, the gathering of the carbon black particle is few, so the conductive path is hard to form; space between congeries is large, the electronics between particles is hard to generate transition, then the composite shows the high resistivity; usually resistivity is higher than $10^{10}\Omega \cdot$ cm. B zone: percolation zone, along with continuous increase of the carbon black contents in the conductive silicon rubber, carbon black particle gathers link, starts forming conductive path; space between congeries continuously decrease, two adjacent particle distance decrease to 1.5-10nm. Probability of the electronic transition raise greatly, as a result sharply increase the conductivity of the composite. C zone: conductive zone, along with the continuous increase of the carbon black filled, carbon black content would exceed percolation threshold value, the space between carbon black particles also would further shorten(it is generally agreed to be 0.35-1.5nm); congeries link gradually stabilizes, the quantity of conductive path basically reach the maximum value. Along with further shortening of the neighboring carbon black particle space, when particle distance is smaller than 0.35nm, it is assumed that contact between the particles could be approximately regarded as pure particle contact, in addition current and voltage show the linear relationship, show the obvious Ohm characteristics.

Summarize the above discussion, composite resistivity in the percolation zone has the maximal change along with different carbon black content. Elastic deformation of the pressure-sensitive composite under the external forces would lead volume fraction variance of the carbon black [6], consequently result in variation of the composite conductivity, which exactly satisfies the design requirement of the tactile sensor, therefore research on the conductive characteristics of the flexible tactile sensing material in this paper is mainly in the percolation zone, and gives priority to quantum tunnel effect theory.

3.2 Piezoresistive effect

Resistance of the pressure-sensitive composite shows the regular variation under the pressure, which possesses good piezoresistive characteristics, could establish piezoresistive characteristics model of the pressure-sensitive composite based on quantum tunnel effect theory. On the assumption that initial thickness of the conductive composite is h_0, initial resistance value is R_0, k_1 is the elastic coefficient of the silicon rubber matrix. When pressure-sensitive composite is under pressure F, the material thickness is h, then within the elastic deformation there is:

$$F = k_1(h_0 - h) \tag{4}$$

Secondly, suppose ω is directly proportional to h, proportion coefficient is k_2 namely:

$$\omega = k_2 h \tag{5}$$

Electric field ε between spaces is inverse proportional when other condition unchanged:

$$\varepsilon / \varepsilon_0 = \omega_0 / \omega = h_0 / h \tag{6}$$

According to Ohm's law, bring the (3) ~ (6) to the resistance equation, then deduce the relationship of resistance value change along with external forces:

$$
\begin{aligned}
R(F) &= h/\sigma S = Jh/ES \\
&= \left(\frac{h_0 k_1 - F}{\sigma_0 S k_1}\right)\left(\exp\left(\frac{\pi\chi k_2 F^2}{2k_1(h_0 k_1 - F)}\right)\right) \\
&\approx \left(\frac{h_0 k_1 - F}{\sigma_0 S k_1}\right)\left[1 + \frac{\pi\chi k_2 F^2}{2k_1(h_0 k_1 - F)} + \frac{\pi^2\chi^2 k_2^2 F^4}{8k_1^2(h_0 k_1 - F)^2} + \cdots\right] \\
&= R_0 - \frac{R_0}{k_1 h_0}\cdot F + \frac{\pi\chi k_2 R_0}{2k_1^2 h_0}\cdot F_2 + \frac{\pi^2\chi^2 k_2^2 R_0}{8k_1^3 h_0 (h_0 k_1 - F)}\cdot F^4 + \cdots
\end{aligned} \tag{7}
$$

It is accurate to quadratic term generally, therefore:

$$R(F) = R_0 - AF + BF^2 \tag{8}$$

Among this, $A = R_0/k_1 h_0$, $B = (\pi\chi k_2 R_0)/(2h_0 k_1^2)$,above equation reflects the functional relationship that the conductivity of pressure-sensitive conductive silicon rubber changes along with external force variation; calculate the value of A and B by least-squares method according to test data R(F)~ F and (8).

3.3 Influencing mechanism of the temperature to conductivity

Because influence of the temperature to matrix material is considerable, hence changes the conductivity of the pressure-sensitive conductive composite , so research on the influence of the temperature to composite conductivity is necessary. Research on the temperature characteristics for the pressure-sensitive conductive composite would mainly gain two properties: positive temperature coefficient characteristics (PTC) and negative temperature coefficient characteristics (NTC). From the discussion of the temperature characteristics of the pressure-sensitive conductive silicon rubber on conductive mechanism will find it is possible that these two properties appear in one sample heating process [7].

According to quantum tunnel effect theory by Sheng P and others, use ε_T to represent the electric field change caused by the thermal disturbance when heating up, ε_A represents the extra electric field, then:

$$\varepsilon = \varepsilon_T + \varepsilon_A \tag{9}$$

When extra electric field ε_A is very low, namely when $\varepsilon_A < \varepsilon_T$, then the actual tunnel current is:

$$\Delta J = J(\varepsilon_A + \varepsilon_T) - J(\varepsilon_A - \varepsilon_T) \tag{10}$$

Here use the equation below to define $\sum(\varepsilon_T)$:

$$\sum(\varepsilon_T) = \frac{\Delta J}{\varepsilon_A} = 2\frac{dJ(\varepsilon_T)}{d\varepsilon_T} \tag{11}$$

Average is obtained as the equation below:

$$\sigma = \int_0^\infty P(\varepsilon_T)\Sigma(\varepsilon_T)d\varepsilon_T \tag{12}$$

In it: $P(\varepsilon_T) = \kappa \exp\left(-\frac{\omega A\varepsilon_T}{8\kappa T^2}\right)$ represents the appearing probability for ε_T.

Whereas fundamental equation of the quantum tunnel effect theory is (3)
Bring the (3) and (10) to the (11); ultimately get the following equation:

$$\sigma = \sigma_0 \exp\left[\frac{\frac{-A\varepsilon_0^2\omega}{8\pi\kappa}}{T + \frac{A\varepsilon_0^2}{4\pi^2\chi\kappa}}\right] \tag{13}$$

In the equation: $\chi = \sqrt{2mV_0/\hbar^2}$, whereas κ is Boltzmann constant, m is the electronic mass, e is the electron charge, V_0 is the barrier height, A is carbon black particle space areas of the capacitance, they are all the constant. There are only two variants ω and T.

$$M = A\varepsilon_0^2 / 8\pi\kappa \,, N = A\varepsilon_0^2 / 4\pi^2\chi\kappa$$

Then (13) becomes:

$$\rho = \rho_0 \exp[M\omega / (T + N)] \tag{14}$$

It could be found from the above equation that conductivity is related to temperature T and ω; ρ decreases along with the increase of T and increases along with the increase of the ω. Regularity could be summarized as when temperature of the pressure-sensitive conductive composite increases, silicon rubber matrix generates thermal expansion that results in space ω expansion and raise the resistivity; whereas temperature rise also raise the electronic transition probability between the carbon black particles (namely thermal disturbance) that results in fall of the resistivity. It could be said that resistivity changes along with the temperature mainly subject to the two kind functions of the thermal expansion and thermal disturbance, and the characteristics showed mainly depends on which takes the leading role. If thermal expansion takes the leading role, and then PTC characteristics appears, if thermal disturbance takes the leading role, then NTC characteristics appears.

4. Study and analysis of piezoresistive and temperature properties

In the above, we have put forward the mechanism of piezoresistive characteristics and temperature properties of flexible tactile sensor material, and in the following, we will analyze its piezoresistive characteristics and temperature characteristics related with the experiment result.

4.1 Piezoresistive effect experiment and result analysis

Piezoresistive effect experiment is intended to test the relationship between pressure and resistance, hence to further analyze the conductivity of the pressure-sensitive conductive composite under pressure, which adopt the measuring method as shown in Fig.3. Among it the upper and lower pole are circular copper poles, and sample and pole are bonded by conductive silver adhesive in order to guarantee good metallic connection. Piezoresistive experiment respectively for sample S3, S4 and S5, records are drawn into curve as shown in Fig.4.

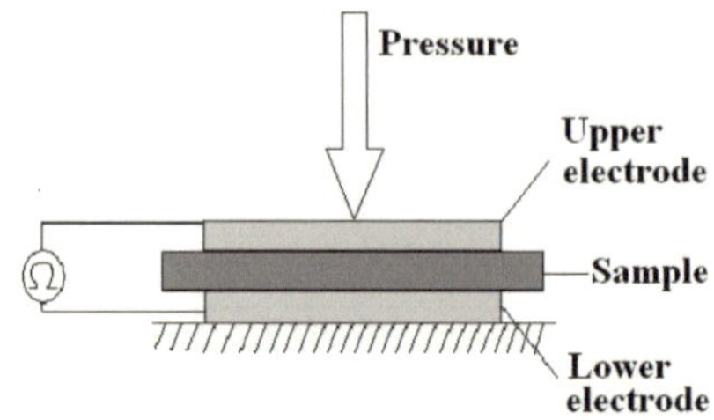

Fig. 3. Piezoresistive effects measuring device

Fig.4 reflects the relationship of the pressure-sensitive conductive composites' resistivity changing along with pressure. According to (8) based on experiment curve, we could fit the value A, B each parameter value of three samples is shown as Table 3.

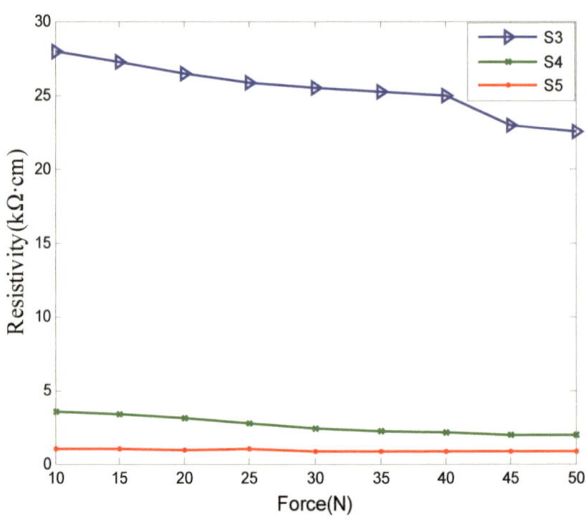

Fig. 4. Piezoresistive characteristics with different carbon black content

Parameters	A(Ω/N)	B(Ω/N)	R_0 (Ω)
Sample s3	11.518	0.9288	915.33
Sample s4	13.573	0.6572	132.27
Sample s5	1.7962	0.0909	35.777

Table 3. Parameters estimated value of the three samples

It is observed from Table 3 that B value decreases along with the increase of the carbon-black content, the linearity of the sample piezoresistive characteristics would be better, which also provides reference for this material satisfying the design requirement of the flexible tactile sensor. The value ω can be estimated from the value of A and B, but is inconsistent with the theoretical value. The reason may be due to the simplified model which did not consider the microscopic structure of carbon black particles. Carbon black particles are not uniform distribution in the matrix material, but uneven distribution.

4.2 Temperature characteristics experiment and result analysis

Place the sample into the high and low temperature experiment box with the model number of GDW-100 produced by Sunan Laboratory Apparatus Factory, introduce conductor through electrode, measure resistance with Ohmmeter, respectively take temperature test to sample S3, S4, S5, heat from 30°C to 90°C, record 20 minutes constant temperature after heating 10°C, gain resistance temperature characteristics of the pressure-sensitive conductive rubber with different carbon black content that is shown in Fig.5.

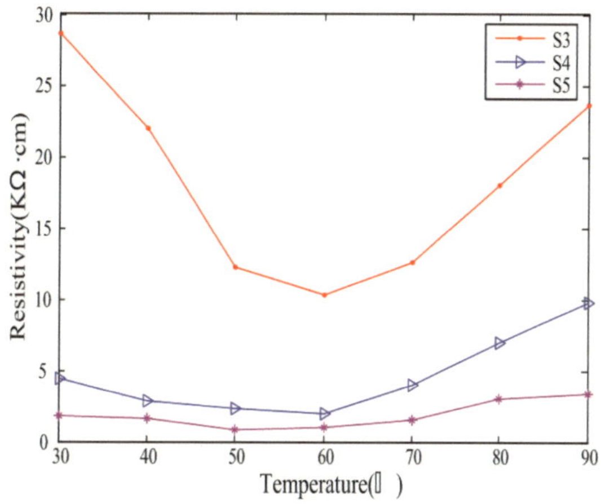

Fig. 5. The resistivity temperature characteristics with different carbon black content

In Fig.5, resistivity of sample S3, S4, S5 all have the phenomenon of continuously rise after decreasing. Major analysis is in the percolation zone, temperature's influence to quantum tunnel effect is comparatively apparent when heating freshly, which means the thermal disturbance takes leading role, then the sample resistivity appears the downward trend, shows the negative temperature coefficient. After temperature reaching certain value, quantum tunnel effect is stabilized, whereas different thermal expansion rates of the carbon black particle and silicon rubber result in the enlarged space of the carbon black particle. Sample resistivity increases, which means the thermal expansion takes leading role, shows the positive temperature coefficient characteristics.

It could also be discovered from Fig.5 that along with increase of the carbon black content, resistivity of the pressure-sensitive conductive rubber decreases; furthermore temperature stability is also better. This is because along with the increase of the carbon black content, average space between carbon black particles become smaller, quantum tunnel effect is increasingly obvious, resistivity value continuously falls. But increase of the carbon black content stabilizes the quantum tunnel effect, reduces influence of the thermal disturbance, whereas combined function of thermal disturbance and thermal expansion stabilize the sample resistivity temperature characteristics much more.

As materials of the flexible tactile sensing, temperature stability influences its operating temperature range. In order to increase the temperature stability of such material, several literatures put forward the way of adding other filled materials such as graphite particle and nanometer particles etc [8].

5. Nano-modification technology of flexible tactile sensor material

5.1 Characteristics and modification mechanism of nanomaterials

Because of high relevance of quantum mechanics, nanomaterials show their different properties from macro and micro's. In the field of polymer materials modification study, the nanoparticles have been considered to be good filling material to improve the mechanical and physical characteristics of high molecular polymer. The specific surface area as huge as $1000m^2/g$ at nanometer level and its unique performance have aroused extensive research in various fields of natural sciences and engineering sciences.

In the process of mixing nano powder and high molecular polymer, agglomerate will be broken, and then become nanoparticles chain. Through study, Friedlander finds that [9-10] inorganic oxide nanoparticles chain aggregates have feature of flexibility, which can lengthen under extension and the nanoparticles chain curl and reunite into agglomerate after removing tension. That is, nanoparticles chain aggregates have feature of reaggregation, which shows property of elasticity mechanics similar to polymer chains, and at the same level with ordinary high polymer molecular chain, as shown in Fig.6. Another important contribution of nanoparticles chain is three-dimensional network formed through nanoparticles chains' self-contact or by nanometer elastic layer absorbed in the surface of the polymer particles chains. The study found that network formed by nanoparticles has property of elasticity mechanics similar to polymer, which makes the particles network and polymer network extend, intervein and mix together, effectively eliminates the interfacial energy difference, and reduces stress concentration, so as to enhance the strength of high polymer complex, thereby, nanoparticles chains with flexibility have a significant impact on nano powder's dispersion and toughening in high polymer. [11].

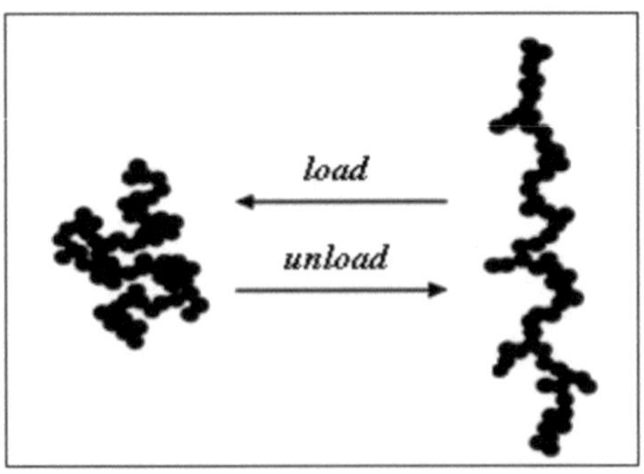

Fig. 6. Elastic behaviors of nanoparticle chain are similar to polymer chain

5.2 Test and analysis of nano composites

5.2.1 Modification of mechanical properties of sensor

In the sensors' measuring, good static and dynamic characteristics are requested, of which, linearity, repeatability and hysteresis quality is particularly important. Nano composites, as a new type of sensor sensitive material, should also have the corresponding characteristics. The following experiment is adopted to measure the effect on the static characteristics of sensitive materials by adding different content of nano-SiO_2 or adding no nano-SiO_2. Fig. 7 shows that area of the hysteresis loop of samples A1 with no nano-SiO_2 added is far larger than that of samples A2, A3 and A4 with nano-SiO_2 added, which means that the addition of nano-SiO_2 improves the hysteresis quality of samples.

Four times loading and unloading cycle experiments are conducted in sample A1with no Nano-SiO_2 added and sample A3 with nano-SiO_2 added, in order to test linearity and repeatability of composites used for the tactile sensor. The results in Fig. 8 show that the repeatability of sample A1 with no nano-SiO_2 added is relatively poor, but sample A3 with nano-SiO_2 added shows good repeatability in the four times experiments. This is because structure with completely different mechanical properties is formed with nano-SiO_2 added. Nano SiO_2 particles, like rigid chain, enhance the toughness and strength of polymer, change the toughness, strength and many other mechanical properties of composites [12]. Therefore, the addition of nano-SiO_2 enhances toughness, repeatability, and hysteresis of tactile sensor to different degrees.

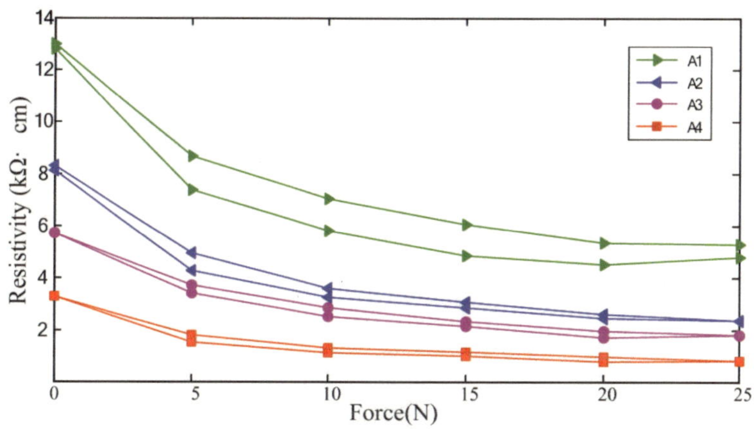

Fig. 7. Linearity and hysteresis curves

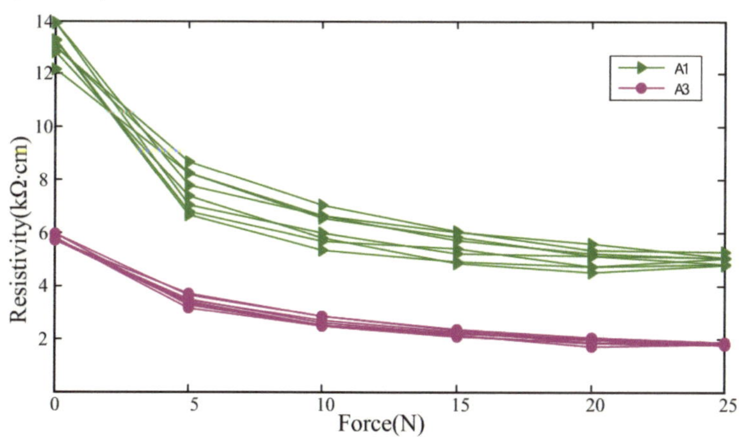

Fig. 8. Reproducibility curves

5.2.2 Characteristic of conductivity

From the experiment of mechanics above, we can also see that with the increase of nano-SiO_2 content, the resistivity of composites shows the trend of gradual decreasing. This is because when the nanoparticles and conductive carbon black are well-mixed in silane coupling agent, network formed by nanoparticles chains can effectively make particle network and polymer network extend, intervein and mix together.

Fig. 9 is the SEM image of 20000 times of sample with 3% nano-SiO_2 added and sample with no nano-SiO_2 added. In the image we can see that the addition of nano-SiO_2 significantly increases the dispersion of conductive carbon black in silicon rubber matrix, improves the uniformity of the sample, thereby increases the conductivity of the material, which is also reflected in the specific resistance in characteristic curve of the four samples in Fig. 7.

(a)

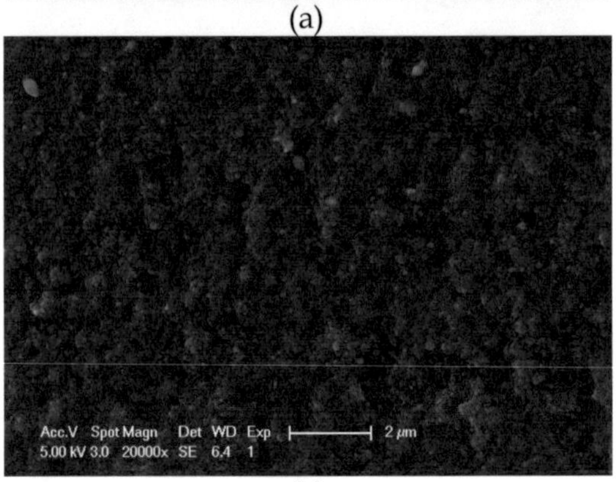

(b)

Fig. 9. Effect of 20,000 times of SEM microstructure of composites(a. no filling nano-SiO_2, b. filling nano-SiO_2)

5.2.3 Effect on resistance creep

As for sensor sensitive material, not only good piezoresistive characteristics and elastic property are required, but also good time stability. In this paper, because substrate material of the flexible sensitive composites is colloid, it shows mechanical behavior of high polymer materials, with both mechanical properties: glutinosity and elasticity. At the same time, the conductive path of the composites varies with the deformation of colloid. These phenomena are mainly embodied in the time stability of the resistance of material, such as the creep of resistance, which directly affects the sensor measurement error.

Fig.10 is the experimental results of resistance creep of sample A1 with no nano-SiO_2 added and A3 with nano-SiO_2 added under different loading. From the Figure, we can see that the relative change of specific resistance $\Delta\rho/\rho_0$ of sample A3 with nano-SiO_2 added is lower than that of sample A1, that is, the resistance creep is relatively small. Therefore, sample with nano-SiO_2 added has a better time response as tactile sensor.
Due to the addition of nano SiO_2 particles, nano SiO_2 particles, like beaded chain, strongly absorb the surrounding silicon rubber molecules with its high surface energy and surface binding energy, and create a strong chemical bonds and physical adsorption with silicone rubber molecules, thereby form a relatively complete crosslinking network structure with nano SiO_2 particle as nodes, which disperses the force to other polymer chain as common burden, prevents the rapid breakage of polymer and effectively improves the resistance creep property of composites under loading.

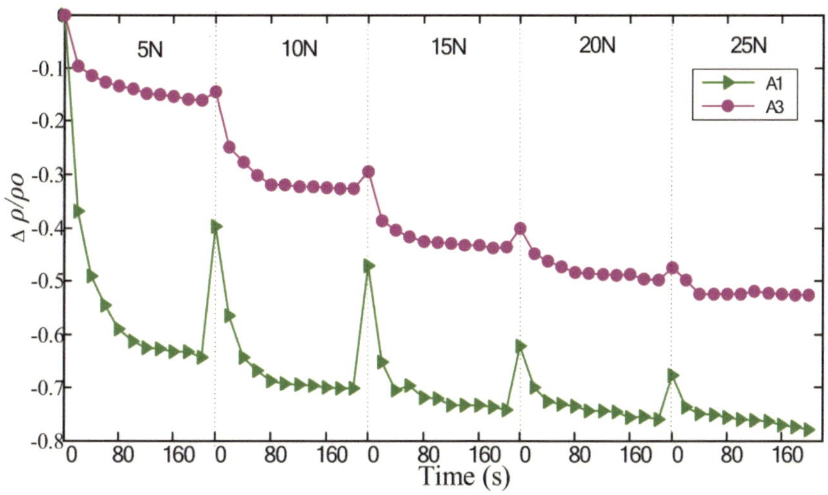

Fig. 10. Resistance creep behavior at different applied loads

5.2.4 Effect on resistance temperature properties

Resistance temperature properties of flexible tactile sensor based on the pressure-sensitive conductive rubber is effect of coactions of material's thermal disturbance and thermal expansion [8]. In the beginning of warming up, the tunnel effect is obviously influenced by temperature, that is, the thermal disturbance plays a leading role, the resistivity of the sample decreases, showing properties of negative temperature coefficient. When the temperature reaches a certain value, the tunnel effect tends to be stable. However, because the thermal expansion rate of carbon black particles and silicon rubber is different, the space between the carbon black particles increases, and the resistivity of sample increases, the thermal expansion at this time plays a leading role, showing properties of positive temperature coefficient. Refer to the Fig.11, which is the temperature experimental curve of sample A1. This phenomenon demonstrating positive and negative temperature properties and large resistance temperature coefficient is disadvantageous to the produced flexible tactile sensor.
Fig.11 shows the resistance temperature properties of samples A1, A2, A3 and A4. The resistance rate change of A2, A3 and A4 is smaller than that of A1, and the variation trend

basically shows monotonicity. We can see that the addition of nano-SiO_2 effectively improves the time stability of the samples. Further more, with the increase of the content of nano-SiO_2, the temperature stability of the samples is gradually improving, which plays an important role in the practical application and temperature properties compensation of flexible tactile sensors.

The reason of the addition nano-SiO_2 changes the temperature properties of the material is mainly related with the micro characteristics of nano-SiO_2. In the process of temperature changing, the distribution and decentralization state of filling material in the matrix decide the resistance-temperature relations. If explained in conductive pathway theory, it is because stable conductive pathway is formed, which improves the temperature stability. Therefore, the appropriate amount of nanoparticles SiO_2 prevents the agglomeration of conductive carbon black in the polymer, and plays a very good role in dispersion, which promotes the formation of a conductive circuit, and increases temperature stability and consistency of pressure-sensitive conductive rubber.

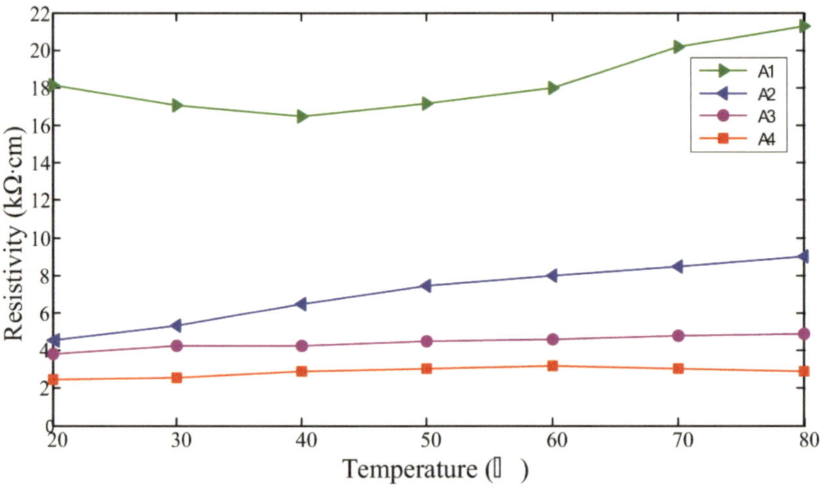

Fig. 11. The resistance temperature characteristic of with different nano-SiO_2 content

5.2.5 Performance Comparison

Of the above four samples' performance, Table 4 shows the comparison of the performance indicators of sample A1 with no nano-materials added and sample A3 with 3% nano-SiO_2 added. From Table 4 we can see that the linearity, repeatability and hysteresis as well as conductivity, resistance creep and temperature properties of sample A3 with nano-SiO_2 added have improved, of which, linearity, repeatability and hysteresis are acquired through curve fitting in the fourth loading and unloading of the above experiment. From this, we

Sample	Linearity (%)	Hysteresis (%)	Repeatability (%)	Conductivity ρ_0 (kΩ· cm)	Creep (10N loading) $\Delta\rho/\rho_0/t$	Temperature $\Delta\rho/\rho_0/T$
A1	10.9	15.5	12.4	13.9	0.0015/s	0.00588/℃
A3	6.5	8.4	7.5	5.7	0.0009/s	0.00273/℃

Table 4. Samples A1 and A3 performance comparison

can see that all the performance of sample A3 with 3% nano-SiO_2 added have been improved, which is the best filling share. It proves that a certain content of nano materials has a modification function on flexible tactile sensor.

6. Conclusion

This paper studies conductive mechanism of the new pressure-sensitive conductive composite based on flexible tactile sensing, believes that conductive mechanism of the composite within percolation zone gives priority to quantum tunnel effect theory. Starting from the quantum tunnel effect theory it studies the relationship between pressure and resistance, and piezoresistive model could be established by using the quantum tunnel effect. Based on theory and experiment, through the analysis and research on the conductive characteristics of such new sensitive material with different content conductive fillers, the conductive mechanism could be obtained within the pressure sensitive range, which provides theoretical foundation for the study of such new flexible tactile sensing material. It also discusses the influence mechanism of the temperature to the conductivity of the pressure-sensitive conductive composite. Positive temperature coefficient (PTC) and negative temperature coefficient (NTC) characteristics of the pressure-sensitive conductive silicon rubber will appear in the heating process of one sample, which provides reference for the practical application of such new sensitive material.

Whereas there is still a certain distance between the performances of the above flexible material compared with the requirements of sensor sensitive materials, therefore, this paper further studies the modification function of nano-SiO_2 by adopting the special nature of nano materials. Since nowadays, the study on nanoparticles modified materials and nano materials' mechanical behavior is only in the early stage, which only detects the external manifestations of mechanical phenomena, far from the study on its internal mechanism [13]. Therefore, this paper mainly conducts experimental study and measurement on modification of pressure-sensitive conductive rubber after the addition of nano-SiO_2 materials. The results show that the unique properties of nano-SiO_2, the mechanical properties of pressure-sensitive conductive rubber, temperature performance and stability are significantly improved. As for the modified composites, we can improve the technical indicators of the produced tactile sensors to meet the requirement of tactile sensing. Further more, the compounding of nanoparticles and polymer modifies the heat-resistant and ageing resistance, and improves the properties of shock resistance and anti-fatigue.

Flexible tactile sensor material, presented in this chapter, has both good tactile sensing function and flexibility, its preparation method is simple. Flexible tactile sensor based on this material can measure various tactile information of contact surface, which is expected to provide a new way of thinking and research methods of the flexible tactile sensor.

7.References

McLachlan, DS; Blaszciewicz, M & Newnham, RE. (1990). Electrical resistivity of composites. Journal of American Ceramic Society, pp.2187-2203, 73(8),1990.

Zhudi Zhao; Wenxue Yu; Xiujuan He & Xinfang Chen. (2003). The conduction mechanism of carbon black-filled poly(vinylidene fluoride) composite. Materials Letters, pp.3082–3088, 57, 2003. (In China)

Sheng, P; Sichel, EK & Gittleman, JI. (1978). Fluctuation-Induced Tunneling Conductionin Carbon-Polyvinylchloride composites. PhysRevLett, 40, 1978.

Sheng, P. (1980). Fluctuation-Induced Tunneling Conductionin Disordered Materials. PhysRev, pp.2180-2195, B21, 1980.

Jaime C G. (2001). Carbon black-filled polymer composites : Property optimization with segregated microstructures. University of Minnesota, 2001.

Wang Peng, Ding Tianhuai, Xu Feng.Piezor (2004)"Resistivity of Conductive Composites filled by Carbon Black Particles," J. Acta Materiae Compositae Sinica, pp.34-38, 21, 2004. (In China)

Shekhar, S; Prasad, V & Subramanyam, S, V. (2006). Transport properties of conducting amorphous carbon–poly(vinyl chloride) composite. Carbon, pp.334–344, 4, 2006.

Xie Quan; Luo Jiao-lian & Gan Fu-xi. (1999). Effect of Filler on the Resistance Temperature Propertier of the complex Conductive Silicone Rubber. Journal of Functional Polymers, pp.414-418, 12, 1999. (In China)

Friedlander S K; Jand H D & Ryu K H. (1998). Elastic behavior of nanoparticle chain aggregates.App Phy Lett, pp.173-175,72,1998.

Sheldon K; Friedlander ; Kenya Ogawa & Marc Ullmann. (2001). Elasticity of nanoparticle chain aggregates: implications for polymer fillers and surface coatings. Powder Technology, pp.90–96.118, 2001.

Zhuang Qingping. (2005). Study on interface state and mechanical behavior between NCA/polymer chain. Polymer Materials Science & Engineering, pp.76-79,Volume 25, No. 1, 2005.

Zhou Hong; Qiao Xueguang & Wang Hongliang. (2005). Study on Property Improvement of Polymer Coated Material for Optical fiber bragg gratings by Using nmSiO_2. Acta Photonica Sinica, pp.1332-1335,Vol.34, No. 9, 2005.

Jia Shaojin; Jiang Pingkai; Zhang Zhicheng & Wang Zhong guang. (2006). Effect of carbon-black treatment by radiation emulsion polymerization on temperature dependence of resistivity of carbon-black-filled polymer blends. Radiation Physics and Chemistry, pp.524-531, 75, 2006.

19

A Principle and Characteristics of a Flexible and Stretchable Tactile Sensor Based on Static Electricity

Yasunori Tada[1], Masahiro Inoue[2], Toshimi Kawasaki[3], Yasushi Kawahito[3], Hiroshi Ishiguro[2] and Katsuaki Suganuma[2]
[1]*Nagoya Institute of Technology,*
[2]*Osaka University,*
[3]*Togawa Rubber Co., Ltd.*
Japan

1. Introduction

Tactile sensors are required for various robots such as robot hands and humanoid robots. In some humanoid robots, the tactile sensor is utilized for interacting with the human. For example, when many humans interact with the humanoid robot such as Robovie, they often try to stroke the robot's head, to hug the robot, or to shake hands (Tajika et al., 2006). To recognize such communications, some humanoid robots have the tactile sensor over the robot's whole body.

Existing tactile sensors for humanoid robots were proposed in the following studies. Tajima et al. (Tajima et al., 2002) proposed a distributed tactile sensor made of conductive fabric and gel. Ohmura et al. (Ohmura et al., 2006) proposed a tactile sensor that the sensor elements were mounted on the flexible printed circuit board. Robovie utilizes the robot skin that piezoelectric films were embedded in the silicone rubber (Miyashita et al., 2007). These tactile sensors have flexibility. The flexibility is important feature. If the surface of the humanoid robots is flexible, the uncomfortable feeling when the human hugs the humanoid robot will be reduced. Moreover, the flexibility is expected to improve the safeness.

Moreover, the stretch ability of the tactile sensor has advantages that the sensor is not frangible and that the sensor can be easily mounted on curved surfaces or deformable parts such as joints. However, many existing tactile sensors (Shan et al., 2005; Mizuuchi et al., 2006; Koterba & Matsuoka, 2006) have the flexibility, but do not have the stretch ability. Hoshi and Shinoda (Hoshi & Shinoda, 2006) proposed flexible and stretchable robot skin made of conductive fabric and urethane foam. However, they did not discuss the sensor output when the skin is stretched.

This paper proposes a flexible and stretchable tactile sensor which is expected to be suited for humanoid robots. The sensor can detect touch and release by using the static electricity and the electrostatic induction. The static electricity and the electrostatic induction are well-known phenomena from old times. However, we do not know a sensor that these phenomena were utilized as the tactile sensor. However, there is a sensor to distinguish materials which utilized the static electricity (Shida et al., 2007). By using the static electricity, the sensor consists of only the charged bodies and the conductive material, and has very simple structure. In this paper, the sensor is made of the soft silicone rubber because the sensor is expected flexible and stretchable. The characteristics of the sensor are investigated by some experiments.

The remainder of this paper is organized as follows. First, a principle of the sensor is introduced. The sensor utilizes the static electricity and the electrostatic induction phenomenon. In a next session, some experiments are shown to demonstrate characteristics of the sensor. A first experiment shows sensor responses when several objects touch the sensor. A second experiment shows the sensor output when the sensor is stretched. A final experiment shows the influence of a cross-talk noise. In a final session, we conclude this paper.

2. Principle of the sensor

A principle of the proposed tactile sensor is shown in Fig. 1. Fig. 1(a) shows a structure of a sensor system. The sensor consists of a base sheet at the bottom of the sensor, a conductive wire on the base sheet, and a charged body on the conductive wire. The conductive wire is connected to the ground through a resistor. A voltmeter measures voltage between the wire and the ground. The structure of the sensor is similar to capacitive tactile sensors. However, the principle of the proposed sensor is different from the existing tactile sensors. The existing capacitive tactile sensors measure the transition of the capacitance whereas the proposed sensor measures the static electricity. In the proposed sensor, when an object touches the charged body of the sensor, the static electricity occurs in the object and the sensor. Moreover, at this time, the static electricity generates the electrostatic induction in the wire. Therefore, the charged body and the conductive wire are performed as the tactile sensor by measuring the voltage of the electrostatic induction.

Each step of the principle is described in the following. The sensor and an object are discharged on initial condition shown in Fig. 1(a). A first step is shown in Fig. 1(b). The static electricity occurs in the sensor and the object when the object touches the sensor. However, the electrostatic induction does not occur in this step, and the voltmeter continues to indicate 0 [V] because the quantities of the positive and negative electric charges are the same. In the figure, the object and the sensor are charged to the positive and the negative, respectively. Fig. 1(c) shows a second step. When the object is released from the sensor, the electrostatic induction occurs in the wire because of effect of the negative charge in the charged body. In other words, the positive charges are collected under the charged body, and the negative charges are collected at near the resistor. Therefore, the voltmeter indicates negative value. A third step is shown in Fig. 1(d). The charges move to the ground because the wire is connected to the ground through the resistor, and the voltmeter indicates 0 [V]. A fourth step is shown in Fig. 1(e). When the positive charged object approximates to the sensor, the electrostatic induction occurs in the wire again. At this step, the positive charges are collected at near the resistor, and the voltmeter indicates the positive value. Additionally, the reason that both the charged body and the wire are charged to the negative is the difference of the quantity of charge between the object and the sensor. Fig. 1(f) shows a final step. The charges move to the ground, and the voltmeter indicates 0 [V]. Therefore, the proposed sensor can detect touch and release by the principle.

We consider a characteristic of the static electricity. The work functions of each object are the important factor to decide whether the object is charged to positive or negative. The electrons in the low work function object move to another object easily. In other words, the electrons in the low work function object reduce, and the object is charged to positive. If the objects have a higher or lower work function than the proposed sensor, the sensor output when the objects touch the sensor is unstable; the positive or negative value is observed when the objects touch the sensor. Therefore, the material of the charged body of the sensor should have the very low or high work function for avoiding unstable sensor output.

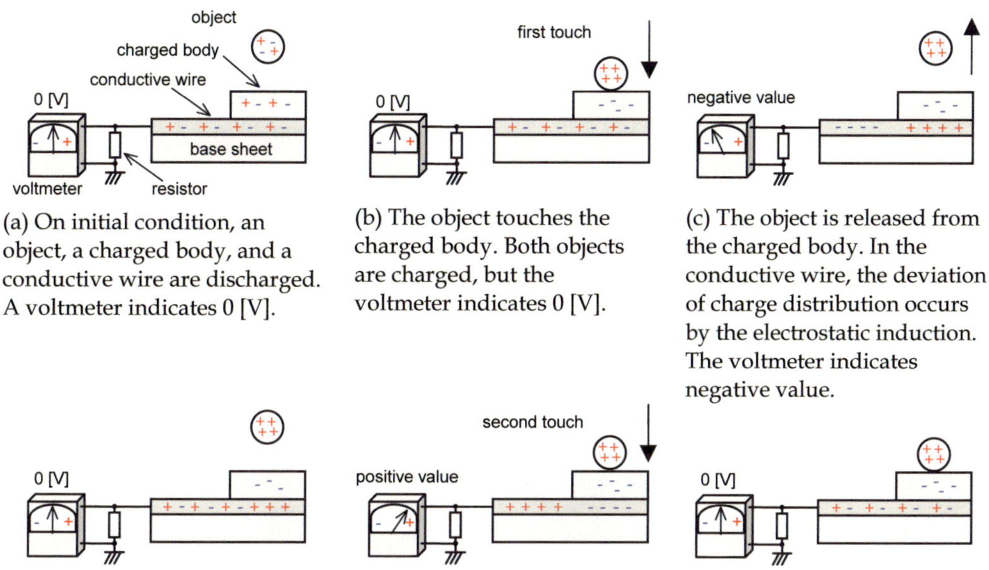

(a) On initial condition, an object, a charged body, and a conductive wire are discharged. A voltmeter indicates 0 [V].

(b) The object touches the charged body. Both objects are charged, but the voltmeter indicates 0 [V].

(c) The object is released from the charged body. In the conductive wire, the deviation of charge distribution occurs by the electrostatic induction. The voltmeter indicates negative value.

(d) The charge moves to the ground through a resistor immediately. The voltmeter indicates 0 [V] again.

(e) The object touches the charged body again. The voltmeter indicates positive value by the electrostatic induction.

(f) The charge moves to the ground. The voltmeter indicates 0 [V].

Fig. 1. A principle of the proposed sensor

3. Experiment

3.1 Experimental setup

We develop a sensor sheet which has a simple structure shown in Fig. 2. Fig. 2(a) shows a top view of the sensor. A square sensor element and a wire for connecting the element to an amplifier are printed on a sheet made of the insulated silicone rubber. The shape of the sensor element is arbitrary. A cross sectional view of the sensor is shown in Fig. 2(b). The sensor element is sealed in the insulated silicone rubber. In addition, the sensor element can be made of arbitrary conductive materials as is clear from the principle of the sensor. For example, if the aluminium foil is sealed in the insulated silicone rubber, it performs as the tactile sensor. In this study, we use the conductive silicone rubber (Fujikura Kasei Co., Ltd., XA-819A) as the sensor element and the conductive wire which can be stretched to nominal 180% (Inoue et al., 2007).

A developed tactile sensor is shown in Fig. 3. The size of the sensor element is square 10 [mm] on a side. The sensor element is connected to an amplifier shown in Fig. 4. The amplifier is a non-inverted amplifier used LF-411 and is connected to a data logger (Keyence Corporation, NR-500). A sampling frequency is 1 [kHz]. Resistors R_s and R_f equal 10 [kΩ]. Therefore, the amplifier doubles input signals. A resistor R_i is the input impedance and equals 1 [MΩ].

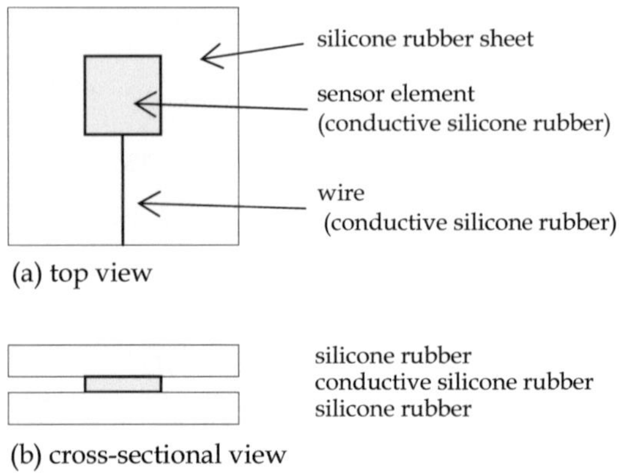

Fig. 2. A structure of the proposed sensor

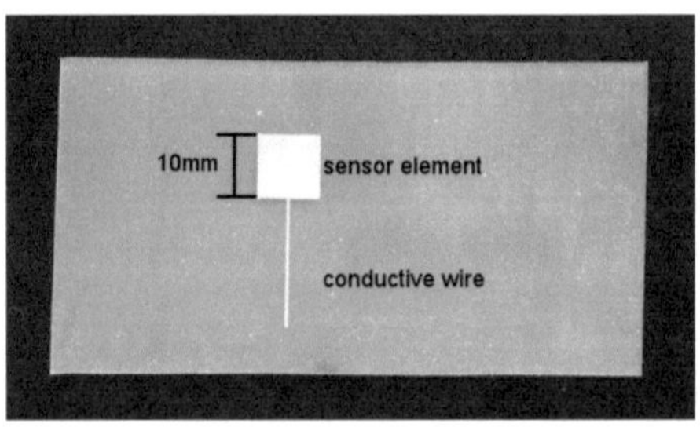

Fig. 3. A photograph of the developed tactile sensor

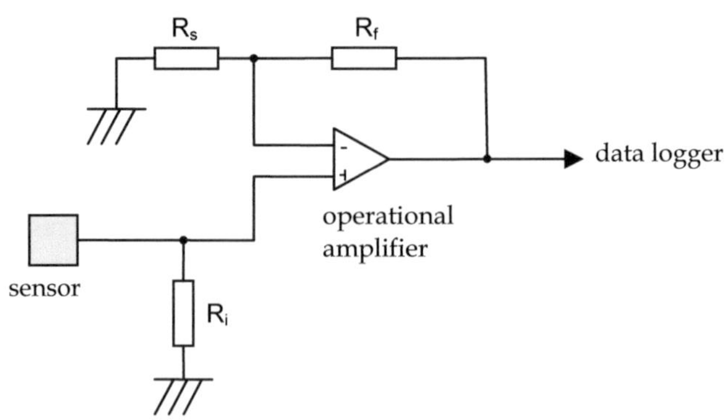

Fig. 4. A diagram of an amplifier circuit

3.2 Characteristics

Fig. 5 shows an example of the sensor output when a human finger touches the sensor. From this figure, the positive value is observed when the finger touches the sensor. The sensor output returns to 0 [V] shortly after the touch. In contrast, the negative value is observed when the finger leaves the sensor. This output characteristic is observed at some objects which are made of different materials. Additionally, the characteristic will change as is pointed out in the section 2 if the sensor is made of different materials. The silicone rubber used in this study is charged to negative very strongly. Therefore, the sensor output shows positive value when almost all the objects touch the sensor.

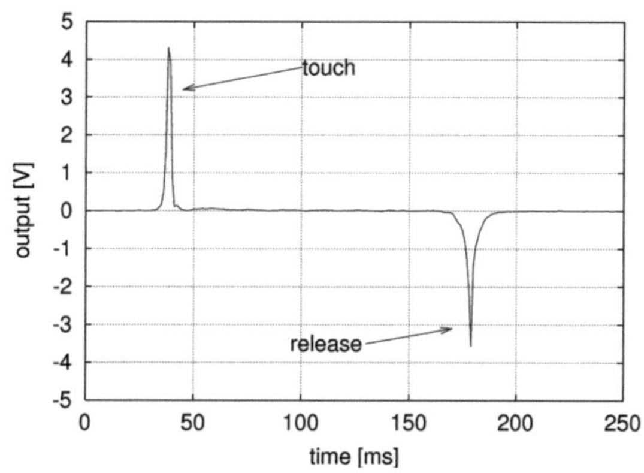

Fig. 5. An example of the sensor output when the human touches the sensor

We investigate the output characteristics of the developed sensor by four experiments: 1) a response to some weights, 2) a response to some areas of the sensor element, 3) a response to some touch velocities, and 4) a response to some materials of the touching object.

In 1) to 3) experiments, we use an object N for touching the sensor. It is a cylinder made of acrylic resin (diameter = 20 [mm], height = 40 [mm]), and a natural rubber sheet is pasted on the bottom surface. In addition, the experiments 2) to 4) are iterated 50 times for each experimental condition.

The first experiment shows a response to some weights. The object N is put on the sensor element on initial condition, and the weight is changed by a compression testing machine. Time courses of the weight and the sensor output are shown in Fig. 6(a). The weight varies 0 [N] to 12 [N]. However, the sensor output continues to indicate 0 [V] because the static electricity does not occur if the contact condition is not changed. Therefore, the sensor does not respond to the weight. In addition, the vibration is observed in the time course of the weight. The reason is that the noise from a motor in the compression testing machine affects a load transducer.

The second experiment shows a response to some areas of the sensor element. The areas of the sensor element are 25, 50, 100, 150, and 225 [mm^2]. The object N is dropped from 30 [mm] height and the sensor output is observed. Because of the bounce of the object, the sensor output is vibratory signal. Thus, we consider a first peak of the sensor output in each trial as a trial result. Fig. 6(b) shows the averages and the standard deviations of peaks of all trials. The figure shows that amplitude of the sensor output is varied by the area of the sensor element. A reason that the sensor output is related to the area of the sensor element is as follows. The current is found by equation (1).

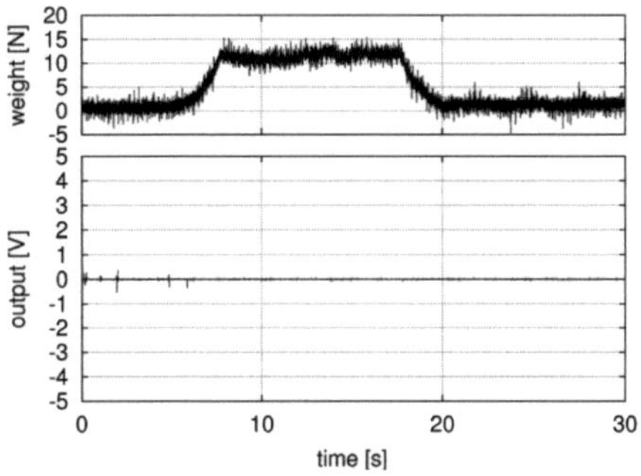

(a) The sensor output when the weight is changed

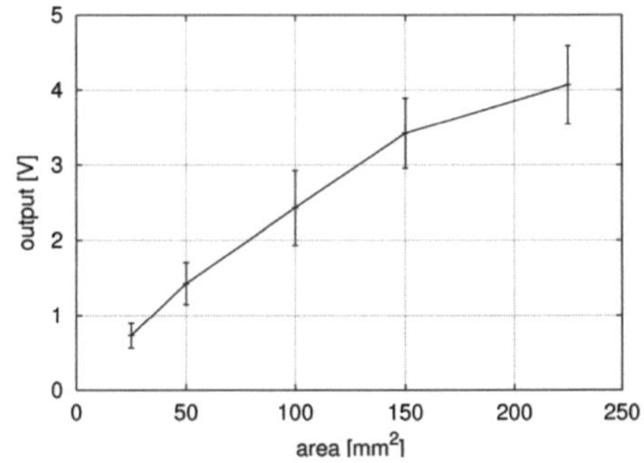

(b) The relationship between the sensor output and the area of the sensor element

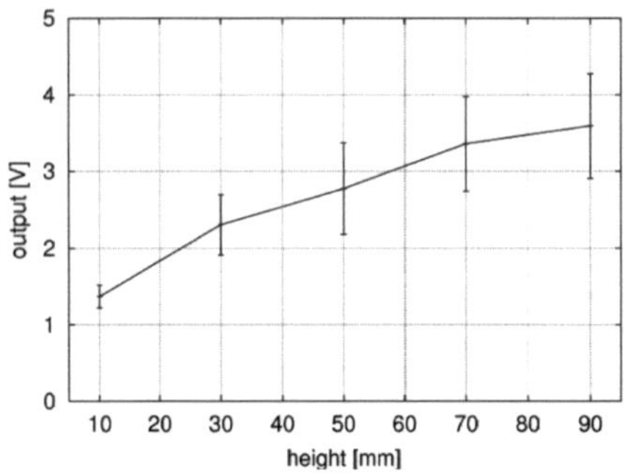

(c) The relationship between the sensor output and the height of fall

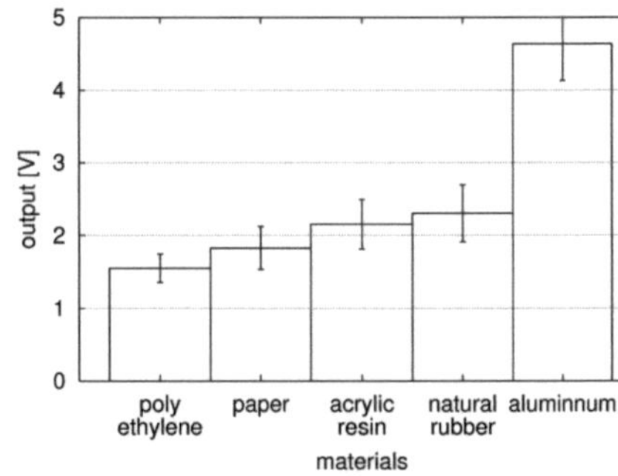

(d) The sensor output in each material

Fig. 6. Characteristics of the proposed sensor

$$I=Q/t \quad (1)$$

where, *I* is the current [A], *Q* is the electric charge [C], *t* is time [s]. The electric charges flow to the ground per unit time because a quantity of the electric charge is very few. Therefore, if the area of the sensor element is large, the quantity of the electric charge by the electrostatic induction increases and the current is large. Moreover, the relationship between the voltage and the current is proportional by Ohm's law. As a result, the sensor output is large if the area of the sensor element is large.

The third experiment shows a response to some touch velocities. The touch velocity is changed by modifying the drop height to 10, 30, 50, 70, and 90 [mm]. In addition, the area of the sensor element is 100 [mm^2]. Fig. 6(c) shows the averages and the standard deviations of peaks of all trials. The result shows that the sensor output is varied by the drop height. In other words, the sensor output is varied by the touch velocity. A reason that the sensor output is related to the touch velocity is similar to the reason described in the previous paragraph. If the touch velocity is slow, the electric charges by the electrostatic induction flow to the ground before observing the sensor output, and the sensor output is small. In contrast, if the touch velocity is fast, much time is required for that the charges flow to the ground. As a result, the sensor output is large.

The fourth experiment shows a response to some materials of the touching object. In this experiment, the area of the sensor element and the drop height is 100 [mm^2] and 30 [mm], respectively. The dropped objects are an aluminium cylinder (diameter = 20 [mm], height = 30[mm]), an acrylic resin cylinder (diameter = 20 [mm], height = 40 [mm]), a polyethylene sheet, a copier paper, and a natural rubber sheet. To make easy the experiment, the polyethylene sheet, the copier paper, and the natural rubber sheet are pasted on the bottom surface of the acrylic resin cylinder (diameter = 20 [mm], height = 40 [mm]). Fig. 6(d) shows the averages and the standard deviations of peaks of all trials. The figure shows that the sensor output is varied by the material of the touching object. It is caused by that the quantity of the static electricity is different by the material of the object.

3.3 Stretchable tactile sensor

In this study, the stretchable conductive silicone rubber is utilized as the tactile sensor element. The silicone rubber keeps electrical conductivity when the rubber is stretched to triple length. Therefore, even if the element is stretched by the external force, the sensor is expected to perform as the tactile sensor. The sensor output when the sensor is stretched is investigated by experiments. We show two experiments in this section. A first experiment shows sensor output when the sensor element is stretching. Another experiment shows sensor output when an object touches the stretched sensor element.

The first experiment shows sensor output when a sensor element is stretching to one way. A shape of the sensor element is square, and its size is 10 [mm] on a side. A tension tester is utilized for stretching the sensor element. Elongation speed is 1000 [mm/min] and elongation length is 40 [mm]. However, the actual elongation length of the sensor element is 10 [mm] because of allowance of fixing jig and deformation characteristics of the silicone rubber. Fig. 7 shows an experimental result. The tension tester is operating from 1 to 3.4 [s]. However, the sensor output is not varied. A reason is that the static electricity does not occur only if the sensor element is stretched. In addition, the sensor output is vibratory as compared to other results because of noise from the tension tester.

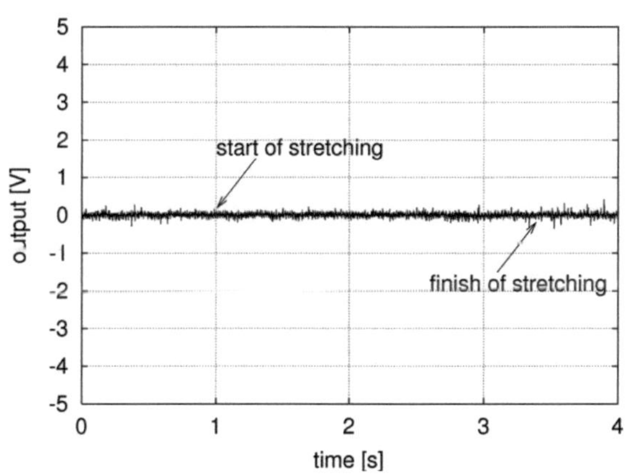

Fig. 7. The sensor output when the sensor is stretching.

The second experiment shows sensor output when an object touches the stretched sensor element. Fig. 8 shows a tactile sensor used in the experiment. An upper photograph shows an unstretched tactile sensor that the size is 5×5 [mm]. The sensor shown in a lower photograph is the same sensor but is stretched to triple length. As a result, the size of the sensor element is changed to 15×3.5 [mm].

In the experiment, two sensors are tested as the stretchable sensor. The size of a tactile sensor element A is 5×5 [mm]. The elongation percentages of the element A are 100% and 200%. The size of another sensor element B is 10×10 [mm]. The elongation percentages of the element B are 25% and 50%. At before and after the stretch, the sizes and the areas of the sensor elements are shown in Table 1. The sensor output when the object N used in the previous section is dropped from 30 [mm] height is observed. Fig. 9 shows an example of the sensor output when the object touches the stretched element A. The figure shows that the sensor performs as the tactile sensor even if the sensor is stretched. The experiment is iterated 50 times for each elongation percentage. The averages and the standard deviations of peaks of all trials are calculated and compared with Fig. 6(b). A result is shown in Fig. 10. The figure shows that the sensor output depends on the area of the sensor element even if the sensor is stretched.

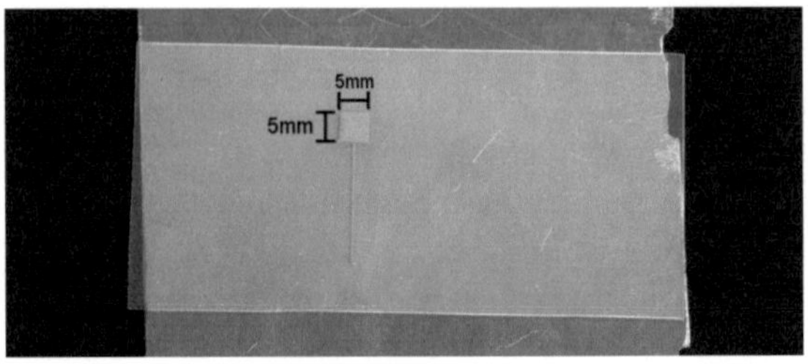

(a) 5×5 [mm]sensor elemnt

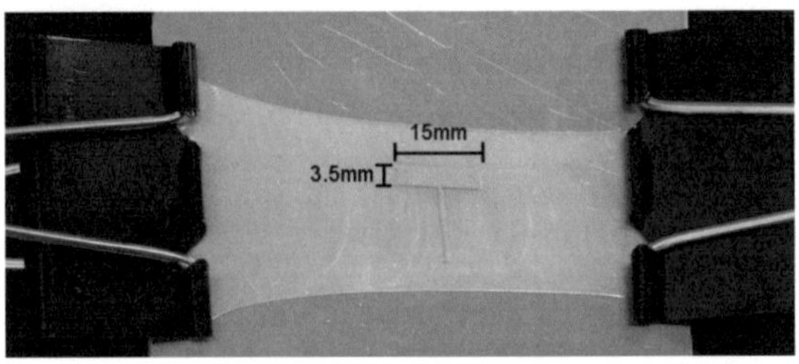

(b) The same sensor elemnt is stretched to triple lentgh.

Fig. 8. Photographs show the same sensor element. The element shown in an upper photograph is unstretched. A lower photograph shows the same sensor element which is stretched to triple length to the lateral direction.

	unstretched	100% stretched	200% stretched
element A	5×5 [mm] 25 [mm^2]	10×4 [mm] 40 [mm^2]	15×3.5 [mm] 52.5 [mm^2]
	unstretched	25% stretched	50% stretched
element B	10×10 [mm] 100[mm^2]	12.5×9.25 [mm] 117.75 [mm^2]	15×8 [mm] 120 [mm^2]

Table 1. The sizes and areas of the sensor elements

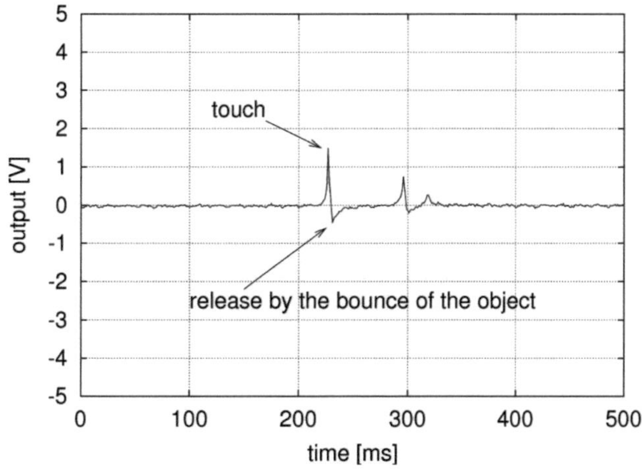

Fig. 9. A figure shows an example of the sensor output when the object touches the stretched element A. Even if the sensor is stretched, the sensor can detect touch and release of the object.

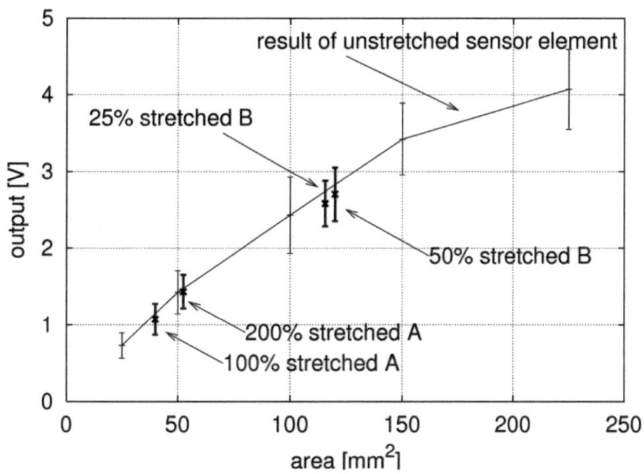

Fig. 10. The results in the section 3.3 are superimposed on Fig. 6(b). The sensor output depends on the area of the sensor element even if the sensor is stretched.

3.4 Investigating cross-talk noise for a multi-element sensor sheet

Many tactile sensor elements are required for the robot skin of humanoid robots. To place many elements, a distance of wiring between the elements has to be approximated. However, if the wiring is close, there is concern that the sensor output is affected by cross-talk noise. In particular, the proposed sensor is easily affected by various noises because the sensor included the amplifier consists of a high impedance circuit. We investigate the amplitude of the cross-talk noise when a distance of wiring is approximated. A sensor sheet used in an experiment is shown in Fig. 11. The sheet consists of two parallel wires. A distance of these wires is 0.2 [mm]. A wire has a sensor element, but another wire has no sensor element. The area of the sensor element is 100 [mm^2]. The sensor output when a finger touches the sensor element is shown in Fig. 12. In Fig. 12(a), a red and a green line show the output of the sensor element and the wiring, respectively. Fig. 12(b) shows a magnified view of Fig. 12(a). In the figure, blue broken lines show a standard deviation of the output of the wiring. The figure shows that the output of the wiring is not affected by the cross-talk noise.

From this result, we developed a sensor sheet which has six sensor elements shown in Fig. 13. In the figure, a square which has vertical lines is a sensor element. A flexible connector is placed on the sheet. The shortest distance between each wire is 0.2 [mm] at the flexible connector. However, the sensor output is not affected by the cross-talk noise.

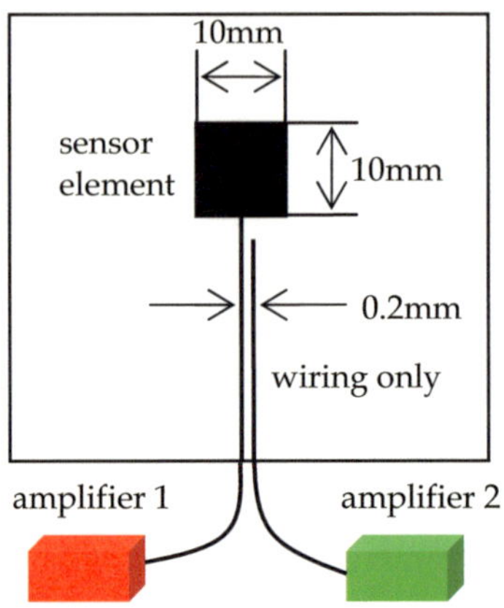

Fig. 11. A figure shows a sensor sheet to investigate the cross-talk noise. The sheet consists of two conductive wires. A wire has a sensor element, but another wire has no sensor element.

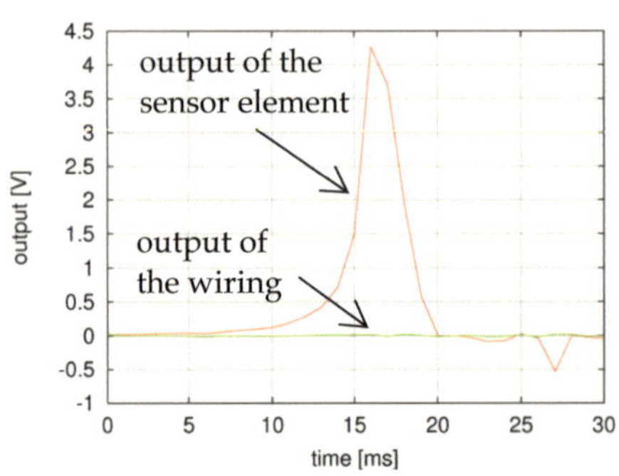

(a) A figure shows that the sensor output of the element and the wiring.

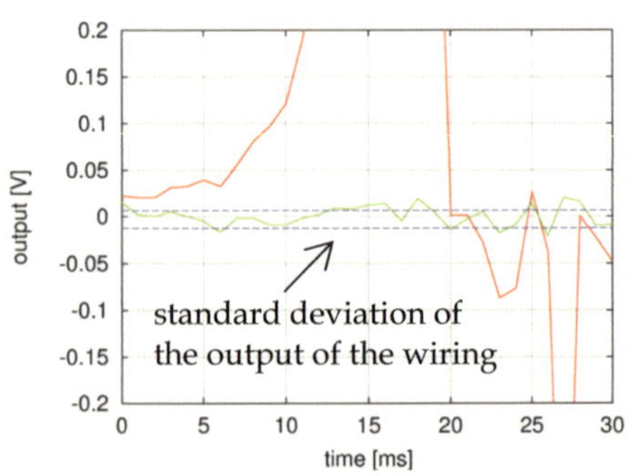

(b) A y-axis of the left figure is magnified. Moreover, a standard deviation of the wiring output is shown by blue broken lines.

Fig. 12. Figures show the sensor output when the finger touches the sensor element. The output of the sensor element is very large whereas the output of the wiring is very small. From this figure, the proposed sensor is not affected by cross-talk noise.

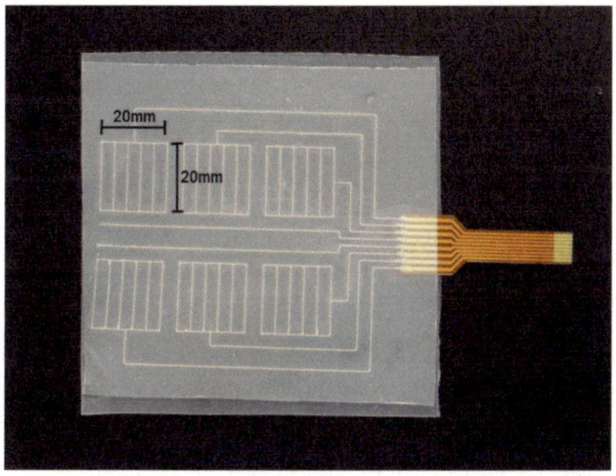

Fig. 13. A photograph shows a developed multi-element sensor sheet. The sensor consists of the six sensor elements and a flexible connector. A square which has vertical lines is a sensor element.

4. Conclusion

This paper proposed the novel tactile sensor utilized the static electricity and the electrostatic induction phenomenon. The sensor consisted of the charged body and the conductive sensor element only. Additionally, the structure of the sensor was very simple. The arbitrary materials were utilized as the sensor. In this paper, the sensor element and the charged body were made of the conductive and the insulated silicone rubber, respectively. Therefore, the developed sensor was flexible and stretchable. Some experiments were performed to investigate the characteristics of the sensor. The first experimental results showed that the sensor output depends on the area of the sensor element, the velocity of the touch, and the material of the object. However, the sensor did not response to the weight. In the second experiments, the sensor output when the sensor was stretched was investigated. The result showed that even if the sensor had been stretched, the sensor could detect touch and release. Additionally, the stretched sensor output also depended on the area of the sensor element. The final experiment showed the proposed sensor is not affected by the cross-talk noise.

The proposed sensor is not to respond to the weight as described above. In general, the characteristic is seemed as a disadvantage. However, it is expected as an advantage if a designer wants to detect just touch and release. Additionally, the sensor is expected to distinguish materials of the objects because the sensor output is varied by the material. We expect that the sensor has advantages in some applications.

On the other hand, there are some disadvantages. To detect the touch of the object, the object has to be charged. Therefore, the proposed sensor should not apply to the collision detection which requires the reliability. Additionally, if the sensor has a dusty or humidity surface, the static electricity does not occur. As a result, the sensing ability is reduced.

In the future work, we investigate the other characteristics of the sensor: rubbing, rolling, and vibration. Moreover, to avoid the depending on the condition of the sensor surface, the other structure of the sensor is proposed.

5. References

Hoshi, T. & Shinoda, H. (2006). Robot Skin Based on Touch-Area-Sensitive Tactile Element. *Proceedings of the 2006 IEEE International Conference on Robotics and Automation*, pp. 3463-3468, Orland, USA, May, 2006

Inoue, M.; Kawahito, Y.; Tada, Y.; Hondo, T.; Kawasaki, T.; Suganuma, K. & Ishiguro, H. (2007), A super-flexible sensor system for humanoid robots and related applications. *Proceedings of the International Conference on Electronics Packaging*, pp. 114–119, Tokyo, Japan, April, 2007

Koterba, S. & Matsuoka, Y. (2006). A Triaxial Force Discernment Algorithm for Flexible, High Density, Artificial Skin, *Proceedings of the 2006 IEEE International Conference on Robotics and Automation*, pp. 1359-1364, Orland, USA, May, 2006

Miyashita, T.; Tajika, T.; Ishiguro, H.; Kogure, K. & Hagita, N. (2007), Haptic communication between humans and robots. *Robotics Research*, ser. Springer Tracts in Advanced Robotics, Thrun, Brooks, & Durrant-Whyte, (Ed.), vol. 28, pp. 525–536, Springer, 2007.

Mizuuchi, I.; Yoshikai, T.; Sodeyama, T.; Nakanishi, T.; Miyadera, A.; Yamamoto, T.; Niemelä, T.; Hayashi, M.; Urata, Y. N. J. ; Nishino, T. & Inaba, M (2006). Development of musculoskeletal humanoid Kotaro, *Proceedings of the 2006 IEEE International Conference on Robotics and Automation*, pp.82-87, Orland, USA, May, 2006

Ohmura, Y.; Kuniyoshi, Y. & Nagakubo, A. (2006), Conformable and scalable tactile sensor skin for curved surfaces. *Proceedings of the 2006 IEEE International Conference on Robotics and Automation*, pp. 1348–1353, Orland, USA, May, 2006

Shan, J. H.; Mei, T.; Sun, L.; Y. Kong, D.; Zhang, Z. Y.; Ni, L.; Meng, M. & Chu, J. R. (2005). The design and fabrication of a flexible three-dimensional force sensor skin, *Proceedings of the 2005 IEEE International Conference on Robots and Systems*, pp. 1965-1970, Edmonton Canada, October, 2005

Shida, K.; Kimoto, A. & Ichinose, Y. (2007), The device to distinguish materials. *Patent in Japan*, No. 2007-57276, 2007.

Tajika, T; Miyashita, T; Ishiguro, H & Hagita, N (2006). Automatic categorization of haptic interactions -what are typical haptic interactions between a human and a robot?, *Proceedings of the 2006 IEEE-RAS International Conference on Humanoid Robots*, pp. 490–496, Genova, Italy, December, 2006

Tajima, R.; Kagami, S.; Inaba, M. & Inoue, H. (2002). Development of soft and distributed tactile sensors and the application to a humanoid robot. *Advanced Robotics*, vol. 16, no. 4, pp. 381–397, 2002

20

Design Considerations for Multimodal "Sensitive Skins" for Robotic Companions

Walter Dan Stiehl
MIT Media Lab, Personal Robots Group
USA

1. Introduction

The sense of touch is the earliest sensory system to become fully functional in all species. Even as early as nine weeks in the womb, the fetus will close its fingers in a gripping motion if its palm is touched (Montagu 1986). The skin, our largest sensory organ, is capable of detecting skin indentations as small as 1μm (Nolte 2002). It is our sense of touch that helps to protect us by allowing us to feel pain so we do not damage our bodies. We manipulate objects using the tactile sensors of our hands. Additionally, touch has an affective and social component - expressed through hugs, handshakes, and pats on the back. One only needs to look at the interactions between humans and their pets to see these expressions of affective touch in action.

To help motivate this discussion of the importance of touch let's conduct a thought experiment. Imagine you wake up in bed and have to start your day without using your sense of touch. How far would you get just relying on vision and sound processing alone? Most likely, you would have trouble leaving your room. Now replace yourself with an autonomous robot and you can see why tactile sensing systems are important.

One major goal of robotics research is to produce robotic systems which can exist in the very complex human world. Touch sensing systems can add a great deal to allowing the robot to function autonomously in this world. First, vision alone is not sufficient to function in unstructured environments due to problems of occlusion (Lumelsky et al. 2001). Second, tactile information can be combined with vision and auditory information to form stronger multi-modal percepts. Third, a full body tactile system can help to convey the "illusion of life" in the robot - as no matter how lifelike the motion of the robot may appear, if the robot is touched and it does not respond, that illusion is instantly broken. Fourth, in the context of human-robot interaction (HRI), the sense of touch can carry social, affective, or other information. In addition to these qualities, a well designed skin should feel pleasant to touch and not distract from the interaction.

In many cases, the field of robotics today has focused much of its attention on vision and auditory sensing. If tactile sensors are used, they have primarily been confined to discrete locations, such as those of Sony's AIBO robotic dog (Sony Product Literature) or have been primarily used in grippers for manipulation tasks, such as NASA's Robonaut (Martin, et al. 2004). While these uses are important, there exists much potential for full-body, multi-modal, tactile sensing systems - or "sensitive skin" as defined by (Lumelsky, et al. 2001) .

Such "sensitive skins" consist of a large variety of sensors with processing capabilities that cover the entire surface of the robot.
Today, there are a few "sensitive skin" systems currently in development. There are fabric sensor based systems such as (DeRossi et al. 2002; Sergio et al. 2002) which are capable of potentially being used for full-body sensitive skins, through currently they are not multi-modal. Other approaches have tried to use inductive coupling to eliminate the wires of such "sensitive skins" (Hakozaki et al. 2001) or have looked to use optics to eliminate the wires in the skin (Yamada et al. 2002). Others such as (Someya et al. 2004) have created flexible skins which combine temperature and pressure sensors into a single flexible skin design with the potential for a large number of sensors. In the realm of robotic companions, the traditional approach has been to use a set of discrete touch sensors in specific locations, and thus there is not usually full-body coverage in such systems. Some examples are the Omron Necoro (Omron Corporation Product Literature) and the Aibo (Sony Product Literature). The one exception is the design of the Paro (Shibata et al. 2001) which features a nearly full body "sensitive skin" but has a small number of sensors which are primarily pressure based. Finally, a good review of the processing capabilities of various "sensitive skins," treating such systems as sensor networks can be found in (Paradiso et al. 2004).
Unlike many previous systems, our approach looks to create a "sensitive skin" designed not just as a patch of skin, but rather to be a fully functional full-body skin on a small robotic companion. Additionally, our emphasis is on social and affective touch and not manipulation. Thus, our desire is to create a multi-modal sensitive skin with a large density of sensors. We believe that the biological design of the human and animal somatosensory cortex provides a great design template for such as skin.
In this chapter I will first present a brief overview of the somatosensory system in humans and animals. Next, using the human and animal system as inspiration I will describe our conceptual approach - the "somatic alphabet." Next I will spend the bulk of this chapter describing our current sensitive skin design for the Huggable, a new teddy bear robotic companion being developed at the MIT Media Lab. Finally, I will end this chapter with a discussion of future areas of research wide open in the field of tactile sensing for robotic systems.

2. Biological inspiration – the human and animal somatosensory system

The human and animal somatosensory system is a great template for how a robotic somatosensory system should be designed. In this section, I will briefly describe some of the important concepts behind the biological system and I encourage the reader to further study these systems if they are interested by looking at the following references in much more detail - (Heller & Schiff 1991; Kandel et al. 2000; Nolte 2002; Rosenzweig et al. 2002; Sekuler & Blake 2002). In this chapter I will work my way from the skin, through the sensors, and then up the spinal cord, finally completing this discussion in the somatosensory cortex of the brain.

2.1 The mechanoreceptors – the "sensors" of the somatosensory system

There are two major types of skin found in the body of humans and other animals. The first type is glaborous, the smooth skin found on the palm of the hand and the sole of the foot.

The second type, hairy, is found on almost every other part of the body. In addition to the two types of skin, the skin can also be divided into two layers - the epidermis above and the dermis below. The actual sensors of the somatic system are referred to as the mechanoreceptors. Each of these receptors has a specific location and stimulus response. Thus, these different mechanoreceptors are the building blocks of our somatic sensation. What becomes instantly clear is that there is not one single sensor for somatic information in the body, but rather, a set of specific receptors which are combined together encode the wide variety of somatic sensations we feel.

These mechanoreceptors can be first divided into the four modality types of somatic sensation as shown in Table 1. For purposes of our discussion we will focus on the mechanoreceptors of glaborous skin in the touch modality – the Meissner's corpuscle, the Merkel disc receptor, the Pacinian corpuscle, and the Ruffini ending. For those interested a good discussion of the pain receptors can be found in (Rollman 1991; Kandel et al. 2000). More information on temperature sensing can be found in (Stevens 1991; Craig & Rollman 1999). Finally, a further discussion of muscle and skeletal mechanoreceptors can be found in (Clark & Horch 1986).

The glaborous skin mechanoreceptors for touch can be arranged in a 2x2 grid as shown in Table 2. One axis of this grid corresponds to adaption, which is how the receptor responds to a sustained stimulus. There are two types of adaptation – slowly adapting (SA) and rapidly adapting (RA). Slowly adapting receptors encode static position and can do so over a period of several minutes. This firing rate first indicates how rapidly the pressure is applied to the skin initially and then in steady state the firing rate is proportional to the skin indentation (Kandel et al. 2000). The rapidly adapting receptors fire at a rate proportional to the speed of motion and their duration corresponds to the duration of the motion (Kandel et al. 2000).

The second axis describes the size of the receptive field, i.e. the how big the area on the skin that receptor is sensitive to. This receptive field size is due to both the location and the depth of the mechanoreceptor in the skin. The mechanoreceptors right below the surface of the epidermis have a small receptive field, while those deep in the dermis have a larger receptive field. The terms punctate (small receptive field with sharply defined boundary) and diffuse (larger receptive field with less definition of boundary) can be used to describe each type (Cholewiak & Collins 1991). A more detailed discussion of how the physical properties of each mechanoreceptor influence their stimulus response can be found in (Kandel et al. 2000; Nolte 2002).

The number and type of mechanoreceptors vary throughout the body. Psychophysical tests such as the two-point limen and the error of localization test have determined that there is spatial perception difference on different parts of the body (Cholewiak & Collins 1991). For example, the fingertips can distinguish a difference in location as small as 2mm on the pad while the back can only distinguish a difference as large as 70mm (Sekuler & Blake 2002). Additionally studies have shown that touch also has temporal properties as well and the limit at which tactile stimuli can be perceived as two separate events if spaced in time was 5ms or more, which is slower than hearing (0.1 ms) but faster than vision (25 ms) (Pohja 1996). A further discussion of other findings from psychophysical studies can be found in (Cholewiak & Collins 1991).

Receptor Type	Fiber Name	Modality
Cutaneous and subcutaneous mechanoreceptors		Touch
Meissner's corpuscle	RA	Stroking, fluttering
Merkel disc receptor	SAI	Pressure, texture
Pacinian corpuscle	PC	Vibration
Ruffini ending	SAII	Skin Stretch
Hair-tylotrich, hair-guard	G1, G2	Stroking, fluttering
Hair-down	D	Light Stroking
Field	F	Skin Stretch
Thermal Receptors		Temperature
Cool receptors	III	Skin cooling (25°C)
Warm receptors	IV	Skin warming (41°C)
Heat nociceptors	III	Hot temperatures (>45°C)
Cold nociceptors	IV	Cold temperatures (<5°C)
Nociceptors		Pain
Mechanical	III	Sharp, pricking pain
Thermal-mechanical	III	Burning pain
Thermal-mechanical	IV	Freezing pain
Polymodal	IV	Slow, burning pain
Muscle and skeletal mechanoreceptors		Limb proprioception
Muscle spindle primary	Ia	Muscle length and speed
Muscle spindle secondary	II	Muscle stretch
Golgi tendon organ	Ib	Muscle contraction
Joint capsule mechanoreceptors	II	Joint angle
Stretch-sensitive free endings	III	Excess stretch or force

Table 1. Somatic Sensation Receptor Types adapted from (Kandel et al. 2000)

The four main types of mechanoreceptors of glaborous skin respond to stimuli in different ways. Numerous studies applying stimuli to the finger pad of monkeys and recording from the individual receptors have shown that texture and roughness can be encoded in different ways at the level of the mechanoreceptor. Some examples of stimuli presented have been dots (Johnson & Lamb 1981; Lamb 1983; Connor et al. 1990; Johnson et al. 1991; Connor & Johnson 1992; Johnson & Hsiao 1992; Johnson & Hsiao 1994), raised letters (Vega-Bermudez et al. 1991), and grooved surfaces (Lederman 1974). From these studies, it has been shown that the encoding of form and texture is done primarily by the Merkel disc receptors with the possibility of some encoding by the Meissner's corpuscles (Johnson & Hsiao 1992).

Other studies have been conducted by recording the response of Merkel disc receptors and Meissner's corpuscles to the encoding of shape (LaMotte & Srinivasan 1993) and curvature (LaMotte & Srinivasan 1987; LaMotte & Srinivasan 1987; Srinivasan & LaMotte 1987). These results have shown that the Merkel disc receptors show an increase in firing rate after indentation as the curvature of the presented stimuli increased.

	Slowly Adapting (SA)	Rapidly Adapting (RA)
Small Receptive Field (Superficial Layers)	Merkel disk receptors	Meissner's corpuscles
Large Receptive Field (Deep Layers)	Ruffini endings	Pacinian corpuscles

Table 2. Glaborous Skin Somatic Sensation Receptor Types adapted from (Kandel et al. 2000)

Finally, studies of vibration have shown that three of the mechanoreceptors of glaborous skin fire an action potential at a rate of one spike per cycle of the sinusoidal wave of a presented stimulus. In addition, these three receptors are "tuned" to different frequencies with the Merkel disc receptors showing a preference for signals between 5-15 Hz, the Meissner's corpuscles preferring a range of 20-50 Hz, and the Pacinian corpuscles are active for a range of 60-400 Hz (Kandel et al. 2000).

Thus from this brief presentation is becomes clear that the mechanoreceptors in the skin of humans and animals are the first building blocks of our somatic sensation. Now we will turn our discussion from the individual receptors to the path that each output signal takes as it travels from the mechanoreceptor to the somatosensory cortex.

2.2 The pathway from receptor to cortex – the "wires" of the somatosensory system

A single nerve fiber will receive input from a cluster of Merkel disc receptors or Meissner's corpuscles. Those mechanoreceptors with a larger receptive field such as the Pacinian corpuscle and the Ruffini ending will be connected to a single nerve fiber. Just as there are different types of receptors in human skin as shown in Table 1, there also exist four different types of peripheral nerve fibers each with different diameters, conduction speeds, and sheath. A further discussion of these fibers can be found in (Rosenzweig et al. 2002) and a mathematical model for the conduction of an action potential along these fibers is found in chapter 6 of (Dayan & Abbott 2001).

What is important to note is that there is an organizational structure as to how the sensory information from each type of somatic receptor enters into the central nervous system through the spinal cord. The division of the spinal cord in humans into 8 cervical sections (neck), 12 thoracic sections (trunk), 5 lumbar sections (lower back), 5 sacral sections (pelvic), and 1 coccygeal section (bottom) maps to specific regions of the skin (known as dematomes) that the spinal nerves innervate (Rosenzweig et al. 2002). Each dermatome actually overlaps the surrounding dermatomes and this helps with higher level processing in the somatosensory cortex, discussed in the next sub-section. The actual specifics of how the sensory nerves synapse, and information travels up the spinal cord to the somatosensory cortex is beyond the scope of this chapter. The reader is encouraged to look at (Kandel et al. 2000; Nolte 2002) for a much more in-depth discussion. Additionally, reflexes, which do not leave the spinal cord level, are further described in (Sekuler & Blake 2002).

2.3 The somatosensory cortex – the "high-level processing" of the somatosensory system

The somatosensory cortex has three major divisions - the primary somatosensory cortex (SI), the secondary somatosensory cortex (SII), and the posterior parietal cortex (Kandel et al. 2000). A full in-depth discussion of the actual physical layout of these systems and their interconnections is beyond the scope of this section and the reader is encouraged to read (Kandel et al. 2000; Nolte 2002) for a greater detail.

The study of the somatosensory cortex has yielded a few interesting results which are particularly important for the design of "sensitive skins" for robotic systems. First, each region of the primary somatosensory cortex is arranged in a somatotopic map. This map shows the number of receptors in each region with areas such as the fingers and lips which have a high density of receptors having more cortical area than those such as the trunk which have a lower density. This map has also been presented in a different fashion, known as the Homunculous, or "little man," where the body form is distorted to reflect the number of receptors present in that region.

A second important fact is that cortical neurons receive input from a large number of mechanoreceptive fibers and inhibition methods (such as surround or lateral) are used to help result in finer modes of discrimination and feature detection (Kandel et al. 2000). In addition, as information moves from lower level to higher level cortical regions, the size of the receptive field increases, but the processing becomes more complex. For example a single lower level receptive field might encompass only a single fingertip, but a higher-level would include the finger pads of all four fingers. Thus, these higher level neurons have been shown to encode such properties as motion, direction, and orientation (Hyvarinen & Poranen 1978). In addition, beyond the somatosensory cortex, other brain regions are integrated with touch to yield multi-modal forms of processing. This discussion is beyond the scope of this chapter, but the reader is encouraged to see (Calvert et al. 2004) for a further discussion.

3. The "Somatic Alphabet Approach"

In the previous section, the biological somatosensory system in humans and animals was described briefly. The goal of this section is to present one approach we have created to help abstract a design methodology for the field of robotics in creating "sensitive skins." We call this approach the "Somatic Alphabet" (Stiehl 2003; Stiehl & Breazeal 2004; Stiehl et al. 2004). Figure 1 shows a diagram of this approach.

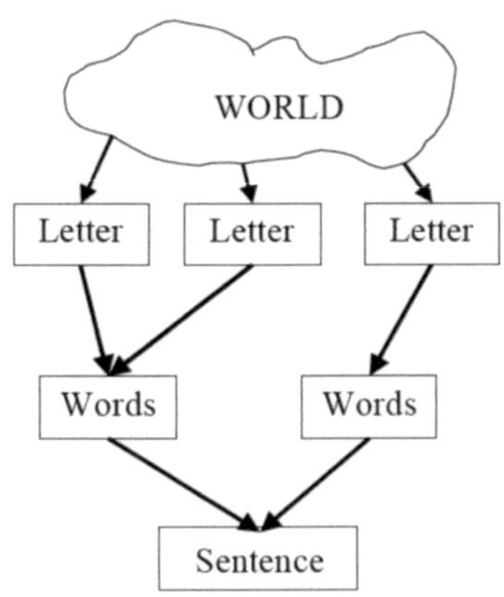

Fig. 1. The Diagram of the "Somatic Alphabet Approach"

The first building blocks of our somatic perception of the world around us are the actual skin receptors in the four modalities of pain, proprioception, touch, and temperature as shown in Table 1. These receptors are designed to encode a specific aspect of a specific type of somatic sensation. Thus nature has created a set of primitives. In the "Somatic Alphabet" approach these primitives are the "letters" of the alphabet.

Just as there is not a single somatic sensor in our skin, thus we should not build "sensitive skins" for robots relying on only one sensor to encode all the properties of the world encountered through touch. From a practical standpoint, it is possible to build a "sensitive skin" using commercially available sensors. For example temperature can be sensed using thermistors, RTDs, or thermocouples. Potentiometers can be used to measure kinesthetic information of joint angle motion. Force sensitive resistors, capacitive sensors, load cells, or vibration sensors can measure touch information. Finally, pain can be encoded as the extremes of temperature or touch information. Additionally, each sensor can be tuned using analog electronics to detect different levels of sensitivity to different input ranges. For example, a voltage divider can be used to change the response of a force sensing resistor (FSR) to different applied inputs. Thus, these hardware sensors act as the "letters" of the "Somatic Alphabet." This approach is similiar to the original concept of a "sensitive skin" consisting of a collection of different sensors as proposed by (Lumelsky et al. 2001).

The next building level of the "Somatic Alphabet" combines the individual "letters" into "words" which now hold a specific meaning. In the biological system, data from the receptors in the skin travels up through the spinal cord to the somatosensory cortex. Here in this cortex, these receptors are combined into receptive fields for higher order cortical process such as orientation, direction of motion, etc as was previously discussed.

In our robotic "sensitive skin" the location and type of the sensors are known due to the specific design layout of the skin. The information from each of these low-level sensors travels from the periphery of the robot to a central processing computer and at each time step this information is aggregated to ascribe somatic meaning to the interaction. This higher level processing, i.e. the creation of the "words" of the "Somatic Alphabet," is the first step towards building meaning about the interaction the robot is having with people and its environment.

Our perception of the world around us consists on not only touch, but rather we combine data from all of our senses to created a much richer understanding of the world around us. For example the smooth surface texture, the color, and the scent all tell us that the object we are holding is an apple and not an orange. Thus the "sentences" of perception can be formed by combining "words" from different modalities in our robotic system. In the next section we will show how we are applying this approach to the design of a multi-modal "sensitive skin" for a robotic companion called the Huggable.

4. The design of a multi-modal "sensitive skin" for a robotic companion

4.1 The huggable robot platform

In January of 2005, our research group began the development of a new type of robotic companion called the Huggable (Stiehl et al. 2005). The Huggable, shown in Figure 2 is designed to function both as a fully autonomous robot as well as a semi-autonomous robot avatar (Lee et al. 2008). We are particularly interested in using this robot to explore how robotic companions for healthcare, education, family communication, and entertainment should be developed. In addition we are developing this system to function as a research platform that can be used by experts in robotics as well as other fields such as eldercare (Stiehl et al. 2008).

The Huggable is being designed with a pair of cameras in the eyes (one color and one black and white), an array of microphones in the head based on (Miaw 2008), and a speaker in the mouth. In the body, the Huggable features an inertial measurement unit based upon (Morris 2004), passive potentiometers in the hips and ankles for joint angle position detection, and an embedded PC with wireless networking. Additionally, the Huggable features a set of 8 quiet, backdrivable actuators using a hybrid belt/gear drive system. These degrees of freedom (DOF) are a 3 DOF neck (nod, tilt, and rotate), a 2 DOF shoulder mechanism in each arm (rotate and in/out), and a 1 DOF ear mechanism. Currently, the Huggable robot runs tethered to a 12V power supply and will ultimately be wireless. The huggable is also being designed with a full body, multi-modal, sensitive skin with over 1500 sensors all over the surface of the body underneath a layer of soft silicone rubber and fur (Stiehl & Breazeal 2006). This skin is the subject of the following sections.

A full discussion of the hardware and software architecture of the Huggable is beyond the scope of this chapter and the reader is encouraged to see our other publications for more information (Stiehl et al. 2005; Lee 2008; Lee et al. 2008; Stiehl et al. 2008; Toscano 2008). Before transitioning to describing the "sensitive skin" system being developed for the Huggable robot, it is important to first motivate this discussion by mentioning a few important design considerations for the Huggable which directly impact the "sensitive skin" design.

Fig. 2. The Current Huggable V3.0 Prototype (in development) (right) and the Concept Plush by Tammy Hendricks (left). In the current prototype only the underlying mechanics of the robot are shown. The sensitive skin system, soft silicone rubber synthetic skin, and final cosmetic fur exterior are not shown in this photo. When fully finished the Huggable robot will look like the concept plush at left.

4.2 Important design considerations for the huggable's "sensitive skin"

The Huggable is designed to look and feel like a soft teddy bear. Thus, in addition to its furry look, the Huggable when touched must feel soft and organic as to not distract from the interaction. Other robotic companions such as the Omron NeCoRo (Omron Corporation Product Literature) place fur over hard plastic coverings or use only plastic covers without fur such as Sony's Aibo (Sony Product Literature). While these are approaches, the feeling is very different for the person interacting with these robots than that of a real animal, soft teddy bear, or other soft skinned robots such as the Huggable and the Paro (Shibata et al. 2003). In the next section we will discuss how we are creating a soft silicone rubber synthetic skin for the Huggable to accomplish this lifelike feel.

Another important design consideration is that the Huggable must feature a full body "sensitive skin" to help promote the illusion of life. No matter how lifelike the Huggable moves or interacts, if a person touches the robot and it does not respond that illusion of life is instantly broken.

Ultimately, we want the "sensitive skin" of the Huggable to be capable of distinguishing a large variety of tactile interactions. Thus we have employed a "Somatic Alphabet Approach" in our design. Our current skin design is capable of detecting aspects from the four modalities of somatic sensation. We use electric field sensing and quantum tunneling composite (QTC) force sensors (Peratech 2004) to detect touch, thermistors to detect temperature, potentimeters to detect joint angle position, and finally pain can be determined as a high value above a determined threshold in each of the previous sensors. In addition to the multi-modal combination of sensors in the skin, we have a high spatial resolution of sensors in the skin - with QTC sensors the width of 1 finger tip in close proximity to one another. This size (width of 1 fingertip) was selected as most of the tactile interactions people will have with the Huggable will involve human touch on the robot in a petting, tickling, or scratching gesture, to name a few. Thus the smallest unit in size is approximately the width of one finger tip.

Finally, from a practical standpoint, the size of the Huggable, approximately 18 inches tall, means that whatever skin that is designed must be developed in a small form factor with much care given to wire management. A bundle of 400 wires would be larger than the arms of the Huggable and thus is not an option. To help reduce the number of wires and make the system as compact as possible, the Huggable features a tree like structure which will be discussed in the next section.

4.3 The electromechanical design of the "sensitive skin" of the huggable

As was previously described in Section 2, the human and animal somatosensory system provides a great template for the design of robotic sensitive skins. In Section 3, the "Somatic Alphabet" approach was presented as a way to create such sensor systems. In this section we will discussion how we have applied lessons from the human and animal somatosensory system and the "Somatic Alphabet" approach to design the "sensitive skin" of the Huggable robot.

As is shown in Figure 2, the current version 3.0 prototype of the Huggable has it's "sensitive skin" design currently under development. For purposes of discussion, in this section we will be presenting the design of the "sensitive skin" of the previous Huggable robot prototype - version 2.0. Lessons learned from this design are currently being applied to the design of the somatic system for our current version 3.0 prototype.

Our "sensitive skin" design decouples the sensors from the actual soft synthetic skin material. This design decision allows for us to not have to make the sensing system flexible, as the silicone skin is capable of large elongations greater than 100%. A further discussion of the synthetic silicone skin appears later in this chapter.

Figure 3 shows two different examples of the "sensitive skin" for two different sections of the Huggable - a section of the arm and the body. Each green circuit board is the first level of the sensing tree and is analogous the skin of the human body. On these sensor circuit boards, the three main sensor types of the Huggable's "sensitive skin" system are found. The white rectangles and squares are the QTC material used for force sensing. These Peratech Quantum Tunneling Composite sensors were chosen for their wide resistance range (10 M-ohm to less than 1 ohm) and low cost (Peratech Product Literature). Additionally the switch substrate material came in A4 sheets which could be cut to any specific size which allowed for much flexibility in our design. There are close to 1400 of these sensors in the current version 3.0 skin design and they are used primarily for high spatial resolution force detection.

The silver circles above the arm sensor circuit board are the Thermometrics NTC thermistors (Thermometrics Product Literature) used for temperature sensing. Ultimately it was decided to replace these through hole sensors with surface mount sensors in the rest of the sections of the "sensitive skin" due to the risk of the small sensor leads breaking during prolonged interaction. Currently we are exploring methods of improving the thermal conductivity of the silicone rubber to get better performance from the surface mount thermistors. However, the slow time constant of these temperature sensors is not a huge

Fig. 3. Two Examples of the "Sensitive Skin" designed for the Version 2.0 Huggable Robot. At left is a section of the arm with units shown in inches. At right is one of the four body region panels with a human finger shown for scale. In each case the silicone skin and soft fur coverings are removed to show the sensor circuit boards.

problem. We do not require fast temperature sensing but rather monitor the temperature rise in a region of contact with a person's body due to body heat.

Finally, the electrodes for the electric field sensors are embedded inside each of the sensor circuit boards on an internal layer. A Motorola/Freescale Semiconductor 33794 Electric Field Sensing IC is used for processing the electric field signal connected to each electrode. This IC was selected as it allowed for 9 input electrodes with a driven shield. The driven shield is connected both to the shield of the coaxial cable carrying the signal as well as a copper shield on the backside of each sensor circuit board (not shown). Additionally, the use of a single IC package solution allowed for the measurement process in one body region to run in parallel while other processes were running on the microcontroller. We currently tune these electric field sensors to be preferential to human contact and thus use them to measure the proximity of a human hand to the surface of the Huggable. This has great human robot applications, as using this sensor we can distinguish between the Huggable sitting in a person's lap or sitting on a table top due to the differences in dielectric material properties and conductivity between a plastic or wood table top and human skin. Another benefit of using this sensor in our skin is that our robot has a soft fur covering of approximately ½" in thickness. Thus, if a child or adult were to gently stroke their hand across the top surface of the fur, they would actually not be applying a force to the skin, but rather just displacing the fur. Ideally, we would want a similar receptor to the hair mechanoreceptors of Table 1. While some have used small piezo sensors with mounted fibers as used in the Tribble project (Paradiso et al. 2004), we felt that given the size limitations and potential places for people to gently brush the fur, it was better to use a proximity sensor for this purpose. The electric field sensor allows us to detect such proximity to about 1" above the surface of the sensor circuit boards and can be read through the silicone rubber sensitive skin. For those who are interested in further reading, a much further discussion of electric field sensing can be found in (Smith 1999).

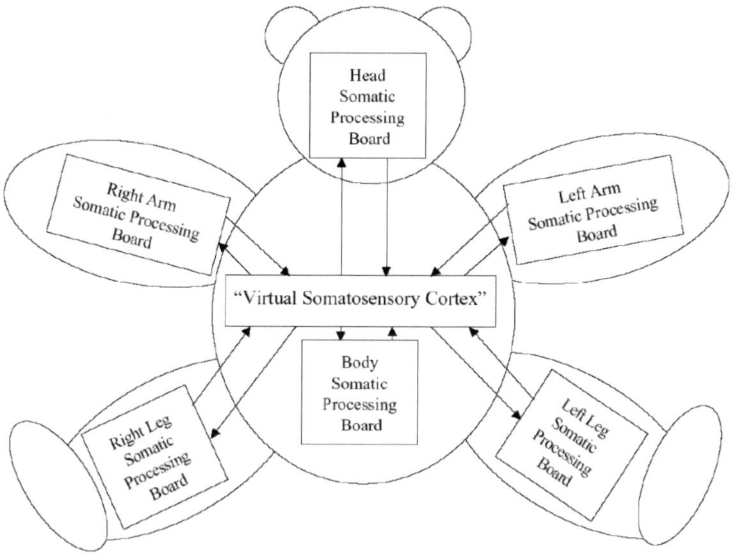

Fig. 4. The Division of the Huggable Robot into Body Regions. As shown in the figure, each body region has its own Somatic Processing Board with the output of each of these circuit boards being sent to the "Virtual Somatosensory Cortex" software system running on the Embedded PC inside the Huggable robot.

Just as the human and animal somatosensory system divides up the skin into dermatome regions, we divide the Huggable's "sensitive skin" into body regions and then sub-regions. First the Huggable is divided into six body regions as shown in Figure 4 with each body region having its own somatic processing circuit board. This circuit board will be discussed in more detail later in the section. This division allows faster processing time for the entire skin system. Additionally, each body region is further divided into independent body sections. Each body section is further divided into a set of sensor circuit boards with each circuit board having a discrete set of temperature, QTC, and electric field sensors. For example, the arm section shown in Figure 3 is one of five arm sections which make up the arm region. As shown in the figure, the arm section shown consists of eight sensor circuit boards with each sensor circuit board consisting of eight QTC, three temperature, and one electrod for the electric field sensor inside the circuit board (not shown). This tree-like structure mirrors the design of the nerves of our somatosensory system.

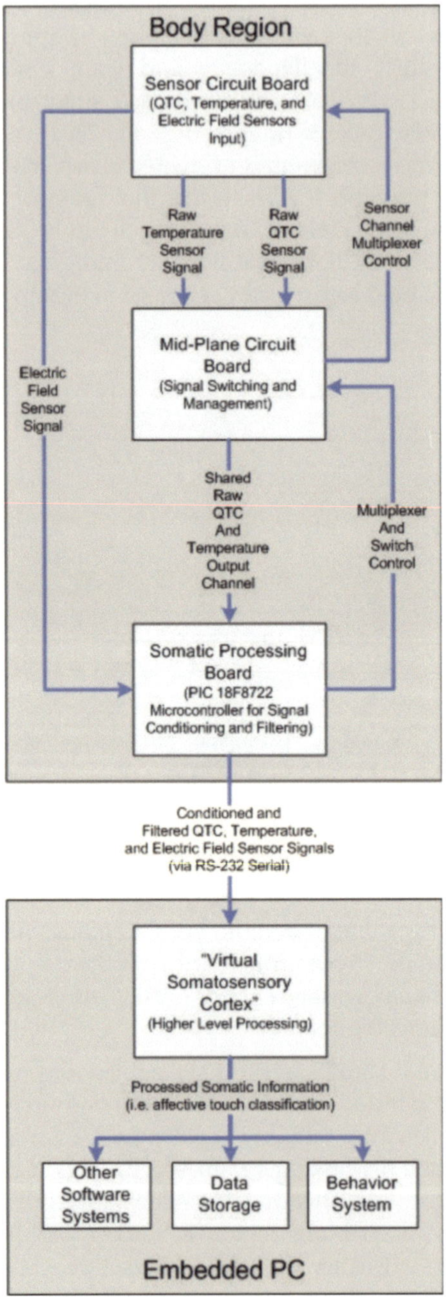

Fig. 5. The Flow of Information in the Huggable's "Sensitive Skin."

To help reduce the overall number of wires we employ a three layered approach of circuit boards in each body region as shown in Figure 5. The first layer is the actual sensor circuit board shown in Figure 3. A pair of on-board multiplexers is used to switch between each QTC and temperature sensor. The electric field sensors have their own path direct to the somatic processing circuit board via a shielded cable to prevent interference. Additionally, in the arm each arm section's set of eight sensor circuit boards are divided into two electrodes with each set of four sensor circuit boards forming one electrode.

The QTC and temperature output of the multiplexers on the sensor circuit board travels to the next layer - the mid-plane circuit board. The mid-plane circuit board has a dual purpose. First, this circuit board mutiplexes the QTC and temperature outputs of each sensor circuit board in a given body region to further reduce the overall wire count.

Second, the mid-plane circuit board features a series of on-board switches to isolate each sensor circuit board from the rest of the system during electric field sensor readings. This isolation is needed to prevent the electric field signal from coupling into the other wires of the system and which carry power, multiplexer control, and ground to the sensor circuit boards. A more in-depth discussion of this design and the isolation process can be found in (Stiehl 2005; Stiehl & Breazeal 2006). Finally, the mid-plane circuit board also acts as a pass through for power, ground, and multiplexer channel select to each of the sensor circuit boards in that body section.

The final layer in each body region is the somatic processing board which is responsible for the control of the entire "sensitive skin" body region, the signal conditioning of the raw sensor values for the temperature, QTC, and electric field sensors, the A/D conversion of these sensor values, and finally the output of these digitized values to the Embedded PC's "Virtual Somatosensory Cortex" software system for higher level processing. Figure 6 shows a schematic overview of the somatic processing circuit board. As shown in this diagram, the QTC sensors are processed in three different ways - light, moderate, and hard. Each processing method was done using a different voltage divider resistor value to sense a different part of the exponential sensor range of the QTC switch substrate (Peratech Product Literature). In addition, though not shown in Figure 6, each signal conditioning system featured a set of digital potentiometers which were designed to be used as part of an auto calibration function. Further work has been done on the auto calibration approach and the description can be found in (Wang 2008). A more detailed description of the somatic processing board is beyond the scope of this chapter, but the reader is encouraged to look at (Stiehl 2005; Stiehl & Breazeal 2006) for more information.

4.4 Synthetic silicone skin

As shown in Figure 7, the Huggable features a soft silicone rubber skin between the soft fur exterior and the hard sensor circuit boards of Figure 3. This synthetic skin helps to provide the Huggable with a more life-like organic feel while helping to protect the sensors. As we wanted to use a soft silicone rubber, a series of tests using professional special fx silicone rubbers were conducted. The details of these tests can be found in Chapter 13 of (Stiehl 2005). The final formulation chosen was Silicone's Inc XT-298 with 20% silicone fluid. This rubber was cast into 3-D printed molds to ensure a proper thickness of 1/8" to ¼" depending on the body location. A full discussion of the casting and pigmentation of silicone rubber is beyond the scope of this chapter, but the reader is encouraged to see (McLaughlin 1999) for more information.

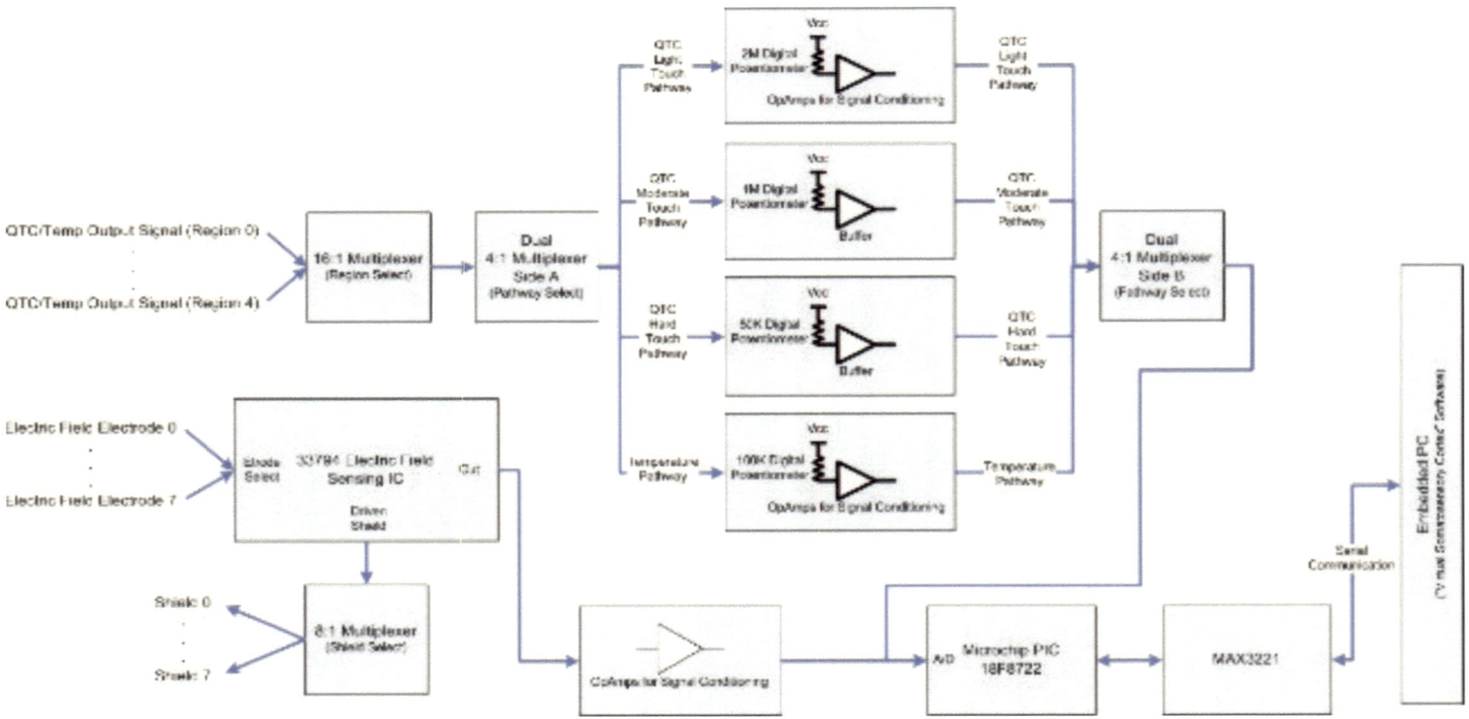

Fig. 6 The Schematic Overview of the Somatic Processing Circuit Board. Note the auto calibration functionality is not shown.

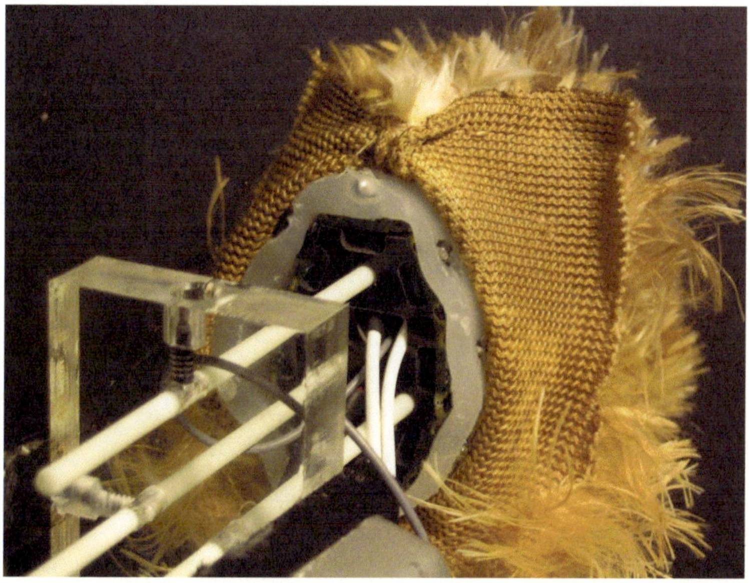

Fig. 7. The Arm Section of the Huggable "Sensitive Skin" showing the layered structure of soft silicone rubber sandwiched between a fur exterior and the sensor circuit boards.

4.5 The "virtual somatosensory cortex"

The somatosensory cortex of humans and animals contains cells which use population coding to arrive at higher levels of processing, as was previously described in section 2. In the population coding approach cells respond to a specific feature with in a given receptive field. This receptive field is formed by combining a set of individual receptors. The outputs of these cortical cells form the inputs to higher levels of processing further on in the somatosensory cortex. In the "Somatic Alphabet Approach," the "words" are formed by this process.

We employ a similar approach in the design of our "Virtual Somatosensory Cortex," which is the name of our somatic processing software system which runs on the embedded PC of the Huggable robot. Our current approach is an extension of work originally done in the early design of a pair of tactile sensing hands for the Leonardo robot at the MIT Media Lab (Stiehl 2003; Stiehl & Breazeal 2004; Stiehl et al. 2004; Stiehl 2005). Just as with the biological system, we employ a population coding approach. Here individual sensor outputs which have been conditioned, filtered, and digitized by the Somatic Processing Circuit Board are combined into receptive fields which are formed in a hierarchical tree. This tree has the smallest unit of a single sensor circuit board, and these sensor circuit boards are combined into larger and larger receptive fields for higher levels of processing, as is the case in the human and animal somatosensory system. From a computational approach, this ability to have multiple levels of receptive fields allows for different interactions to be encoded. For example, a "look at where touched" behavior would only need to know the rough location where the robot was touched, but a more complex behavior needs to not only know where, but also how the robot is being touched, for example giggling if tickled. Finally, the processed outputs of the "Virtual Somatosensory Cortex" system are then fed to the behavior system and integrated with other sensor data, such as the IMU, vision, or sound.

A detailed discussion of the various methods we currently employ in the calculation of our sensitive skin is beyond the scope of this chapter and the reader is encouraged to read the indicated references for more information. Our current system design first normalizes all sensor values with respect to a baseline to account for small changes in individual sensors. Then a set of low-level features, including determining the centroid of force in a given receptive field, the orientation, the direction of motion, timing characteristics, and other information is calculated following the steps outlined in (Stiehl & Breazeal 2004; Stiehl 2005; Stiehl & Breazeal 2005). As a first test to verify that affective processing was possible using our skin, an off-line neural network approach using MATLAB was done as described in (Stiehl & Breazeal 2005). This approach focused on just a small section of the arm and

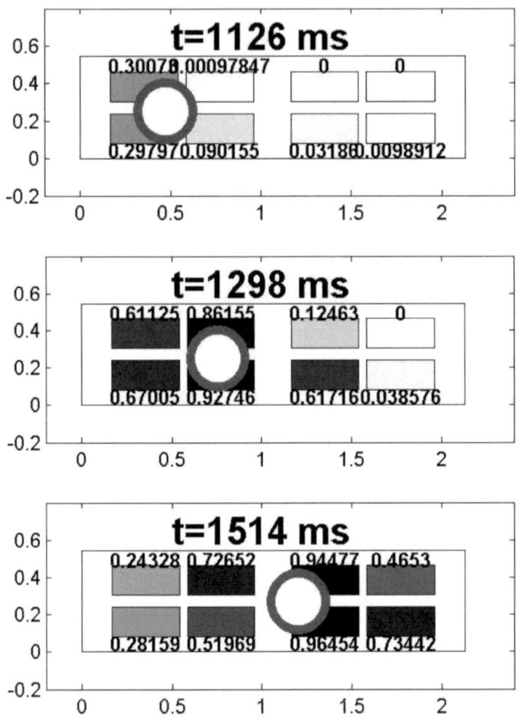

Fig. 8. The calculation of the QTC sensor centroid location for a petting gesture applied by human hand to the top surface of a sensor circuit board. The sequence in time is shown from top to bottom. Each sensor value is indicated both numerically above or below the sensor, and in greyscale with black as maximum value. The white circle with grey edge is the centroid location.

we are currently expanding our methodology to the full skin and we are currently experimenting with various new pattern classification techniques such as HMMs in the design of our new "sensitive skin" for the Huggable Version 3.0, currently in development. A discussion of these new techniques can be found in (Knight 2008).

4.6 Results

To help illustrate the performance of our sensitive skin we include the following figures originally published in (Stiehl 2005; Stiehl & Breazeal 2006). A detailed discussion of these results is beyond the scope of this chapter, but we highlight only a few figures to show the power of a multi-modal sensitive skin. Figure 8 shows the calculated centroid of force during a petting gesture. Figure 9 shows the raw sensor response to two types affective

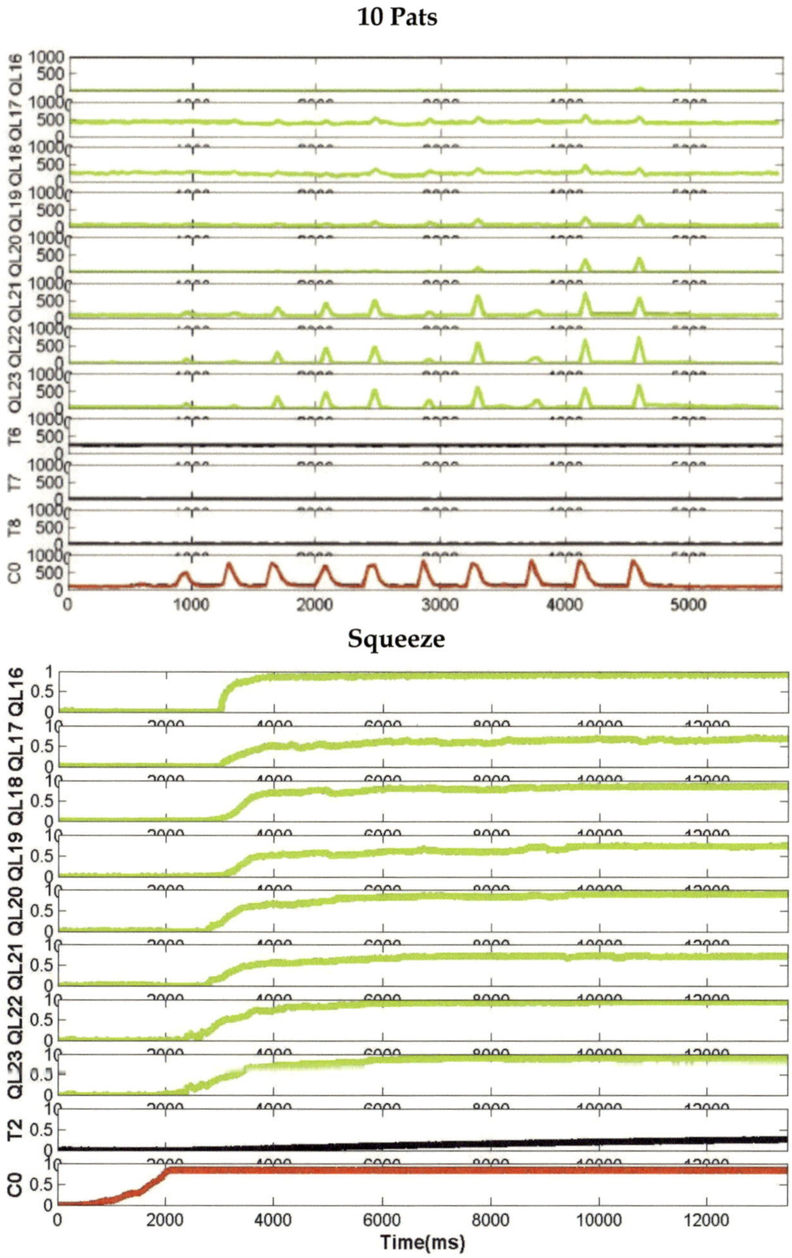

Fig. 9. The response of the QTC (QL, green), temperature (T, black), and electric field sensor (C, red) of one sensor circuit board to two different interactions. At top is a series of 10 pats with sensor values shown as their raw 10-bit value. At bottom is a response to a squeeze with sensor values normalized from 0 to 1.

touch - a series of 10 pats and a squeeze. In both figures, the stimulus was applied by a human hand to the top sensor circuit board in the arm section shown in Figures 3 and 7.

As shown in Figure 8, it is clear that across the time sequence the centroid location is moving in a direction from left to right, as is the case for a petting gesture. In the case of the 10 pats gesture of Figure 9, a few interesting aspects emerge from the figure. First, the QTC sensors directly under the patting gesture (QL21-23) show a larger response than the other sensors, which convey the location that the gesture was applied. Next, it becomes clear that the patting gesture did not have constant force, but all ten peaks are visible in time. Third, the temperature sensors, with their long time constant, do not show any effect. Finally, the electric field sensor (C0) clearly shows all 10 pats as it is a measure of proximity and not force.

The squeeze gesture of Figure 9 demonstrates a different behavior. With the squeeze, the entire sensor circuit board indicates a maximum value. In this case, the QTC light pathway is only shown, and thus all QTC sensors are close to being fully saturated. What is different in this case is that unlike the short interaction of the patting gesture, the squeeze has a much longer interaction time. This longer length allows the temperature sensors to show a gradual increase over time due to the body heat from the person being sense by these sensors. Thus, for longer interactions where the electric field and force sensors may saturate, the slow time constant of the temperature sensors may yield additional information. Finally, the electric field sensor actually responds prior to the onset of touch. As this sensor measures the proximity of contact, the anticipation of touch can be recorded.

While we currently have much more work to do with this system, these early results from the Huggable Version 2.0 "sensitive skin" are encouraging. We believe that the multimodal nature of these skins allow for more types of interactions to be sensed. With a larger potential feature set due to the different types of sensors and methods of processing, we hope that our pattern recognition algorithms will show improved results.

5. Conclusion

The goal of this chapter has been to present the reader with a brief introduction to some of the design considerations for multi-modal "sensitive skins" for robotic companions. In order to motivate how such systems should be designed, a brief discussion of the human and animal somatosensory system was presented. Central to the design of the biological system is the notion that our sense of tactile perception consists of four main modalities - pain, proprioception, touch, and temperature. Within each of these modalities are a set of receptors specifically designed to respond to and encode a specific set of stimuli. These building blocks are then grouped into receptive fields in the somatosensory cortex of the brain where higher lever processing is conducted. Finally, this somatic information is integrated with other sensory systems to form the perception of the world around us.

Using this biological model, the "Somatic Alphabet Approach" was then presented. This approach treats specific sensors as the building blocks or "letters" of our perception. These "letters" are then combined through higher level processing into the "words" which hold the somatic meaning. Finally, these somatic "words" are combined with "words" from other sensory systems to form the "sentences" of the perception of the world around us.

Next, using the "Somatic Alphabet Approach" as inspiration, the design of the multi-modal, "sensitive skin" for the Huggable robotic companion was presented. This "sensitive skin" system featured sensors across the four modalities of somatic sensation - touch, temperature, proprioception, and pain. In addition this system consisted of a dermatome like mapping of the individual QTC, temperature, and electric field sensors into specific body regions, sections, and specific sensor circuit boards. A brief discussion of the electromechanical design and some of the processing methods used in this system followed to show how the design of this system was inspired by the somatosensory system of humans and animals. Finally, the brief results presented show that such multi-modal "sensitive skins" hold great potential for improved interaction processing capabilities and thus could allow for a much richer human robot interaction.

6. Future work

The Huggable is very much an active ongoing project. Over the past year, our focus has been on the final mechanical design of the Huggable Version 3.0 robot as well as the development of the semi-autonomous robot avatar system (Lee et al. 2008; Toscano 2008). As we move back to the "sensitive skin" design for this new robot, we expect to explore new methods to improve this current design. This includes the potential for fully implementing the entire "sensitive skin" with auto calibration functionality, following the method proposed by (Wang 2008). Additionally, Heather Knight in the Personal Robots Group has been exploring the potential for inferring social touch as part of her upcoming MEng Thesis (Knight 2008). Unlike the affective touch approach which looked primarily at analyzing the tactile interaction in a specific region, the approach Heather is taking will integrate tactile information from all across the body of the robot to infer things such as hugs and other types of social touch. Finally, we are going to be expanding the "Virtual Somatosensory Cortex" software system. One current goal will be to allow the Huggable to integrate full body touch with joint angle information and the inertial measurement unit to understand how it is being held, where the person is in relation to its body, and then using this information allow the Huggable to show relational touch behaviors such as nuzzling into the persons body or looking up at them.

Beyond the Huggable, the design of "sensitive skins" for robotic companions holds many open research questions. As shown in Table 1, our human skin features a large number of receptor types. Our current robotic skin with three sensors is a small fraction of the potential a more sensor rich, multi-modal skin could achieve. Thus, there are the engineering challenges of densely packing a large number of different sensor types into one system. Additionally, if such skins were created, they hold a series of processing challenges. Finally, would it be possible to include sensors not in the human system that could benefit robots where they have deficiencies, even by adding features from other sensory systems into the somatic skin. For example, can a "sensitive skin" be designed which functions as a crude vision system to help improve upon a robot's capabilities in the human world?

Lastly, when robots with full body, multi-modal, "sensitive skins" become common place, what new applications could emerge? Could these systems be used for therapy applications where the type of touch imparted to the robot or object is measured and recorded? One example is the notion of play therapy, if a child plays with a toy in the therapy session and it records how that toy is being interacted with through its "sensitive skin," could new methodologies and treatments emerge? Can these "sensitive skins" be combined with prosthetic systems and potentially give an amputee a synthetic sense of touch? Can these skins be applied to furniture to create large surface covers that detect many things about their environment and expand the notion of robot to include furniture? One can clearly see that when we give robots a sense of touch that begins to approach or exceed our own human system, entirely new doors will open in human robot interaction.

7. Acknolwedgements

The author would like to thank the other members of the Personal Robots Group at the MIT Media Lab, specifically Professor Cynthia Breazeal who heads the research group, and the other members of the research group over the past four years who have worked on the Huggable project with Dan. In addition, the author would like to thank Professor Chris Moore of the MIT Brain and Cognitive Science Department who provided the author with help in understanding the human and animal somatosensory cortex, and Professor Joe Paradiso of the MIT Media Lab who provided some early help with the design of the "sensitive skin" described in this chapter. This work is funded in part by a Microsoft iCampus Grant, the MIT Media Lab Things That Think and Digital Life Consortia, and the NSF Center for Bits and Atoms Contract No. CCR-0122419. We also would like to thank Don Steinbrecher for his support of this project. Further information about the Huggable project can be found on our group website at http://robotic.media.mit.edu/projects /robots/huggable/overview/overview.html

8. References

Calvert, G., C. Spence, et al., Eds. (2004). The Handbook of Multisensory Processes. Cambridge, Massachusetts, The MIT Press.

Cholewiak, R. W. and A. A. Collins (1991). Sensory and Physiological Bases of Touch. The Psychology of Touch. M. A. Heller and W. Schiff. Hillsdale, NJ, Lawrence Erlbaum Associates, Inc.: 23-60.

Clark, F. J. and K. W. Horch (1986). Kinesthesia. 1986. K. R. Boff, L. Kaufman and J. P. Thomas. New York, Wiley & Sons. **1:** 1-62.

Connor, C. E., S. S. Hsiao, et al. (1990). "Tactile Roughness: neural codes that account for psychophysical magnitude estimates." Journal of Neuroscience 10: 3823-3836.

Connor, C. E. and K. O. Johnson (1992). "Neural coding of tactile texture: comparisons of spatial and temporal mechanisms for roughness perception." Journal of Neuroscience 12: 3414-3426.

Craig, J. C. and G. B. Rollman (1999). "Somesthesis." Annual Review Psychology 50: 305-31.

Dayan, P. and L. F. Abbott (2001). Theoretical Neuroscience: Computational and Mathematical Modeling of Neural Systems. Cambridge, Massachusetts, MIT Press.

DeRossi, D., F. Carpi, et al. (2002). Electroactive Fabrics for Distributed, Conformable and Interactive Systems. Sensors 2002, Orlando, Florida, IEEE.

Hakozaki, M., A. Hatori, et al. (2001). A Sensitive Skin Using Wireless Tactile Sensing Elements. 18th Sensor Symposium.

Heller, M. A. and W. Schiff, Eds. (1991). The Psychology of Touch. Hillsdale, NJ, Lawrence Erlbaum Associates.

Hyvarinen, J. and A. Poranen (1978). "Movement-sensitive and direction and orientation-selective cutaneous receptive fields in the hand area of the postcentral gyrus in monkeys." Journal of Physiology (London) 283: 523-537.

Johnson, K. O. and S. S. Hsiao (1992). "Neural Mechanisms of Tactual Form and Texture Perception." Annual Review of Neuroscience 15: 227-50.

Johnson, K. O. and S. S. Hsiao (1994). "Evaluation of the relative roles of slowly and rapidly adapting afferent fibers in roughness perception." Canadian Journal of Physiology and Pharmacology 72: 488-497.

Johnson, K. O. and G. D. Lamb (1981). "Neural Mechanisms of spatial tactile discrimination: Neural patterns evoked by Braille-like dot patterns in the monkey." Journal of Physiology (London) 310: 117-44.

Johnson, K. O., J. R. Phillips, et al. (1991). Tactile pattern recognition. Information Processing in the Somatosensory System. O. Franzen and J. Westman. London, Macmillan: 308-18.

Kandel, E. R., J. H. Schwartz, et al. (2000). Principles of Neuroscience, 4th Edition. New York, McGraw-Hill Health Professions Division.

Knight, H. (2008). The Role of Touch in Social Communication for Robots. MIT EECS MEng Thesis *forthcoming*.

Lamb, G. D. (1983). "Tactile discrimination of textured surfaces: peripheral neural coding in the monkey." Journal of Physiology (London) 338: 567-587.

LaMotte, R. H. and M. A. Srinivasan (1987). "Tactile Discrimination of Shape: Responses of Rapidly Adapting Mechanoreceptive Afferents to a Step Stroked Across the Monkey Fingerpad." The Journal of Neuroscience 7(6): 1672-1681.

LaMotte, R. H. and M. A. Srinivasan (1987). "Tactile Discrimination of Shape: Responses of Slowly Adapting Mechanoreceptive Afferents to a Step Stroked Across the Monkey Fingerpad." The Journal of Neuroscience 7(6): 1655-1671.

LaMotte, R. H. and M. A. Srinivasan (1993). "Responses of cutaneous mechanoreceptors to the shape of objects applied to the primate fingerpad." Acta Psychologica 84: 41-51.

Lederman, S. J. (1974). "Tactile roughness of grooved surfaces: The touching process and effects of macro- and microsurface structure." Perception & Psychophysics 16(2): 385-395.

Lee, J. K. (2008). Affordable Gesture Recognition Based Avatar Control System. MIT Media Arts and Sciences, Master Of Science Thesis *forthcoming*.

Lee, J. K., R. L. Toscano, et al. (2008). The Design of a Semi-Autonomous Robot Avatary for Family Communication and Education IEEE International Workshop on Robot and Human Interactive Communication (RO-MAN 2008) (to appear).

Lumelsky, V. J., M. S. Shur, et al. (2001). "Sensitive Skin." IEEE Sensors Journal 1(1): 41-51.

Martin, T. B., R. O. Ambrose, et al. (2004). Tactile Gloves for Autonomous Grasping with the NASA/DARPA Robonaut. International Conference on Robotics and Automation (ICRA'04), New Orleans, LA.

McLaughlin, T. (1999). Silicone Art: Materials, Methods & Techniques for Coloring, Painting, and Finishing Silicone Rubber Version 1.1. Sherman Oaks, California, McLaughlin Productions Publishing.

Miaw, D. (2008). The Huggable Four-Channel Microphone Input Board. MIT EECS 6.UAP Final Report.

Montagu, A. (1986). Touching. The Human Significance of the Skin, 3rd Edition. New York, Harper & Row.

Morris, S. J. (2004). A Shoe-Integrated Sensor System for Wireless Gait Analysis and Real-Time Therapeutic Feedback. Health Sciences and Technology Sc.D. Thesis. Cambridge, MIT.

Nolte, J. (2002). The Human Brain: An Introduction to Its Functional Anatomy 5th Edition. St. Louis, Missouri, Mosby Inc.

Omron Corporation Product Literature. "NeCoRo website: http//www.necoro.com."

Paradiso, J. A., J. Lifton, et al. (2004). "Sensate Media - Multimodal Electronic Skins as Dense Sensor Networks." BT Technology Journal 22(4).

Peratech (2004). The Science of QTC: http://www.peratech.co.uk/science.htm.

Peratech Product Literature http://peratech.co.uk/.

Pohja, S. (1996). Survey of Studies on Tactile Sensing: R96:02. Kista, Sweden, RWCP Neuro SICS Laboratory: 1-13.

Rollman, G. B. (1991). Pain Responsiveness. The Psychology of Touch. M. A. Heller and W. Schiff. Hillsdale, NJ, Lawrence Erlbaum Associates, Inc.: 91-114.

Rosenzweig, M. R., S. M. Breedlove, et al. (2002). Biological Psychology, 3rd Edition. Sunderland, Massachusetts, Sinauer Associates.

Sekuler, R. and R. Blake (2002). Perception, 4th Edition. New York, McGraw-Hill.

Sergio, M., N. Manaresi, et al. (2002). A Textile Based Capacitive Pressure Sensor. Sensors 2002, Orlando, Florida, IEEE.

Shibata, T., T. Mitsui, et al. (2001). Mental Commit Robot and its Application to Therapy of Children. International Conference on Advanced Intelligent Mechatronics, Como, Italy.

Shibata, T., K. Wada, et al. (2003). Subjective Evaluation of a Seal Robot at the National Museum of Science and Technology in Stockholm. IEEE International Workshop on Robot and Human Interactive Communication, Millbrae, CA, USA.

Smith, J. R. (1999). Electric Field Imaging. MIT Media Lab Ph. D Thesis. Cambridge, MIT Media Lab.

Someya, T., T. Sekitani, et al. (2004). "A large-area, flexible pressure sensor matrix with organic field-effect transistors for artificial skin applications." Proceedings of the National Academy of Science USA 101(27): 9966-9970.

Sony Product Literature "AIBO ERS-7M2 website: http://www.sonystyle.com/is-bin/INTERSHOP.enfinity/eCS/Store/en/-/USD/LC_BrowseCatalog-Start;sid=5PwkGv1dhiQkIb8nCd8uEbJRgVs-2uQV4R0=?CategoryName=lc_AIBO."

Srinivasan, M. A. and R. H. LaMotte (1987). "Tactile Discrimination of Shape: Responses of Slowly and Rapidly Adapting Mechanoreceptive Afferents to a Step Indented into the Monkey Fingerpad." The Journal of Neuroscience 7(6): 1682-1697.

Stevens, J. C. (1991). Thermal Sensibility. The Psychology of Touch. M. A. Heller and W. Schiff. Hillsdale, NJ, Lawrence Erlbaum Associates, Inc.: 61-90.

Stiehl, W. D. (2003). Tactile Perception in Robots: From the Somatic Alphabet to the Realization of a Fully "Sensitive Skin". MIT Mechanical Engineering Bachelor's Thesis. Cambridge, MIT: 174.

Stiehl, W. D. (2005). Sensitive Skins and Somatic Processing for Affective and Sociable Robots based upon a Somatic Alphabet Approach. MIT Media Lab M.S. Thesis. Cambridge, MA, MIT: 251.

Stiehl, W. D. and C. Breazeal (2004). Applying a "Somatic Alphabet" Approach to Inferring Orientation, Motion, and Direction in Clusters of Force Sensing Resistors. IEEE/RSJ International Conference on Intelligent Robots and Systems (IROS04).

Stiehl, W. D. and C. Breazeal (2005). Affective Touch for Robotic Companions. First International Conference on Affective Computing and Intelligent Interaction, Beijing, China.

Stiehl, W. D. and C. Breazeal (2006). A "Sensitive Skin" for Robotic Companions Featuring Temperature, Force, and Electric Field Sensors. IEEE/RSJ International Conference on Intelligent Robots and Systems (IROS 2006), Beijing, China.

Stiehl, W. D., L. Lalla, et al. (2004). A "Somatic Alphabet" Approach to "Sensitive Skin". International Conference on Robotics and Automation, New Orleans, Louisiana, IEEE.

Stiehl, W. D., J. K. Lee, et al. (2008). The Huggable: A Platform for Research in Robotic Companions for Eldercare. AAAI Fall Symposium: AI in Eldercare (to appear).

Stiehl, W. D., J. Lieberman, et al. (2005). Design of a Therapeutic Robotic Companion for Relational, Affective Touch. IEEE International Workshop on Robot and Human Interactive Communication (RO-MAN 2005), Nashville, TN.

Thermometrics Product Literature NTC Thermistors: http://www.thermometrics.com/assets/images/ntcnotes.pdf.

Toscano, R. L. (2008). Building a Semi-Autonomous Sociable Robot Platform for Robust Interpersonal Telecommunication. MIT EECS MEng.

Vega-Bermudez, F., K. O. Johnson, et al. (1991). "Human tactile pattern recognition: Active versus passive touch, velocity effects and patterns of confusion." Journal of Neurophysiology 65: 531-46.

Wang, Y. (2008). Auto-Calibration of Sensor Signal Conditioning Pathways in the Context of the Sensate Skin of a Social Robot. MIT EECS 6.UAP Final Report.

Yamada, K., K. Goto, et al. (2002). A Sensor Skin Using Wire-Free Tactile Sensing Elements Based on Optical Connection. SICE 2002, Osaka.

21

Compliant Tactile Sensors for High-Aspect-Ratio Form Metrology

Erwin Peiner
Technische Universität Carolo-Wilhelmina zu Braunschweig
Germany

1. Introduction

Robust manufacturing of micro parts at high throughput needs tailored concepts and methods to assure unchanging work piece quality. An actual challenge is the dimensional metrology with meso-to-microscopic high-aspect ratio features of micro-system components, e.g. micro gears for watches and micro motors, ferrules of optical fibers, hollow microneedles for painless drug delivery into living cells, planar reflective lenses for X-ray optics, densily packed arrays of square channels in lobster-eye optics and through silicon vias (TSV) for vertical electrical interconnects of 3D integrated micro systems and circuits (Baron et al., 2008; Engelke et al., 2007; Hon et al., 2005; Li et al., 2007; Meyer et al., 2008; Muralikrishnan et al., 2005; Simon et al., 2008; Vora et al., 2008; Weckenmann et al., 2006). Form and roughness of deep and narrow micro holes play an important role for the function of nozzles designed for liquid injection. The inlet roundness of the hole, e. g., mainly determines whether flow is detached from the wall resulting in a constricted liquid jet or the fluid is atomized into small droplets. Detached flow can be observed with hydroentangling nozzles of 80 to 150 µm in diameter and 1 mm in depth (Anantharamaiah et al., 2007) while in the case of fuel injection nozzles atomization of the fluid occurs. The form of the spray holes was found to be crucial for optimal fuel injection, i.e. for combustion at low consumption and exhaust gas emission (Jung et al., 2008). Diameter, surface roughness, and inlet chamfer radius have a direct influence on fuel spray atomization, air entrainment and fuel-air mixing. Improved fuel economy, better soot oxidation and reduced NOx emission are benefits observed at optimized hole machining conditions. Furthermore, fuel penetration and spray angle were found to depend on the conicity of the hole which is attributed to the fuel atomization near the hole exit (Bae et al. 2002).

While several techniques exist for micro deep hole machining at high throughput, e. g., micro drilling, electro-discharge machining (EDM), laser beam machining (LBM) and deep reactive ion etching (DRIE) (Diver et al., 2004; Hon et al., 2005; Jung et al., 2007; Krüger et al., 2007; Li et al., 2007; Seo et al., 2008), access to the inside surface of deep and narrow micro holes for metrology is very limited. In Fig. 1 the schematic and scanning electron microscopy (SEM) photographs of the seat of a VCO (valve covered orifice) of direct injection (DI) Diesel engine nozzle is shown comprising six spray holes which are 1.1-mm-deep and have a diameter of 165 µm. In the right part of this figure a close-up of one hole by SEM as well as by digital microscopy are shown which display the nozzle surface around the outlet edge of the hole but give only a vague impression of form and roughness of the near-edge inner surface. Form measurement of the inside surface profile or topography requires either sectioning of the nozzle along its axis or impressing the hole with a polymer, which is destructive and/or time-consuming (Anantharamaiah et al., 2007; Baron et al., 2008; Cusanelli et al., 2007; Diver et al., 2004; Hon et al., 2005; Jung et al., 2007; Jung et al., 2008; Krüger et al., 2007). Novel metrological tools are required for fast and non-destructive form and roughness measurements of spray holes, which enable an in-process control of the hole machining parameters under the constraints of batch fabrication at high throughput. Measurement resolution and uncertainty in the nanometer range are needed over a millimeter length scale within a laterally constricted volume of hundred microns in diameter showing that form measurement inside spray holes is a typical example for micro-/nano metrology.

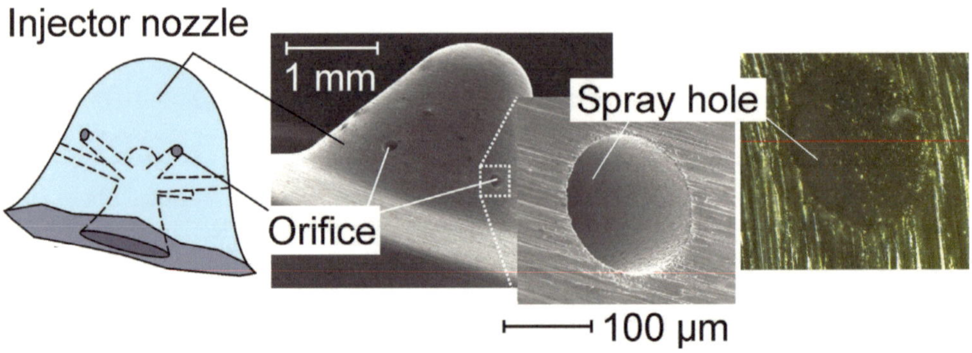

Fig. 1. Schematic (left) and scanning electron microscopy (SEM) photograph (middle) of the round end of a VCO DI Diesel nozzle. In the right part the SEM photograph and a digital optical microphotograph of one spray hole are shown at higher magnifications.

In recent years coordinate measuring machines (CMM) have been constructed for direct metrological characterization of high aspect-ratio microstructures (Bos et al., 2004; Bos et al., 2007; Küng et al., 2007; Peggs et al., 1999; Seitz, 2005; Tibrewala et al., 2008). For narrow holes actively sensing probing elements were developed which were scaled down to diameters of 120 - 300 μm. A schematic of such a type of 3D micro probe is displayed in Fig. 2a. The membrane supporting the shank comprises the stress sensing elements which indicate the position of the probing sphere. Further reduction of the probe diameter is only feasible with simultaneous reduction of the shaft length, i. e. to avoid bending of the probing shaft, its length-to-diameter ratio is limited.

Alternative concepts of tactile micro probe sensors have been proposed and tested based on a compliant instead of a stiff shaft (Table 1). Passive-sensing fiber-optic and vibrating carbon probes comprising either a small glass sphere (Fig. 2b, Cusanelli et al., 2007; Kao & Shih, 2007; Rauh, 2005; UMAP, 2008) or a "virtual tip" (Fig. 2c, Bauza et al., 2005; Woody & Bauza, 2007) as the probing element offer very low probing forces around 1 μN and less. However, the specified uncertainty is rather limited, only for the virtual-tip probe a repeatability of ~ 10 nm is reported (Woody & Bauza, 2007). A severe disadvantage is the probing speed of few tens of pts/s or few tens of μm/s which is too slow for high-throughput form and roughness measurements (Cusanelli et al., 2007). Furthermore, the diameter of the probing sphere, which is 74 μm in the case of the glass fiber probe (Kao & Shih, 2007), is yet a limitation for the accessible hole diameter.

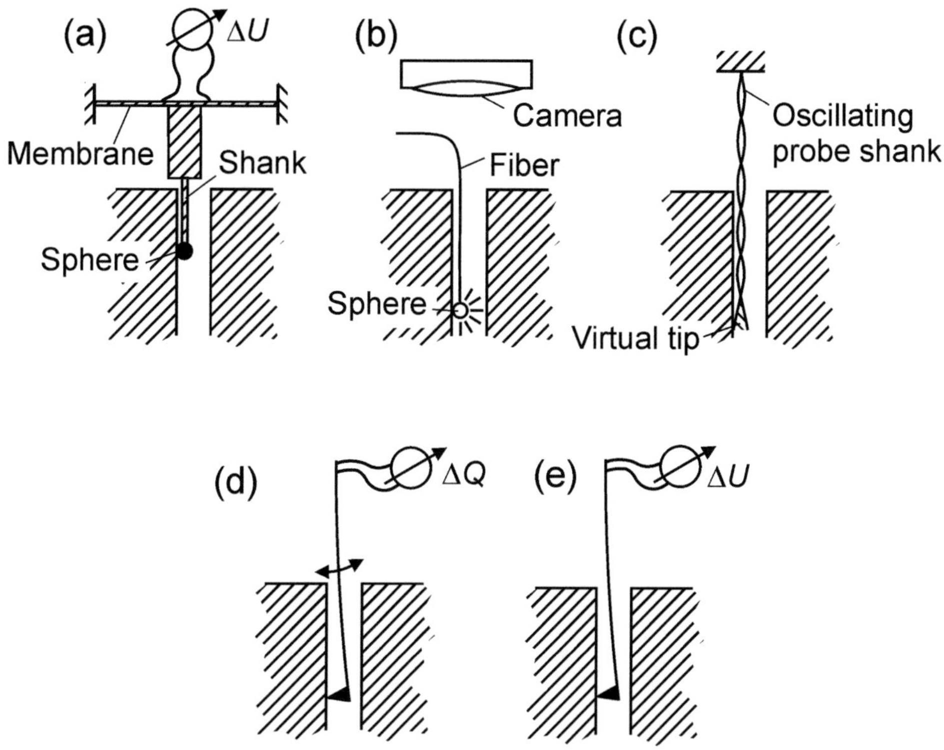

Fig. 2. Schematics of tactile probe sensors with stiff shaft (a) passive-sensing (b, c) and active-sensing (d, e) compliant shaft.

A compliant cantilever which actively senses its physical contact with a surface appears to be the most feasible concept to meet the present lack of metrology capabilities for contour and roughness measurement inside deep, narrow holes (Figs. 2d and e). Both, cantilevers using piezoelectric and piezoresistive signal transduction have been proposed in recent years (Behrens et al., 2003; Lebrasseur et al., 2002; Peiner et al., 2004; Peiner et al., 2007; Peiner et al., 2008; Pourciel et al., 2003; Yamamoto et al., 2000). Piezoelectric cantilevers of very small shaft diameter were fabricated from tungsten carbide with an integrated PZT layer for piezoelectric read-out (Yamamoto et al., 2000). However, its accuracy is limited (~ 0.1 µm) and electro discharge machining (EDM) is needed to realize the shaft.

The potential of cost-effective batch processing using bulk micromachining is provided by silicon cantilevers with piezoresistive read-out (Fig. 2e). They can be operated in both static and dynamic modes for surface scanning either at high speed or at low contact force. Furthermore, piezoresistive cantilevers can be miniaturized and integrated in an array for parallel probing applications (Choudhury et al., 2007; Mathieu et al. 2007). A first prototype of a high-aspect-ratio piezoresistive cantilever was developed and tested for inner-surface profiling of micro holes of 80 µm in diameter and 200 µm in depth showing an uncertainty of ~ 35 nm (Lebrasseur et al., 2002; Pourciel et al., 2003). Tiny tips of height and radius of curvature of 7.2 µm and less than 50 nm, respectively, were described, which are likely to be prone to damage or wear, especially during high-speed operation. Furthermore, larger tip heights are required to get access to undercut or concave surface profiles without needing a tilt of the cantilever (Pourciel et al.,2003).

Parameter	Glass / carbon fiber probe	Cantilever probe	
		Tungsten carbide	Silicon
Shaft length	2 – 200 mm	1 mm	1 – 5 mm
Shaft diameter	7 – 300 µm	10 µm (min.)	20 – 200 µm
Probing element	glass sphere, virtual tip	integrated tip, radius 0.5 – 2 µm	integrated tip, radius 0.5 – 2 µm
Probing force	≤ 1 µN	< 50 µN	1 - 100 µN
Stiffness	few N/m	500 N/m	1 – 10 N/m
Read out	camera / piezoelectric vibration damping	piezoelectric	piezoresistive
Fabrication	piecewise	batch + piecewise	batch
Uncertainty/Repeatability	0.01 - 0.5 µm	0.1 µm	± 35 nm
Probing speed	few tens of pts/s or few tens of µm/s	-	10 – 630 µm/s

Table 1. Comparison of tactile probing sensors for contour and roughness metrology with deep, narrow holes (Bauza et al., 2005; Behrens et al., 2003; Cusanelli et al., 2007; Kao & Shih, 2007; Lebrasseur et al., 2002, Peiner et al., 2004; Peiner et al., 2007; Peiner et al., 2008; Pourciel et al., 2003; Rauh; 2005; Woody & Bauza, 2007; Yamamoto et al., 2000; UMAP, 2008).

In this chapter, tailored slender and compliant cantilevers with integrated tips are addressed which are designed for tactile probing along high-aspect-ratio microstructures, e.g. inside deep and narrow spray holes of DI fuel nozzles (Fig. 1). A full-bridge piezoresistive strain gauge is located on the cantilever designed to be read out using a standard instrumentation amplifier. In the following, design and fabrication of slender cantilever sensors are described as well as their characterization and test in form and roughness measurements with micro holes. A schematic of the probe head and the silicon sensor chip comprising the cantilever with a probing tip and a piezoresistive strain gauge as well as contact pads for wire bonding and die testing/calibration is shown in Fig. 3. Furthermore, the circuit diagram of the strain gauge resistors connected into a Wheatstone bridge is displayed as well as an integrated resistor/diode for temperature control.

2. Tactile sensor

2.1 Layout

Design and fabrication of the slender cantilever sensor was determined by the specific requirements of tactile probing within narrow and deep micro holes:

- A high degree of freedom was required for the geometrical design of the cantilever in combination with high mechanical stability. From this point of view bulk silicon wafers were preferred over silicon-on-insulator wafers (SOI) which have a fixed thickness and are costly.
- The probing tip was integrated at the cantilever's free end. A conical shape of an angle of 60°/90° and a radius of curvature of 2 µm is recommended for measurements of good quality surfaces according to ISO 3274. Therefore we employed anisotropic wet tip etching instead of reactive ion etching often used for sharp and tiny atomic force

microscopy (AFM) tips. Furthermore, in contrast to AFM probes we located the tip at the bottom side of the cantilever. So, robust tips of heights ranging from 10 to 200 μm could be realized while leaving the upper chip surface for a planar integration of the strain gauge.

- The cantilever deflection was measured using a balanced Wheatstone bridge located close to the cantilever clamping. Cantilever dimensions and bridge layout are displayed in Fig. 3. The large contact pads were provided for die testing and calibration. During the back-end processing of the probe head this area was used for the deposition of glue for the fixing of the sensor to the steel finger. Electrical connection was realized via wire bonds from the small pads on the sensor chip to a flexible circuit board glued onto the steel finger.

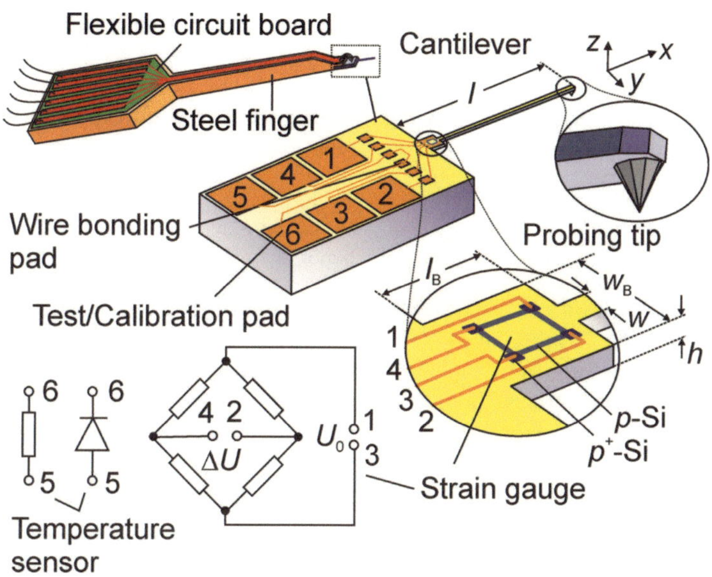

Fig. 3. Schematic of probe head based on a tactile cantilever sensor with enlarged representations of the probing tip and the Wheatstone bridge as well as the circuit diagram of the bridge and a temperature sensing device.

Slender cantilevers of low stiffness as required for probing inside narrow and deep micro holes generate only small strain values upon tip deflection. Therefore, a high gauge factor and an optimum location of the gauge on the cantilever were necessary to meet the sensitivity requirements. Simultaneously, temperature drift, susceptibility to ambient light, power consumption, and noise had to be kept as low as possible. As a trade-off we designed a full Wheatstone bridge of four *p*-type resistors of a sheet carrier concentration of 3×10^{14} cm^{-2} to obtain a bridge resistance of 2.5 kΩ for which we could expect a gauge factor of $K \approx 80$, a temperature drift of ~ (1 – 2) × 10^{-3}/K, noise of ~ 1 μV in a bandwidth of 20 kHz and a power consumption of 0.4 mW at $U_0 = 1$ V.

2.2 Vertical loading

Using the cantilever sensor in Fig. 3 as a tactile sensor three directions of force application to the cantilever free end can be distinguished with respect to the cantilever axis: vertical, lateral and axial loading. In the case of vertical loading, i. e. the normal loading case, a force F_z acts onto the probing tip perpendicularly to the chip plane. It results in a deflection of the cantilever of:

$$\delta_z = \frac{4l^3\left(1-\nu^2\right)}{Ewh^3} c_B F_z = \frac{1}{k} F_z \tag{1}$$

with the plate modulus $E/(1-\nu^2) = 170$ GPa of a (001) silicon cantilever aligned to the [110] crystal direction and l, w and h as defined in Fig. 3. Plane strain is assumed. The cantilever stiffness is denoted by k_z. The widening of the cantilever at its clamped end (w_B, l_B cf. Fig. 3) is taken into account by the factor

$$c_B = 1 - \frac{3l_B\left(l - l_B\right)}{l^2}\left(1 - \frac{w}{w_B}\right). \tag{2}$$

At the cantilever surface uniaxial strain is generated along the cantilever axis depending on the tip deflection δ_z, which has its maximum at the cantilever clamping amounting to:

$$\varepsilon_B = \frac{3h}{2l^2}\frac{w}{w_B c_B} \times \delta_z \tag{3}$$

Stiffness and strain values calculated for the given cantilever geometries ($l = 1.5 - 5$ mm, w = 30 – 200 µm, $h = 25 - 50$ µm, $w_B = 100 - 200$ µm, $l_B = 250$ µm) using eqs. (1) to (3) were compared with the data obtained by finite element modelling (FEM) using ANSYS 8.1. We found an agreement within 2-3 % for the stiffness and 8-10 % for the strain.
Four resistors R_{ij} (indices denote the numbers assigned to each resistor contact) are connected into a full Wheatstone bridge (Fig. 3). Assuming for simplicity that each of the four legs of the bridge, which are aligned either in parallel (longitudinal: R_{14} and R_{23}) or perpendicularly (transversal: R_{12} and R_{34}) to the cantilever, is uniformly strained by ε_B we observe resistance changes of almost identical absolute value but opposite sign. At a constant voltage supply to the bridge of U_0 we find:

$$\frac{\Delta U}{U_0} = K\varepsilon_B \tag{4}$$

with the piezoresistive gauge factor K. Either an additional resistor or a diode is integrated close to the strain-sensing Wheatstone bridge and can be connected via the contacts 5 and 6 for on-chip temperature sensing.

2.3 Lateral loading

In general, during scanning over a not ideally flat work piece surface the cantilever may be deflected not only in vertical but also in lateral direction, i.e. the probing force acting on the tip is a superposition of vertical and lateral contributions. A lateral force F_y applied to the tip caused e.g. by friction forces emerging during scanning the cantilever over a surface in the direction perpendicular to the cantilever axis lead to a lateral cantilever deflection. Simultaneously, a moment about the cantilever axis is exerted causing an additional tip deflection. In total we obtain:

$$\delta_y = \left[\frac{4l^3(1-\nu^2)}{Ehw^3} + c_{torsion}\frac{(h_t + h/2)^2 l}{Gwh^3}\right]F_y = \frac{1}{k_y}F_y \tag{5}$$

with the shear modulus $G = E/(1 + \nu) = 80$ GPa ($\nu = 0.064$) and $c_{torsion} = 3.6$ for $h/w = 4$ and $c_{torsion} = 7.1$ for $h/w = 1$ (Bao 2000). In the present case of slender cantilevers, i. e. h . w . $0.02l$ and a tip height of h_t # h the torsional contribution is more than two orders of magnitude smaller than the lateral bending. This was confirmed by FEM. Non-uniform uniaxial strain across the Wheatstone bridge is induced: At a lateral deflection δ_y the longitudinally oriented resistors (R_{14} and R_{23}) are strained at equal absolute value but at opposite sign

$$\varepsilon_B = \frac{3w}{4l^2} \times \delta_y$$

while strain across the transversally oriented resistors (R_{12} and R_{34}) averages to zero. The longitudinal resistors are located at $\pm w/2$ from the neutral axis. Connecting both longitudinal resistors into a half bridge (hb) we obtain an output signal of:

$$\left(\frac{\Delta U}{U_0}\right)_{hb} = \frac{1}{2}K\frac{3w}{4l^2} \times \delta_y \tag{6}$$

2.4 Axial loading

Axial loading results from moving the cantilever with its free end against a fixed body. Three modes of deflection of an axially loaded cantilever which have been implemented in MEMS technology (Beyeler et al., 2008; Ruther et al. 2007; Samuel et al., 2006) are schematically shown in Fig. 4 where cantilevers fixed to a support by clamping (left), a hinge (middle) and a spring (right) are depicted. Due to its slender shape the first one is best suited for probing the bottom surface of deep and narrow blind holes, e.g. through silicon vias (TSV) for 3D interconnects. Under an axial load F_x a cantilever beam is uniformly compressed until buckling occurs, when F_x exceeds a critical value:

$$F_c = \beta\left(\frac{\pi}{l}\right)^2 \frac{Ewh^3}{12} \tag{7}$$

In this case a uniform rectangular cross section was assumed. The constant β depends on the boundary conditions, i. e. $\beta = 1/4$ for a beam with one end clamped and the other free (type I) and $\beta = 1/(0.7)^2$ for a beam with one end clamped and the other pinned (type II). In the case of slight initial cantilever bending buckling occurs gradually as the load approaches F_c. Below F_c the cantilever is uniformly compressed leading to a strain at the piezoresistive bridge of:

$$\varepsilon_B = \frac{k_x}{Ewh} \times \delta_x \tag{8}$$

with the axial stiffness of the cantilever:

$$k_x = \frac{Ewh}{l} \tag{9}$$

For a 5-mm-long cantilever and a gauge factor of the piezoresistive bridge of 80 we find a sensitivity of 16 mV/µm.

Fig. 4. Schematic of tactile sensing using axially loaded cantilevers.

2.5 Fabrication

Sensor prototypes were realized using a bulk micromachining process which is schematically shown with a sectioned piece of the silicon chip in Fig. 5:

- An *n*-doped (100) silicon wafer (300 ± 3 µm) was thinned in a time-controlled process using either deep reactive ion etching (DRIE, SF_6/O_2) at cryogenic temperature (75 °C to (-95 °C) or wet anisotropical etching in TMAH (tetra methyl ammonium hydroxide, 20 %, 80 °C) solution through a mask of photo resist or thermal oxide, respectively. Etching was stopped at a residual thickness corresponding to the desired cantilever thickness plus the tip height (Fig. 5a). The standard deviation of the thickness measured with the generated membranes was typically less than 1 %. An advantage of cryogenic DRIE over anisotropic wet etching is the considerably higher etch rate of ~ 4 µm/min vs. 0.7 µm for TMAH. Thus, the time consumed for this process step is drastically reduced from ~6 h to ~1 h. Furthermore, a photo resist mask can be employed instead of thermal oxide needed for TMAH etching.
- Subsequently, *p*-type stripes arranged in a square geometry were designed as the resistor legs of a full Wheatstone bridge (Fig. 5a). They were realized by boron diffusion from a spin-on silica emulsion source (Emulsitone Borofilm 100) or by boron implantation. Contact formation to the *p*-type silicon was improved by an extra boron diffusion/implantation dose in the corner regions of the bridge square (Fig. 5b). The standard deviation of the measured resistivity about the target value was 4.1 % and 0.6 % for the diffused and implanted wafers, respectively. The doping profile was measured during various stages of the process with monitor wafers using electrochemical capacitance-voltage profiling (ECV). Subsequent to the final high-temperature step we found a junction depth of 4.5 µm and a surface concentration of $1.5\text{-}3.0 \times 10^{18}$ cm^{-3} which is a tradeoff to obtain a high piezoresistive coefficient around $\pi_{44} \approx 1$ GPa^{-1} and a low temperature coefficient around 1×10^{-3} K^{-1} (Cho et al., 2006).
- A probing tip was generated at the cantilever bottom side by undercut etching of a circular or square oxide (nitride) mask using either TMAH or KOH (Fig. 5c). In this case

photolithography had to be performed within the backside-etched depression shown in Figs. 5a and b. Its depth was determined by the desired tip height, i.e., it has a maximum value of ~ 250 µm for the smallest tips. Using single exposure of positive resist (Shipley, S1818) we realized squares of an edge length of ~ 70 µm as the smallest structures showing deviations from the desired length of typically less then 10 %. During anisotropic etching a micro pyramid with an octagonal base developed underneath the mask with its angle of apex determined by the emerging sidewall facets. We used TMAH (20 %, 80 °C) and KOH (45 %, 80 °C) to generate tip angles of ~ 90° and ~ 40°, respectively. SEM photographs of tips in the backside-etched groove before and just after complete under etching of a square oxide mask using KOH (45 %, 80 °C) are depicted in Figs. 6a and b, respectively.

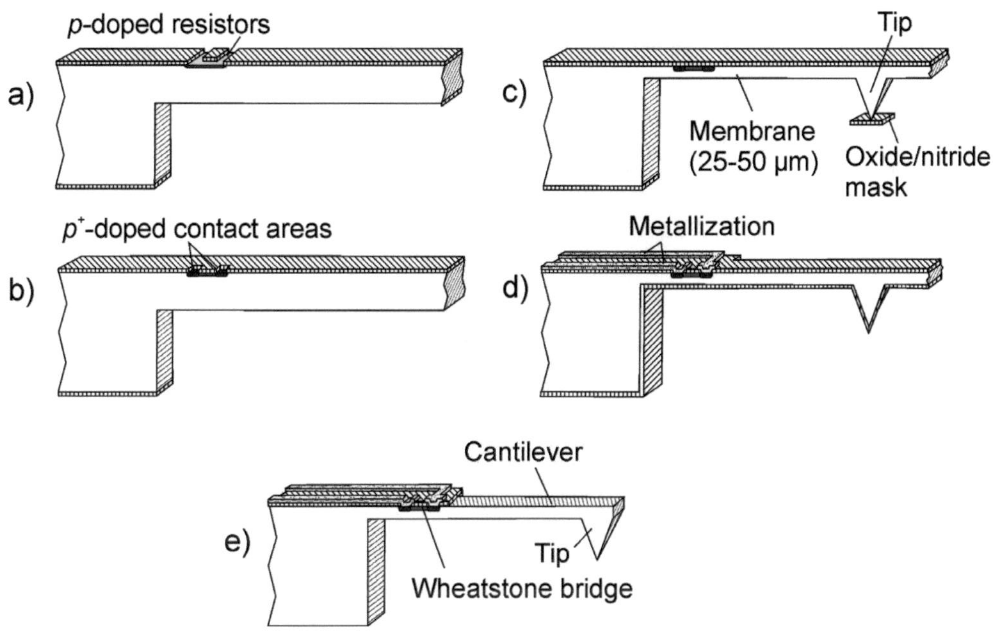

Fig. 5. Schematic of the sensor fabrication process: Membrane etching (a), boron doping (a, b), tip etching (c), metallization (d) and cantilever etching (e).

- After tip formation the wafer was oxidized and patterned for contact holes to the Wheatstone bridge. Either a gold/chromium or an aluminum metallization was used (Fig. 5d).
- Finally, the cantilever was released by either DRIE at cryogenic temperatures using SF_6/O_2 or anisotropic wet etching using KOH (30 %, 60 °C) (Fig. 5e). In both cases a protection of the Au/Cr metallization was not necessary. In the case of DRIE we could employ a photo resist mask and a CMOS compatible Al metallization while an oxide mask and an Au/Cr metallization were used for the KOH process. Samples of the cantilever sensor of 1.5-5 mm in length, 30-200 µm in width and 25-50 µm in height are shown in Fig. 7.

Figure 8 shows a realized probe head comprising the cantilever sensor mounted on a steal finger, a retractable plastic cover protecting the cantilever during transport and mounting bracket.

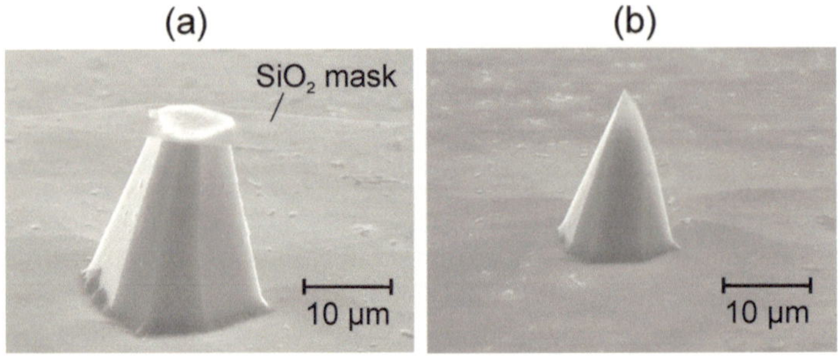

Fig. 6. SEM photographs of tips in the backside-etched groove before (a) and just after complete under etching (b) of a square oxide mask using KOH (45 %, 80 °C).

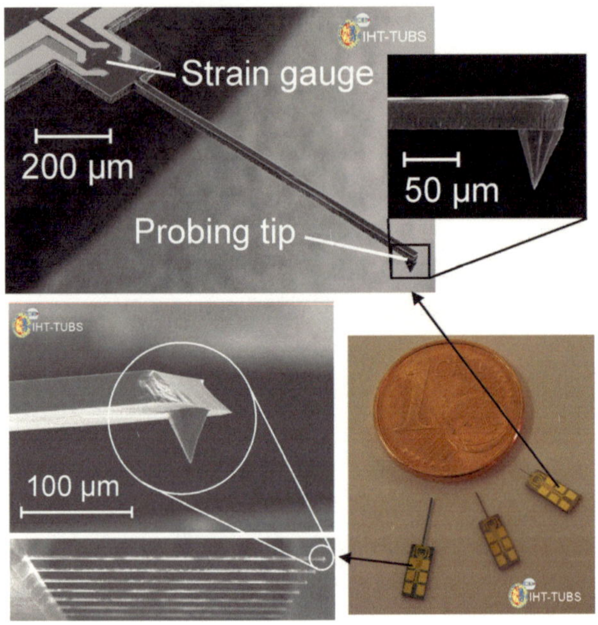

Fig. 7. Samples of slender piezoresistive cantilever sensors with integrated probing tip. Either DRIE at cryogenic temperatures (upper) or wet etching using KOH (lower) was employed for the final release of the cantilevers.

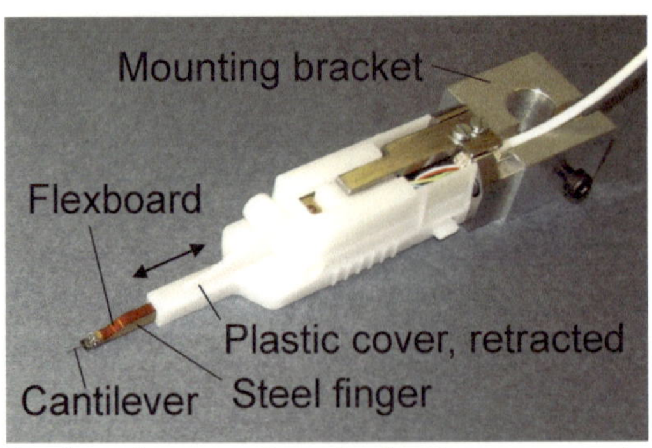

Fig. 8. Probe head after back-end processing. A plastic cover serving as a protection of the cantilever during transport and mounting into a scanning unit is retracted.

3. Sensor performance

Realized sensors were calibrated using a nanonewton force testing setup (Behrens et al., 2003; Peiner et al., 2007; Peiner et al., 2008). For this purpose the sensor dies were mounted into a custom-made metal case. Electrical connection was provided using contact pins which were pressed against the large contact pads shown in Fig. 3 serving as the counterparts for a temporary, easily detachable connection. Temperature and relative humidity in the calibration box were stabilized within 21.4 - 23.5 °C and 23 - 39 %, respectively. The output signal of the full Wheatstone bridge operated at a supply voltage of $U_0 = 1$ V was connected to an instrumentation amplifier (HBM, ML 10B) via a shielded cable. During a typical calibration measurement the cantilever was incrementally moved with its tip against the weighing pan of an electronic balance (Sartorius, SC 2). Simultaneously with the force measurement the output signal ΔU of the integrated piezoresistive gauge was recorded. A calibration curve typically comprised ~ 100 sample increments and was repeated ~ 500 × for each sensor device.
The complete setup was mounted on a platform comprising stabilizer pneumatic isolators with automatic leveling for vibration damping to cancel ground vibrations and acoustic noise. A shielded cable was used to protect the bridge output signal against electromagnetic interference.

3.1 Vertical loading

The typically measured performance data of realized 5-mm-long silicon cantilever sensors were listed in Table 2. Cantilevers show linear load-deflection characteristics in a range up to 200 µm with a fracture limit exceeding 1.6 mm. For the stiffness we found ~ 12 N/m at a repeatability of 2.5 %. The stiffness of the balance of > 10 kN/m is by far higher. Therefore, it was not taken into account for the analysis of the cantilever stiffness. The main source for a deviation from the design value was the cantilever height given by the etching in a time-controlled process. Measurements across the wafer showed a typical variation of ~ 1 - 2 % about the target height.

Parameter	Value
Range δ_{max}	200 µm @ FS, fracture limit: $\delta_z > 1.6$ mm, $\delta_y > 0.3$ mm
Stiffness	11.9 N/m @ repeatability of 2.5 %
Sensitivity S	0.25 µV/nm @ without amplification, repeatability of 1 %
Non-linearity	0.3 %FS @ 200 µm, 0.2 %FS @ 20 µm
Gauge factor K	76 ± 2
Switch-on delay	~ 1 s
Temperature coefficient of R	- 0. 2 %/K
Temperature drift	10 nm/K @ referred to vertical deflection
Light sensitivity	4-10 nm @ neon light: 100 µW/cm^2, referred to vertical deflection
Long-term stability	6 nm @ 70 h, $\Delta T < 1$ K, referred to vertical deflection
Resolution δ_{min}	1.8 nm @ $f_{max} = 1.6$ kHz, $f_{min} = 0.003$ Hz 1.3 nm @ $f_{max} = 800$ Hz, $f_{min} = 0.003$ Hz 0.6 nm @ $f_{max} = 100$ Hz, $f_{min} = 0.003$ Hz
Uncertainty ($k = 2$)	30 nm @ 1 µm

Table 2. Sensor performance ($l = 5$ mm, $w = 200$ µm, $h = 50$ µm, $U_0 = 1$ V, $T = 21.4 - 23.5$ °C, $rH = 23 - 39$ %.)

A deflection sensitivity of 0.25 μV/nm and a non-linearity of 0.3 %FS was measured at a repeatability of 1 % in an exceptionally large deflection range up to 200 μm. Using eqs. (3) and (4) we could calculate from these results a gauge factor of $K = 76 \pm 2$ which is close to the desired value of 80. The resistivity showed a temperature coefficient of - 0.2 %/K. A stable read-out signal was achieved typically within one second after switch-on of the voltage supply. The cross sensitivity against temperature and ambient light was below 10 nm at $\Delta T = 1$ K and an illumination intensity of 100 μW/cm², respectively. The input-referred stability of the strain-gauge output signal amounted to 6 nm over 70 h at $\Delta T < 1$ K.
Noise of the complete system including sensor and amplifier measured in a frequency range from 10^{-3} Hz to 20 kHz showed characteristic $1/f$ and white noise regimes below and above ~ 10 Hz, respectively (Peiner et al., 2007). White noise of 5.8×10^{-11} mV²/Hz can be calculated for a symmetric Wheatstone bridge of a resistance of 2.5 kΩ of each leg. This corresponds very well to the measured value of 6×10^{-11} mV²/Hz obtained as the difference of measured total and amplifier noise in the white noise regime. $1/f$ noise comprises contributions from both the Wheatstone bridge and the amplifier according to:

$$\frac{\overline{U_{\mathrm{H}}^2}}{f} = \left(\frac{\alpha U_0^2}{2N} + \overline{U_{\mathrm{A}}^2} \right) / f \tag{10}$$

where α, N and $\overline{U_{\mathrm{A}}^2}$ denote the Hooge constant, the total number of carriers in a single resistor, and the amplifier noise, respectively (Nesterov & Brand, 2006). With the bridge supply voltage of 1 V and a total number of carriers of 2.5×10^9 within each resistor we calculate a Hooge constant of $\alpha = 1.3 \times 10^{-6}$.
Integration of $1/f$ noise and white noise (Johnson noise: $\overline{U_{\mathrm{J}}^2} / \Delta f$) from f_1 to f_2 yields:

$$\overline{U_{\mathrm{noise}}^2} = \overline{U_{\mathrm{H}}^2} \ln\left(\frac{f_2}{f_1} \right) + \frac{\overline{U_{\mathrm{J}}^2}}{\Delta f} (f_2 - f_1) \tag{11}$$

With $\overline{U_{\mathrm{H}}^2} = 8 \times 10^{-10}\ (\mathrm{mV})^2$, $\overline{U_{\mathrm{J}}^2} / \Delta f = 1.5 \times 10^{-10}\ (\mathrm{mV})^2 / \mathrm{Hz}$, $f_1 = 10^{-3}$ Hz and $f_2 = 1.6$ kHz we we find $\sqrt{\overline{U_{\mathrm{noise}}^2}} = 0.5\ \mu\mathrm{V}$, which at a sensitivity of 0.25 μV/nm corresponds to a resolution of 2 nm.
A high sampling rate is required for scanning at high levels of speed (> 1 mm/s) and lateral resolution (< 1 μm), i.e. the ratio of scanning speed to upper cutoff frequency should be on one hand considerably lower than the minimum lateral structure width which has to be resolved. On the other hand, however, for nanometer vertical resolution high-frequency noise has to be cancelled by reducing the upper cutoff frequency. As a tradeoff we selected an upper cutoff frequency of 100 Hz and reduced the probing speed to around 10 μm/s, if nanometer vertical resolution and sub-micrometer lateral resolution were required. If a lower vertical resolution around 10 nm was acceptable we could operate the amplifier at 19.2 kHz and increase the probing speed to around 1 mm/s.
We tested the vertical and lateral scanning resolution using a photolithography mask comprising 60-nm-high and 1-to-10-μm-wide stripes of chromium on a glass substrate. Scanning of the entire test area of 310 – 100 μm² in the fast modus, i.e. within < 3 min reveals all stripes clearly resolved. High-resolution scans were then performed with the 1-, 2-, and

3-μm-wide stripes at a speed of 15 μm/s and an upper cutoff frequency of 100 Hz. According to the noise analysis we can expect lateral and vertical resolutions of 0.15 μm and 0.5 nm, respectively. Measured stripe heights and widths are summarized in Table 3 showing deviations from the nominal height and width of less than 16 % and 6 %, respectively. The measured stripe width corresponds to the distance between the raising and falling flanks at 90 % of the height. The variances measured for the heights of 1.6 to 2.5 nm are higher than expected. They can be assigned to a 50 Hz interference due to not perfect shielding of the signal transmission. Measurement uncertainty was determined within the deflection range of 0.3 - 10 μm using a depth setting standard (EN 48200). We found a value of 30 nm (k = 2) for a deflection of 1 μm. These results confirm the potential of the described slender piezoresistive cantilever sensor for contour and roughness measurements of structured surfaces at sub-micron lateral and nanometer vertical resolutions.

nominal stripe width (μm)	measured stripe width (μm)	measured stripe height (nm)
1	1.06 ± 0.01	51.8 ± 1.9
2	1.96 ± 0.01	53.2 ± 2.5
3	3.03 ± 0.01	50.7 ± 1.6

Table 3. Scanning across Cr stripes on a glass carrier using a slender cantilever sensor (l = 3 mm, w = 100 μm, w = 50 μm, U_0 = 1V, scanning speed = 15 μm/s, probing force = 80 μN, f_2 = 100 Hz).

3.2 Lateral loading

We investigated the behaviour of a cantilever of uniform cross section of l = 5 mm, w = 200 μm, and h = 40 μm under combined vertical and lateral loading. Combining eqs. (4) and (6) we find a ratio of lateral-to-vertical sensitivity of $w/(4h)$ = 1.25. Measurements were performed of the output signals of the strain gauge resistors under tilted loading conditions, i.e. by moving the tip against a flat work piece inclined by - 30° and 45° about an axis parallel to the cantilever axis. We find values of 0.84-0.93 for the ratio of lateral-to-vertical sensitivity which in fair agreement with the expected value of 1.25. Thus, vertical and lateral signals can be decoupled by analyzing the responses of all four resistors in the conventional full bridge arrangement and the longitudinal resistors alone connected into a half bridge, respectively.

3.3 Axial loading

Moving a cantilever with its free end axially against a fixed body can lead to three different stages of deformation as schematically displayed in Fig. 9a. Initially, it is uniformly compressed. After exceeding a critical load F_c (eq. (7)) buckling occurs which may be either of type I or II depending on whether the cantilever free end can move or is pinned on the probed surface. Axial loading tests were performed with cantilever sensors below and above the critical load for buckling F_c. The photographs in Fig. 9b to d show an axially loaded 5-mm-long cantilever at the initial surface contact (b) and at axial displacements of δ_x = 2 μm and δ_x = 80 μm, respectively. In Fig. 9d the type-II buckling form of a beam is exhibited which typically occurs under the boundary conditions of the cantilever of one end clamped and the other pinned.

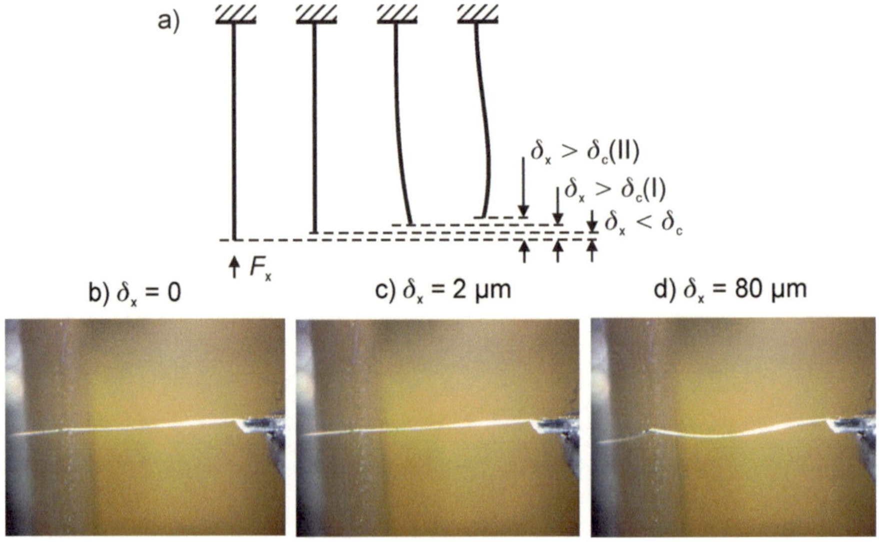

Fig. 9. Schematic (a) and photographs of an axially loaded 5-mm-long cantilever (b-d) at different stages of axial displacement.

The sensor response measured with the cantilever depicted in Figs. 9b-d during axial loading is shown in Fig. 10a where the sensor signal is displayed in dependence on the axial displacement of the cantilever moved against a fixed body. Two probing speeds were selected: 0.25 and 8 µm/s. Up to a maximum displacement of 40 µm an almost linear increase of the output signal amplitude with δ_x is observed at 0.25 µm/s with a buckling form of type I (Fig. 9b). At δ_x . 50 µm the signal drastically increased corresponding to the transition from type-I to type-II buckling (Fig. 9d). At a probing speed of 8 µm/s this transition occurred much earlier, i. e. at an axial displacement between 10 and 20 µm indicating the dynamic-loading effect. The sensitivities of ~ 0.5 mV/µm and ~ 4 mV/µm observed under the conditions of type-I and type-II buckling, respectively, are lower than the sensitivity of 16 mV/µm expected for uniform compression.

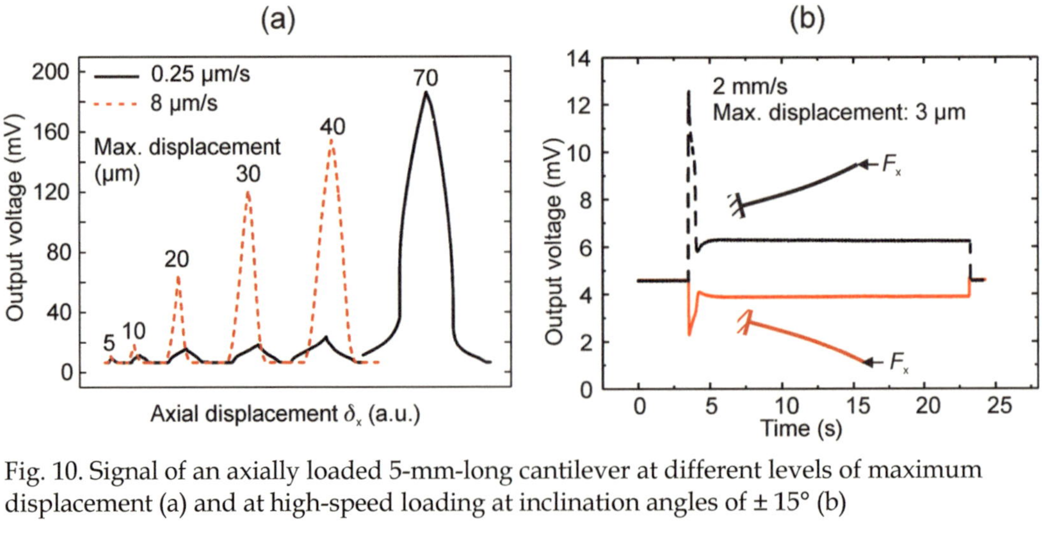

Fig. 10. Signal of an axially loaded 5-mm-long cantilever at different levels of maximum displacement (a) and at high-speed loading at inclination angles of ± 15° (b)

In Fig. 10b the sensor response on high-speed axial probing (2 mm/s) against a fixed body at a maximum displacement of 3 µm and an inclination angle of ± 15° is shown. Type-I buckling is observed in both cases with positive and negative signs of signal change indicating compressive and tensile strain, respectively, to the Wheatstone bridge. After an initial sharp increase the signals rapidly decayed towards constant amplitudes.
Finally, axial loading tests below F_c were performed using the nanonewton force testing setup described above. Under these conditions the balance stiffness is much less than the axial cantilever stiffness k_x calculated using the cantilever dimensions and eq. (9). We find values of typically > 100 kN/m. Therefore, the balance stiffness had to be considered when the measured load-deflection curves were analyzed.

Parameter	Value
Vertical sensitivity S_z	0.1953 ± 0.0008 µV/nm @ without amplification
Axial sensitivity S_x	11.15 µV/nm @ without amplification
	8.5 µV/nm @ calculated using eqs. (4) and (8)
Average residual from linearity	± 5 µV @ 0.7 mV FS
Stiffness k_z	16.33 ± 0.04 N/m
Stiffness k_x	261.6 kN/m @ calculated using eq. (9)
Axial deflection before buckling	1.2 µm @ calculated using k_x and eq. (8)
Fracture limit	580 ± 58 µm @ vertical deflection
	300 ± 30 µm @ lateral deflection
	260 ± 26 µm @ axial deflection

Table 4. Sensor performance (l = 3 mm, w = 100 µm, U_0 = 1V, T = 22.0 – 22.1 °C, rH = 40.1 – 41.5 %.)

In Table 4 the results obtained from the calibration of a slender cantilever sensor are summarized. Vertical stiffness k_z and sensitivity S_z were measured and analyzed using eqs. (1) to (5) yielding a cantilever thickness of h = 46.2 µm and a gauge factor of K = 25.4. Under axial loading we observed a stiffness of 11.46 kN/m which is much less than the axial cantilever stiffness k_x = 261.6 kN/m calculated using eq. (9), i. e. it nearly corresponds to the balance stiffness. For the axial sensitivity we measured a value of 0.488 µV/nm which had to be corrected by multiplying with the ratio of measured stiffness to k_x yielding S_x = 11.15 µV/nm. Using eqs. (4) and (8) we obtain a axial sensitivity of S_x = 8.5 µV/nm which is in fair agreement with the measurement.

4. Form measurement

Silicon wafers patterned by deep reactive ion etching (DRIE) and spray holes manufactured using electro discharge machining (EDM) were used as artefacts of form and roughness measurements using the described slender cantilever sensor.

4.1 Micro sac hole

The results of the previous chapter show that a front-side loaded slender cantilever can be used to measure the depth of a micro sac hole. The highest sensitivity occured at uniform compression but even at $F > F_c$ we observed values around 1 µV/nm leading to sub nanometer resolution. We checked the measurement uncertainty by repeatedly measuring the height Δh of a step fabricated on a silicon wafer using DRIE. The results are displayed in

Fig. 11 where the measured values of the step height are plotted. A typical trace of the sensor signal in dependence on axial cantilever position is shown in the inset. The contact position x_0 was defined as the position where the signal exceeded the average zero signal (offset voltage) by the fivefold of its standard deviation. For the step we found a mean value of 252.407µm at a standard deviation of σ = 82 nm.

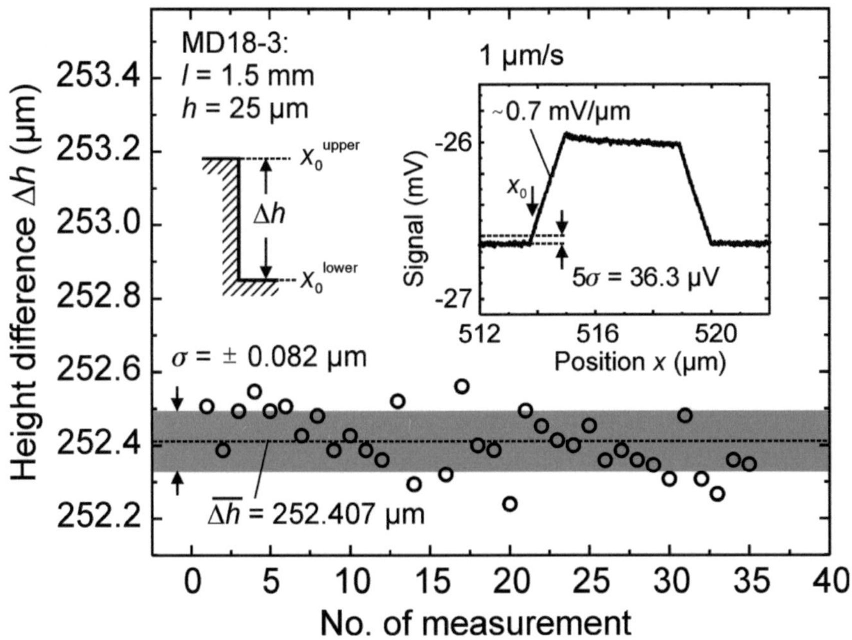

Fig. 11. Step height measured with a DRIE patterned silicon wafer using an axially loaded cantilever.

4.2 Injector nozzle

VCO (valve covered orifice) direct injection (DI) Diesel nozzles with six spray holes of 110 – 170 µm in diameter fabricated by (EDM) were investigated using realized prototype sensors to check the capability of slender piezoresistive cantilevers for in-hole form and roughness measurements. For these experiments we used 1.5-mm-long, 30-µm-wide, and 36-to-41-µm high cantilever sensors with 50-µm-high tips of a radius of curvature of 1.5 µm and a cone angle of 40°. Calibration of the sensors yielded a vertical sensitivity of $\Delta U/\delta_z$ = 0.25 - 0.31 µV/nm and a vertical stiffness of k_z = 19.2 - 29.1 N/m.

A photograph and a schematic of the measurement setup are depicted in Fig. 12. The cantilever sensor with the piezoresistive Wheatstone bridge was connected via Au wire bonding to a printed circuit board and then via unshielded cables to an instrumentation amplifier (HBM ML 10B). This experimental probe head was then mounted on a 2D piezo positioning stage featuring a travel range of 800 µm at sub-nanometer resolution (P-628.2CD with digital piezo controller PI 710, Physik Instrumente, Germany) which was fixed for rough positioning to an *x-/y-/z*-table. The nozzle was arranged on a rotating/tilting stage. Before starting the scanning process the cantilever and hole axes were carefully adjusted.

The schematic in Fig. 13 shows the movement of a cantilever sensor along the inner contour of a spray hole. Before moving into the hole the cantilever had to be carefully aligned to the hole axis. Digital optical micrographs of a VCO nozzle and a slender cantilever into one the six spray holes are depicted in Fig. 14.

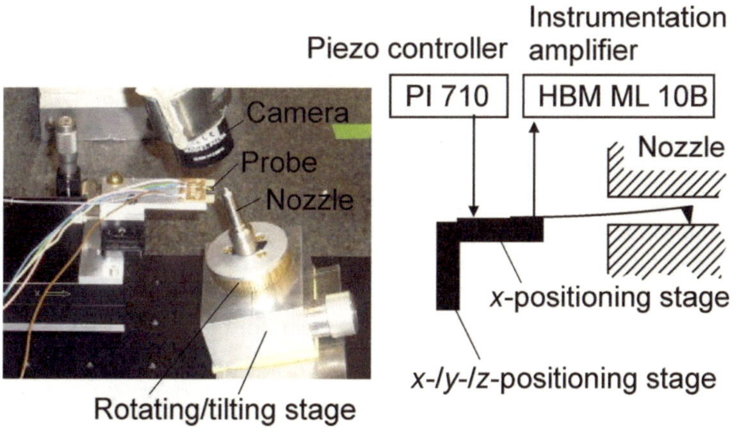

Fig. 12. Photograph (left) and schematic (right) of a setup for surface scanning inside spray holes of DI nozzles.

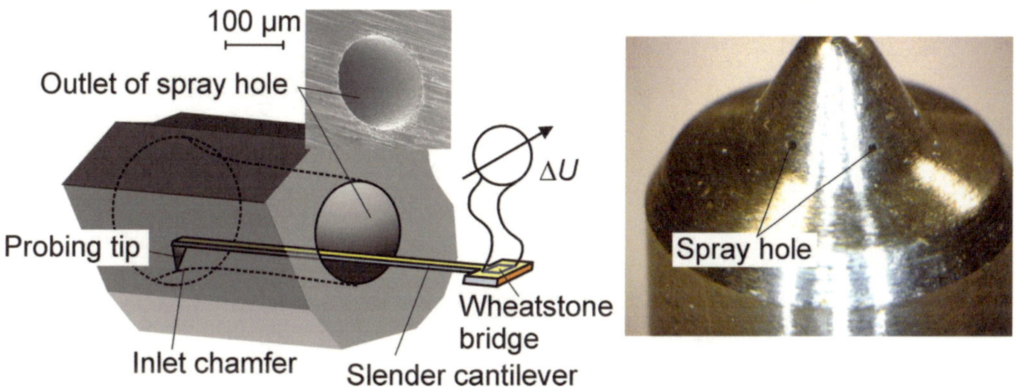

Fig. 13. Schematic of a slender cantilever sensor during scanning along the inner surface profile of deep narrow micro hole.

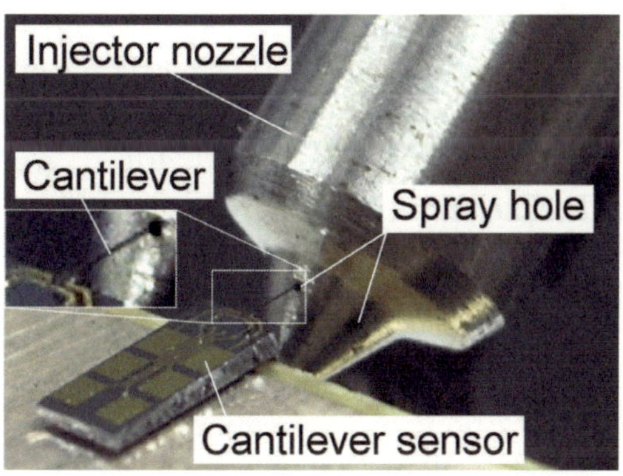

Fig. 14. Optical micrographs of a VCO Diesel injector nozzle with a slender cantilever sensor probing inside a spray hole.

Fig. 15 shows the complete inner-surface profile of the spray hole measured using a slender cantilever sensor. The scans performed at a constant speed of 2 – 200 μm/s were started at the inner hole edge, i.e. at the entry of fuel flow. A step of 50 μm in height corresponding to the tip height was measured during the initial scanning stage which can be assigned to the transition from shaft contact at the beginning of the scan to tip contact (outer left schematic). Then the tip touched the injector wall with its side facet and was moved along the hole edge until the hole wall was reached (inner left schematic). Linear slopes of 30° and 23° appeared at the rising and the trailing flanks, respectively, which are close to the half of the tip angle. Thus both the rising and the trailing flanks of the profiles represent superpositions of the shapes of the tip and the hole edge, respectively.
Regular probing conditions were achieved when the inlet edge of the hole was reached (inner right schematic) and the tip is moved further (outer right schematic). The profile in Fig. 15 corresponds to a not optimal form of a micro hole by EDM indicating a neck at the hole inlet. Necking is related to the loss of erosion particles occurring at the end of the drilling operation, leading to a constricted diameter of the hole at the inlet (Diver et al., 2004). For the surface roughness we found values of 0.4-0.8 μm which is a typical range for micro holes fabricated by EDM (Li et al., 2007; Cusanelli et al., 2007; Diver et al., 2004). Abrasive flow machining (AFM) can be used subsequent to the EDM process to improve surface finish and chamfer radius (Jung et al., 2008).

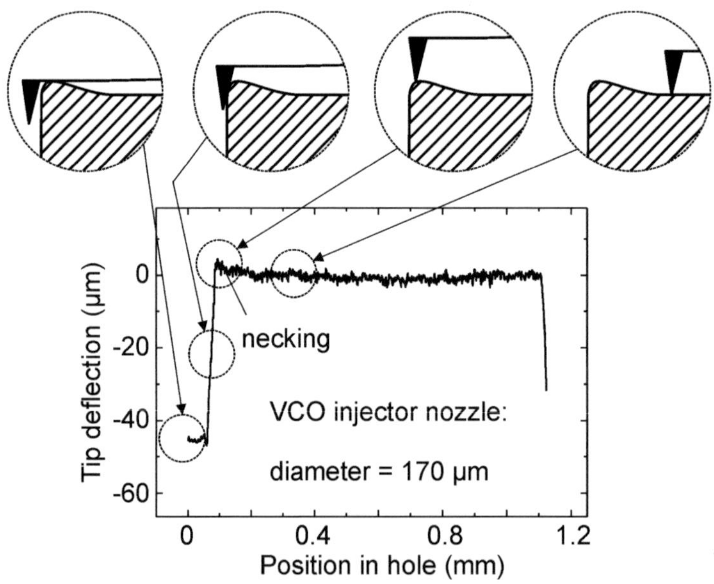

Fig. 15. Typical surface profile measured by scanning within a spray hole of a VCO Diesel injector nozzle using a slender cantilever sensor (lower). The schematics represent the different contact scenarios of the probe about the inlet edge.

In Fig. 16a the profiles from subsequent in-hole scans along identical traces are shown revealing good agreement as indicated by the occurrence of characteristic signatures at identical positions. The profiles provide information on roughness and waviness of the profiles being a measure of the quality of tool and the machine, respectively. Exemplarily, roughness parameters and waviness profile were determined from one the measured profiles according to ISO 4287 and displayed in Table 5 and Fig. 16b, respectively.

Parameter	Value
R_p	1859.27 nm
R_V	2716.97 nm
R_{max}	4692.09 nm
R_z	4576.24 nm
R_a	637.08 nm
R_q	744.48 nm
R_c	1778.09 nm
R_{Sm}	16.72 μm
$R_{\Delta q}$	0.4318
R_{sk}	-0.321
R_{ku}	2.782
R_t	4460.39 nm

Table 5. Roughness parameters according to ISO 4287 extracted from the profile of the inner-wall surface of a VCO nozzle spray hole displayed in Fig. 14.

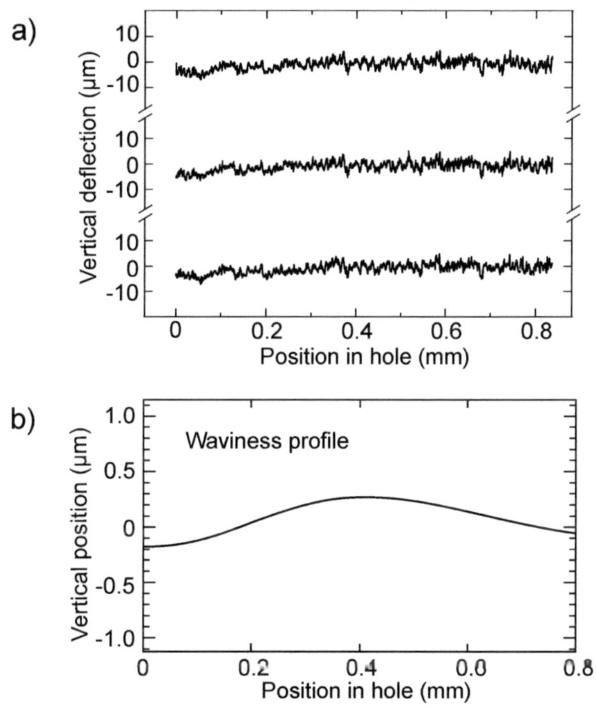

Fig. 16. Surface profiles measured repeatedly along the same trace within a spray hole of a VCO Diesel injector nozzle (a) and waviness profile calculated according to ISO 4287 (b).

Scanning measurements within spray holes were repeated using different sensors at various scanning speeds (2 – 200 μm/s) and probing forces. We found good agreement of the signatures in the profiles. Furthermore, the roughness values 0.80, 0.73, 0.76 and 0.74 μm determined at probing speeds of 2, 20, 100, and 200 μm/s over a scan distance of 300 μm did not show a dependence on probing speed. We conclude that the described piezoresistive cantilever sensors have the potential for fast and non-destructive contour and roughness measurement within spray holes.

5. Conclusion

Construction, fabrication and testing of slender piezoresistive cantilever probes were addressed which were designed for tactile shape and roughness measurements with high-aspect-ratio micro components. In the normal cantilever-bending mode the sensor could be operated within an exceptionally large deflection range (hundreds of µm) at high scanning speeds (> 1 mm/s) and low probing forces (< 100 µN). Vertical and lateral resolution amounted to ~ 10 nm and ~ 1 µm, respectively which fulfils the requirements of form and roughness measurements with machined surfaces. Cross sensitivity vs. temperature and ambient light was typically less than 10 nm at measurement conditions of temperature and light intensity variations of 1 K and 0.1 mW/cm^2, respectively. Sensor response on axial loading could be used for probing the bottom of deep and narrow sac holes. In this case cantilever buckling was normally observed which was monitored by the bridge output to measure the structure heights of 3D-patterned silicon. For the first time form and roughness measurement inside spray holes of injector nozzles could be demonstrated with not sectioned holes. Using tailored prototypes of slender piezoresistive cantilever sensors good reproducibility was obtained at different scanning speeds and loading forces. We could demonstrate the feasibility of slender piezoresistive cantilever sensors for fast, non-destructive and high-performance form and roughness measurements with deep and narrow micro holes.

6. Acknowledgements

The author is grateful to the valuable technical assistance by Nadine Beckmann, Doris Rümmler, Julian Kähler and Stefan Kahmann. This work was supported in part by the German Federal Ministry of Education and Research (BMBF) in the framework of the collaborative project "Prüfung und Bewertung geometrischer Merkmale in der Mikrosystemtechnik (µgeoMess)" within the cluster MSTPrüf under no. 16SV1944.

7. References

Anantharamaiah, N.; Vahedi Tafreshi, H. & Pourdeyhimi, B. (2007). A simple expression for predicting the inlet roundness of micro-nozzles. *J. Micromech. Microeng.*, Vol. 17, N31–N39

Bae, C; Yu, J.; Kang, J.; Kong,J. & Lee, K.-O. (2002). Effect of Nozzle Geometry on the Common-Rail Diesel Spray. *SAE Techn. Pap. Ser.* 2002-01-1625

Bao, M.-H. (2000). *Micro Mechanical Transducers*, Elsevier, ISBN 0-444-50558-X, Amsterdam

Baron, N.; Passave, J.; Guichardaz, B. & Cabodevila, G. (2008). Investigations of development process of high hollow bevelled microneedles using a combination of ICP RIE and dicing saw. *Microsyst. Technol.* DOI 10.1007/s00542-008-0596-1

Bauza, M. B.; Hocken, R. J.; Smith, S. T. & Woody, S. C. (2005). Development of a virtual probe tip with an application to high aspect ratio microscale features. *Rev. Sci. Instrum.*, Vol. 76, 095112

Behrens, I.; Doering, L. & Peiner, E. (2003). Piezoresistive cantilever as portable micro force calibration standard. *J. Micromech. Microeng.*, Vol. 13, 1279-1288

Beyeler, F.; Muntwyler, S.; Nagy, Z.; Graetzel, C.; Moser, M; & Nelson, B. J. (2008). Design and calibration of a MEMS sensor for measuring the force and torque acting on a magnetic microrobot. *J. Micromech. Microeng.* Vol. 18, 025004 (7pp)

Bos, E. J. C. ; Heldens, R. W. P.; Delbressine, F. L. M.; Schellekens, P. H. J. & Dietzel, A. (2007) Compensation of the anisotropic behavior of single crystalline silicon in a 3D tactile sensor. *Sens. Actuat. A* , Vol.134, 374–381

Bos, E. J. C.; Delbressine, F.L.M. & Haitjema, H. (2004). High-Accuracy CMM Metrology for Micro Systems, *Proceedings of 8th International Symposium on Measurement and Quality Control in Production, ISMQ2004*, pp. 511-522, Erlangen, Germany, October, 2004, VDI, Düsseldorf, Germany, No. 1860

Cho, C. H.; Jaeger, R. C. & Suhling, J. C. (2006). Experimental Characterization of the Temperature Dependence of the Piezeresistive Coefficients of Silicon, *Proceedings of. ITherm 2006*, pp. 928–935, San Diego, CA , May, 2006

Choudhury, A.; Hesketh, P. J.; Thundat, T. G.; Hu, Z. & Vujanic, R. (2007). Design and Testing of Single and Double Sided Cantilevers for Chemical Sensing", *Proceedings of IEEE Sensors 2007 Conf.*, pp. 1432-143, Atlanta, GA, October, 2007

Cusanelli, G.; Minello, M.; Torchia, F.; Ammann, W. & Grize, P. E. (2007). Properties of Micro-Holes for Nozzle by Micro-EDM, *Proceedings of 15th Intern. Symp. Electromachining (ISEM XV)*, pp. 241-245, Pittsburgh, PA, April, 2007, ISBN (13): 978-0-9794977-0-4

Diver, C.; Atkinson, J.; Helml, H. J. & Li, L. (2004). Micro-EDM drilling of tapered holes for industrial applications. *J. Mater. Process. Technol.*, Vol. 149, 296–303

Engelke, R.; Ahrens, G.; Arndt-Staufenbiehl, N.; Kopetz, S.; Wiesauer, K.; Löchel, B.; Schröder, H.; Kastner, J.; Neyer, A.; Stifter, D. & Grützner, G. (2007). Investigations on possibilities of inline inspection of high aspect ratio microstructures. *Microsyst. Technol.*, 13, Vol. 319–325

Hon, R.; Lee, S. W. R.; Zhang, S. X. & Wong; C. K. (2005). Multi-Stack Flip Chip 3D Packaging with Copper Plated Through-Silicon Vertical Interconnection, Proceedings of 7th Electron. Packaging Technol. Conf., (EPTC 2005), Vol. 2, pp. 384-389, Singapore, December, 2005

Jung, J.-W.; Ko, S. H.; Jeong, Y. H.; Min, B.-K. & Lee, S. J. (2007). Estimation of Material Removal Volume of a Micro-EDM Drilled Hole Using Discharge Pulse Monitoring. *Int. J. Precision Eng. Manufact.* Vol. 8, 45-49

Jung, D.; Wang, W. L.; Knafl, A.; Jacobs, T. J.; Hu, S. J. & Assanis, D. N. (2008). Experimental investigation of abrasive flow machining effects on injector nozzle geometries, engine performance, and emissions in a DI Diesel engine. *Intern. J. Automotive Technol.*, Vol. 9, No. 1, 9-15

Kao, C.-C. & Shih, A. J. (2007). Form measurements of micro-holes. *Meas. Sci. Technol.*, Vol. 18, 3603–3611

Kiuchi, M.; Matsui; S. & Isono, Y. (2008) The piezoresistance effect of FIB-deposited carbon nanowires under severe strain. *J. Micromech. Microeng.*, Vol, 18, 065011 (10pp).

Krüger, O.; Schöne, G.; Wernicke, T.; John, W.; Würfl, J. & Tränkle, G. (2007). UV laser drilling of SiC for semiconductor device fabrication. *J. Phys.: Conf. Ser.*, Vol. 59, 740–744

Küng, A.; Meli, F. & Thalmann, R. (2007). Ultraprecision micro-CMM using a low force 3D touch probe. *Meas. Sci. Technol.*, Vol. 18, 319–327

Lebrasseur, E.; Pourciel, J. P.; Bourouina, T.; Masuzawa, T. & Fujita, H. (2002). A new characterization tool for vertical profile measurement of high-aspect-ratio microstructures. *J. Micromech. Microeng.*, Vol. 12, 280-285

Li, X.; Wang, J. & Li W. (2007). Current State and Prospect of Micro-Machining, *Proceedings of IEEE Intern. Conf. Automation and Logistics*, pp. 1414-1419, Jinan, China, August, 2007

Muralikrishnan, B.; Stone, J. A. & Stoup, J. R. (2006). Fiber deflection probe for small hole metrology. *Precision Engineering* Vol. 30, 154-164.

Mathieu, F.; Saya, D.; Bergaud, C. & Nicu, L. (2007). Parallel Detection of Si-Based Microcantilevers Resonant Frequencies Using Piezoresistive Signals Downmixing Scheme. *IEEE Sens. J.*, Vol. 7, 172–178.

Nesterov, V. & Brand, U. (2006). Modelling and investigation of the mechanical and electrical characteristics of the silicon 3D-boss microprobe for force and deflection measurements. *J. Micromech. Microeng.*, Vol. 16, 1116–1127

Peggs, G.; Lewis A. & Oldfield S. (1999). Design for a compact high-accuracy CMM. *CIRP Annals - Manufacturing Technology*, Vol. 48, No. 1, 417-420

Peiner, E.; Balke, M. & Doering, L. (2007). Slender Tactile Sensor for High-Aspect-Ratio Micro Metrology, pp. 760-763, Proceedings of IEEE Sensors 2007 Conf., Atlanta, GA, October, 2007

Peiner, E.; Doering, L. & Nesterov, V. (2004). Tactile Probes for High Aspect Ratio Micrometrology. *mstnews*, No. 02/2004, 41,42

Peiner, E.; Doering, L.; Balke, M. & Christ A. (2008). Silicon Cantilever Sensor for Micro-/Nanoscale Dimension and Force Metrology. *Microsyst. Technol.*, Vol. 14, 441–451

Pourciel, J.-B.; Jalabert, L. & Masuzawa, T. (2003). Profile and Surface Measurement Tool for High Aspect-Ratio Microstructures. *JSME Intern. J. Series C*, Vol. 46, 916-922

Rauh, W. (2005). Präzision mit gläserner Faser. *Mikroproduktion*, No. 1/2005, 1-6, Carl Hanser, München, Germany

Ruther, P.; Spinner, S.; Cornils, M. & Paul, O (2007). Cantilever-based tactile sensor with improved sensitivity for dimensional metrology of deep narrow drillings. *Proceedings of 14th Intern. Conf. Solid-State Sensors, Actuators and Microsystems, (Transducers & Eurosensors '07)*, pp. 1469-1472, Lyon, France, June, 2007

Samuel, B. A.; Desai, A. V. & Haque, M. A. (2006). Design and modeling of a MEMS pico-Newton loading/sensing device. *Sens. Actuators A*, Vol. 127, 155–162

Seitz, K. (2005). Qualitätsprüfung in der Mikrosystemtechnik. *inno*, Vol. 29, No. 03/05, 19, IVAM, Dortmund, Germany

Seo, T.-I..; Kim, D.-W.; Cho, M.-W. & Lee, E.-S. (2008). A Study on the Characteristics of Micro Deep Hole Machining Processes. *Key Engineering Materials*, Vols. 364-366, 566-571, Trans Tech Publications, Switzerland

Simon, M.; Reznikova, E.; Nazmov, V. & Last, A. (2008). Measurement of side walls of high aspect ratio microstructures. *Microsyst. Technol.* DOI 10.1007/s00542-008-0618-z

Tibrewala, A.; Phataralaoha, A.; & Büttgenbach, S. (2008). Analysis of full and cross-shaped boss membranes with piezoresistors in transversal strain configuration. *J. Micromech. Microeng.*, Vol. 18, 055001 (6pp)

UMAP (2008) Vision System, Catalog No. E4257-361, Mitutoyo Corporation, Japan. Available: http://www.mitutoyo.co.jp

Vora, K. D.; Shew, B.-Y.; Lochel, B.; Harvey, E. C.; Hayes, J. P.& Peele, A. G. (2008). Sidewall slopes and roughness of SU-8 HARMST. *Microsyst. Technol.*, DOI 10.1007/s00542-007-0506-y

Weckenmann, A.; Peggs, G. & Hoffmann, J. (2006). Probing systems for dimensional micro- and nano-metrology. *Meas. Sci. Technol.*, Vol. 17, 504–509

Woody, S. C. & Bauza, M. B. (2007). High Aspect Ratio Sensors with an application Application to microscale Microscale metrologyMetrology, *Proc. 2007 NCSL Intern. Workshop & Symp. on Metrology's Impact on Products and Services.* Available: http://www.x-cd.com/ncsli07/search.html

Yamamoto, M.; Takeuchi, H. & Aoki, S. (2000). Dimensional measurement of high aspect ratio micro structures with a resonating micro cantilever probe. *Microsyst. Technol.*, Vol. 6, 179-83

22

Tactile Sensor Without Wire and Sensing Element in the Tactile Region using New Rubber Material

Yo Kato and Toshiharu Mukai
Bio-Mimetic Control Research Center, RIKEN
Japan

1. Introduction

Recently the idea of covering a robot's surface with a 'skin' of soft tactile sensors has attracted the attention of researchers, and some human-interactive robots covered with such sensors have actually been made (Tajima et al., 2002; Kanda et al., 2002). However, most conventional tactile sensors need a large number of sensing elements and wires because every detection point needs one sensing element and wiring to an A/D converter. There are some studies aiming to overcome this wiring problem by using 2D surface communication or wireless communication (Shinoda & Oasa, 2000; Ohmura et al., 2006), but these are very complicated and expensive solutions.

We have developed a soft areal tactile sensor made of pressure-sensitive conductive rubber without any wire or sensing element in the tactile region. The distribution of applied pressure, relating to the resistivity change of the pressure-sensitive rubber, can be estimated by using inverse problem theory. We employed electrical impedance tomography (EIT) to reconstruct the resistivity distribution from information obtained by electrodes placed around the region.

EIT is an established method in medical and industrial applications (Holder, 2005), but it has not been applied to tactile sensors until recently. Nagakubo and Alirezaei proposed a tactile sensor using an EIT algorithm operating with commonly used EIT software and commercially available pressure-sensitive rubber (Nagakubo & Kuniyoshi, 2006; Alirezaei et al., 2006). Their method is based on the same principle as ours, but their pressure-sensitive conductive rubber is not suitable for this method. We have newly developed special pressure-sensitive conductive rubber for this sensor, and adopted a new computation technique suitable for this rubber. We have also developed a prototype sensor system that can measure pressure distribution in real-time.

In this paper, we describe basic structure and computation technique of our sensor system, as well as experimental results obtained using our prototype sensor system.

2. Device design

2.1 Basic structure

We have developed, in collaboration with Tokai rubber industries, Ltd., a new type of pressure-sensitive conductive rubber, the resistivity of which increases when pressure is

applied, unlike ordinary conductive rubber. This new rubber is suitable for a tactile sensor using EIT, because its initial resistivity can be set low, which leads to the accurate reconstruction of pressure distribution from voltage measurements with relatively low noise. Fig. 1 shows the dimensions of the pressure-sensitive conductive rubber that we used in experiments. We formed the rubber into a 1-mm-thick, 195-mm-diameter disc and put 16 electrodes around the disc by vulcanization gluing at regular intervals.
This rubber also has the important feature that its resistivity increases in either case where compressive or tensile strain is applied to it. Fig. 2 shows the relationship between the distortion factor and the resistivity of the rubber. The filled circles are for the compressive case, and the open triangles are for the tensile case. We confirmed that the resistivity increases regardless of the type of strain (i.e. any type of stress).

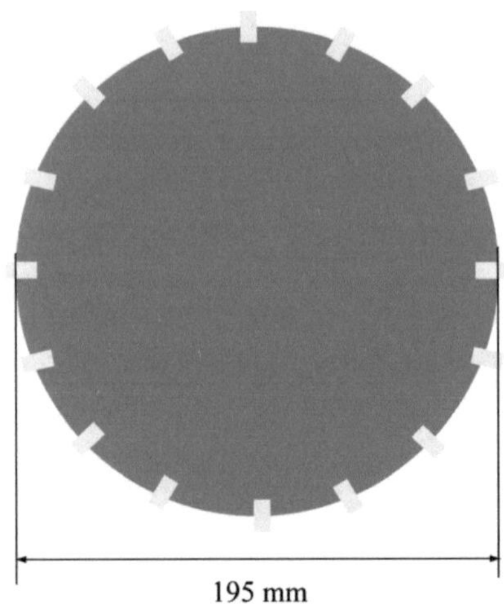

Fig. 1. Pressure-sensitive conductive rubber sheet with 16 electrodes

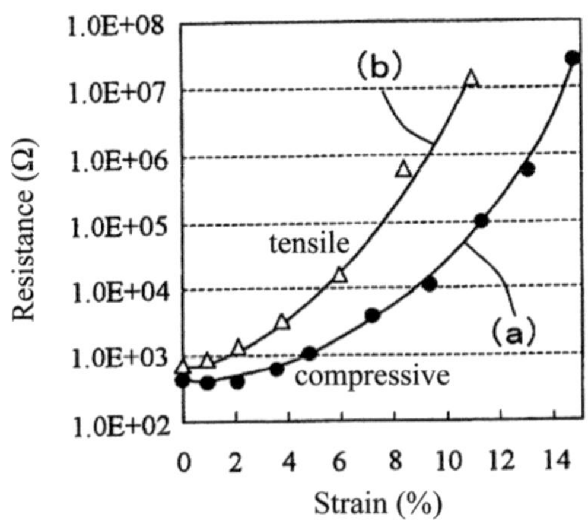

Fig. 2. Resistance changes depends on a) compressive, and b) tensile strains

2.2 Measurement system

The processing of our tactile sensor consists of two steps. The first step is the measurement of conductivities using the 16 electrodes placed around the rubber sheet at even intervals.

An AC voltage is applied, and resistance is calculated using the four-electrode method from the differential voltage between neighboring electrodes, for every possible combination (Fig. 3). In the figure, alternating voltage is supplied between the first and second electrodes and the differential voltage is measured between the 10th and 11th electrodes. The number of possible combinations is 208. However, the number of the independent measured data is actually 104, because the measurements when the supply and the measuring electrode pairs are symmetrically positioned are not independent.

This basic measurement procedure of EIT was developed at Sheffield University (In the original method, current source is used, but we use voltage source instead, for circuit simplicity). The main advantages of this procedure are that it is relatively precise and easy. We can measure at quite a high rate by switching measurement electrodes with a higher frequency AC voltage. In the experiment, we used a 3.685kHz, 4.5 V_{pp} AC voltage. This frequency was determined from the conversion speed of an A/D converter in our sensor controller, dsPIC board, described below.

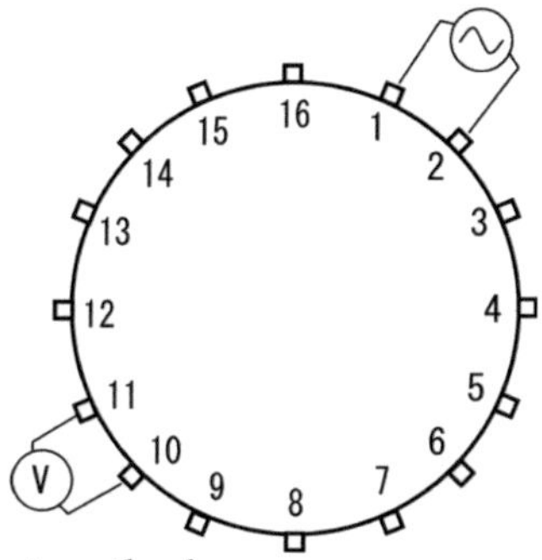

Fig. 3. Conductivity measurement method

2.3 Reconstruction algorithm

In the second step, the distribution of resistivity change is reconstructed from the measurements. This distribution also indicates the pressure distribution because the resistivity change is caused by the change of the pressure applied to the sheet. To get a reconstruction image, we use a sensitivity matrix (Kotre, 1989) that is the Jacobian matrix between δV and $\delta\rho$.

$$\delta V\,(m,n) = S_{m,n,x,y}\,\delta\rho\,(x,y) \tag{1}$$

Here, m and n indicate the position of the applying electrode pair and the measuring electrode pair, respectively, and x and y indicate the position coordinates of the discrete domain in the rubber plate.

The sensitivity matrix can be obtained by numerically solving a forward problem. Then solving (1) by using a least squares (Lawson & Hanson, 1974) or generalized inverse matrix method, the distribution of resistivity change $\delta\rho\,(x,y)$ is obtained. However this is inverse problem. The inverse solution is apt to become instable and, in our case, the ordinary least squares method does not work because of this instability. To overcome this problem, inverse problem theory suggests various normalization methods. We employed the positive value constraint, because we can assume that the resistivity change is always positive thanks to our new pressure-sensitive conductive rubber, unlike ordinary EIT method that uses the regularization technique.

Equation (2) shows the least squares method with a non-negative constraint. In the next section, we show that the constraint successfully stabilizes the solution. In the case of EIT, A, x, b are corresponding to S, $\delta\rho$, δV.

$$\min_x \|Ax - b\|^2 \quad (x \geq 0) \tag{2}$$

This method needs the sensitivity matrix in advance; we calculate the matrix using (3).

$$S_{m,n,x,y} = \int_a \nabla\phi_m \cdot \nabla\phi_n \, da \tag{3}$$

Here, $\nabla\phi_m$ and $\nabla\phi_n$ are the electric potential gradient, which is derived from the electric supply from the electrode pair of m or n, respectively. In other words, we can calculate each component of the sensitivity matrix by integrating the inner products of the potential gradients that are caused by the electric current from m and n electrode pairs in the area indicated by x and y. This equation can be derived from electromagnetic theory.

Fig. 4 shows the example calculation results of the inner products. These solutions are obtained by using the Partial Differential Equation Toolbox on MATLAB (The MathWorks Inc.).

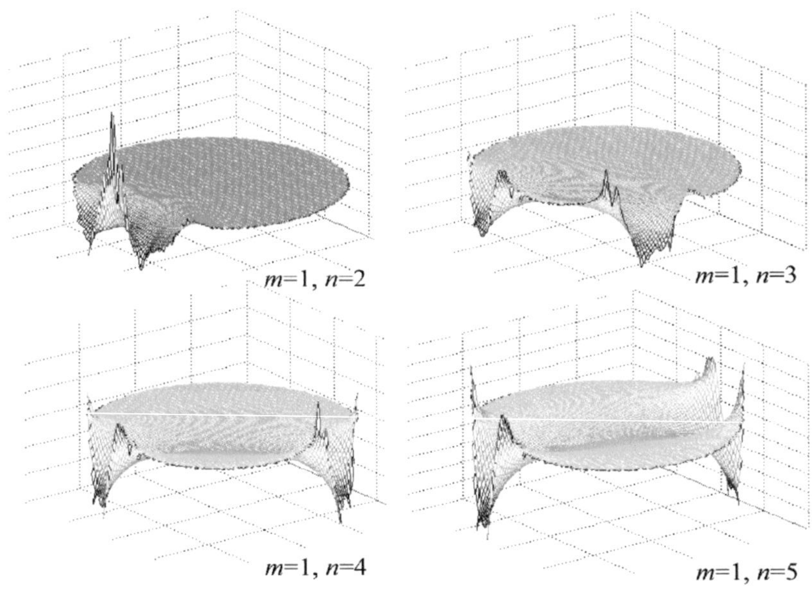

Fig. 4. Example calculation results of the inner product of the potential gradients that caused by m and n electrode pairs

3. Reconstruction experiments

We conducted experiments to determine the performance of our tactile sensor. The measurement conditions were as follows (Kato et al., 2008).

We measured the potential of all electrodes by using sensor controller and driver board (Fig. 5). The main controller is dsPIC board that is a general-purpose 35x50 mm^2

sensing/controlling board we developed, having a dsPIC (Microchip technology 30F6012A) as the CPU and a USB interface IC (FTDI FT232RL) for 1 Mbps communication.
It has many connectors through which most pins are accessed. It also has a stacking connector for an extension board. A tactile sensor controller we developed based on the dsPIC board uses a 12 bit A/D converter and digital I/Os in the 30F6012A. Its program is written in C and downloaded in flash memory of the 30F6012A, as firmware.
Within dsPIC board, all measurement is done using Lock-in amp method. Lock-in amp method is very durable to noise and can be effectively treated by DSP module in dsPIC. The collected data are sending to PC via USB. The reconstruction algorithm is treated by PC.

Fig. 5. Sensor controller and driver consisting of dsPIC board and extension board

The reconstruction algorithm produced the intensity distribution image of the change in resistivity from 208 measured values (Redundant pairs were measured for noise tolerance). The relation between the change in resistivity and pressure distribution is not perfectly clear, but we can conjecture it from Fig. 2. To display the reconstruction image in gray scale, we smoothed the image by inserting interpolation pixels among the pixels that represented actual data. We found that this interpolation successfully displayed a natural distribution image through the anti-aliasing effect when the pressure point was spread over more than one reconstruction domain.

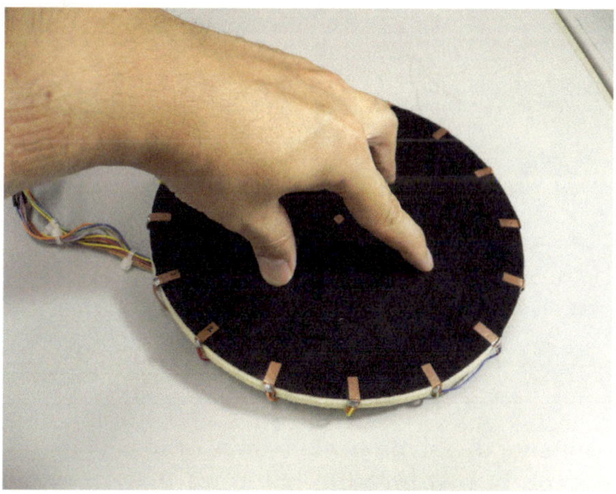

Fig. 6. Point pressures are applying by fingers

First, we tested the stabilization ability of our non-negative least squares algorithm. We applied point pressures by finger as shown in Fig. 6. We compared the reconstruction images generated using a generalized inverse matrix method and those generated using the non-negative least squares method from the same data. The non-negative constraint successfully stabilized the solution, as shown in Fig. 7, while the generalized inverse matrix method failed. The image obtained by non-negative least squares was very stable, and we found that areal tactile sensor functioned well enough for practical usage.

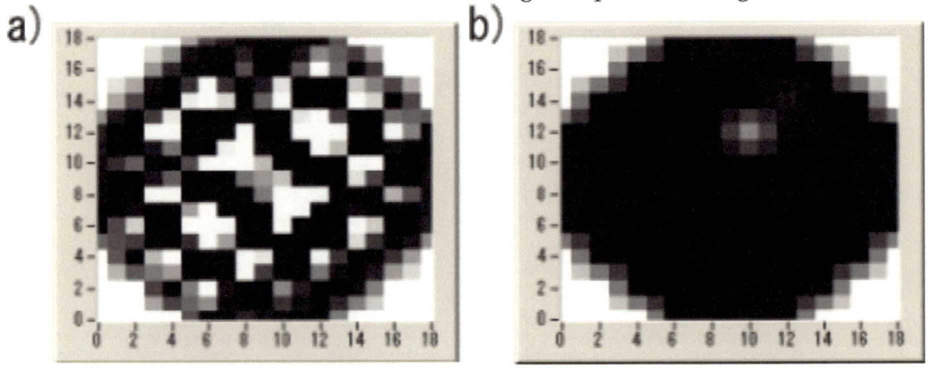

Fig. 7. Comparison between ordinal least square solution and non-negative least square solution from the same input data a) ordinal least square solution, b) non-negative least square solution

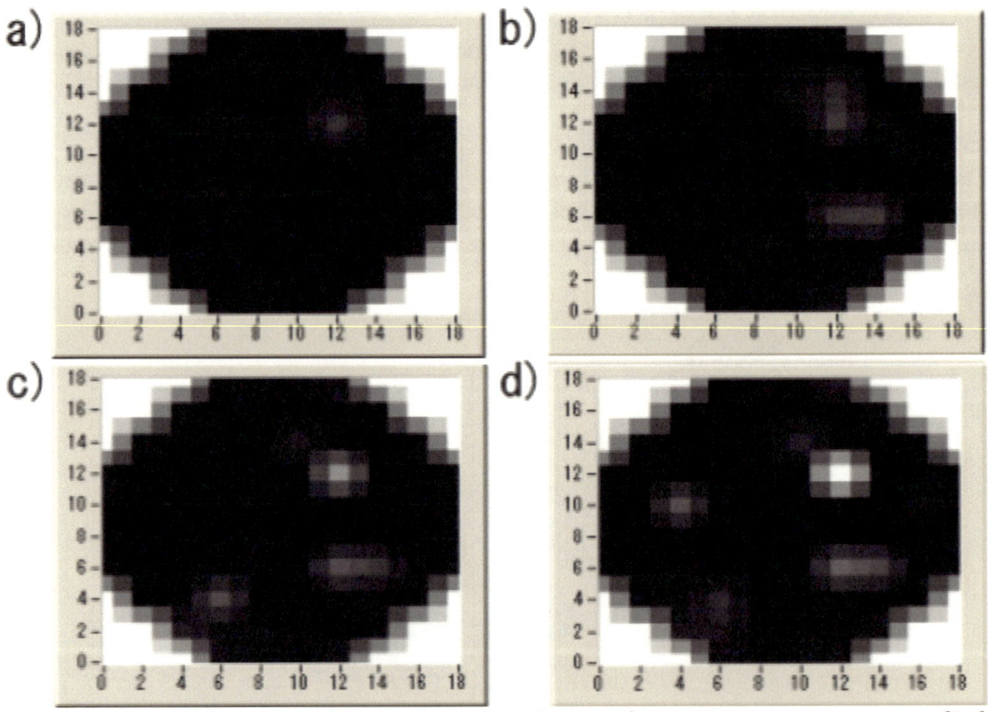

Fig. 8. Reconstructed pressure distribution when three or four point pressures are applied at a time a) a pressure point, b) two pressure points, c) three pressure points and d) four pressure points

Next, we tested multiple pressure points. Fig. 8 shows reconstructed images when applying multiple pressure points by fingers. The images, from the upper left to the lower right, show the reconstruction results when pressure applied to one, two, three or four points. We found that the non-negative least squares algorithm makes it possible to recognize individual pressure points.
Lastly, we measured the update time including the measurement time and reconstruction time. The update time was about 360ms. This is relatively slow; however, we can improve the time by optimizing measurement process.

4. Confirmation of basic performance

To examine basic tactile sensor performance, we compared experiments using a digital force gauge (Shimpo FGC-2B) as shown in Fig. 9 under the same conditions as in previous sections. A disk-like tip of 25mm in diameter was attached to the digital force gauge to apply force to conductive rubber. Data was collected in parallel by tactile sensors.

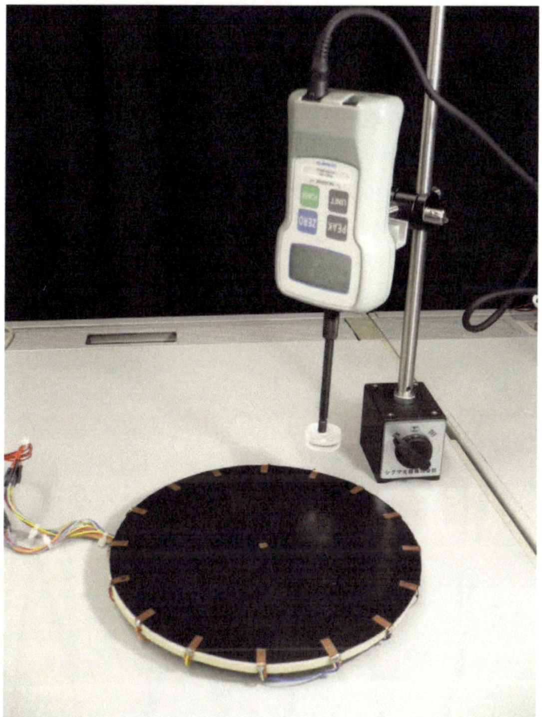

Fig. 9. Experimental setup using digital force gauge

Fig. 10 shows time-averaged reconstructed values by adding time-changing pressures by the force gauge at one point near the center. Since the tip of the force gauge is almost same size as the integration area -- 20mm in diameter -- corresponding to an element of the sensitivity matrix, responses were expected to appear on only one or a few adjacent elements, which was confirmed by results. When the location for adding force was changed, the area expressing the peak also changed. When peaks appeared in multiple areas, they were in areas adjacent to each other.

We extracted time profile of the value in the area showing the peak (Fig. 10) and compared it with measurement from the digital force gauge (Fig. 11). We found good tracking of pressure changes at an early stage, but output is reduced with repeated pressure changes, possibly due to the present characteristic of the material in which deformation tends to remain longer than the force recovery to the original state when large force is applied. We plan to improve response by increasing restoration force and speed by optimizing characteristics such as Young's modulus.

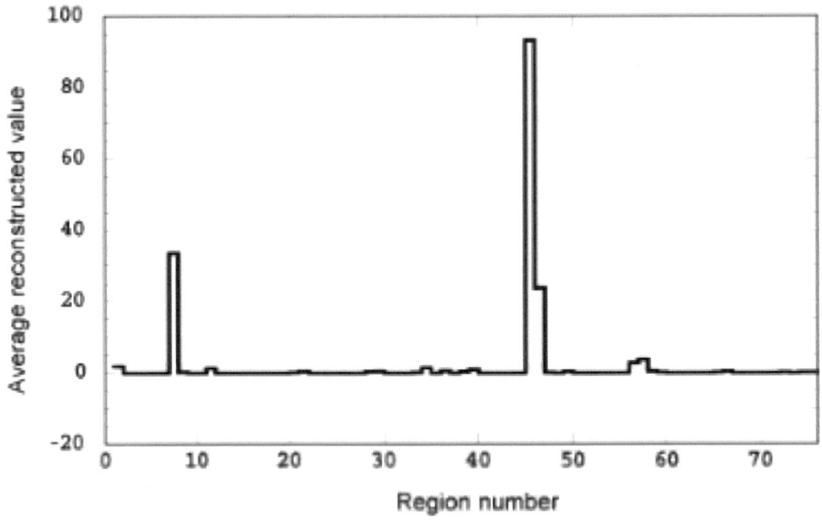

Fig. 10. Average of reconstructed value of each region when pressure changing over time was applied to 1 point

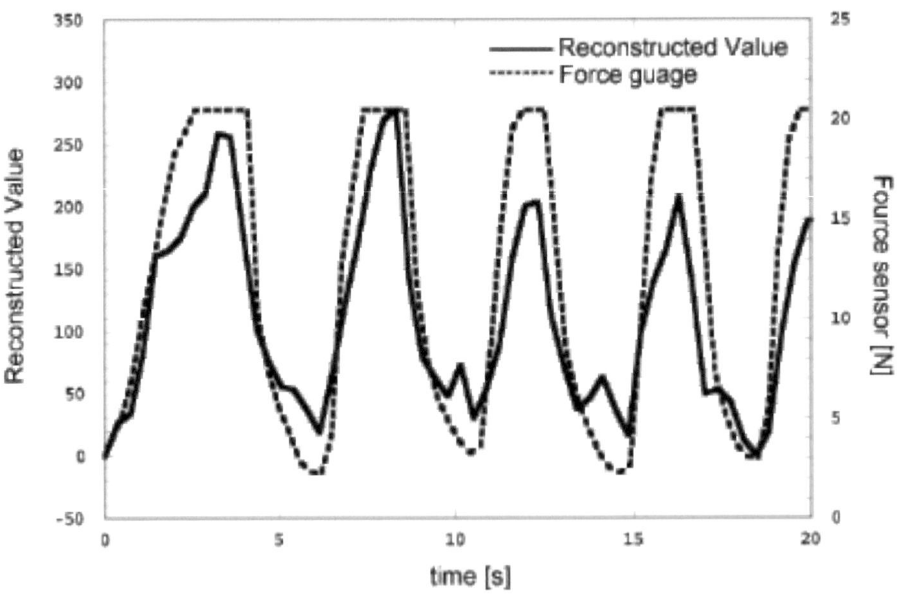

Fig. 11. Comparison of time traces between reconstructed value of the tactile sensor and force measurement of the digital force gauge when changing force is applied

To estimate noise, we measured signal to noise ratio, assuming 2-norm of the voltage measurement vector in the unloaded condition as noise and 2-norm of the voltage vector approximated by the least squares method with a force of 25N applied to one point as a signal. The ratio was found to be 27dB.

5. Conclusion

We developed a tactile sensor system based on the EIT method that has no wire or sensing elements in the tactile region. We believe that the sensor system is extremely durable to shocks and large transformations, and is low cost because there is no wire or sensing elements. We obtained good results from the experiments investigating the basic performance of our sensor system. The ordinary EIT algorithm assumes not only positive reconstructed values but also negative values, which leads to instability in reconstruction with noisy data. We restrict the reconstructed values to positive, which is justified by using a new kind of rubber whose electrical resistivity always increases in response to applied deformation.

In the future, we will improve the accuracy of the sensor and shorten the measurement time. We will also mount sensors as artificial sensitive skin on various objects (e.g. robots, seats and bumpers of automobile).

6. References

Alirezaei, H.; Nagakubo, A. & Kuniyoshi, Y. (2006). A tactile sensor based on inverse problem theory: Basic experiments, in *Proc. the 24th Annual Conference of the Robotics Society of Japan*, 1C32 (in Japanese)

Holder, S. D. (2005). *Electrical Impedance Tomography: Methods, History and Applications*, Inst. of Physics Pub. Inc.

Kanda, T.; Ishiguro, H.; Ono, T.; Imai, M. & Nakatsu, R. (2002). Development and evaluation of an interactive humanoid robot 'Robovie', in *Proc. ICRA2002*, pp. 4166-4173

Kato, Y.; Hayakawa, T. & Mukai, T. (2008). Soft areal tactile sensor using tomography algorithm, *Journal of Robotics and Mechatronics*, Vol. 20, No. 4 (In press)

Kotre, C. J. (1989). A sensitivity coefficient method for the reconstruction of electrical impedance tomograms, *Clinical physics and physiological measurement*, Vol. 10, pp. 275-281

Lawson, C. L. & Hanson, R. J. (1974). *Solving Least-Squares Problems*, Prentice-Hall, Chapter 23

Nagakubo, A. & Kuniyoshi, Y. (2006). A tactile sensor based on inverse problem theory: The principle, in *Proc. the 24th Annual Conference of the Robotics Society of Japan*, 1C31 (in Japanese)

Ohmura, Y.; Kuniyoshi, Y. & Nagakubo, A. (2006). Conformable and scalable tactile sensor skin for curved surfaces, in *Proc. ICRA2006*, , pp. 1348–1353

Shinoda, H. & Oasa, H. (2000). Wireless tactile sensing element using stress-sensitive resonator, *IEEE/ASME Trans. on Mechatronics*, Vol. 5, No. 3, pp.258-265

Tajima, R.; Kagami, S.; Inaba, M. & Inoue, H. (2002). Development of soft and distributed tactile sensors and the application to a humanoid robot, *Advanced Robotics,* Vol. 16, No. 4, pp.381-397

23

Recognition of Contact State of Four Layers Arrayed Type Tactile Sensor by Using Neural Networks

Seiji Aoyagi
Kansai University
Japan

1. Introduction

An advanced tactile sensor is desired for the purpose of realizing complicated assembly tasks, recognizing objects in the space where vision sensor cannot be used (in the darkness, etc.), and so on. For example, development of a robot hand is important for realizing a practical humanoid robot. In this hand, a tactile sensor of micro size is desired to be developed to give a vivid sense of touch like a human being. There are two subjects in developing such sensor: one is fabrication of a practical sensor having numerous force sensing elements, and another is information processing of data obtained from these elements.

The human skin contains numerous force-sensing receptors distributed horizontally and vertically at intervals of about 1 mm (Maeno, 2000). Many artificial tactile sensors are proposed (Kinoshita, 1981; Lee & Nicholls, 1999; Shinoda, 2000; Ishikawa & Shimojo, 1988): however, none among them approaches the human equivalent because it is technically difficult to fabricate a sensor having numerous force sensing elements. To overcome this problem, several researches are carried out, which obtain and analyze stress distribution optically (Kamiyama et al., 2003; Hiraishi et al., 1989; Maekawa et al., 1994). However, it is still desirable to detect the stress distribution by setting many sensing elements and obtaining electrical signals from these elements like the human tactile sensing mechanism, which directly processes stress data from receptors (i.e., does not indirectly process once optically transformed data). This direct processing method would be preferable in the industrial viewpoint of sensing accuracy, and moreover in the academic viewpoint of human mimetics.

Micromachining based on semiconductor manufacturing is being widely researched (Kovacs, 1998). Numerous arrayed miniature force sensors with uniform performance are fabricated on a silicon wafer with fine resolution of several microns, which may make it possible to fabricate a practical tactile sensor. Tactile sensors proposed and developed based on micromachining are classified to piezoresistive (Kobayashi & Sagisawa, 1991; Ohka et al., 1990; Horie et al., 1995; Nguyen et al., 2004; Takao et al., 2005) and capacitive (Esashi et al., 1990; Kane et al., 2000; Suzuki et al., 1990; Lee et al., 2006). In dynamic measurement, piezoelectric type is also applicable (Yagi, 1991; Matsushita et al., 2004). Author is fabricating a multiaxis piezoresistive type tactile sensor (Maeda et al., 2004; Izutani et al., 2004; Aoyagi et al., 2005) and a flexible tactile sensor of arrayed capacitive type (Aoyagi & Tanaka, 2007; Ono et al., 2008), which are described in the next section in detail.

Table 1 shows the comparison of characteristics of these sensors. The capacitive type has characteristics of nonlinear output, easy fabrication, and excellent sensitivity. The piezoelectric type has characteristics of acting also as an actuator, not applicable for the static sensing because of charge dissipation as time elapse. The piezoresistive type has characteristics of a linear output, and a wide range of strain detection.

	Capacitive	Piezoresistive	Piezoelectric
Detecting range	Narrow	Wide	Narrow
Sensitivity	Good	Normal	Normal
Power consumption	Small	Large	Small
Temperature dependence	Small	Big	Small
Digital procesing	Pulse	A/D converter necessary	A/D converter necessary
Linearity	Bad	Good	Good
CMOS compatible	Good	Good	Bad
Fabrication cost	Low	High	High
Sensing area	Small	Large	Small

Table 1. Comparison between characteristics of micromachined sensor types

In the reported micromachined tactile sensors, the force sensing elements are distributed two dimensionally on a surface. However, in the human skin, four kinds of tactile receptors (Meissner corpuscle, Merkel cell, Ruffini ending, and Pacinian corpuscle) are distributed three dimensionally, i.e., not only on the skin surface but also in the four corresponding depths of the skin (Maeno, 2000) (Fig. 1).

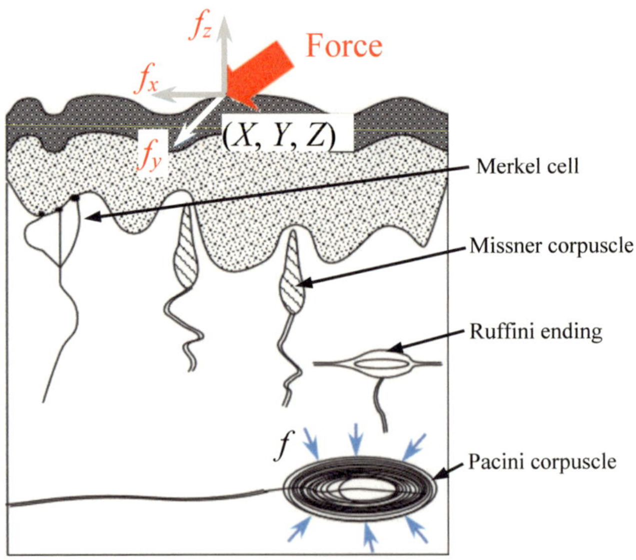

Fig. 1. Schematic view of human tactile receptors

In the human tactile sensing, the brain synthesizes nerve signals from many receptors and obtains cutaneous stress distribution to finally recognize the contact state. This human information processing mechanism has not been cleared yet: therefore, many artificial intelligence methods are proposed and evaluated. As one of the methods of processing information from many sensing elements, neural networks (referred to herein as NN) are well known (Wasserman, 1993; Watanabe & Yoneyama, 1992). As for the pattern recognition by vision sensors, there are many researches applying NN for processing image pixel data (Marr, 1982; Sugie, 2000). However, there are few reports applying NN for tactile sensors (Aoyagi et al, 2005; Aoyagi & Tanaka, 2007), since a practical, inexpensive, and widely used tactile sensor composed of many sensing elements has not been established mainly because of fabrication difficulties.

The following of this chapter is constructed as follows:

1. The micromachined force sensing elements under development by the author's group are introduced. One has the silicon structure having a pillar on a diaphragm, on which four piezoresistors are fabricated to detect the distortion caused by a force input to the pillar. Another has the polymer PDMS structure having a concave area inside, on top and bottom surfaces of which aluminum electrodes are deposited, realizing a capacitor.
2. Since a practical arrayed tactile sensor composed of many of the force sensing element is under development, the output of an assumed arrayed type tactile sensor is simulated by the finite element method (FEM). The FEM-simulated stress distribution data are assigned to each assumed stress sensing element of the array. Then, all data of these elements are processed by NN.
3. Imitating the human skin, an arrayed type tactile sensor comprising four layers is proposed and assumed. The information processing method of this sensor is investigated by FEM simulation. A recognizing method of force and its direction is proposed by using two stages NN. A recognizing method of object shape, which is contacted with the sensor surface, is also investigated by a simulation.

2. Example of micromachined force sensing element

2.1 Piezoresistive type

A structure having a pillar and a diaphragm has been developed by authors using micromachining technology. The schematic structure of one sensing element is shown in Fig. 2 (Izutani et al., 2004). Piezoresistors are fabricated on a silicon diaphragm to detect the distortion which is caused by a force input to a pillar on the diaphragm. Three components of force in x, y, and z direction can be simultaneously detected in this sensing element. The principle of measurement is shown in Fig. 3.

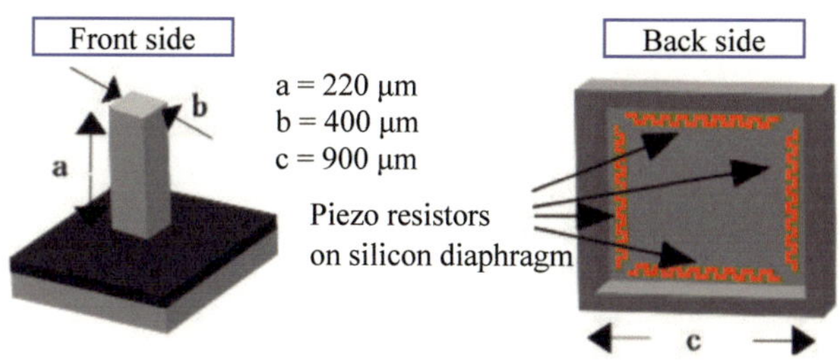

Fig. 2. Schematic structure of sensing element of piezoresistive type

In order to determine the arrangement of piezoresistors, FEM analysis was carried out. The distribution of strain in horizontal direction on the diaphragm when the force of 10 gf is applied vertically to the pillar tip is shown in Fig. 4(a). The distribution when the force is applied horizontally is shown in Fig. 4(b). It is proven that the strain is maximal at the edge of the diaphragm. Therefore, the piezoresistors were arranged near the edge of the diaphragm as far as possible.

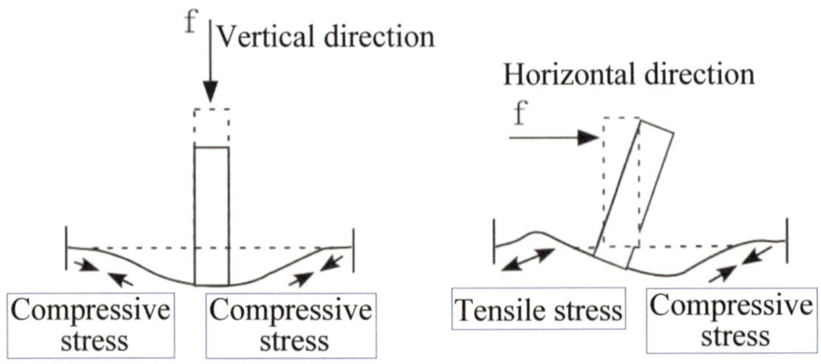

Fig. 3. Principle of force measurement for 3 axes

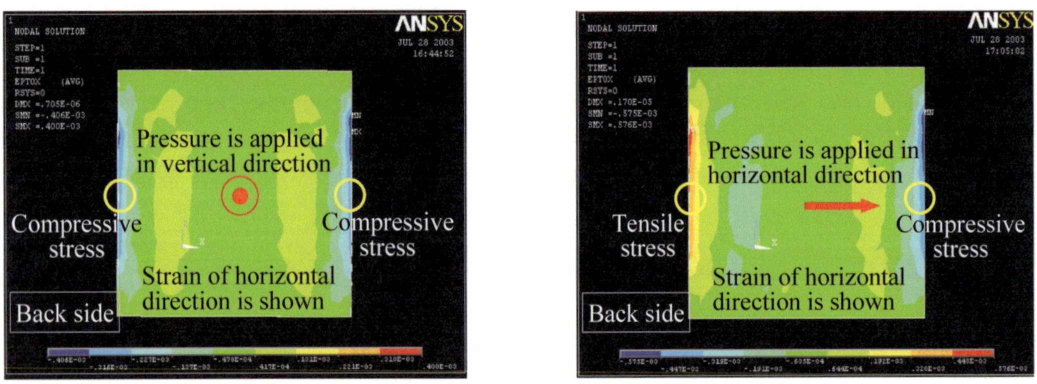

(a) Vertical case (b) Horizontal case

Fig. 4. FEM simulation result of distortion of a diaphragm

The micromachining fabrication process of this sensing element is shown in Fig. 5. The SEM image of a fabricated sensing element is shown in Fig. 6. In z direction, it is experimentally proven this element can detect the input force with good linearity within the range from 0 to 200 gf, as shown in Fig. 7. Characterization of performance of force detection in x and y direction, and fabrication of an arrayed type micro tactile sensor by using many sensing elements are ongoing. Furthermore, coating a polymer Parylene (Tai, 2003) film on arrayed elements is planned in future, as shown in Fig. 8. Chemical Vapor Deposition (CVD) can realize a conformal deposition (that is, the deposition is performed not only on the top surface of a target object but also on the back/side surface of it). Four of coated sheets are stacked one by one and bonded to each other, finally forming an arrayed tactile sensor having four layers.

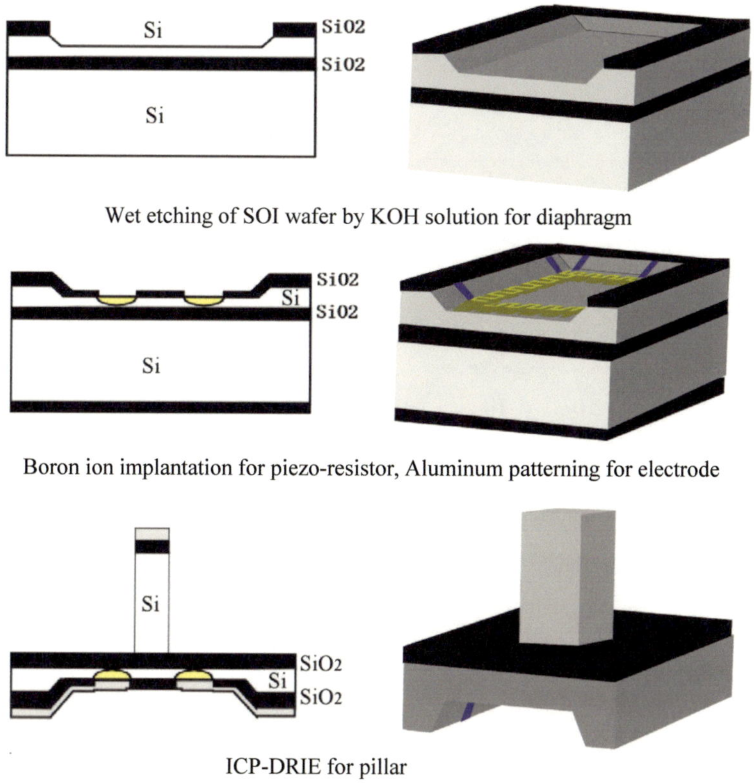

Fig. 5. Microfabrication process of a force sensing element of Piezoresistive type

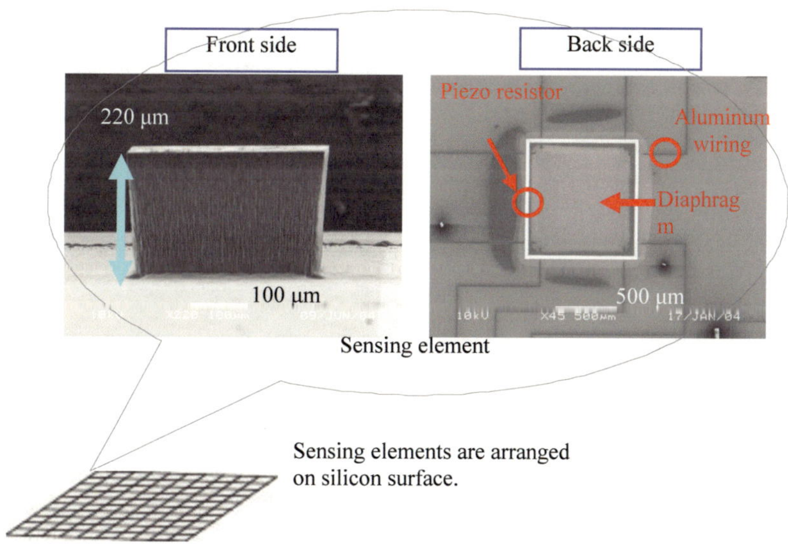

Fig. 6. SEM image of a fabricated sensing element and its application to an array type tactile sensor

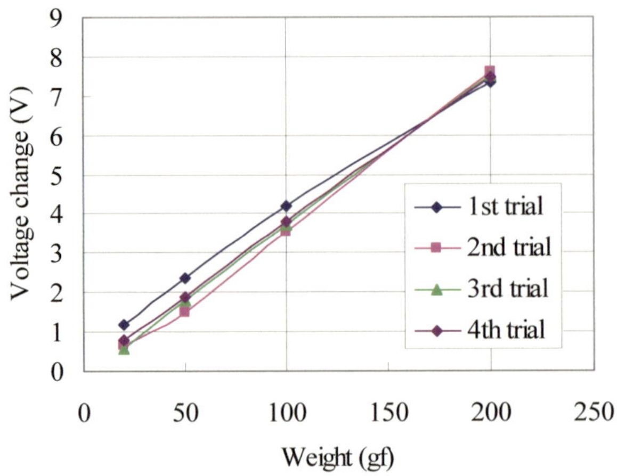

Fig. 7. Output voltage change with respect to applied weight

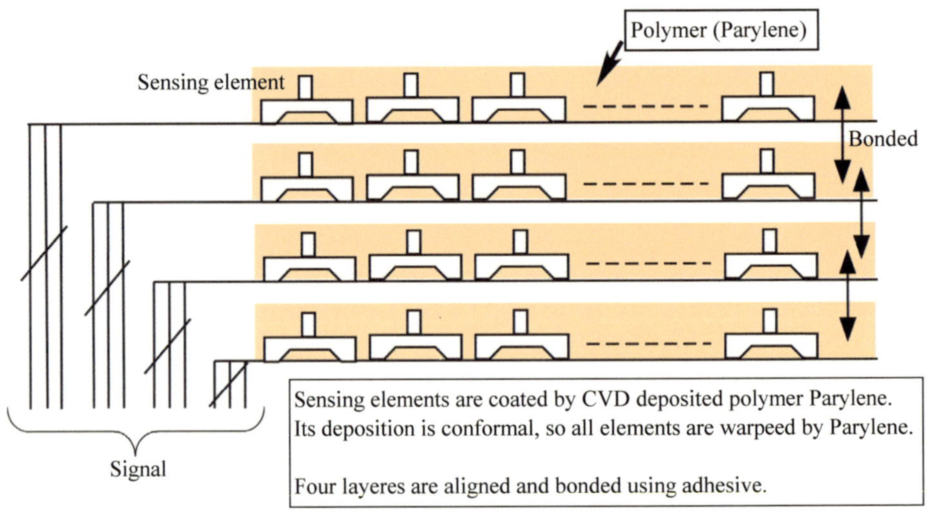

Fig. 8. Four layers tactile sensors comprising polymer sheets deposited on sensing elements (under planning at present)

2.2 Capacitive type

Imitating the human skin structure, a flexible arrayed type tactile sensor having four layers is under development using micromachining technology (Aoyagi & Tanaka, 2007; Ono et al., 2008). The fabrication process of this sensor is shown in Fig. 9. As the material of a layer, polydimethylsiloxane (PDMS), which is a kind of flexible silicone rubber, is used. This process is summarized as follows: one PDMS layer having electrodes is fabricated by a spin-coated method. Another PDMS layer having electrodes is fabricated by a casting method, on which a number of concave space is formed as negative of patterned sacrificial photoresist. These two layers are bonded with each other by applying heat and pressure (see detailed condition in this figure).

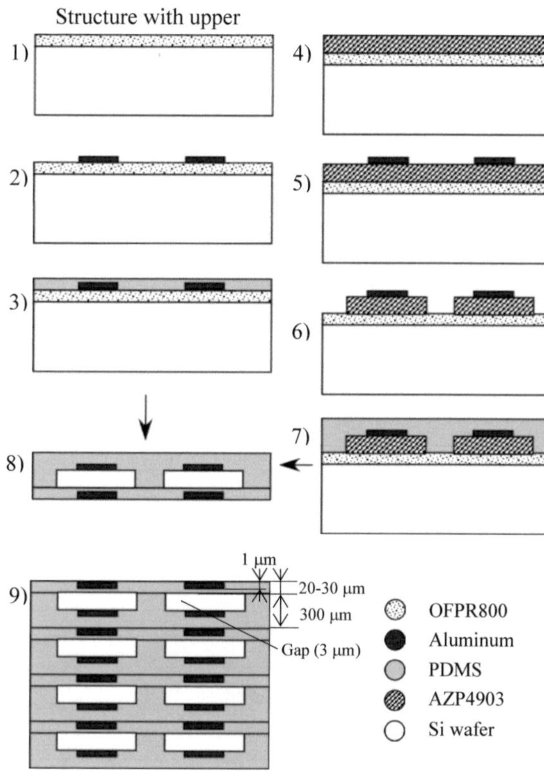

1) Photoresist (OFPR800) is spin-coated on Si substrate for sacrificial layer.
2) Aluminum (1 μm) is deposited and patterned for upper electrodes.
3) PDMS (20-30 μm) is spin-coated as structural material.
4) Photoresist (OFPR800) is spin-coated on Si wafer for the 1st sacrificial layer, followed by hard bake for giving resistivity to O_2 plasma etching afterward.
Thick photoresit (AZP4903) is spin-coated for the 2nd sacrificial layer (10 μm).
5) Aluminum (1 μm) is deposited and patterned for lower electrodes.
6) The 2nd sacrificial layer is patterned by O_2 plasma.
7) PDMS (300 μm) is cast and cured in air.
The 1st and 2nd sacrificial layers are wet etched away using acetone, consequently PDMS structure is pealed off from Si substrate.
8) PDMS structure with lower electrodes is turned over, and bonded to that with upper electrodes under condition as follows: baking temp. is 120°C, pressure is 20 kPa.
Bonded PDMS structure is pealed off

Fig. 9. Fabrication process of micro tactile sensor composed of many capacitive sensing elements distributed in four PDMS layers

Each sealed concave space has lower and upper electrodes, forming a capacitance. This capacitance changes as the distance between electrodes changes when the structure is deformed based on applied force, i.e., a capacitive force sensing element is realized. The obtained structure having many sensing elements forms one layer, four of which are stacked one by one and bonded to each other, finally forming a tactile sensor having four layers.

A structure of one layer has been fabricated at the moment. An optical image of this structure is shown in Fig. 10(a), of which layout of capacitive sensing elements is shown in Fig. 10(b). Including a 5 by 5 array, many types of arrays are designed on trial. Wiring in one direction, and that in its perpendicular direction are formed, on the crossing areas of which, capacitive sensing elements exist. By selecting corresponding two bonding pads for these two directions, detecting the capacitance of the target sensing element is possible.

The performance of one capacitive force sensing element and that of an arrayed sensor composed of 3×3 elements are characterized. First, a weight was set on the surface of the fabricated sensor having one layer. Then, the capacitance change of one sensing element (1 mm square, 3 μm gap) was detected with the aid of a CV converter IC (MicroSensors Inc., MS3110), the programmable gain of which was set to 0.1 pF/V. Four weights of 5, 10, 20, and 50 gf were employed, of which radii are 5.5, 6.5, 7.5, and 10 mm, respectively. Namely, whole area of one sensing element was covered by each weight and was applied pressure of 516, 738, 1,109, and 1,560 Pa, respectively.

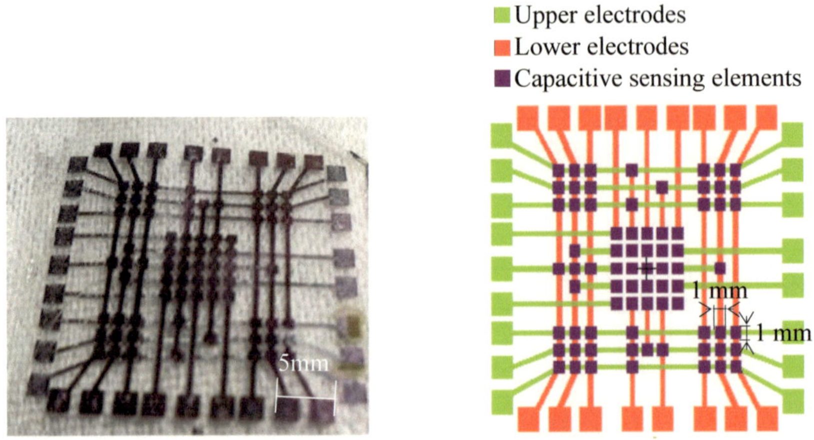

(a) Optical image of one layer sensor (b) Layout of capacitive sensing elements

Fig. 10. Fabricated sensor having one layer composed of many capacitive sensing elements

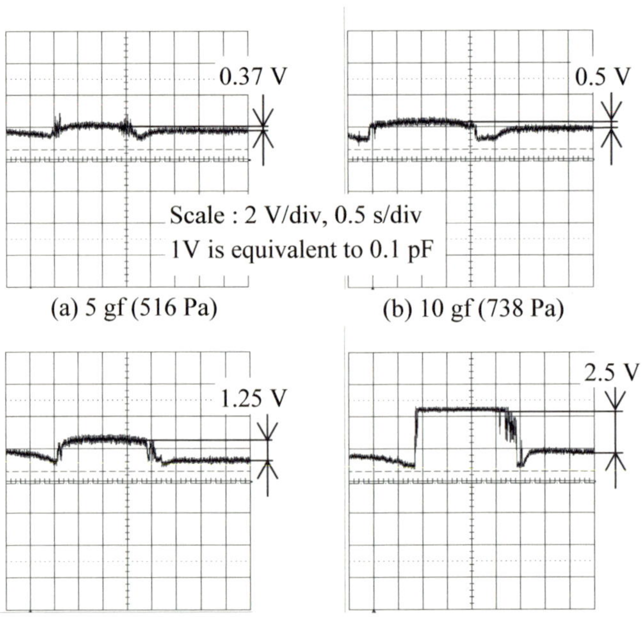

(a) 5 gf (516 Pa) (b) 10 gf (738 Pa)

(c) 20 gf (1,109 Pa) (d) 50 gf (1,560 Pa)

Fig. 11. Capacitance change with respect to applied force

Experimental results of output voltage of the IC for several applied force, which are observed by an oscilloscope, are shown in Fig. 11. It is confirmed that the capacitance surely changes by applying force. The results are arranged in Fig. 12, which shows the relationship between the applied pressure and the capacitance change of one sensing element. It is proven that the capacitance increases as the pressure increases. In this figure, the theoretical value is based on the FEM multiphysics simulation, which analyzes the capacitance under the boundary condition defined by the mechanical deformation of the sensor structure. Measured and theoretical curves have similar trends, although the error is rather large at the pressure of 1,560 Pa.

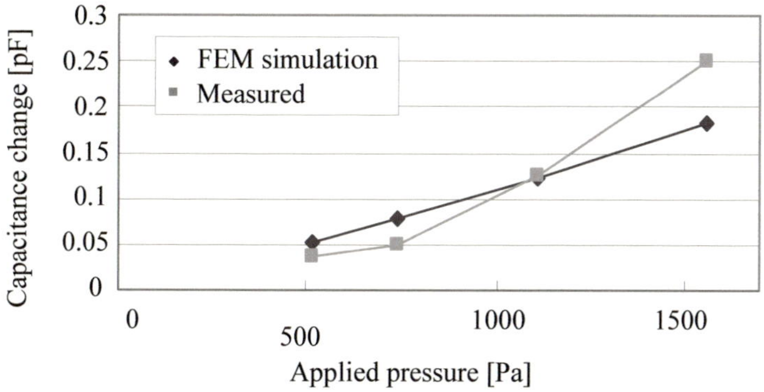

Fig. 12. Relationship between capacitance change and pressure

Next, a distributed load was preliminarily detected using the developed arrayed sensor having one layer. A weight of 5 gf was set, i.e., the pressure of 516 Pa was applied, under two conditions: one is that the weight completely covers the surface area of an arrayed sensor consisting of 3×3 sensing elements (see Fig. 13(a), the sensor exists in the lower right corner of this figure), and another is that the weight partially covers the arrayed sensor, leaving some uncovered elements near the corner of the sensor (see Fig. 13(b)). Then the capacitance change of each sensing element was detected one by one. The results for these cases are shown in Figs. 14(a) and (b), respectively. Looking at these figures, in the former case, almost the constant capacitance changes for all the sensing elements are obtained: while in the latter case, the comparatively lower capacitance changes are obtained at the sensing elements near the corner of the fabricated sensor, where the sensing elements are not covered completely by the weight. These results imply the possibility of this sensor to detect a distributed load.

Weight: 5 gf, Pressure: 516 Pa

(a) A weight completely covers an arrayed sensor.

(b) A weight partially covers an arrayed sensor.

Fig. 13. Experimental condition for distributed load measurement by the arrayed sensor with 3×3 elements

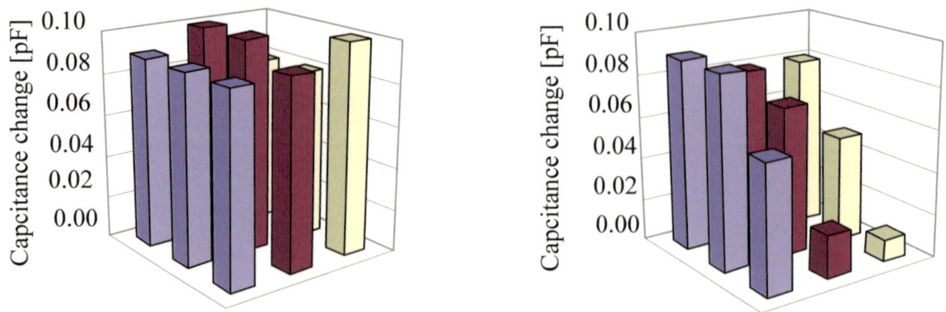

(a) Under the condition shown in Fig. 13(a)

(b) Under the condition shown in Fig. 13(b)

Fig. 14. Result of distributed load measurement

3. FEM simulation on data processing of arrayed tactile sensor having four layers

3.1 Acquisition of contact data by FEM

Since a practical tactile sensor composed of many force sensing elements distributed on four layers is under development, FEM simulation is employed to simulate the data from these sensing elements. As a tactile sensor, an elastic sheet is assumed of which side is 15.0 mm and thickness is 5.0 mm, as shown in Fig. 15. Sensing elements are horizontally distributed in 1.25 mm pitch, and vertically distributed in 1.0 mm pitch. That is, the sensor has four layers, which are positioned at 1 mm, 2mm, 3mm, and 4 mm in depth from the surface. The number of sensing elements is 13×13×4=676 in total. Furthermore, to show the effectiveness of the sensor having four layers, a sensor having one layer is assumed for the reference, of which sensing elements are positioned at 1 mm in depth from the surface, and the number of sensing elements of which is 13×13×1=169 in total.

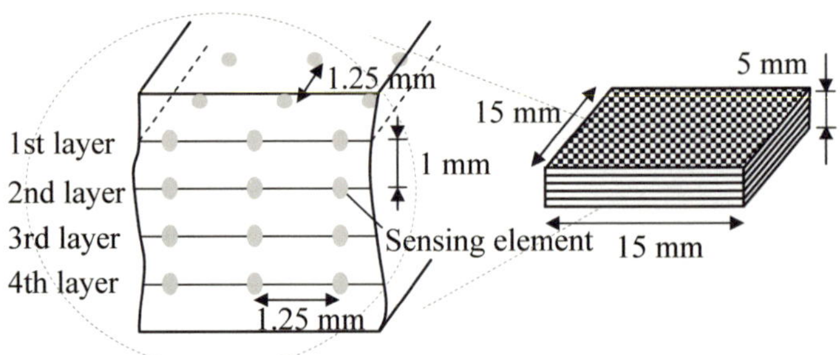

Fig. 15. Assumed model of four layers arrayed type tactile sensor

In case of recognizing force magnitude and its direction using NN (details are explained later in Chapter 4), the stress distribution inside the sensor sheet is simulated under the condition shown in Fig. 16. ANSYS (ANSYS, Inc.) is used as simulation software. As a material of composition, PDMS (Young's modulus: 3.0 MPa) is assumed. Distributed load is applied to the circle of 3 mm in radius on the sheet surface. An object that cuts diagonally a cylinder is used to apply the force, because this software is difficult to deal with a diagonal load to a sheet surface. The friction of coefficient between the sheet surface and the bottom of object is assumed to be 1.0. Under this condition, stress distribution inside the sheet is simulated for many times, changing the force magnitude and its direction. Considering the sensing range of the practical arrayed tactile sensor under development, the applied force magnitude is changed within the range from 10 to 200 gf. Figure 17 shows a simulated example of distribution of Mises stress σ_{mises} , when θ is 15° and force is 10 gf.

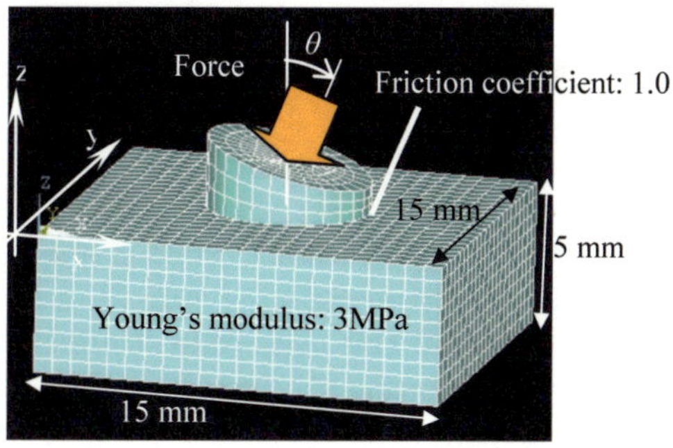

Fig. 16. FEM simulation condition of stress distribution for contact force recognition

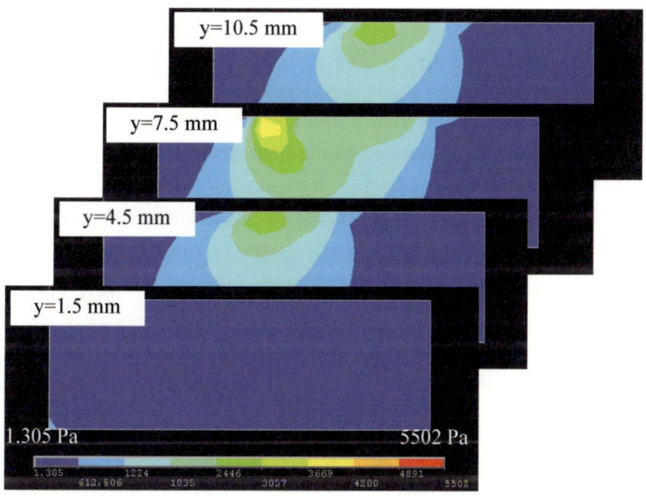

Fig. 17. FEM simulation result of stress distribution for contact force recognition (in case of θ=15 degree)

In case of recognizing the shape of contact object using NN (details are explained later in Chapter 5), the stress distribution in the sensor sheet is simulated under the condition shown in Fig. 18 (a). The contact objects having various bottom shapes are employed. Each object is pressed vertically, i.e., under θ =0°, against the assumed tactile sensor, being applied force of which magnitude is 10 gf. Figure 18(b) shows a simulated example of distribution of σ_{mises} , where the bottom shape of object is circle.

10gf is applied vertically.

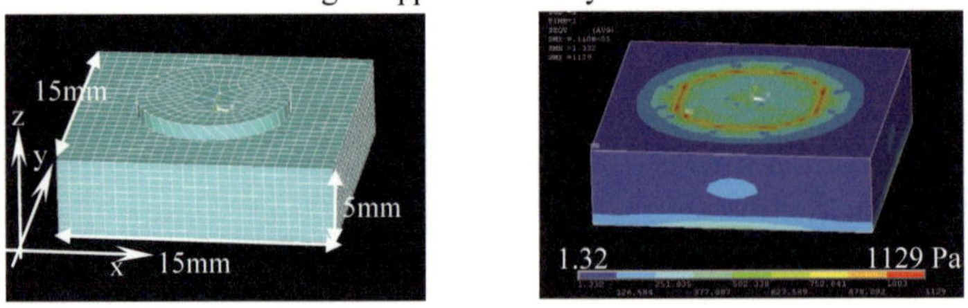

(a) FEM simulation (b) Simulation result

Fig. 18. FEM simulation of stress distribution for object shape recognition (in case of circle shape)

3.2 Assignment of FEM data to sensing elements

It is necessary to assign σ_{mises} at each node on FEM meshed element to each sensing element of the tactile sensor (Fig. 15). A sampling area of 0.625 mm in radius, of which center is the position of a sensing element, is assumed. The σ_{mises} data of FEM nodes within this area are averaged, being assigned to the corresponding sensing element as its output.

4. Recognition of contact force

4.1 Recognition method of force magnitude and its direction using two stages neural networks

In usual NN researches, several features, such as area, surrounding length, color, etc., are extracted from raw data, and they are input to NN. On the other hand, in this research, all raw data are directly input to NN at the first step, considering that the information processing mechanism in the human brain has not been cleared, i.e., whether some features are extracted or not, and what features are extracted if so.

In usual researches, single NN is used for pattern recognition. In case of tactile sensing, single NN may be possible, to which stress data of sensing elements are input, and from which three components f_x, f_y, f_z of force vector are output. However, in case of recognizing both magnitude and its direction with practical high precision by single NN, numerous training data and long training time would be necessary. On the other hand, in this case, as far as the force direction is kept to be identical, the aspect of stress distribution does not change, whereas the stress value at each sensing element changes linearly in proportion to the input force magnitude. Therefore, force direction could be detected irrespective of force magnitude by normalizing stress data of all sensing elements from 0 to 1, and inputting them to the first stage NN (Fig. 19). Then, the direction information, i.e., three components of the normalized unit force vector, and the maximum stress value of each layer, are input to the second stage NN for detecting the force magnitude (Fig. 20). Since NN of each stage perform its own allotted recognition processing, the number of training data and training time are expected to be much reduced, keeping high detecting precision.

As a learning method of network's internal state that decreases the error between NN outputs and training data, RPROP method (Riedmiller & Braun, 1993) modifying the well-known back propagation method is adopted. Stress distribution data of unknown force vectors are input to the learned two stages NN, and its direction and magnitude are recognized. From these results, the generalization ability of the NN is investigated.

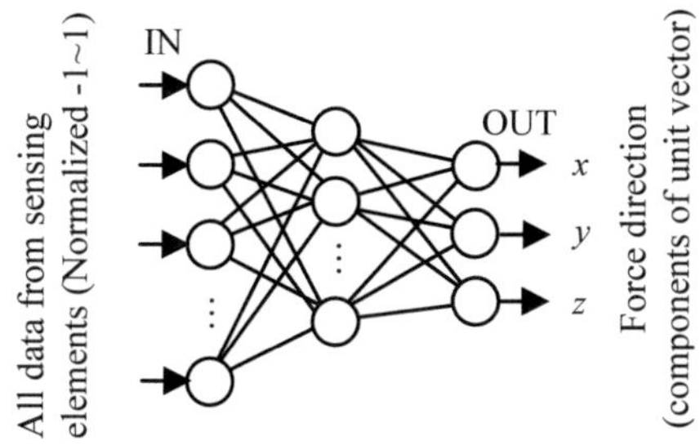

Fig. 19. First stage neural networks for force direction recognition

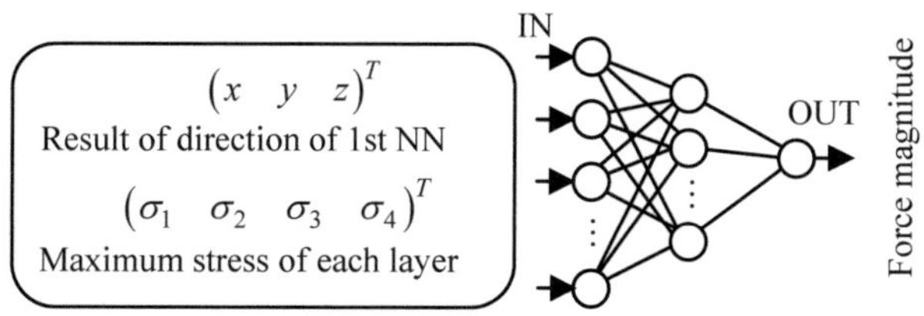

Fig. 20. Second stage neural networks for force magnitude recognition

4.2 Results of force direction recognition

The number of neurons of the first stage NN (see Fig. 19) is as follows: 676 for input group in case of the four layers sensor (this is 169 in case of the one layer sensor), 20 for hidden group, and 3 for output group. Stress information of all the sensing elements is input to the neurons of input group. The neurons of output group determine the unit vector of applied force (3 outputs). Training data are 8 kinds of stress distribution, of which force direction θ (see the definition of θ in Fig. 16) ranges from 0 to 35° in 5° intervals. The convergence of learning of NN is good for both of the one layer sensor and the four layers sensor, of which training error is equivalent to 0.04°, as shown in the second line of Table 2.

As the unknown test data, four kinds of stress distribution, of which force directions θ are 1, 13, 18, and 27°, are input to the learned NN. The output of NN is converted to θ, which is shown in Table 2. It is proven that the recognition accuracy of the four layers sensor is slightly better than that of the one layer sensor. The errors are within 0.2° for both cases.

	NN of one layer sensor	NN of four layers sensor
Training error	0.04°	0.04°
Unknown input 1°	0.8°	0.9°
Unknown input 13°	12.8°	13.0°
Unknown input 18°	18.0°	18.0°
Unknown input 27°	26.8°	27.2°

Table 2. Results of force direction recognition

4.3 Results of force magnitude recognition

The number of neurons of the second stage NN (see Fig. 20) is as follows: 7 for input group in case of the four layers sensor (this is 4 in case of the one layer sensor), 169 for hidden group, and 1 for output. The output of NN is from 0 to 1, normalizing the full range of sensor output, which is from 0 to 200 gf. Training data are 160 kinds of stress distribution, i.e., 8 kinds of degree ranging from 0 to 35° in 5° intervals, 20 kinds of force magnitude ranging from 0 to 200 gf in 10 gf intervals, then 8×20=160 kinds in total. Contrary to the case of force direction recognition, the convergence of learning the NN is not so good, depending on initial connection weights of neurons. Therefore, ten kinds of initial connection weights are tested, from which the NN is learned, setting the limit of iteration number to 100,000. Obtained training errors for them are averaged, and described in the second line of Table 3, showing that the training error for the one layer sensor is inferior to that for the four layers sensor.

One NN realizing the smallest training error is selected among the ten, the generalization ability of which is estimated. As the unknown test data, 76 kinds of stress distribution are prepared, of which θ and force magnitude are as follows: θ are 1, 13, 18, and 27°, and force magnitudes are from 15 to 195 gf in 10 gf interval. The outputs of the NN for the test data are evaluated by comparing them with true values of force magnitude. The results of absolute errors between them in case of the one layer sensor are shown in Fig. 21. Those in case of the four layers sensor are shown in Fig. 22. The average and the standard deviation of all the absolute errors are calculated for each case, which are shown in the third and forth lines of Table 3.

From these results, it is proven that the accuracy of force magnitude recognition of the four layers sensor is fairly better than that of the one layer sensor. The reason of this advantage of the four layers sensor compared to the one layer sensor would be based on its larger number of sensing elements distributed not only horizontally but also vertically, realizing a fine interpolation of nonlinear characteristics of stress distribution caused by applied force, which does not contradict the better convergence of learning NN (see the second line of Table 3).

	NN of one layer sensor	NN of four layers sensor
Average of training error	0.003 [gf]	0.001 [gf]
Absolute average error	0.53 [gf]	0.23 [gf]
Standard deviation	0.17 [gf]	0.03 [gf]

Table 3. Results of force magnitude recognition

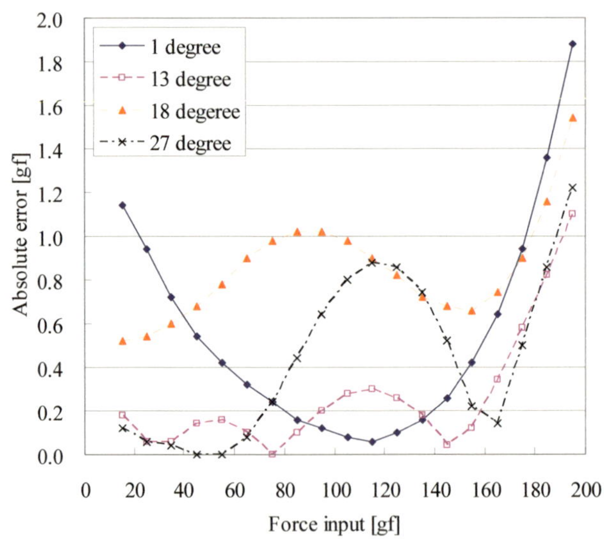

Fig. 21. Absolute errors between NN outputs and true values (in case of one layer sensor)

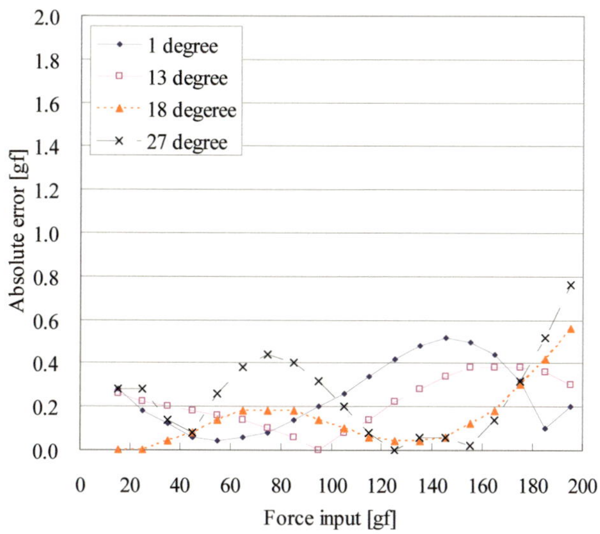

Fig. 22. Absolute errors between NN outputs and true values (in case of four layers sensor)

5. Recognition of object shape

5.1 Recognition method of object shape using neural networks

The shape of contact object is recognized by applying NN to the FEM simulated data of force sensing elements. It is assumed that only approximate contact position is known by some recognition method. Then, the important point is to recognize the shape with robustness to unwanted shift of the object from the reference position, where the template for the recognition was constructed. The method using NN for object shape recognition is schematically shown in Fig. 23.

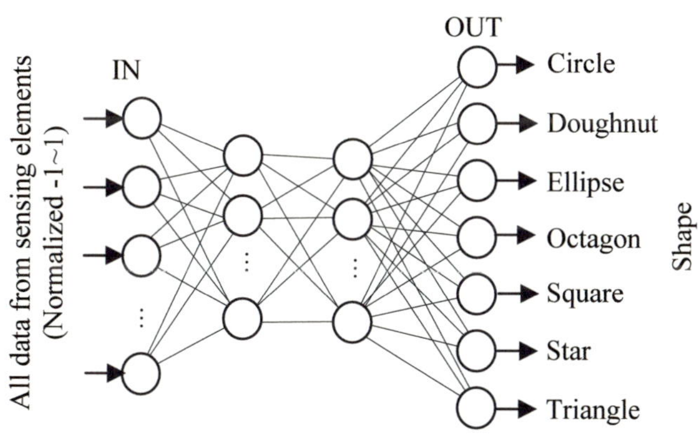

Fig. 23. Neural networks for object shape recognition

As the object shape, seven kinds of circle, doughnut, ellipse, octagon, square, star, and triangle are employed, which are circumscribed for a 10 mm square. As the training data, the stress distributions are simulated by FEM, when the objects are positioned precisely in

the center of the sensor surface of 15 mm square, and pressed vertically by applying 10 gf force. As a learning method of network's internal state, RPROP method (explained in Section 4.1) is adopted.
The unknown test data are prepared, which are obtained from the stress distributions when the objects are shifted from the center of the sensor surface by 1.25 mm. This shift is beyond 10% of the object side, which is comparatively large. Using these data, the generalization ability of NN is investigated, and the effectiveness of using four layers is estimated.

5.2 Results of object shape recognition

The number of neurons of the NN (see Fig. 23) in case of the four layers sensor is as follows: 676 for input group, 676 for the first hidden group, 20 for the second hidden group, and 7 for output group. That in case of the one layer sensor is as follows: 169 for input group, 169 for the first hidden group, 13 for the second hidden group, and 7 for output group. The employment of two hidden groups, and the definition of the number of neurons of them are based on the adjustment by trial and error. Note that the adjustment in case of the four layers sensor was much easier than that in case of the one layer sensor, implying the good interpolating ability of using four layers.
The results of object shape recognition for unknown objects are shown in Table 4 and Fig. 24(a) in case of the one layer sensor. Those in case of the four layers sensor are shown in Table 5 and Fig. 24(b). The shaded values in these tables are the maximum NN's output value among the seven candidates. Seeing Table 4, in case of the one layer sensor, the circle is mistaken for the ellipse, whereas the doughnut and the octagon are mistaken for the circle. By contrast, seeing Table 5, all objects are finely recognized as the correct shapes in the case of the four layers sensor.

Unknown input	Output of NN in case of the one story sensor						
	Circle	Doughnut	Ellipse	Octagon	Square	Star	Triangle
Circle	0.62	0.00	0.99	0.00	0.00	0.00	0.00
Doughnut	0.89	0.00	0.00	0.00	0.00	0.00	0.00
Ellipse	0.00	0.00	1.00	0.00	0.00	0.00	0.00
Octagon	0.62	0.00	0.38	0.13	0.00	0.00	0.00
Square	0.00	0.00	0.00	0.00	0.97	0.00	0.00
Star	0.00	0.00	0.00	0.16	0.00	0.98	0.00
Triangle	0.00	0.00	0.00	0.00	0.00	0.00	1.00

Table 4. Results of object shape recognition by NN in case of the one layer tactile sensor

Unknown input	Output of NN in case of the four stories sensor						
	Circle	Doughnut	Ellipse	Octagon	Square	Star	Triangle
Circle	0.95	0.00	0.00	0.00	0.01	0.00	0.00
Doughnut	0.00	0.99	0.00	0.00	0.00	0.00	0.00
Ellipse	0.00	0.00	1.00	0.00	0.01	0.00	0.00
Octagon	0.01	0.00	0.00	0.81	0.00	0.00	0.00
Square	0.00	0.00	0.01	0.00	1.00	0.00	0.00
Star	0.00	0.00	0.00	0.00	0.00	0.99	0.00
Triangle	0.00	0.00	0.00	0.00	0.00	0.01	1.00

Unknown objects shifted from the center of sensor by 1.25 mm are recognized.

Table 5. Results of object shape recognition by NN in case of the four layers tactile sensor

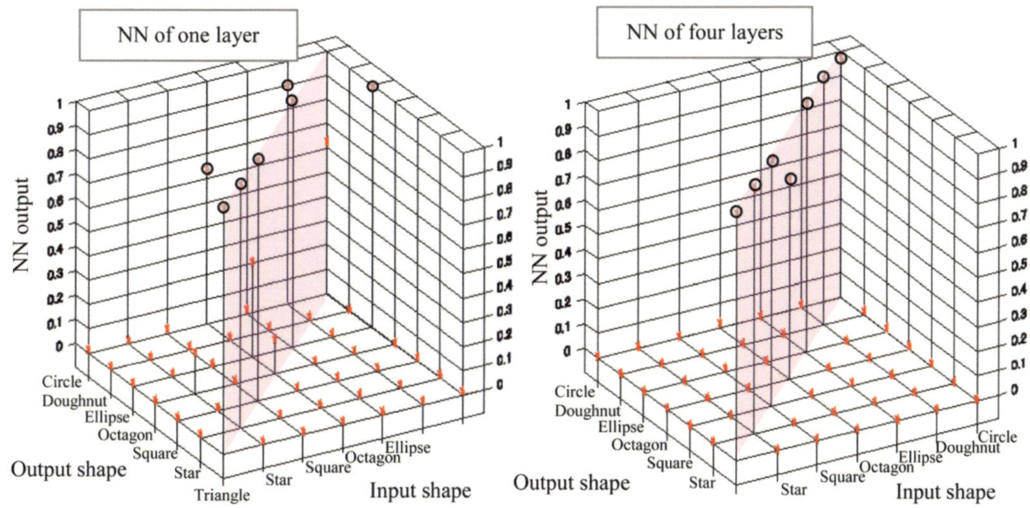

(a) In case of the one layer tactile sensor (b) In case of the four layers tactile sensor

Fig. 24. NN output of tactile sensor for unknown object shape

The stress distributions on each surface of the four layers are shown in Fig. 25. Seeing this figure, the contour edge of stress distribution becomes obscure as the depth becomes large, which means the influence of the object shift on the stress distribution change becomes smaller. If four layers are employed, the stress information of deeper layers, which is robust to the object shift, is available, which would be one of the reasons for the higher recognition ability of using four layers compared to that of using only one layer.

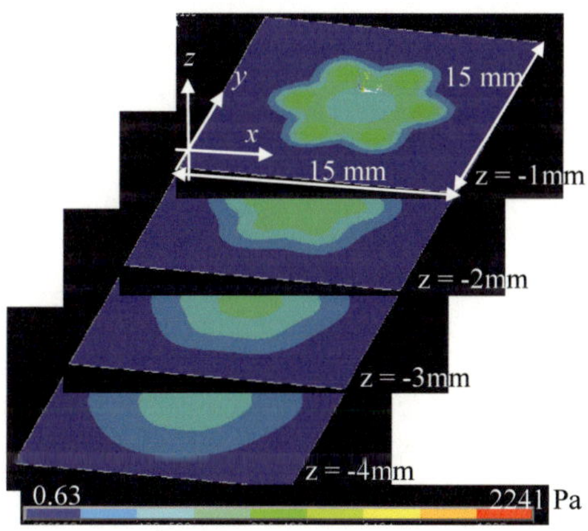

Fig. 25. Stress distribution on each surface of the four stories (in case of star shape)

6. Comparison with human tactile sensing

The density of tactile receptors in the human finger is very high and some optimum information processing may be carried out in the human brain. To compare the ability of artificial NN with that of a human being, the experiment is carried out in which a human senses object shape by finger touch, while his eye is occluded by a bandage. The situation of this experiment is shown in Fig. 26. Three subjects of circle, triangle, and square are adopted. Two human testers are employed. Each of them touches randomly 30 subjects. The result is shown in Table 6. Also, NN simulation is applied to this case, and the result is shown in the same table.

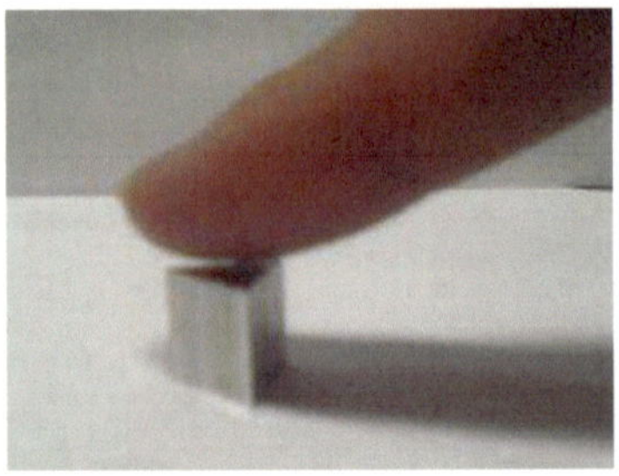

Fig. 26. Situation of object shape recognition by human being

	Circle	Triangle	Square
Human (testers A and B)	100.0%	100.0%	100.0%
Neural networks	96.8%	100.0%	98.2%

Table 6. Comparison between human and neural networks

Seeing this result, human ability is fairly good, whereas NN ability is a little bit inferior to human ability, but not so bad. The mechanism of tactile sensing of human being is not clear; however, NN is a good candidate imitating human tactile ability, and authors think improving the recognition rate of NN is possible by increasing the number of sensing elements and selecting appropriate parameters of NN, such as the number of neurons, the number of hidden groups, etc.

7. Module networks for detecting total contact state

On the basis of this research, constructing networks (not restricted to neural networks) which can recognize not only force direction/magnitude and object shape but also size, contact position, orientation, texture, etc., totally is a projected work. The authors are planning to use module type networks, of which concept is schematically shown in Fig. 27. In this type networks, for example, if an object of different shape is added, the system can cope with this case only by re-learning a network for object shape recognition (not re-learning all networks), which reduces training data number and learning time. Modulation makes it possible to select effective and proper recognition methods not limited to NN, such as Support Vector Machine (SVM) (Cristianini & Taylor, 2000) etc., which would improve the recognition ability of the whole system.

In our previous research (Aoyagi et al., 2005), single NN recognize the object shape contacted with a sensor sheet despite size and contact position: however, enormous training data of stress distribution changing size and contact position variously were needed. As for this single NN, in order to cope with adding kinds of recognized objects, the learning process must be carried out again from the start, which needs more training data and training time as far as keeping a practical recognition rate. To solve this problem, in the projected work, two network modules which determine size and contact position respectively are used (Fig. 27). From the output of these two modules, the most suitable NN is selected to recognize the object shape. By using the modulation method described here, reducing training data number, increasing recognition rate, increasing generalization ability, etc., are expected in future.

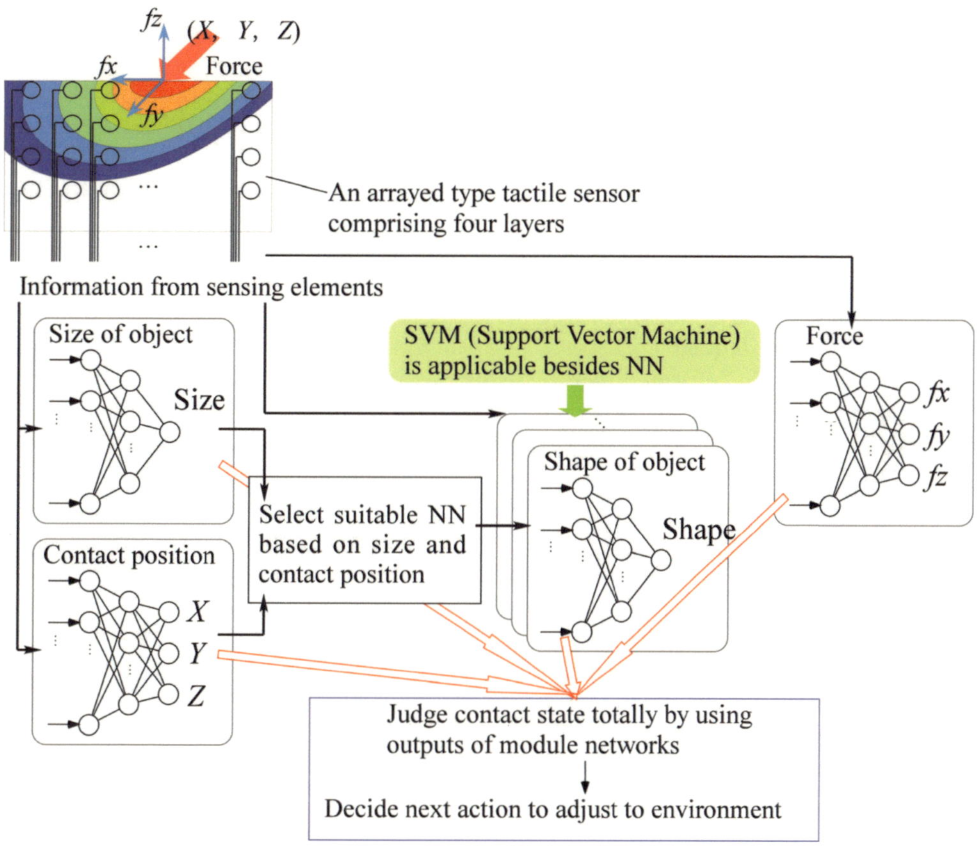

Fig. 27. Concept of module networks for detecting total contact state

8. Conclusions

A tactile sensor having four layers is proposed, and the information processing method for this sensor using neural networks (NN) is investigated. The summary is as follows: 1) Imitating the human skin structure, an arrayed type tactile sensor composed of many force sensing elements distributed on four layers is proposed. 2) The micromachining process for the four layers sensor composed of many capacitive sensing elements is proposed. 3) The output data from the force sensing elements are simulated by FEM. 4) For recognizing the contact force, two stages NN, the first stage of which is for detecting force direction and the second stage of which is for detecting force magnitude, is proposed. The effectiveness of this method is confirmed by simulation. 5) The contact object shape is recognized by simulation, confirming the effectiveness of NN. 6) In both cases of force recognition and shape recognition, the sensor having four layers is confirmed to be superior to that having one layer, in the viewpoint of recognition accuracy. 7) A concept of module networks for detecting total contact state is proposed.

The novelties of this study on tactile sensor are 1) using four layers, 2) using raw data of all sensing elements and inputting them to NN, and 3) proposing two stages NN and module networks.

9. Acknowledgments

This work was supported by Ministry of Education, Culture, Sports, Science and Technology of Japan (MEXT) KAKENHI (17656090).

10. References

Aoyagi, S.; Tanaka, T. & Makihira, K. (2005). Recognition of Contact State by Using Neural Network for Micromachined Array Type Tactile Sensor. *Int. J. Information Acquisition*, Vol. 2, No. 3, pp. 1-10

Aoyagi, S. & Tanaka, T. (2007). Proposal of a Micromachined Tactile Sensor Having Four Stories and Its Information Processing Method Using Module Networks. *Neural Information Processing -Letters and Reviews*, Vol. 11, Nos. 4-6, pp. 147-158

Cristianini, N. & Taylor, J. S. (2000). *An Introduction to Support Vector Machines and Other Kernel-Based Learning Methods*, Cambridge University Press, ISBN 0-521-7809-5, Cambridge, UK

Esashi, M.; Shoji, S.; Yamamoto, A. & Nakamura, K. (1990). Fabrication of Semiconductor Tactile Imager. *Trans. Institute Electronics, Information and Communication Engineers*, Vol. J73-C-II, No. 1, pp. 31-37

Hiraishi, H.; Suzuki, N.; Kaneko, M. & Tanie, K. (1989). Profile Detection of Objects by a High-Resolution Tactile Sensor Using a Light Conductive Plate. *J. Japan Society of Mechanical Engineers*, Vol. 55, No. 516, pp. 2091-2099 (in Japanese)

Horie, M.; Funabashi, H. & Ozawa, T. (1995). Development of Three-Axial Force Sensors for Microrobots. *J. Japan Society of Mechanical Engineers*, Vol. 61, No. 591, pp. 4563-4568 (in Japanese)

Ishikawa, M. & Shimojjo, M. (1988). An Imaging Tactile Sensor with Video Output and Tactile Image Processing. *J. Society of Instrument and Control Engineers*, Vol. 24, No. 7, pp, 662-669

Izutani, J.; Maeda, Y. & Aoyagi, S. (2004). Development of a Micro Tactile Sensor utilizing Piezoresistors and Characterization of its Performance. *Proc. International Conference on Machine Automation (ICMA2004)*, pp. 193-196, Osaka, Japan, November 2004

Kamiyama, K.; Kajimoto, H.; Inami, M.; Kawakami, N. & Tachi, S. (2003). Development of A Vision-based Tactile Sensor. *Trans. Institute of Electrical Engineers of Japan*, Vol. 123, No. 1, pp. 16-22 (in Japanese)

Kane, B. J.; Cutkosky, M. R. & Kovacs, G. A. (2000). A Tactile Stress Sensor Array for Use in High-Resolution Robotic Tactile Imaging. *J. Microelectromechanical Systems*, Vol. 9, No. 4, pp. 425-434

Kinoshita, G. (1981). Overview of the Basic Research needed to advance the Robotic Tactile Sensors. *J. The Robotics Society of Japan*, Vol. 2, No. 5, pp. 430-437 (in Japanese)

Kobayashi, M. & Sagisawa, S. (1991). Three Direction Sensing Silicon Tactile Sensors. *Trans. Institute Electronics, Information and Communication Engineers*, Vol. J74-C-II, No. 5, pp. 427-433

Kovacs, G. T. A. (1998). *Micromachined Transducers Sourcebook*, McGraw-Hill, ISBN 0-07-290722-3, New York, USA

Lee, M. H. & Nicholls, H. R. (1999). Tactile Sensing for Mechatronics -A State of the Art Survey-. *Mechatronics*, Vol. 9, pp. 1-31

Lee, H. K.; Chang, S. I. & Yoon, E. (2006). A Flexible Polymer Tactile Sensor: Fabrication and Modular Expandability for Large Area Deployment. *J. Microelectromechanical Systems*, Vol. 15, No. 6, pp. 1681-1686

Maeda, Y.; Aoyagi, S.; Takano, M. & Hashiguchi, G. (2004). Development of a Micro Tactile Sensor for Robot Application and Its Evaluation. *Proc. Spring Meeting of Japan Society for Precision Engineering*, pp. 1121-1122, Tokyo, Japan, March 2004 (in Japanse)

Maekawa, H.; Tanie, K.; Kaneko, M.; Suzuki, N.; Horiguchi, C. & Sugawara, T. (1994). Development of a Finger-Shaped Tactile Sensor Using a Hemispherical Optical Waveguide. *J. Society of Instrument and Control Engineers*, Vol. 30, No. 5, pp. 499-508

Maeno, T. (2000). Structure and Function of Finger Pad and Tactile Receptors. *J. The Robotics Society of Japan*, Vol. 18, No. 6, pp. 767-771 (in Japanese)

Marr, D. (1982). *Vision: A Computational Investigation into the Human Representation and Processing of Visual Information*. Freeman, W. H. and Company, ISBN 0716712849, New York, USA

Matsushita, T.; Aoyagi, S. & Kozuka, H. (2004). Sol-gel Preparation of a PZT Thick Film from a Solution Containing Polyvinylpyrrolidone and Its Application for a Micro Ultrasonic Sensor. *Proc. Asia-Pacific Conf. Transducers and Micro-Nano Technology 2004 (APCOT2004)*, pp. 159-162, Sapporo, Japan, July 2004

Nguyen, A. T.; Dao, D. V.; Toriyama, T.; Wells, J. C. & Sugiyama, S. (2004). Measurement of Loads Acting on a Near-Wall Particle in Turbulent Water Flow by Using a 6-DOF MEMS-Based Sensor. *Proc. Asia-Pacific Conf. Transducers and Micro-Nano Technology (APCOT2004)*, pp. 508-512, Sapporo, Japan, July 2004

Ohka, M.; Kobayashi, M.; Shinokura, T. & Sagisawa, S. (1990). Data Processing of Tactile Information for Three-Axis Tactile Sensor. *J. Japan Society of Mechanical Engineers*, Vol. 56, No. 531, pp. 2919-2925 (in Japanese)

Ono, D.; Fukutani, T. & Aoyagi, S. (2008). Development of an Arrayed Tactile Sensor Having Four Stories and Recognition of Contact Using Neural Networks, *IEEJ Transaction SM*, Vol. 128, No. 5, pp. 246-251

Riedmiller, M. & Braun, H. (1993). A Direct Adaptive Method for Faster Backpropagation Learning: The RPROP Algorithm. *Proc. IEEE International Conference on Neural Networks*, pp. 586-591, San Francisco, USA, March 1993.

Shinoda, H. (2000). Tactile Sensing for Dexterous Hand. *J. The Robotics Society of Japan*, Vol. 18, No. 6, pp. 772-775 (in Japanese)

Sugie, N. (2000). Sensory Information Processing by Means of Neuro-computing. *J. The Robotics Society of Japan*, Vol. 9, No. 2, pp. 209-216 (in Japanese)

Suzuki, K.; Najafi, K. & Wise, K. D. (1990). A 1024-Element High-Performance Silicon Tactile Imager. *IEEE Trans. Electron Devices*, Vol. 37, No. 8, pp.1852-1860

Tai, Y. C. (2003). Parylene MEMS: Material, Technology and Application. *Proc. 20th Sensor Symp.*, pp. 1-8, Tokyo, Japan, July 2003

Takao, H.; Sawada, K. & Ishida, M. (2005). Multifunctional Smart Tactile-Image Sensor with Integrated Arrays of Strain and Temperature Sensors on Single Air-Pressurized Silicon Diaphragm. *Tech. Digest Transducers '05*, pp. 45-48, Seoul, Korea, June 2005

Wasserman, P. D. (1993). *Advanced Methods in Neural Computing*, Van Nostrand Reinhold, ISBN 0442004613, New York, USA

Watanabe, S. & Yoneyama, M. (1992). An Ultrasonic Visual Sensor for Three-Dimensional Object Recognition Using Neural Networks, *IEEE Trans. Robotics and Automation*, Vol. 8, No. 2, pp. 240-249

Yagi, S. (1991). Array Tactile Sensor using Piezoelectric Film. *J. The Robotics Society of Japan*, Vol. 7, No. 7, pp. 906-907 (in Japanese)

24

Tactile Information Processing for the Orientation Behaviour of Sand Scorpions

DaeEun Kim
Yonsei University, School of Electrical and Electronic Engineering
Corea (South Korea)

1. Introduction

Arachnids including sand scorpions and spiders use their tactile sense organs, called basitarsal compound slit sensilla (BCSS), to detect their prey. The sense organs consisting of mechanoreceptors are located at or near joints in the cuticle, and they can sense a vibrational signal caused by prey movement. The nocturnal sand scorpion *Paruroctonus mesaensis* has a distinguished capability of finding their prey only with these tactile senses. The sand scorpions show their orientation behaviour of positioning themselves directly towards their prey and then run into the prey when there is a vibration disturbance caused by the prey. According to the biological researches (Brownell & Farley, 1979, Brownell, 1984), it is presumed that the sand scorpions respond to Rayleigh waves, surface waves of sand, to detect the direction of a vibration source and possibly longitudinal vibrations to estimate the distance. Especially, the time delay between arrival of a vibration signal at the BCSS sense organs is an important cue to determine the direction of their prey (Brownell & Farley, 1979).

The central nervous system should process stimulus-locked neuron firings of the sense organs on their eight legs to induce the orientation behaviour. How the nervous system is organized to handle the orientation behaviour is still an open question. Only a few studies explain this behaviour mechanism. Stuerzl et al. (2000) introduced a neuronal mechanism to support the orientation behaviour of scorpions, and it is based on the difference of the arrival time of stimulus signals produced at sense organs on their legs. They argued that the brain of sand scorpions receives sensor signals from mechanoreceptors on their eight legs and processes an inhibition mechanism among a set of command neurons projected from the sense organs on each leg. This inhibitory interaction leads to more accumulated firings of the command neurons near the prey source and less firings at the opposite sides. The accumulated neuron firings thus form a tuning curve for a specific resource direction. Then the distribution of neuron firings can vote to determine the resource direction.

In fact, arachnids have sensory projections to the central nervous system for each leg (Babu & Barth, 1989, Anton & Barth, 1993). Previously Brownell and Farley (1979) suggested that eight command neurons (or eight clusters of neurons in the brain) accumulate neuron firings from mechanoreceptors on the eight legs of scorpions, respectively and also interact each other with triad inhibitions. According to the triad inhibition hypothesis, early arrival of vibration stimulus to mechanoreceptors on a leg excites the corresponding command neuron and the command neuron subsequently inhibits three command neurons on the opposite side. This inhibition mechanism forms an appropriate tuning curve of neuron firings for the resource direction. Brownell and Farley (1979) built this conceptual model with triad inhibitions for interaction over the eight receptor neurons (command neurons) and later Stuezl et al., (2001) tested the hypothesis with a neuronal model. Triad inhibition mechanism over a stimulus vibration showed a good agreement with the real orientation data of sand scorpions. Recently Kim (2006b) showed that pentad and heptad inhibitions as well as triad inhibitions can determine the resource direction. Thus, we infer that inhibition processes play a critical role for the orientation behaviour.

Once the distribution of neuron firings is available for the set of receptor neurons, the accurate turning angle can be decided. Population coding (Georgopoulos et al., 1982, Pouget et al., 1998, Deneve et al., 1999) over a set of accumulated neuron firings, that is, voting calculation can determine the final turning angle towards a prey (Stuerzl et al., 2000). The voting contribution of the neuron activations from the eight directions can be simply calculated (Georgopoulos et al., 1982) as follows:

$$ze^{i\phi} = \sum_{k=1}^{m} z_k e^{i\phi_k} \tag{1}$$

where z_k is the firing rate or spike counts for each direction and ϕ_k is the angular position for the *k*-th leg direction (*m=8*). The argument ϕ will represent the direction that the scorpion finally chooses. Each directional unit is assigned the weight proportional to the activation, and a population of neurons can determine the vector direction by the voting procedure.

Triad, pentad, heptad inhibition connections among a set of command neurons can lead to a good agreement with the real orientation data of sand scorpions (Kim, 2006b) However, the effect of weight configurations and synaptic delays among the eight receptor neurons have not been studied in detail. In this paper, we investigate the role of inhibition mechanism and synaptic delays in the network configuration among the command neurons to determine the direction of a vibration source. Also we will re-visit the triad inhibition hypothesis suggested by Brownell and Farley (1979). Relevant works were presented in the paper (Kim, 2006a, 2006b).

2. Command neurons

There are eight legs for sand scorpions and their foot positions form a circle (Brownell et al., 1979), as shown in Figure 1. With this structure, we can formulate the time difference of a vibration stimulus arrival on a set of legs. The time delay of vibration between a pair of legs can be calculated simply (Stuerzl et al., 2000) as

$$\begin{aligned}\Delta t(j,k) &= \frac{1}{v}(d(p_j,\bar{s}) - d(p_k,\bar{s})) \\ &\approx \frac{R}{v}[\cos(\phi-\theta_j) - \cos(\phi-\theta_k)]\end{aligned}$$

where $\Delta t(j,k)$ is the time difference of Rayleigh waves between the *j*-th and *k*-th foot, $d(\alpha, \beta)$ is the distance between position α and β, p_k is the position of the *k*-th foot, $\bar{s}$ is the position of a vibration source, R is the radius of the foot circle (see Fig. 1, R ≅2.5 cm), v is the Rayleigh wave speed in the sand (v ≅ 50 m/sec), ϕ is the angle of vibration source, and θ_k is the angle of the k-th leg from the front direction (θ_k = 18, 54, 90, 140, 220, 270, 306, 342 degree). According to the equation, the maximum time difference of stimulus on a pair of legs is around 1 msec.

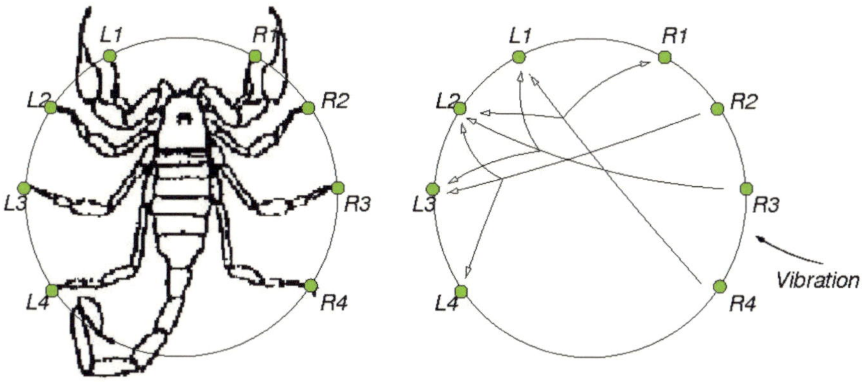

Fig. 1. Foot position of sand scorpions in a circle and triad inhibitions in the eight command neurons (reprinted from (Brownell, 1984; Brownell and Farley, 1979))

The neuron firings in the sense organs on legs depend on the amplitude of vibration signal, which is generated by prey movement. A *command neuron* or *receptor neuron* corresponding to each leg is activated by BCSS, and it accumulates the sensor activations. As shown in Fig. 1, there are eight receptor neurons in the brain and the command neuron for each leg receives inhibitory signals from three neurons on the opposite side, which is called triad inhibitions among command neurons. Stuerzl et al. (2000) assumed that the inhibitions have synaptic delay about 0.7 msec. Once sense organs on a leg first detect a vibration signal, the signal is transmitted to the corresponding command neuron. Then the inhibitory neuron signals from the command neuron often arrive at the three receptor neurons on the opposite side earlier than the vibration waves stimulate the three receptor neurons. Therefore, the neurons on the opposite side are deactivated for the vibration signals whilst the receptor neurons near the vibration source have intense neuron firings. The distribution of these accumulated neuron firings or the averaged firing rates on the eight directions will determine the orientation direction of scorpions capturing their prey. Here, the time delay from sense organs to the command neuron is common for each leg, and so what command neurons receive stimulus signals in an earlier time or later time is an important cue to determine the orientation.

For the neural mechanism, the perception vector for the orientation behaviour is a distribution of the weighted average of the sensor activations in the preferred directions, responding to a given stimulus. We can assume the stimulus direction follows population coding (Georgopoulos et al., 1982). It is represented as a sinusoidal array of eight elements for the sensor activations. The phase of each element follows the angular position of the corresponding leg (see Fig. 1).

3. Experiments

We simulate neuronal processes for the orientation behaviour. For our experiments, the command neurons in the brain receive stimulus signals at the same angular position as sense organs of sand scorpions. We modelled the neuronal firings of the eight command neurons with inhibitory interaction by following our previous works (Kim, 2006b). Each receptor neuron has an integrate-and-fire model and the firing rate is proportional to the magnitude of membrane potential. The neural network is similar to the continuous-time recurrent neural networks (CTRNN) (Beer & Gallagher, 1992). Here, the neuron has a membrane potential f_i,

$$
\begin{aligned}
\tau \frac{df_i}{dt} &= -f_i(t) + \sum_{j=1, j \neq i}^{8} \mu_{ji} z_j(t-\delta) + g I_i(t) \\
z_j(t) &= H(1/\{1+\exp[-(f_j(t)+\beta_j)]\} - 0.5) \\
H(x) &= \begin{cases} x & x > 0 \\ 0 & \text{otherwise} \end{cases}
\end{aligned}
$$

where τ is the time constant, z_j is the firing rate of the j-th receptor neuron for a burst of neuron spikes, δ is the synaptic delay of inhibitory connection, μ_{ji} is the weight from the j-th neuron to the i-th neuron, I_i is the intensity of the sensory input for the i-th leg, g is a gain factor, β_j is a bias term and $H(x)$ is a rectifying function to obtain positive firing rates. In the experiments, we set $\beta_j = 0$, $\tau = 0.33$, g=1.

Without any interaction signal among command neurons, each neuron produces neuron firings purely depending on the stimulus amplitude. In our experiments, for each command neuron, we integrate the firing rates for 500 msec to count the number of neuron spikes for the period. The distribution of accumulated activations over the eight receptor neurons will determine the resource direction. The vibration power spectrum of sand has a peak at 300 Hz (Aicher & Tautz, 2000) and so we assume that a burst of neuron firings has 1-2 msec duration. In the test, the input signal I_i has a form of half-wave rectified sinusoid with noise, and the vibration frequency for the input sinusoid is sampled every cycle with a Gaussian distribution with mean 300 Hz and standard deviation 50 Hz. In our model, we apply 10% random noise to the input signal and the vibration power spectrum also varies around the central frequency 300 Hz. Noisy input is given into the receptor neurons and the receptor neuron activation is accumulated for 500 msec. That integration process has a low-pass filter effect on the distribution of the average firing rate or the number of spikes over the eight receptor neurons, and thus produces steady response angles with small variance to a given stimulus direction. It is reasonable to observe the distribution of neuron firings accumulated for a time interval of 500 msec, when we consider the biological test as in Brownell and Farley's experiments (Brownell & Farley, 1979)

We first tested the orientation direction of scorpions with triad inhibitions among the eight receptor neurons. For triad inhibitions shown in Fig. 1, L_k neuron has triad connections with R_{4-k}, R_{5-k}, R_{6-k} neurons, and likewise R_k with L_{4-k}, L_{5-k}, L_{6-k} neurons for k=1,...,4 (for convenience, R_0=L_1, L_0=R_1, R_5=L_4, L_5=R_4). Inhibitions among the receptor neurons greatly influence the decision of resource direction. If we assume there is no inhibition at all, each receptor neuron will have almost the same level of neuron firings and the direction cue cannot be observed. Fig. 2 shows how the temporal difference of sensory afferents with the triad inhibitions changes the neuron firings. *L1* neuron initially produces neuron firings for the onset of stimulus vibration and later receives an inhibitory signal from *R3* neuron after an inhibition delay time. Accordingly, the number of spikes for the neuron *L1* will be depressed by inhibition. The inhibition signal arrives at *L2*, *L3* neurons earlier than the

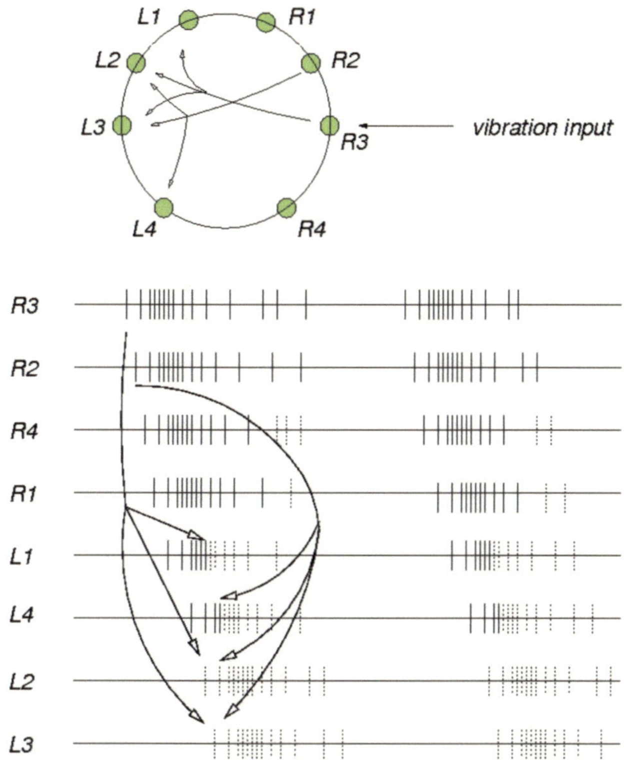

Fig. 2. Sequence of neuron firings with inhibitory actions among the eight receptor neurons (vibration source is given at the direction of 90 degree; solid spikes: actual spikes, dotted spikes: spikes inhibited by other neurons)

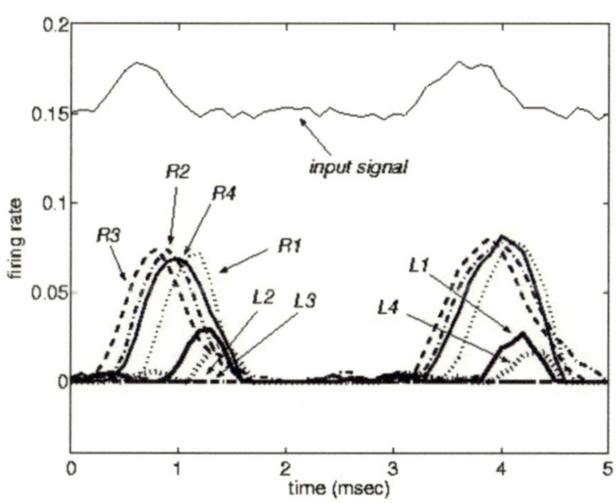

Fig. 3. Firing rate of receptor neurons, z_j with a half-wave sinusoidal input at the direction of 90 degree (*L2*, *L3* neurons have almost no firing by inhibitory signals from *R2*, *R3*, *R4* and *L1*, *L4* have relatively low amplitudes)

vibration stimulus innervates the neurons and thus no neuron firing is detected. Similar inhibition operation is found among *R2* neuron and *L2, L3, L4* neurons on its opposite side. Spikes at *L1, L4* neurons will subsequently trigger another inhibition into *R3, R4, L4* neurons and *L1, R1, R2* neurons, respectively. In Fig. 3, we can observe the transition and distribution of firing rates in time course. We characterize the response of the receptor neurons by the total number of neuron spikes in a fixed time interval, which can be obtained by integrating the firing rates in the interval, say, for 500 msec. It reduces the effect of variance due to the noise from the external stimulus, sense organs or neurons.

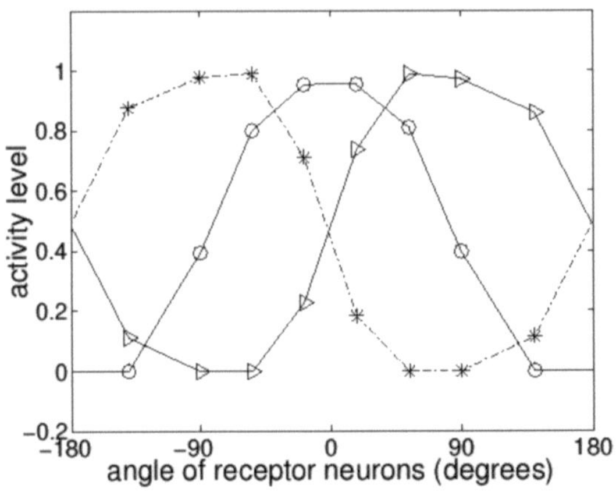

Fig. 4. Activity level of eight receptor neurons; the vibration source is given at the direction of -90, 0, and 90 degrees (reprinted from (Kim, 2006b))

Fig. 4 shows the activities of receptor neurons for a vibration stimulus at the direction of -90 degree, 0 degree and 90 degree, respectively. Interestingly, the activity of the eight receptor neurons roughly has a cosine tuning curve, and if it is normalized, then it follows $(1+\cos(\phi-\theta_k))/2,\ k=1,\ldots,8$ for a given direction ϕ of vibration source. The neuron firings of the receptors are significantly influenced by the temporal difference of arrival of sensory afferent signals. It makes a relative difference of neuron firings among the eight command neurons. It shapes such a tuning curve. To estimate the resource direction from the activity curve of the eight command neurons, we need to apply a decoding method shown in equation (1). Several decoding methods are available to explain the population coding. Here, we use the complex coding, that is, simple voting mechanism with the amplitude for each phase (Georgopoulos et al., 1982, Kim, 2006b) .

To find the influence of synaptic delay on the estimation of a resource direction, we tested a variety of delay time from 0.1 msec up to 1.5 msec on triad inhibitions. The vibration stimulus is given at the direction of 90 degree and the activity levels in the set of eight command neurons change depending on the transmission delays.

We first see how neuron firings fluctuate in time course. Fig. 5 shows the transition of firing rates depending on the transmission delays of inhibitions. If the delay time δ=0.2 msec is taken, the duration of inhibition on the opposite side becomes small while the vibration waves continue to stimulate the neurons, as *R3, R2* neurons cannot suppress the neuron activations of *L2, L3* neurons sufficiently in Fig. 5(a). As a result, such interaction of the inhibition produces a distorted bimodal distribution of activity in the set of command neurons (see Fig. 6). In contrast, a long delay time, for example, δ =1.5 msec allows all the receptor neurons to be activated during the stimulus period. It thus produces a large estimation error for the resource direction.

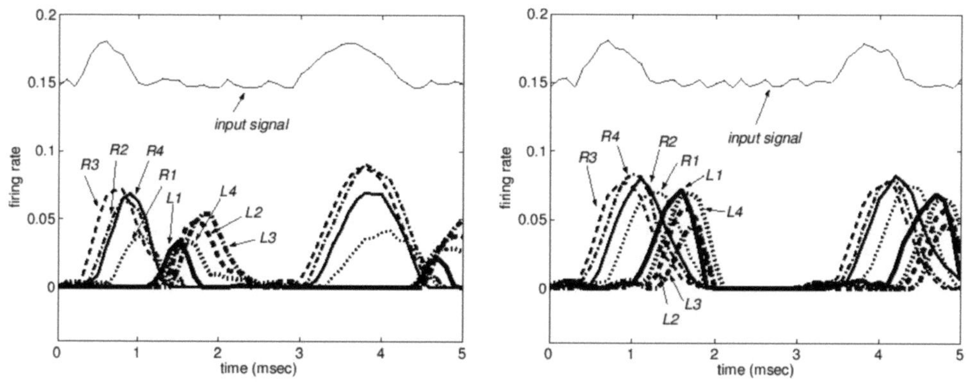

(a) (b)

Fig. 5. Firing rate of receptor neurons with different transmission delays (a) δ=0.2 msec (b) δ =1.2 msec (vibration source is given at the direction of 90 degree)

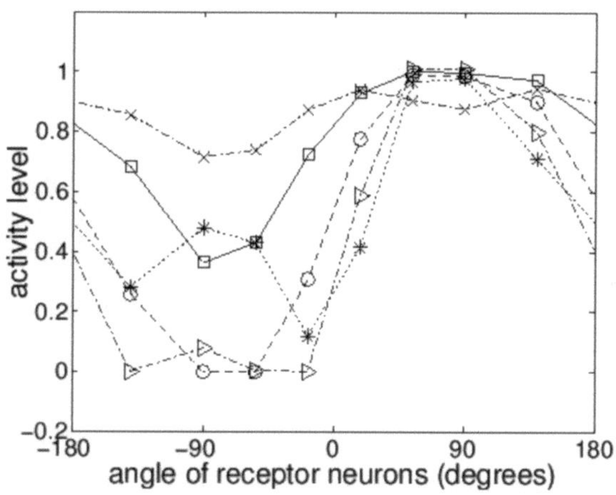

Fig. 6. Activity level of command neurons with various delays on inhibitory signals (vibration source is given at the direction of 90 degree, *: δ= 0.2 msec, triangle: δ=0.5 msec, o: δ=0.8 msec, square: δ=1.2 msec, x: δ=1.5 msec) (reprinted from (Kim, 2006b))

As shown in Fig. 6, especially a small delay time, for instance, δ =0.2 msec will produce a small hill in the tuning curve in addition to the main cosine-like curve. It results from neuron firings which was not inhibited enough on the opposite side. That is, too small delays may not inhibit incoming neuron firings transmitted from sense organs on the opposite side. As a result, it produces a distorted bimodal-like tuning curve for the activity of receptor neurons rather than a cosine-tuning curve. However, too large delays induce a relatively small contrast among the sensory activations, since the inhibition process does not reduce neuron firings of other command neurons sufficiently - see Fig. 5(b). The decoding method in equation (1), irrespective of the distribution of neuron firings, will pinpoint an orientation angle.

Brownell and Farley (1979) showed biological experiments for the orientation of sand scorpions towards their prey. They collected the orientation data of the scorpion for stimulus vibrations at various angles. We tested our orientation simulation over the real data with a neural model described above. Here, the orientation response angles were estimated over distribution of accumulated neuron firings. Initially, we simulated a variety of transmission delays on triad inhibitions, and compared the orientation results with the real orientation data of scorpions to see how much our model matches the real data. The agreement with the real data was reasonably good within the range of 0.5 - 1.1 msec; the performances were not significantly different within that range. However, too small or too large delays show a deviation from desirable estimation of orientation angle (Kim, 2006b).
Brownell and Farley (1979) tested another experiment with ablated slit sensilla on one or a few legs. From the result, they suggested a hypothesis of triad inhibitions among the eight receptor neurons. Stuerzl et al. (2000) used a stochastic model of neuron firings with triad inhibitions to examine the hypothesis, and found its good agreement with the real orientation data of scorpions. In their experiments, monad inhibitions were worse than triad inhibitions in agreement with the real data. However, how many inhibitions are available is still unknown. Thus, we explored monad, dyad, triad, pentad, heptad inhibitions among the eight receptor neurons in the previous work (Kim, 2006b) – see the inhibition arrangements in Fig. 7.

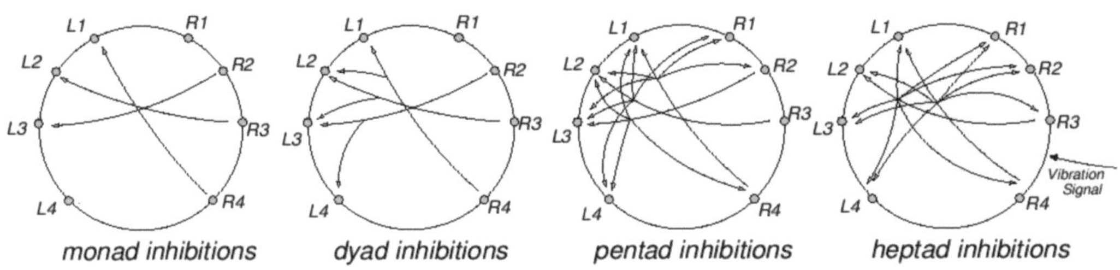

Fig. 7. Inhibitory connections among the receptor neurons (reprinted from (Kim, 2006b))

The inhibitory mechanism was tested on the biological data (Brownell & Farley, 1979, Brownell, 1984) with some slit sensilla ablated. For these experiments, we used δ=0.9 msec for a delay time of inhibitory signal, which was a good delay parameter for the triad inhibition test in the above. For the monad inhibition arrangement, each receptor neuron has an inhibitory connection with a single receptor neuron on the opposite side (L_k with R_{5-k} for $k=1,...,4$ and vice versa), and for the heptad connection if one neuron is activated, its inhibitory signal is transmitted into all the other seven neurons. For dyad inhibitions, L_k neuron has an inhibitory connection to two neurons R_{5-k} and R_{6-k}, on its opposite side, and similarly R_k neuron to two neurons L_{5-k} and L_{6-k}. For pentad inhibitions, each neuron transmits inhibitory signals to five command neurons except its two neighbour neurons.
Fig. 8 shows that the neural network can determine the appropriate orientation. As Stuerzl et al. (2000) showed, triad inhibitions generally produce good performance in agreement test with the real data, but monad inhibitions have poor agreement with the real data, as shown in Fig. 8(b) and 8(d). Three inhibitions or more from one neuron cell can produce good agreement. Pentad or heptad inhibitions show the best agreement in some cases (Kim, 2006b), and possibly more inhibitions rather than three inhibitions might be a plausible idea to support the orientation data. However, pentad or heptad inhibitions need more connections among the receptor neurons just for a slight improvement of the agreement.

Thus, it may be arguable that the experiments still support the triad inhibition hypothesis over the eight receptors, since their performance differences are not much significant.

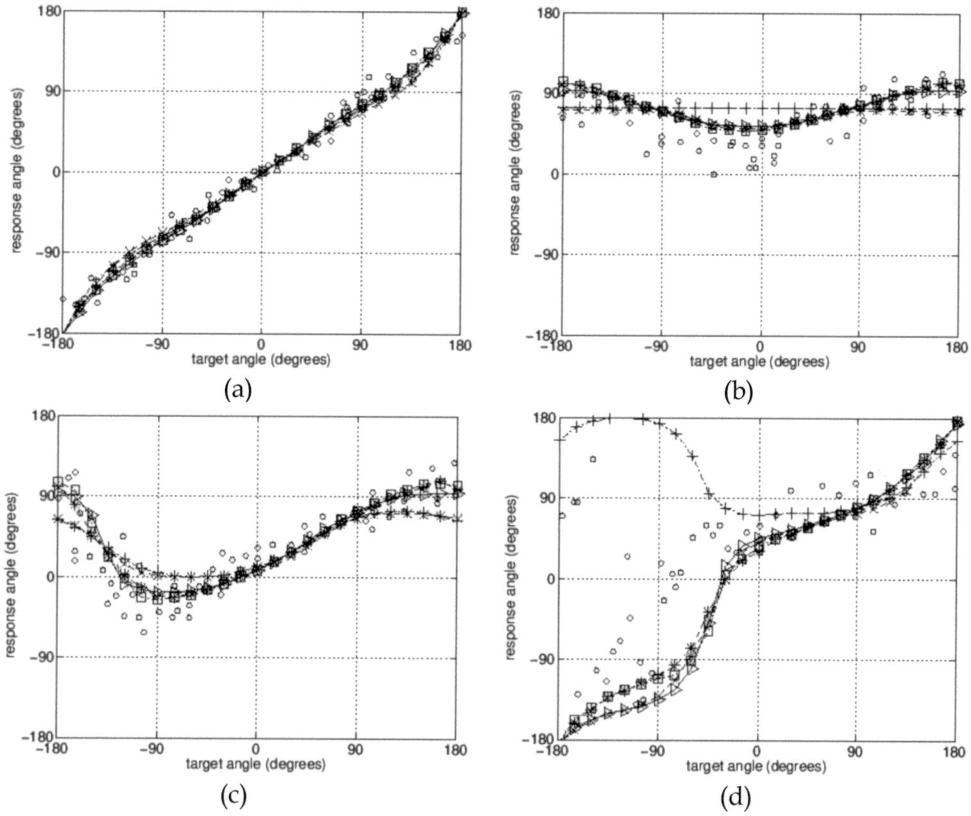

Fig. 8. Test of orientation angles with the real orientation data (a) no ablation (b) L1, L2, L3 and L4 ablated (c) L3 and L4 ablated (d) L1 and L2 ablated (o: real data, +: monad inhibition, x: dyad, triangle: triad, square: pentad, *: heptad); real orientation data of scorpions have been reprinted from the paper (Brownell & Farley, 1979, Brownell, 1984, and Kim, 2006b)

In our experiments, synaptic delays on triad inhibitions significantly influence the estimation of a resource direction. We applied a varying delay time ranging from 0.1 msec up to 1.5 msec to several arrangements of inhibitions, that is, monad, dyad, triad, pentad and heptad connections, and the results were compared with the orientation data of sand scorpions. We measured the estimation error for each sample in the orientation data and calculated the root mean squared error for each set of data. Fig. 8 shows that each arrangement of inhibition needs a little different range of delay time for good performance in the agreement test. The delay time from 0.5 msec to 1.1 msec covers the best performance in Fig. 9(a)-(c), and 1.0-1.4 msec in Fig. 9(d). It seems that triad, pentad or heptad connections produce better performance than the other connection types. From the result, we also infer that sand scorpions need a synaptic delay on the inhibitions within the range of 0.5 - 1.2 msec.

According to the agreement test, we see there is still an error between the theoretical model and the real data. The estimation error is often found in the experiments with ablated sense organs. As shown in Fig. 8(b) and 8(d), even for varying inhibition arrangements, the orientation estimation does not perfectly match the real data. We guess several factors to cause

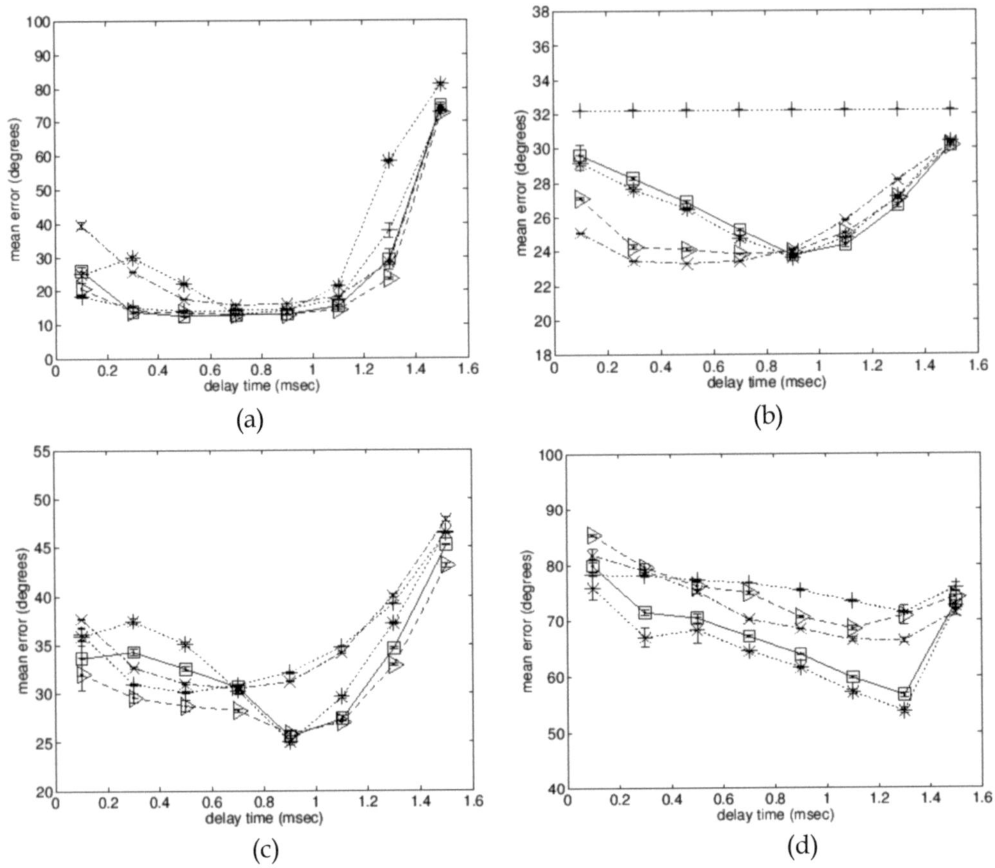

Fig. 9. Estimation error with varying delay time on inhibitory signal (a) no ablation (b) L1, L2, L3 and L4 ablated (c) L3 and L4 ablated (d) L1 and L2 ablated (root mean squared error and standard deviation is measured with 5 trials, o: real data, +: monad inhibition, x: dyad, triangle: triad, square: pentad, *: heptad) (reprinted from (Kim, 2006a))

this mismatch. Possibly sand scorpions may have different activity patterns for the receptor neurons or use a little different decoding method for the orientation. We assumed unbiased sensitivity over the eight receptor neurons for the incoming vibration waves, but it may not be true in reality. Above all, from the agreement test, receptors corresponding to front legs seem to have more contribution in the voting procedure than the others, because a large discrepany between the estimation and the reality is observed near the front direction when the sense organs at a front leg are ablated. Or, there might exist conditional operations or neuromodulations among the receptor activations; for instance, one receptor's firings may change the contribution rate or presynaptic weights of another receptor neuron. Alternatively, different arrangement of inhibitory connections, or partly excitatory connections may be available among the command neurons. Presumably, the command neurons for the front legs may interact each other with strong inhibitions, since ablation of the sense organs on the front leg at one side seems to trigger more neuron firings of the front leg at the other side (left or right). We need further study to explain it more precisely. In fact, there are many factors to influence the agreement performance, such as synaptic delay, neuron firing rate, coupling strengths among the receptor neurons, and decoding methods. More rigorous parameter setting and anatomical / neurophysiological result will be required for close agreement with the real data.

4. Discussion

In this paper, we have shown a simplified neuronal model to explain the orientation mechanism of sand scorpions, which is rather similar to the CTRNN network. Stuerzl et al. (2000) tested another type of neural networks, a population of spiking neurons to see how the arrival time of vibration on each leg influences the neural firings. They found the neuron firings on each leg is a function of the time difference of vibration stimulus among the triad inhibitory interactions. Our neural model is simplified to eight neurons each of which has its own firing rate to approximate the number of neuron spikes. The experimental result shows our neuron model is well fit to the real orientation data, which suggests that the eight neuron model may be sufficient to explain the orientation mechanism of sand scorpions. This simplified model has more potential for robotic application or other vibration sensing application, since it can be easiliy built with artificial devices.

Fig. 10. Mobile robot with multi-legs

Following the suggested neural architecture, we are testing robotic orientation with tactile senses. The sand scorpions use BCSS organs near joints to detect any vibration in the environment. We mount vibration sensors, piezoelectric sensors at the foot and measure the temporal signals of vibrations. We found that the time delay between arrival of a vibration signal on robotic legs is not much significant in an office environment when a vibration disturbance is given near the robot. Rayleigh waves of vibration can be easily recognized in the desert environment for sand scorpions (Brownell, 1984), but the flat surface shown in Fig. 10 is only involved with longitudinal waves, that is, sound waves. The sound speed of the solid surface is much higher than the sound wave in the air. To apply our neural approach to the robotic system in the office environment, we need to measure high resolution of temporal signal to detect the time delay of arrival. As an alternative approach, we can use the amplitude or power spectrum of the vibration signal. We observe that more distant area has more attenuation in the amplitude of sound waves. However, it is questionable if the distribution of amplitude attenuation can exactly pinpoint the direction of vibration source. The orientation with the two types of vibration signals will be tested in the environment similar to the desert. It is still an open question that sand scorpions use both Rayleigh waves and longitudinal waves or either for resource localization. We need further study to find the principle of orientation of sand scorpions and apply it to engineering problems.

The connectivity pattern and the time delay of inhibitory signal have been investigated in this paper. The results are based on our neural model and it seems that the neuron firings on each leg follow an almost linear function of the time difference of vibration stimulus when triad, pentad or heptad connections are used. It implies that our model is a good approximation of the realistic neuron model suggested by Stuerzl et al. (2000). To validate our neuron model, we still need further study of whether more realistic neuronal model produces the same results about the inhibitory pattern or delay.

We suggested neural networks with inhibition mechanism or suitable delay time for the inhibitory connections. However, there is no neurophysiological or anatomical evidence for the neural network of command neurons. From the experiments, it seems that the inhibitory interaction has a significant role on the localization of a prey source at least in the agreement test with the real data. We still need anatomical data on the scorpion brain to support the argument. Also if the information of the recording time of vibration signals for the sand scorpions can be obtained, it may provide the relevance and significance of the integration in our neural network.

To model the decoding process of orientation angle over distribution of neuron firings in the command neurons, we can build a neuronal layer with a lot of directional cells (Kim, 2006b). One hypothesis is that there exist a large number of direction cells to pinpoint the exact direction of the prey source and the direction is determined before the appropriate motor action. As an alternative decoding mechanism, we can build a relatively small number of neurons directly coupled to motor neurons for the orientation behaviour. With that neural structure, appropriate weights may guide the orientation angle accurately.

5. Conclusion

In this paper, we present a neuronal model to process tactile information from the sense organs of sand scorpions. We tried to explain how sand scorpions orient towards their prey when there is any vibration disturbance caused by prey. They have eight clusters of neurons in the brain to process stimulus-locked neuron firings of the sense organs on their eight legs (Anton & Barth, 1993, Babu & Barth, 1989). It is modelled simply as eight command neurons with inhibitory interactions in our neural system.

Our continuous-time recurrent neural networks can encode the temporal difference of tactile vibration on each leg, and also implement inhibitory interactions depending on synaptic delays. Then we compared the neuronal simulation results with the orientation of sand scorpions. According to the agreement test, the eight command neurons show the best performances with triad, pentad or heptad inhibitions, which implies three inhibitory connections or more may be available in the brain system of sand scorpions. Varying synaptic delays also influence the agreement performances. The experiments suggest that they have a synaptic delay of inhibition within the range of 0.5 - 1.2 msec.

The neural model and its experiments showed that a relatively small number of neurons can achieve the orientation mechanism for a vibration source. This study can provide a foundation of tactile information processing to localize a prey source, or to detect the environmental cue with a collection of sensor signals. We still need further study to understand the orientation of sand scorpions. This will be involved with testing various factors including the coupling strength, non-uniform synaptic delays or varying neural connections among the command neurons.

6. Acknowledgements

The work has been supported by the MKE (The Ministry of Knowledge Economy), Korea, under the ITRC (Information Technology Research Center) support program supervised by the IITA (Institute for Information Technology Advancement)" (IITA-2008-C1090-0803-0006)

7. References

Aicher, B. and Tautz, J. (1990). Vibrational communication in the fiddler crab, *Uca pugilator*. I. signal transmission through the substratum. *Journal of Comparative Physiology A*, 166:345--353.

Anton, S. and Barth, F. (1993). Central nervous projection patterns of trichobothria and other cuticular sensilla in the wandering spider *Cupiennius saley* (arachnida, araneae). *Zoomorphology*, 113:21--32.

Babu, K. and Barth, F. (1989). Central nervous projections of mechanoreceptors in the spider *Cupiennius saley* keys. *Cell and Tissue Research*, 258:69--82.

Beer, R. and Gallagher, J. (1992). Evolving dynamic neural networks for adaptive behavior. Adaptive Behavior, 1(1):91--122.

Brownell, P. (1984). Prey detection by the sand scorpion. *Scientific American*, 251(6):94--105.

Brownell, P. and Farley, R. (1979). Orientation to vibrations in sand by the nocturnal scorpion *Paruroctonus mesaensis*: Mechanism of target localization. *Journal of Comparative Physiology A*, 131:23--30.

Deneve, S., Latham, P., and Pouget, A. (2000). Reading population codes: a neural implementation of ideal observers. *Nature Neuroscience*, 2(8):740--745.

Georgopoulos, A., Kalaska, J., Caminiti, R., and Massey, J. (1982). On the relations between the direction of two-dimensional arm movements and cell discharge in primate motor cortex. *Journal of Neuroscience*, 2(11):1527--1537.

Gerstner, W. and Kistler, W. (2002). Spiking Neuron Models. Cambridge University Press.

Kim, D. (2006a). Role of Inhibitions in Neural Networks for the Orientation Behaviour of Sand Scorpions, *Proceedings of Int. Conf. on the Simulation and Synthesis of Living Systems*, pp. 275--281, 2006

Kim, D. (2006b). Neural Network Mechanism for the Orientation Behaviour of Sand Scorpions Towards Prey", *IEEE Transactions on Neural Networks*, vol. 17 (4), pp. 1070-1076, 2006

Pouget, A., Zhang, K., Deneve, S., and Latham, P. (1998). Statistically efficient estimation using population coding. *Neural Comp.*, 10:373--401.

Stuerzl, W., Kempter, R., and van Hemmen, J. (2000). Theory of arachnid prey localization. *Physical Review Letters*, 84(24):5668--5671.